BASIC MATHEMATICS

Martin M. Zuckerman
City College of the City University of New York

D. VAN NOSTRAND COM[PANY]
New York Cincinnati Toron[to]

In memory of Joseph Dimentstein

D. Van Nostrand Company Regional Offices:
New York Cincinnati

D. Van Nostrand Company International Offices:
London Toronto Melbourne

Library of Congress Catalog Card Number: 79-64465
ISBN: 0-442-21911-3

Published by D. Van Nostrand Company
135 West 50th Street, New York, N.Y. 10020

10 9 8 7 6 5 4 3 2 1

PREFACE

Basic Mathematics is a text whose broad coverage of arithmetic, algebra, and geometry helps students acquire the basic mathematical skills necessary for college-level courses in mathematics, science, and technology. Arithmetic and algebraic concepts are integrated throughout the text. Algebraic methods are employed in the final two chapters on geometry, the treatment of which is unusually complete.

The book has been designed for efficient, step-by-step mastery of skills and for maximum adaptability to each student's needs.

Each chapter begins with a Diagnostic Test that helps students determine which sections require study. Answers to the Diagnostic Tests appear at the end of the book.

The textual material is reinforced with an abundance of illustrative examples. Virtually every new idea is followed by an example demonstrating how problems are solved.

The examples are followed by Class Exercises that are similar to the examples. Answers to these exercises are provided at the end of each section and frequently are presented as fully explained, step-by-step solutions.

At the end of each section there is an extensive and varied set of Home Exercises. Many of these exercises concern practical, real-life situations. There is an emphasis on solving verbally stated problems. Answers to the odd-numbered Home Exercises appear at the end of the book.

A comprehensive set of Review Exercises closes each chapter. All the answers to these exercises appear in the back of the book.

There are more than 400 figures to illustrate important concepts.

An Instructor's Manual is available that provides answers to even-numbered Home Exercises and Sample Exams for the entire text.

ACKNOWLEDGMENTS

I wish to thank the following reviewers for their many helpful suggestions:

John Clifton, El Centro Community College
Jan Ford, San Diego State University
David Hares, El Centro Community College
Wei-Jen Harrison, American River College
Albert Liberi, Westchester Community College
Alfonzo Patricelli, Olive-Harvey College
Harold L. Schoen, University of Iowa
Richard Semmler, Northern Virginia Community College at Annandale
Michael Schub, Westchester Community College
Ara B. Sullenberger, Tarrant County Junior College

I gratefully appreciate the efforts and contributions of Ralph DeSoignie and Harriet Serenkin of D. Van Nostrand Company.

CONTENTS

1 INTEGERS

DIAGNOSTIC TEST Perhaps you are already familiar with some of the material in Chapter 1. This test will indicate which sections in Chapter 1 you need to study. The question number refers to the corresponding section in Chapter 1. If you answer *all* parts of the question correctly, you may omit the section. But *if any part of your answer is wrong, you should study the section.* When you have completed the test, turn to page A1 for the answers.

1.1 The Number Line

a. Write the 3-digit number that has 7 as its 100's digit, 4 as its 10's digit, and 0 as its 1's digit.
b. Express 5902 in verbal form, that is, express this number in words.
c. In Figure 1A, which integer is represented by each of the points Q, R, S, T?
d. Round 79,850 to the nearest 1000.

Fig. 1A

1.2 Negative Integers

a. In Figure 1B, which integer is represented by each of the points Q, R, S, T?
b. Fill in "<" or ">". $-7 \square -3$
c. Rearrange the integers so that you can write "<" between any two integers.
$$-7, 7, 0, -3, -10$$
d. Find $|-9|$. e. Which numbers are 9 units from the origin?

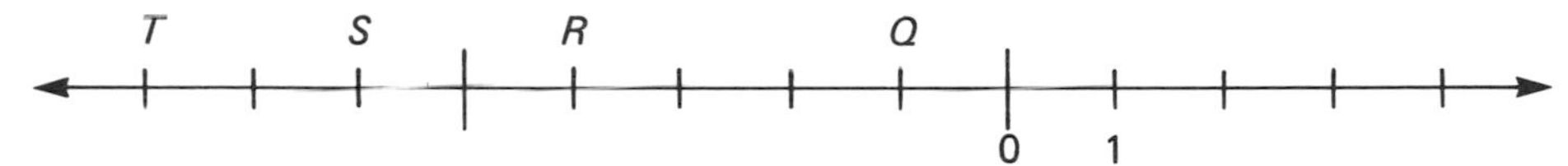

Fig. 1B

1.3 Addition and Subtraction of Positive Integers

a. Add.

$$\begin{array}{r} 4819 \\ 2634 \\ \underline{5036} \end{array}$$

b. Subtract.

$$\begin{array}{r} 842 \\ \underline{-\ 159} \end{array}$$

c. Approximate by first rounding each integer to the nearest 100, and then adding.

$$\begin{array}{r} 4289 \\ 7943 \\ 8449 \\ \underline{7501} \end{array}$$

d. Ann spends \$9 for a hat, \$28 for a pocketbook, and \$6 for a scarf. How much does she spend on her wardrobe?

1.4 Addition and Subtraction of Integers

Add or subtract.

a. $10 + (-7)$ b. $(-5) + (-3)$ c. $10 - (-4)$
d. $6 + 8 - (5 + 1) - (7 - 9)$
e. When the temperature rises from $-7°$ Celsius to $3°$ Celsius, how many degrees Celsius does it rise?

1.5 Multiplication and Exponentiation

Find each value.

a. $\begin{array}{r} 792 \\ \times\ 89 \\ \hline \end{array}$ b. $(-2)(-1)(-3)$ c. 9^2 d. 2^5 e. $(-1)^{19}$

f. If each shirt costs \$9, how much do eight shirts cost?

1.6 Division

Divide, or indicate that division is not defined. Also, check your result.

a. $\frac{0}{-7}$ b. $\frac{-7}{0}$ c. $\frac{-56}{-7}$
d. Divide and find the remainder. $93\overline{)4087}$
e. There are 16 ounces in a pound. 100 pounds = ______ pounds and ______ ounces.

1.7 Order of Operations

Find each value

a. $(7 + 3)4 - 1$ b. $\frac{6 - 2}{2} - (1 - 2)^2$ c. $(5 \cdot 4 + 1) \div 3$

1.1 THE NUMBER LINE

DIGITS The number 10 is the **base** of our number system.

$$0, 1, 2, 3, 4, 5, 6, 7, 8, 9$$

are called **digits**. When you write the **three-digit number**

$$237$$

2 is the 100's digit, 3 is the 10's digit, and 7 is the 1's digit.

100's	10's	1's
2	3	7

Thus, 237 stands for

$$\underbrace{2 \cdot 100}_{200} + \underbrace{3 \cdot 10}_{30} + \underbrace{7 \cdot 1}_{7}$$

You can also consider a large number, such as the **seven-digit number**

$$4{,}307{,}500$$

Example 1 Express 4,307,500 in terms of 1,000,000's, 100,000's, and so on.

Solution

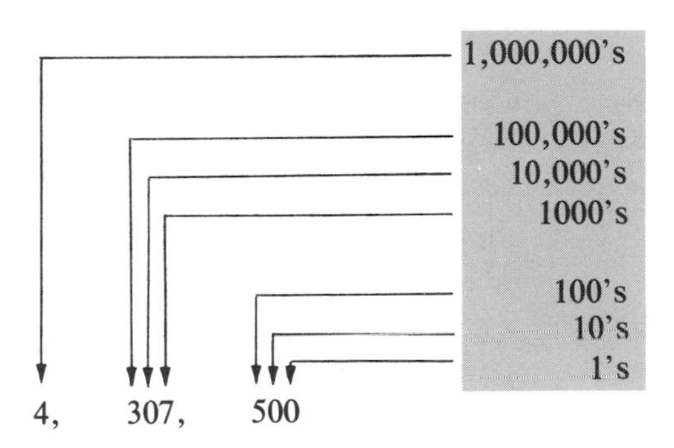

Thus,

$$\begin{aligned} 4{,}307{,}500 &= \underbrace{4 \cdot 1{,}000{,}000} + \underbrace{3 \cdot 100{,}000} + 0 \cdot 10{,}000 + \underbrace{7 \cdot 1000} + \underbrace{5 \cdot 100} + 0 \cdot 10 + 0 \cdot 1 \\ &= 4{,}000{,}000 + 300{,}000 + 7000 + 500 \end{aligned}$$

Observe that the digit 0 appears three times: as the 10,000's digit, as the 10's digit, and as the 1's digit

---

CLASS EXERCISES

1. Consider 396.
   a. What is its 100's digit? **b.** What is its 10's digit?
   c. What is its 1's digit?
2. Consider 400,360.
   a. What is its 100,000's digit? **b.** What is its 100's digit?
   c. Fill in: 0 is its ______'s digit, its ______'s digit, and its ______'s digit.
3. Express 53,742 in terms of 10,000's, 1000's, 100's, 10's, and 1's.

VERBAL FORM OF NUMBERS

Newspapers and books often present numbers in verbal form. You may read that **fifty thousand** fans attended a football game. It is important to know how to translate from words to digits, and vice versa. For example, **fifty thousand** means:

10,000's	1000's	100's	10's	1's
↓	↓	↓	↓	↓
5	0,	0	0	0

Thus, write **50,000**.

Note the correspondence:

1's	⟷	ones
10's	⟷	tens
100's	⟷	hundreds
1000's	⟷	thousands
10,000's	⟷	ten thousands
100,000's	⟷	hundred thousands
1,000,000's	⟷	millions

Example 2 Express each number in terms of digits.
a. three hundred fifty-four **b.** twenty thousand, one hundred thirty-six

Solution

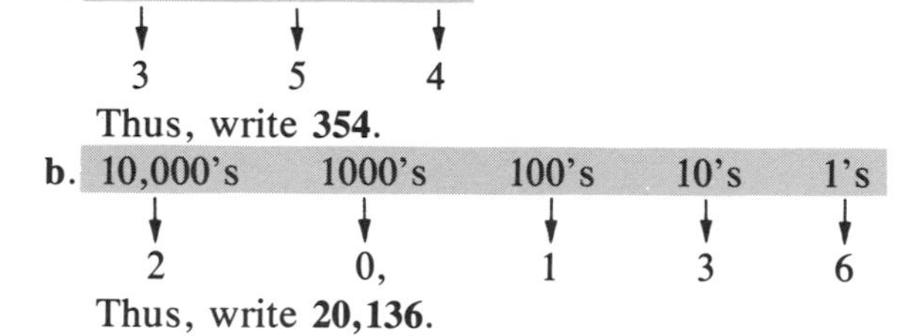

When writing integers with *more than four digits*, begin at the right and use commas to separate the digits into groups of three. For example, consider

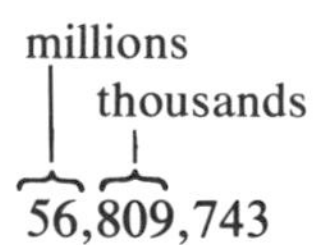

This represents

56 million, 809 thousand, 743

Note that the *left*-most group may have one, two, or three digits.

Example 3 Express each number in verbal form.
a. 3841 **b.** 6,002,187

Solution **a.**

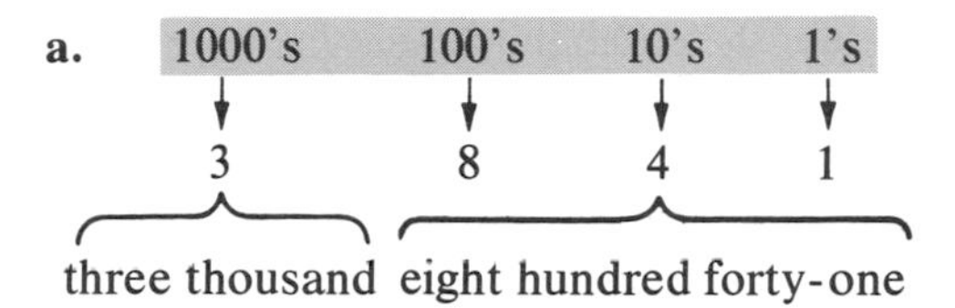

b.

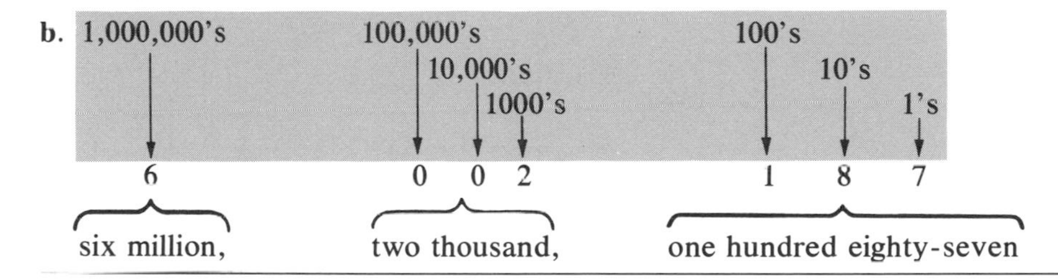

CLASS EXERCISES

4. Express in terms of digits: fifty thousand, one hundred twenty
5. Express in verbal form: 72,404

PICTURING POSITIVE INTEGERS

Definition

> The numbers
>
> 1, 2, 3, 4, . . .
>
> are known as **positive integers.** The three dots to the right are read "and so on."

Among the positive integers are the two-digit number 27, the three-digit number 843, and the six-digit number 100,481.

There is a very convenient way of picturing 0 and the positive integers along a straight line—the so-called **number line**. Consider a *horizontal* line, as in Figure 1.1, and suppose that it extends indefinitely both to the left and to the right. Choose any point on the line, and let the point represent the number

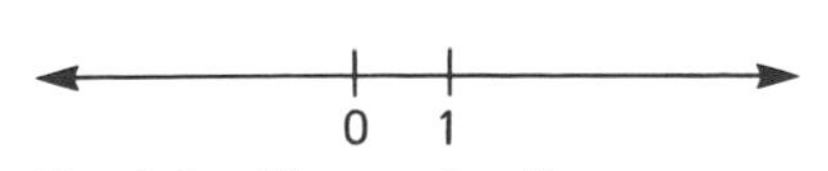

Fig. 1.1. The number line.

0. This point is called the **origin**. Now, choose another point to the *right* of 0. This second point represents the number 1.

A portion of a line that lies between two points (including these "end points") is called a **line segment.** The *length* of the line segment between 0 and 1 determines the basic **unit of distance**. The number 2 is represented one (distance) unit to the right of 1. The number 3 is represented one unit further to the right. Continue this process to represent the numbers 4, 5, 6, and so on. (See Figure 1.2 .)

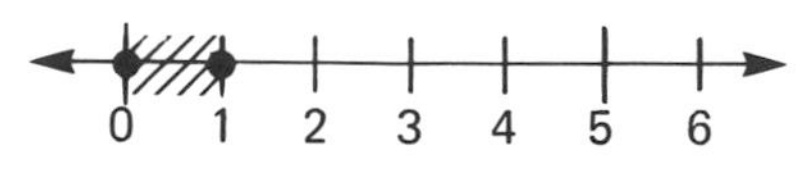

Fig. 1.2. The line segment between 0 and 1 is indicated.

Example 4 Consider the number line shown in Figure 1.3(a). Label the points that represent the following numbers.

a. 4 **b.** 7 **c.** 8 **d.** 10

Fig. 1.3 (a)

Solution

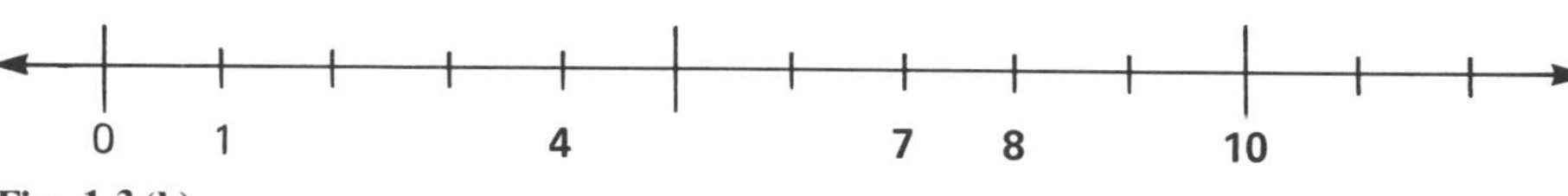

Fig. 1.3 (b)

CLASS EXERCISES

6. Consider the number line shown in Figure 1.4 . Label the points that represent the following numbers.

a. 3 **b.** 6 **c.** 9 **d.** 11

0 1

Fig. 1.4

7. In Figure 1.5, indicate which number is represented by each of the points Q, R, S, T, U.

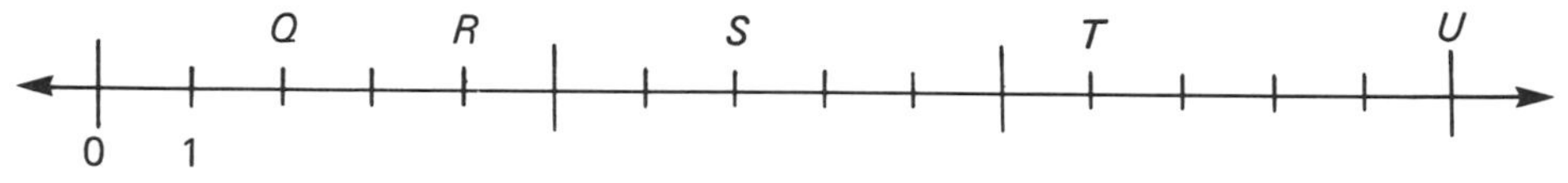

Fig. 1.5

ROUNDING

Numbers are often "rounded" to the nearest thousand, to the nearest million, and so on. For example, the distance between Boston and Los Angeles, which is 3050 miles, may be rounded to 3000 miles.

> To round to the nearest 1000:
> 1. If the 100's digit is 5 or more, increase the 1000's digit by 1.
> 2. Replace the last three digits by 0's (as in 1000).

Example 5 Round to the nearest 1000.

a. 7362 **b.** 8594 **c.** 49,630

Solution **a.** The 100's digit of 7362 is 3, which is less than 5. Thus, round to 7000.

b. The 100's digit of 8594 is 5. Increase the 1000's digit, 8, by 1. Round to 9000. Do you see that 8594 is closer to 9000 than to 8000? [See Figure 1.6 .]

Fig. 1.6

c. The 100's digit of 49,630 is 6 and is therefore more than 5. Increase the 1000's digit from 9 to 10. To do so, you must also consider the 10,000's digit, and increase 49 to 50. Thus,

49,630 rounds to 50,000

The rules for rounding to the nearest 1000 can be modified for rounding to the nearest 100, to the nearest 100,000, to the nearest 10,000,000, and so on.

Example 6 Round 126,342,171 to the nearest

a. 100,000 **b.** 10,000,000 **c.** 100

Solution **a.** The 10,000's digit of 126,342,171 is 4. Round to 126,300,000.

b. The 1,000,000's digit of 126,342,171 is 6, which is more than 5. Increase the 10,000,000's digit from 2 to 3. Round to 130,000,000.

c. The 10's digit of 126,342,171 is 7. Round to 126,342,200.

CLASS EXERCISES

8. Round 8174 to the nearest 1000. **9.** Round 9599 to the nearest 1000.

10. Round 7,391,504 to the nearest 100,000.

SOLUTIONS TO CLASS EXERCISES

1. a. 3 **b.** 9 **c.** 6 **2. a.** 4 **b.** 3 **c.** 10,000's, 1000's, 1's

3. $53{,}742 = \underbrace{5 \cdot 10{,}000}_{} + \underbrace{3 \cdot 1000}_{} + \underbrace{7 \cdot 100}_{} + \underbrace{4 \cdot 10}_{} + \underbrace{2 \cdot 1}_{}$

$= 50{,}000 + 3000 + 700 + 40 + 2$

4. 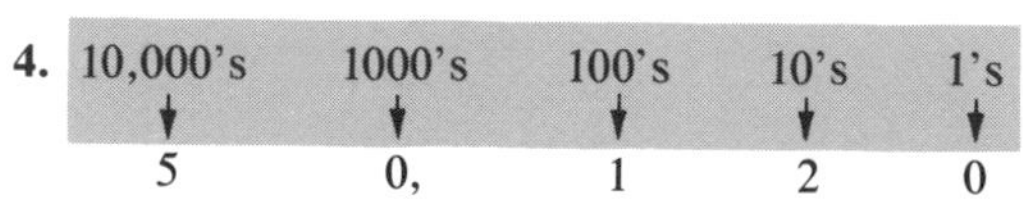

5 0, 1 2 0
Thus, write **50,120**.

5. 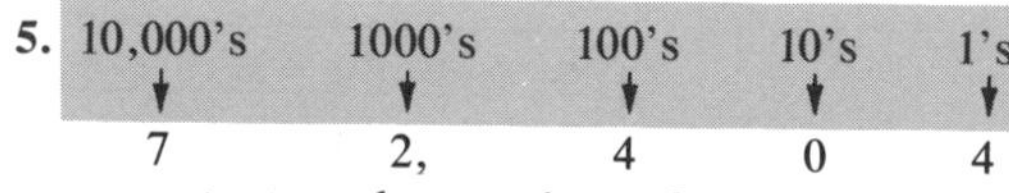

7 2, 4 0 4
seventy-two thousand, four hundred four

6. 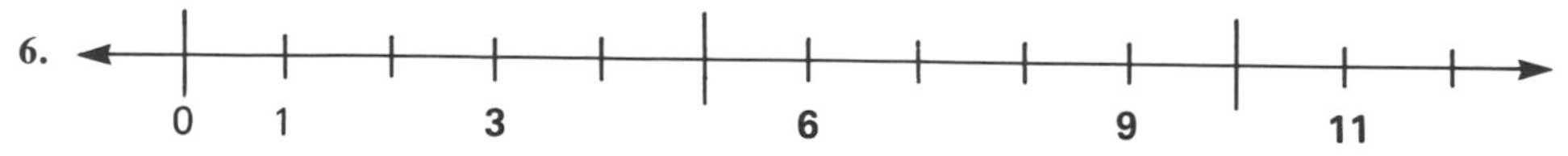

Fig. 1.7

7. 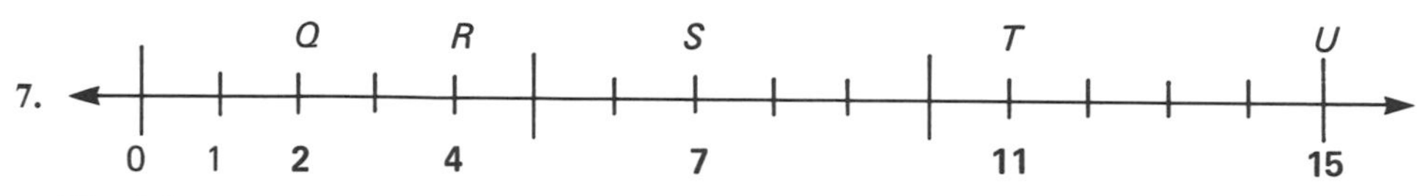

Fig. 1.8

8. 8174 rounds to 8000. **9.** 9599 rounds to 10,000. **10.** 7,391,504 rounds to 7,400,000.

HOME EXERCISES

In Exercises 1–4: **a.** *What is the 100's digit?* **b.** *What is the 10's digit?* **c.** *What is the 1's digit?*

1. 542 **2.** 437 **3.** 905 **4.** 816

In Exercises 5–8: **a.** *What is the 100,000's digit?* **b.** *What is the 100's digit?* **c.** *What is the 10's digit?*

5. 748,261 **6.** 480,300 **7.** 300,008 **8.** 1,472,805

In Exercises 9–12, express each number in terms of 1000's, 100's, 10's, and 1's.

9. 4283 **10.** 3859 **11.** 2670 **12.** 9003

In Exercises 13 and 14, express each number in terms of 1,000,000's, 100,000's, and so on.

13. 5,834,003 **14.** 7,200,000

15. Write the three-digit number that has 5 as its 100's digit, 2 as its 10's digit, and 4 as its 1's digit.

16. Write the five-digit number that has 7 as its 10,000's digit and 1 as its other digits.

17. Write the seven-digit number that has 4 as its 1,000,000's digit, 2 as its 1000's digit, and 0 as its other digits.

In Exercises 18–20, express each number in terms of digits.

18. three thousand, five hundred ninety-eight **19.** sixty thousand, fifty-five

20. five million, one hundred twenty-six thousand, nine hundred sixty

In Exercises 21–24, express each number in verbal form.

21. 409 **22.** 30,802 **23.** 161,284 **24.** 7,004,510

In Exercises 25 and 26, express the number in each sentence in terms of digits.

25. The population of Providence is approximately two hundred fifty thousand.

26. The daily circulation of the New York Times is over one million, nine hundred fifty thousand.

For Exercises 27–32, a horizontal line has been drawn in Figure 1.9 . Locate the following numbers on this number line.

27. 2 **28.** 6 **29.** 10 **30.** 12 **31.** 14 **32.** 15

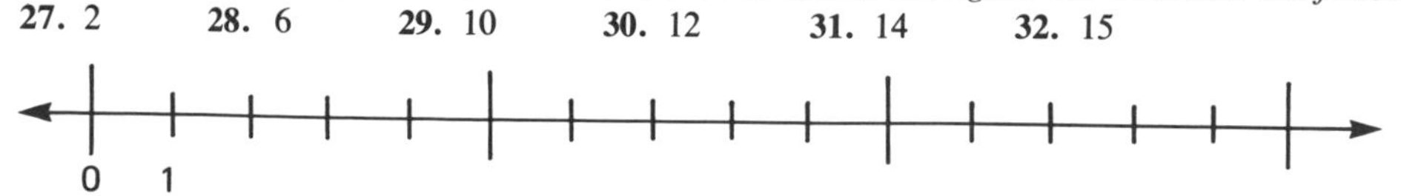

Fig. 1.9

33. In Figure 1.10, which integer is represented by each of the points Q, R, S, T, U?

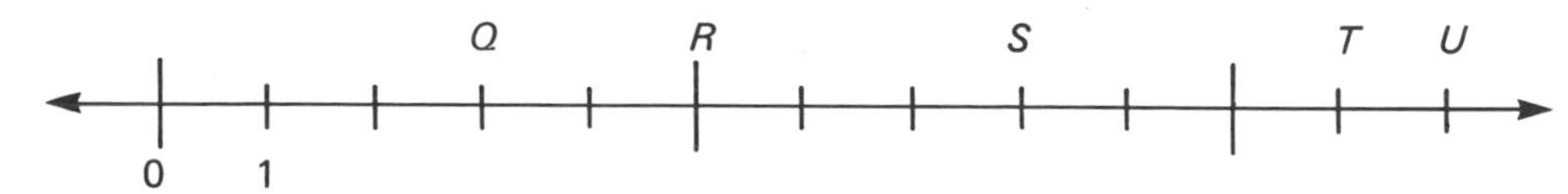

Fig. 1.10

34. In Figure 1.11, which integer is represented by each of the points Q, R, S, T, U?

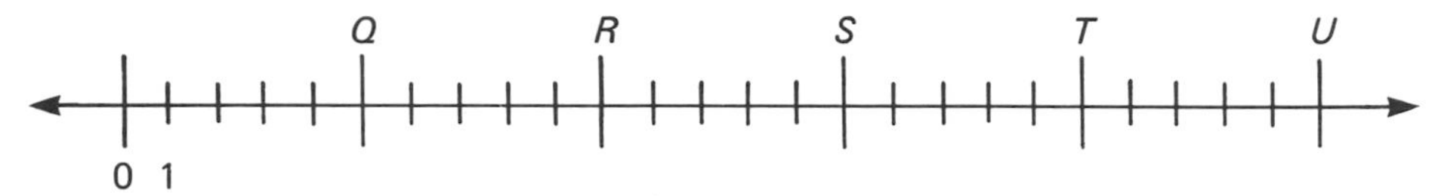

Fig. 1.11

In Exercises 35–38, round to the nearest 1000.

35. 6172 **36.** 6692 **37.** 6500 **38.** 99,950

In Exercises 39–42, round to the nearest 1,000,000.

39. 20,499,532 **40.** 6,666,666 **41.** 19,932,147 **42.** 39,500,000

43. Round 536,215 to the nearest **a.** 100 **b.** 1000 **c.** 100,000

44. Round 165,559,200 to the nearest **a.** 100,000 **b.** 1,000,000 **c.** 10,000,000

In Exercises 45–47, round the populations of the following cities to the nearest 100,000.

45. San Francisco 740,316 **46.** Chicago 3,369,357 **47.** Philadelphia 1,949,996

In Exercises 48–50, round the asking prices of the following houses to the nearest $1000.

48. Split-level $28,898 **49.** Ranch-style $49,500 **50.** Mobile Home $19,999

1.2 NEGATIVE INTEGERS

PICTURING NEGATIVE INTEGERS

At times positive numbers do not suffice. *Negative numbers* must be introduced.

Example 1 When the temperature drops below 0°C (0 degrees Celsius), you express it in terms of a negative number. For example, the temperature in Milwaukee one January morning is expressed by -12°C.

Example 2 A football team *loses* 3 yards on a play. Its *gain* is -3 yards.

DEFINITION

> The numbers $-1, -2, -3, -4, \ldots$ are known as **negative integers**.

Recall that positive integers are pictured to the *right* of 0 along a horizontal line, as in Figure 1.12 .

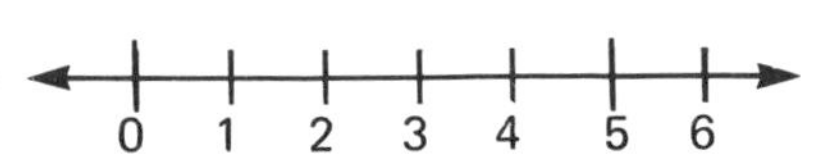

Fig. 1.12

Now, go one distance unit to the *left* of 0. This point represents the integer -1. Go one unit further to the left to represent -2. Continue to the left to represent -3, -4, -5, and so on, as in Figure 1.13 .

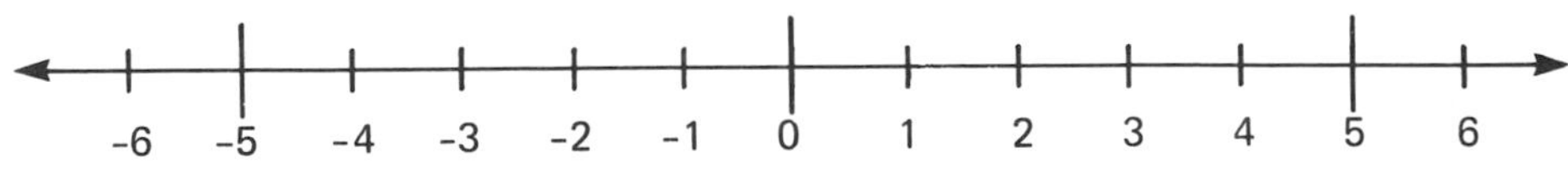

Fig. 1.13

Example 3 Consider the number line. Label the points that represent the following numbers.
a. -4 **b.** -8 **c.** -9 **d.** -12

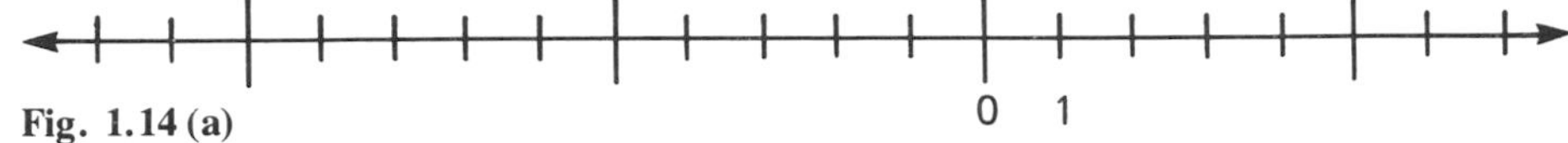

Fig. 1.14 (a)

Solution

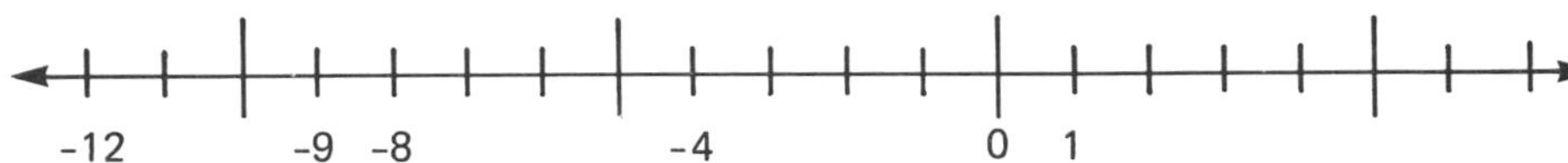

Fig. 1.14 (b)

CLASS EXERCISES

1. Consider the number line. Label the points that represent the following numbers.
   a. -3 **b.** -5 **c.** -7 **d.** -10

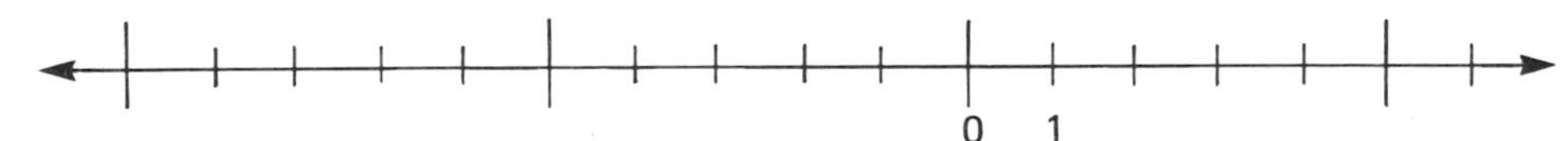

Fig. 1.15

2. In Figure 1.16, indicate which number is represented by each of the points Q, R, S, T, U.

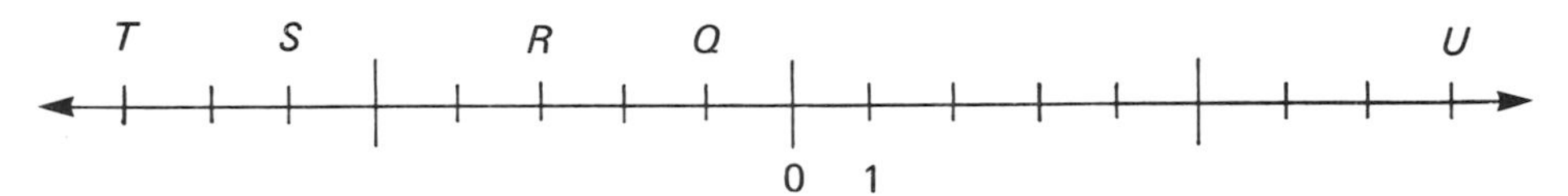

Fig. 1.16

ORDER

Consider Figure 1.17 . If 3 lies to the *left* of 5 on the number line, you say that 3 *is less than* 5. You can indicate "is less than" by the symbol "<". Thus,

"$3 < 5$" is read as "3 is less than 5."

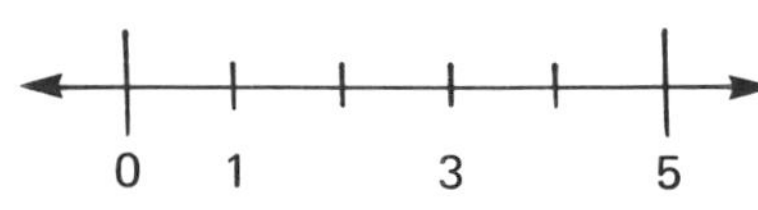

Fig. 1.17

In mathematics, you often use letters to represent the numbers you are discussing. Instead of writing

write

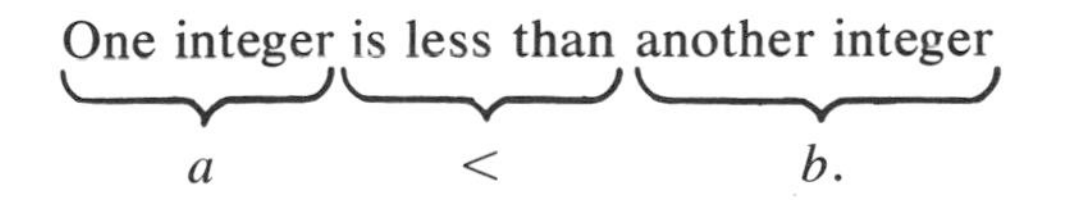

DEFINITION

LESS THAN. Let a and b be any number.* If a lies to the left of b (on the number line), you say a **is less than** b. In this case, you write

$$a < b$$

You can also say that a **is smaller than** b.

When $a < b$, you can also say that b **is greater than** a and write

$$b > a$$

In this case, b lies to the *right* of a. Thus, $5 > 3$ because 5 lies to the right of 3. (See Figure 1.17 .) Note that the symbols

$$< \qquad >$$

each point to the *smaller* number. Also, *if* $a < b$, *then* $b > a$.

Example 4 Fill in "<" or ">". $10 \square 8$

Solution $10 \boxed{>} 8$

In Figure 1.18, note that -5 lies to the left of -2. Thus, -5 must be less than -2.

$$-5 < -2$$

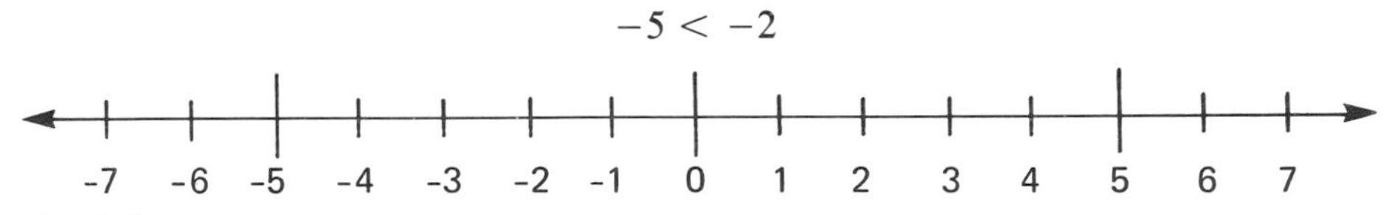

Fig. 1.18

Example 5 Fill in "<" or ">". $-7 \square -4$

Solution (Refer to Figure 1.18 .) $-7 \boxed{<} -4$

*Although only integers have been introduced thus far, much of the discussion on order and on arithmetic applies to (real) numbers, in general. For this reason, many definitions and rules are stated in terms of numbers.

Example 6 Fill in "left" or "right." -1 lies to the ☐ of -6 (on the number line).

Solution (Refer to Figure 1.18.) -1 lies to the [right] of -6 (on the number line).

CLASS EXERCISES *In Exercises 3–5, fill in "<" or ">".*

3. 0 ☐ 2 **4.** -9 ☐ -10 **5.** -2 ☐ 0

In Exercises 6–8, fill in "left" or "right."

6. 1 lies to the ☐ of 4 (on the number line).

7. -3 lies to the ☐ of -12. **8.** 5 lies to the ☐ of -1.

INVERSE

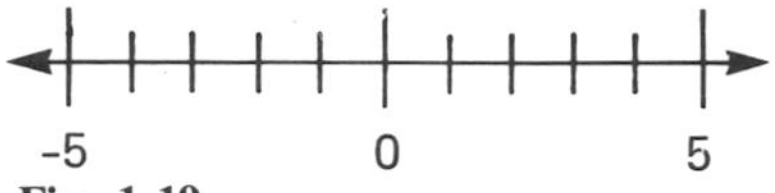

-5 0 5

Fig. 1.19

The numbers 5 and -5 lie on opposite sides of the origin, 0, but are the same distance from 0. The *positive* number 5 lies 5 units to the *right* of the origin, whereas the *negative* number -5 lies 5 units to the *left* of the origin. (See Figure 1.19.)

DEFINITION

> INVERSE. Let a be a number other than 0. The **additive inverse of a**, or for short, the **inverse of a**, is the number that lies on the opposite side of the origin and is the same distance from the origin as is a. The (**additive**) **inverse of *0*** is 0 itself.

For every number a, let $-a$ denote the inverse of a. Thus, -5 is the inverse of 5.

Example 7 **a.** The inverse of 10 is -10. **b.** The inverse of -10 is 10. In symbols, write

$$\underbrace{-}_{\text{the inverse of}}(-10) = 10$$

Observe that *the inverse of a positive number is a negative number. The inverse of a negative number is a positive number.*

CLASS EXERCISES *Write the inverse of each number.* **9.** 3 **10.** -12 **11.** 0

ABSOLUTE VALUE

DEFINITION

> ABSOLUTE VALUE. Let a be any number. The **absolute value of a** is its distance from the origin.

Let $|a|$ denote the absolute value of a.

Example 8 See Figure 1.20. **a.** $|7| = 7$ because 7 lies 7 units from the origin. **b.** $|-7| = 7$ because -7 lies 7 units from the origin. **c.** $|0| = 0$

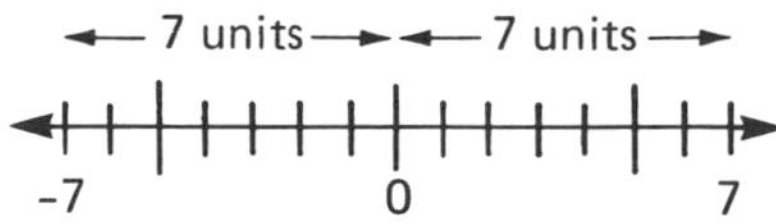

Fig. 1.20

For every number a, $\quad |a| = |-a|$

Both numbers are the same distance from the origin, but on opposite sides of it. Thus, *absolute value measures distance from the origin, but neglects direction.*

Observe that

$$|a| = a, \quad \text{if } a \text{ is positive or } 0$$
$$|a| = -a, \quad \text{if } a \text{ is negative}$$

In fact, if a is *negative*, a lies to the *left* of the origin, and $-a$ lies to the *right* of the origin. Thus, $-a$, which is *positive*, is the distance from a to the origin.

Example 9 **a.** $|4| = 4$ **b.** $|-4| = -(-4) = 4$

Example 10 A Datsun gets 25 miles to the gallon. It uses the same gallon of gas to travel 25 miles east as it would to travel 25 miles west. Thus, gas consumption is measured in terms of absolute value. You consider distance, but neglect direction.

CLASS EXERCISES *Find each absolute value.* **12.** $|9|$ **13.** $|-9|$

SOLUTIONS TO CLASS EXERCISES

1.

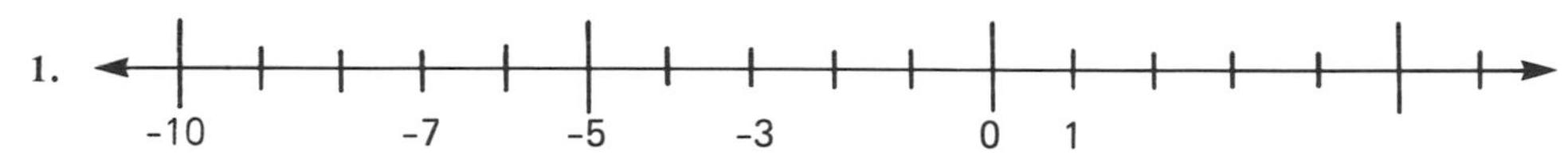

Fig. 1.21

2.

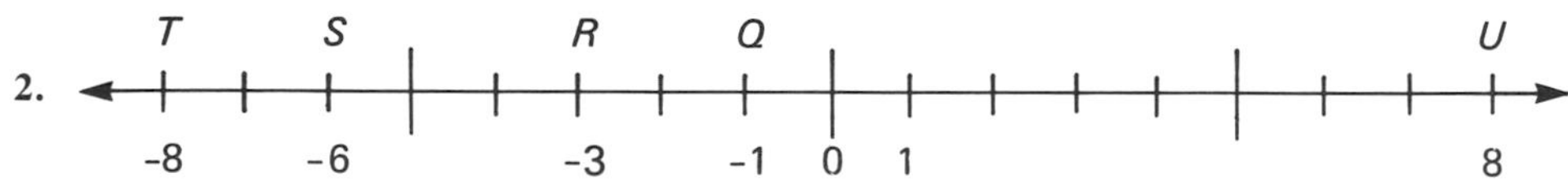

Fig. 1.22

3. $<$ **4.** $>$ **5.** $<$ **6.** left **7.** right
8. right **9.** -3 **10.** 12 **11.** 0 **12.** 9 **13.** 9

HOME EXERCISES

In Exercises 1–6, locate the following numbers on the number line. Use Figure 1.23 .

1. -2 **2.** -5 **3.** 5 **4.** -8 **5.** -10 **6.** -11

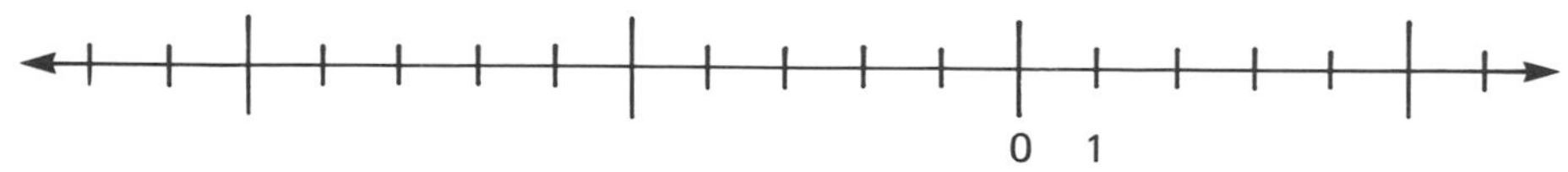

Fig. 1.23

7. In Figure 1.24, which integer is represented by each of the points Q, R, S, T, U?

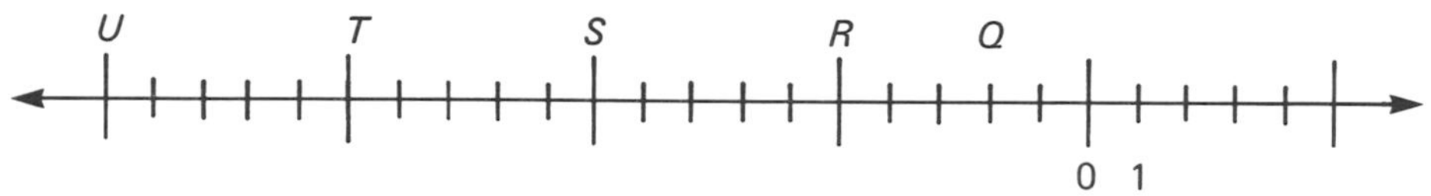

Fig. 1.24

In Exercises 8–14, fill in "<" or ">".

8. 2 □ 8 **9.** 0 □ 4 **10.** 53 □ 48 **11.** -8 □ -3 **12.** -9 □ -12 **13.** 0 □ -6 **14.** -12 □ 4

In Exercises 15–20, fill in "left" or "right."

15. 4 lies to the ☐ of 9 (on the number line).
16. 10 lies to the ☐ of 3.
17. 0 lies to the ☐ of 3.
18. -5 lies to the ☐ of -3.
19. -2 lies to the ☐ of -6.
20. 0 lies to the ☐ of -20.

In Exercises 21–26, rearrange the integers so that you can write "<" between any two integers.

Sample 2, 10, 1, 7	**Solution** $1 < 2 < 7 < 10$

21. 3, 8, 1, 5 **22.** 8, 10, 0, 4, 9 **23.** $-3, -6, 3, 6$ **24.** $-2, -4, 0, 4$ **25.** $-5, -3, 1, -7, 7$
26. $-12, 12, -3, 3, -4, 4$

In Exercises 27–32, write the inverse of each number.

27. 20 **28.** -20 **29.** 0 **30.** 100 **31.** -75 **32.** -1000

In Exercises 33–38, find each absolute value.

33. $|8|$ **34.** $|-8|$ **35.** $|0|$ **36.** $|100|$ **37.** $|-100|$ **38.** $|-5000|$

In Exercises 39–44, simplify.

Sample $-|4|$	**Solution** -4

39. $-|7|$ **40.** $-|-7|$ **41.** $|-7|$ **42.** $-(-7)$ **43.** $-|0|$ **44.** $-(-12)$

Sample (for Exercises 45–48). Which numbers are 5 units from the origin?	**Solution** 5 and -5

45. Which numbers are 8 units from the origin?

46. Which numbers are 15 units from the origin?

47. Which positive number is 11 units from the origin?

48. Which negative number is 13 units from the origin?

49. Which number is further from the origin, 27 or -44?

50. Which number is further from the origin, 32 or -29?

1.3 ADDITION AND SUBTRACTION OF POSITIVE INTEGERS

BASIC RULES OF ADDITION

Let a and b be any numbers. The **sum of *a* and *b*** is denoted by

$$a + b \qquad \text{(Read: ``}a \text{ plus } b\text{'')}$$

It expresses "adding b objects to a objects." For example,

$3 + 2$ expresses adding 2 objects to 3 objects.

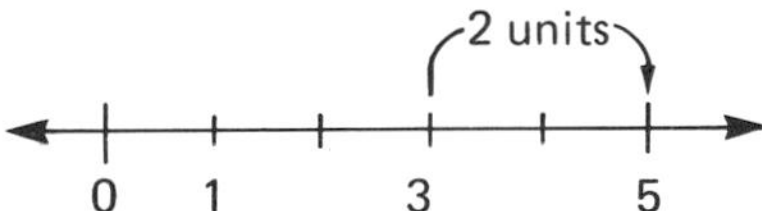

Fig. 1.25. 3 + 2 lies 2 units to the right of 3.

$3 + 2 = 5$

Figure 1.25 depicts the sum of 3 and 2 geometrically. Note that

$3 + 2$ lies 2 units to the *right* of 3.

In general, if p is a *positive* number, $a + p$ lies p units to the *right* of a. *For any number a,*

$$\boxed{a + 0 = a}$$

For example, $6 + 0 = 6$

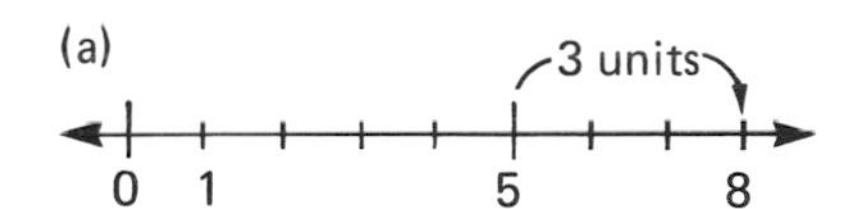

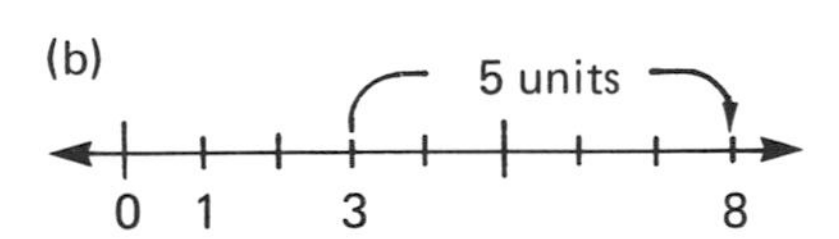

Fig. 1.26(a) $5 + 3 = 8$.
(b) $3 + 5 = 8$.

Figure 1.26 illustrates: $5 + 3 = 3 + 5 = 8$

When you add two numbers, you can add them in either order. Thus, *for any two numbers a and b,*

$$\boxed{a + b = b + a}$$

This is called the **Commutative Law of Addition.**

Next, consider what happens when you add three numbers.

Example 1 **a.** $\underbrace{(4 + 1)}_{5} + 2 = 5 + 2 = 7$

Parentheses ( ) are used here to indicate which sum comes first. Parentheses always come in pairs. Look for groupings within a pair of parentheses. Here, the parentheses indicate that you first add 4 and 1, and then add 2 to the sum, 5.

b. $4 + \underbrace{(1 + 2)}_{3} = 4 + 3 = 7$

Here, the parentheses indicate that you add 3, the sum of 1 and 2, to 4.

In both cases, the sum of the numbers 4, 1, and 2 is 7. Thus,

$$(4 + 1) + 2 = 4 + (1 + 2)$$

When you add three numbers a, b, and c, you get the same result if you first add a and b, and then c, or if you add the sum of b and c to a.

$$\boxed{(a + b) + c = a + (b + c)}$$

This is known as the **Associative Law of Addition.** Note that parentheses indicate which addition occurs first. Because of the Associative Law, you usually write the sum

$$a + b + c$$

without parentheses. Thus,

$$a + b + c = (a + b) + c = a + (b + c)$$

CLASS EXERCISES

1. Show that both sums, **a** and **b**, are the same. **a.** $(5 + 2) + 3$ **b.** $5 + (2 + 3)$

COLUMN ADDITION

Larger numbers are usually added in columns. You add the 1's digits, then the 10's digits, then the 100's digits, and so forth.

The Commutative and Associative Laws enable you to rearrange numbers when adding them. For example, observe that

$$\begin{aligned} 6 + (10 + 3) &= (6 + 10) + 3 && \textit{by the Associative Law} \\ &= (10 + 6) + 3 && \textit{by the Commutative Law} \\ &= 10 + (6 + 3) && \textit{by the Associative Law} \end{aligned} \qquad (*)$$

Thus, if you want to add two numbers such as 26 and 13 you use

$$26 + 13 = \overbrace{(20 + 6)}^{26} + \overbrace{(10 + 3)}^{13}$$

$$= 20 + [6 + (10 + 3)] \quad \textit{in the Associative Law, let } a = 20,\ b = 6,\ c = 10 + 3$$

$$= 20 + [10 + (6 + 3)] \quad \textit{by } (*)$$

$$= \underbrace{(20 + 10)}_{30} + \underbrace{(6 + 3)}_{9} \quad \textit{in the Associative Law, let } a = 20,\ b = 10,\ c = 6 + 3$$

$$= 39$$

In practice, you add in columns.

10's	1's
2	6
1	3
3	9

To add $37 + 16$, use

$$37 + 16 = \overbrace{(30 + 7)}^{37} + \overbrace{(10 + 6)}^{16}$$

$$= (30 + 10) + \underbrace{(7 + 6)}_{13}$$

$$= (30 + 10) + \overbrace{(10 + 3)}^{13}$$

$$= (30 + 10 + 10) + 3$$

When you arrange in columns, you will see that the 1's column sum is 13, or $10 + 3$.

10's	1's
3	7
1	6
	13

Carry the **1** to the 10's column, and add it along with the other numbers in the 10's column.

10's	1's
1	
3	7
1	6
5	3

You may have to carry more than once in an addition example.

Example 2 Add.

```
3089
2682
5189
2977
----
```

Solution

```
  132
 3089
 2682
 5189
 2977
-----
13937
```

The sum is 13,937.

CLASS EXERCISES *Add in columns.*

2. 23 + 51 + 12

3.
```
8469
 638
5943
6927
----
```

APPROXIMATING SUMS Rounding enables you to estimate a sum quickly. This is particularly important in checking for gross errors.

Example 3 Approximate the following sum by first rounding each integer to the nearest 1000, and then adding.

```
12 234
 8 916
 6 102
 5 518
------
```

Solution Round, and then add.

```
12 000
 9 000
 6 000
 6 000
------
33 000
```

The sum is *approximately* 33,000.

The actual sum is obtained as follows.

```
12 234
 8 916
 6 102
 5 518
------
32 770
```

CLASS EXERCISES

4. Approximate the sum

```
18 419 327
12 501 064
11 103 841
 9 712 349
 5 532 849
----------
```

by first rounding each integer to the nearest 1,000,000, and then adding.

SUBTRACTION Let a and b be positive numbers, and suppose that $a > b$. Then the **difference between a and b** (in this order) is denoted by

$$a - b. \qquad (\textit{Read: ``a minus b''})$$

It expresses "subtracting b objects from a objects." Thus,

$6 - 2$ expresses subtracting 2 objects from 6 objects.

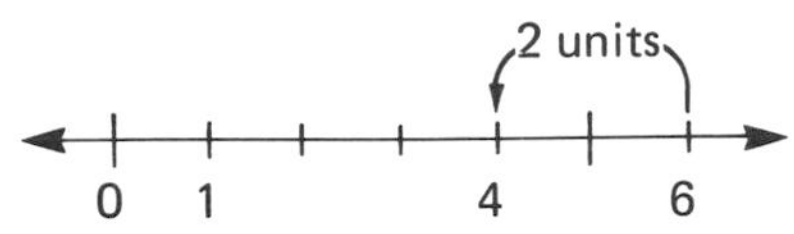

Fig. 1.27. $6 - 2$ lies 2 units to the left of 6.

$6 - 2 = 4$

Here is a geometric way of picturing this difference. Figure 1.27 illustrates:

$6 - 2$ lies 2 units to the *left* of 6.

In general, if a and b are positive numbers and $a > b$, then $a - b$ lies b units to the *left* of a.

Let a be any number. When you subtract 0 from a, the result is a.

$$\boxed{a - 0 = a}$$

On the other hand, when you subtract a from a, the result is 0.

$$\boxed{a - a = 0}$$

(a)

3 units

0 1 5 8

(b)

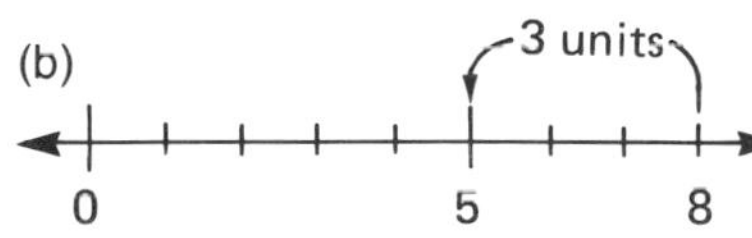

Fig. 1.28(a) $5 + 3$ lies 3 units to the right of 5.

$5 + 3 = 8$

(b) $8 - 3$ lies 3 units to the left of 8.

$8 - 3 = 5$

For example, $7 - 0 = 7$ and $9 - 9 = 0$

Subtraction undoes addition. For example, as illustrated in Figure 1.28,

$$5 + 3 = 8, \quad \text{but} \quad 8 - 3 = 5$$

Thus, you can *check* a *subtraction* example *by adding*. Similarly, you can *check* an *addition* example *by subtracting*.

Usually, larger integers are subtracted in columns.

Example 4 Subtract 23 from 58. Check your result by adding.

Solution Subtract 1's.

$$\begin{array}{r} 58 \\ -\ 23 \\ \hline 5 \end{array}$$

Subtract 10's.

$$\begin{array}{r} 58 \\ -\ 23 \\ \hline 35 \end{array}$$

check

$$\begin{array}{r} 35 \\ +\ 23 \\ \hline 58 \end{array}$$

The difference is 35.

Example 5 Consider subtracting:

$$\begin{array}{r} 31 \\ -\ 19 \\ \hline \end{array}$$

A difficulty arises in the 1's column because $1 < 9$. To overcome this difficulty, observe that

$$\begin{aligned} 31 &= 30 + 1 \\ &= (20 + 10) + 1 \\ &= 20 + (10 + 1) && \textit{by the Associative Law} \\ &= 20 + 11 \end{aligned}$$

Thus, expand as follows:

$$\begin{array}{r} 31 = 30 + 1 = 20 + 11 \\ 19 = 10 + 9 = \underline{10 + \ \ 9} \\ 10 + \ \ 2 \end{array} \qquad \textit{Subtract.}$$

In column form this is shortened to:

$$\begin{array}{rr} \mathbf{2} & \mathbf{11} \\ \not{3} & \not{1} \\ -\,1 & 9 \\ \hline 1 & 2 \end{array}$$

You "borrow" from the 10's column in order to increase 1 to 11 in the 1's column. Then, you subtract 9 from 11 to obtain 2. In the 10's column, you borrowed 1 from 3, leaving 2. Thus, subtract 1 from 2 to obtain 1.

Example 6 Consider subtracting:

$$\begin{array}{r} 402 \\ -\,137 \\ \hline \end{array}$$

You want to borrow from the 10's column in order to subtract

$$\begin{array}{r} 12 \\ -\ \ 7 \\ \hline \end{array}$$

in the 1's column. Because 0 is the 10's digit of 402, you must first borrow from the 100's column.

$$\begin{array}{ccc} \mathbf{3} & \mathbf{10} & \\ \not{4} & \not{0} & 2 \end{array}$$

Now, proceed as before.

$$\begin{array}{rccc} & & \mathbf{9} & \mathbf{12} \\ & \mathbf{3} & \mathbf{10} & \\ & \not{4} & \not{0} & \not{2} \\ - & 1 & 3 & 7 \\ \hline & 2 & 6 & 5 \end{array}$$

CLASS EXERCISES

5. Subtract. Check your result by adding.

$$\begin{array}{r} 5948 \\ -\ 2706 \\ \hline \end{array}$$

6. Subtract.

$$\begin{array}{r} 80\ 004 \\ -\ 50\ 217 \\ \hline \end{array}$$

PROFIT When an item is sold, the **profit** made is its **sale price** minus its **cost**, or the **wholesale price** minus the **retail price**. (We neglect "overhead.") For example, suppose a dress costs a store $25. The store sells the dress for $35. The profit is $10.

$35	-----	*sale price*
− $25	-----	*cost*
$10	-----	*profit*

Example 7 A car dealer pays a wholesale price of $3132 for a Volkswagen. The car is then sold for a retail price of $3698. What is the dealer's profit?

Solution

$3698	-----	*retail price*
− $3132	-----	*wholesale price*
$ 566	-----	*profit*

CLASS EXERCISES

7. It costs a manufacturer $45 to produce a suit. A jobber buys it for $72. What is the manufacturer's profit?

8. A real estate broker buys a house for $27,000. Two months later, she sells it for $35,500. What is her profit?

SOLUTIONS TO CLASS EXERCISES

1. a. $(5 + 2) + 3 = 7 + 3 = 10$ **b.** $5 + (2 + 3) = 5 + 5 = 10$ **2.** 86 **3.** 22,004

4.
$$\begin{array}{r} 18\ 000\ 000 \\ 13\ 000\ 000 \\ 11\ 000\ 000 \\ 10\ 000\ 000 \\ 6\ 000\ 000 \\ \hline 58\ 000\ 000 \end{array}$$

5.
$$\begin{array}{r} 5948 \\ -2706 \\ \hline 3242 \end{array} \qquad \textbf{check} \quad \begin{array}{r} 3242 \\ +\ 2706 \\ \hline 5948 \end{array}$$

6. 29,787 **7.** $27 **8.** $8500

HOME EXERCISES

In Exercises 1 and 2, show that both sums, **a** *and* **b**, *are equal.*

1. a. $(9 + 6) + 5$ **b.** $9 + (6 + 5)$ **2. a.** $(30 + 10) + 20$ **b.** $30 + (10 + 20)$

In Exercises 3–8, add in columns. When adding numbers with five or more digits, instead of using commas, place a slight separation at three-digit intervals. Thus, read 63 479 *as* 63,479

3.	**4.**	**5.**	**6.**	**7.**	**8.**
5123	8736	80 936	63 924	634 897	924 001
1610	9582	74 939	53 875	108 501	309 682
3214	8473	19 259	85 392	219 243	714 093
	5904	74 382	55 555	832 951	217 725
		98 517	39 239		832 043

In Exercises 9 and 10, approximate each sum by first rounding each integer to the nearest 1,000,000 and then adding.

9.	**10.**
16 592 401	14 392 195
18 389 526	13 996 194
21 099 599	11 099 593
14 593 999	10 504 399
	9 984 892

In Exercises 11 and 12: **a.** *Approximate each sum by first rounding each integer to the nearest* 1000 *and then adding.* **b.** *Find the actual sum by adding the integers as given.* **c.** *Round the sum in part* **b** *to the nearest* 1000. **d.** *Do parts* **a** *and* **c** *agree?*

11.	**12.**
6314	5321
2129	3216

13. A donut costs 20 cents and a cup of coffee costs 25 cents. What is the cost of a donut and a cup of coffee?

14. Jose drives 60 miles. After stopping for lunch he drives 50 miles further. Altogether, how many miles does he drive?

15. A carpenter earns $27 on Monday, $32 on Tuesday, $29 on Wednesday, $40 on Thursday, and $36 on Friday. How much does he earn for this five-day week?

16. Maria buys a dress for $37, a hat for $8, a pair of shoes for $26, and a scarf for $4. How much does she spend on her wardrobe?

17. Subtract 53 from 97. **18.** Subtract 204 from 857.

In Exercises 19–24, subtract.

19.	**20.**	**21.**	**22.**	**23.**	**24.**
8847	8145	63 842	20 007	60 507	500 003
− 2943	− 3056	− 49 132	− 10 108	− 10 809	− 400 004

In Exercises 25–28, add or subtract, as indicated. Check your result.

25.	**26.**	**27.**	**28.**
498	584	715	5839
+ 351	− 312	+ 284	− 3882

In Exercises 29–32, approximate each difference by first rounding each number to the nearest 1000.

Sample	**Solution**
Approximate.	Round.
17 021	17,021 rounds to 17,000
− 5 894	5894 rounds to 6000
	17 000
	− 6 000
	11 000

29.	**30.**	**31.**	**32.**
4044	19 420	12 795	77 163
− 3075	− 11 710	− 5 845	− 76 992

33. It costs a craftsman \$18 to make a pair of sandals. He sells the pair for \$36. What is his profit?
34. One week a restaurant owner spends \$835 to operate her restaurant. Total cash receipts for the week are \$1321. What is her profit for that week?
35. It costs a wholesaler \$92 to buy a coat. A retailer buys the coat for \$145. What is the wholesaler's profit?
36. Slick Sid wins \$125 on one race and then loses \$78 on the next race. How much has he won?
37. Of the 2147 men and women enrolled at a college, 1326 are men. How many are women?
38. During one season the Baltimore Orioles won 87 of the 162 games they played. How many games did they lose?
39. The temperature reading at 9 a.m. in Buffalo is 27°F (27 degrees Fahrenheit). At noon the temperature is 42°F. What is the rise in temperature?
40. Rosa has \$420 in her savings account. She withdraws \$150. How much is left in her account?

1.4 ADDITION AND SUBTRACTION OF INTEGERS

$a + (-b)$

We can extend the definitions of addition and subtraction by considering the number line. Throughout this section, the subtraction symbol will be printed in bold to distinguish it from the symbol for inverse. Thus, $4 - 2$ is read: 4 *minus* 2, whereas -5 is read: *the inverse of* 5.

Let a be *any* number—positive, negative, or 0—and let b be a *positive* number or 0. Then, $a + b$ lies b units to the *right* of a (on the number line) and $a - b$ lies b units to *left* of a.

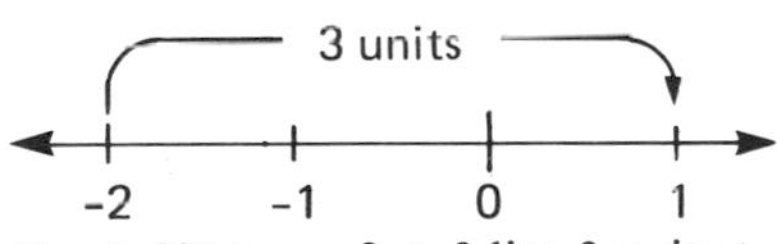

Fig. 1.29(a) $-2 + 3$ lies 3 units to the right of -2.

$$-2 + 3 = 1$$

Example 1

a. $-2 + 3$ lies 3 units to the right of -2. Thus,

$$-2 + 3 = 1 \qquad \text{[See Figure 1.29 (a).]}$$

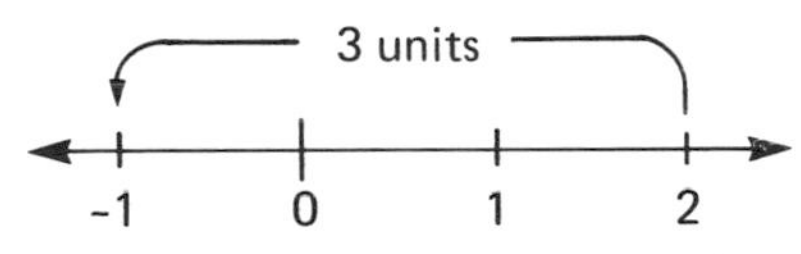

(b) $2 - 3$ lies 3 units to the left of 2.

$$2 - 3 = -1$$

b. $2 - 3$ lies 3 units to the left of 2. Thus,

$$2 - 3 = -1 \qquad \text{[See Figure 1.29 (b).]}$$

Note that

$$\underbrace{2 - 3}_{-1} = -\underbrace{(3 - 2)}_{1}.$$

3 units

-5 -4 -3 -2 -1 0

(c) $-2 - 3$ lies 3 units to the left of -2.

$$-2 - 3 = -5$$

c. $-2 - 3$ lies 3 units to the left of -2. Thus,

$$-2 - 3 = -5 \qquad \text{[See Figure 1.29 (c).]}$$

Let a be any number and let b be a *positive* number. What does it mean to add $-b$, the inverse of b, to a?

DEFINITION

$$a + (-b) = a - b$$

Thus, *to add* $-b$, *simply subtract* b. For example,

$$5 + (-2) \quad \text{means} \quad 5 - 2, \text{ or } 3$$

If b is any positive number,

$$0 - b = 0 + (-b) = -b$$

CLASS EXERCISES *Add.* **1.** $3 + (-2)$ **2.** $3 + (-9)$

ADDING NEGATIVE NUMBERS Negative numbers are added in terms of their inverses, as in the following example.

Example 2 Note that

$$(-5) + (-4) = -5 - 4 = -9$$

$$(-5) + (-4) = -(5 + 4)$$ [See Figure 1.30.]

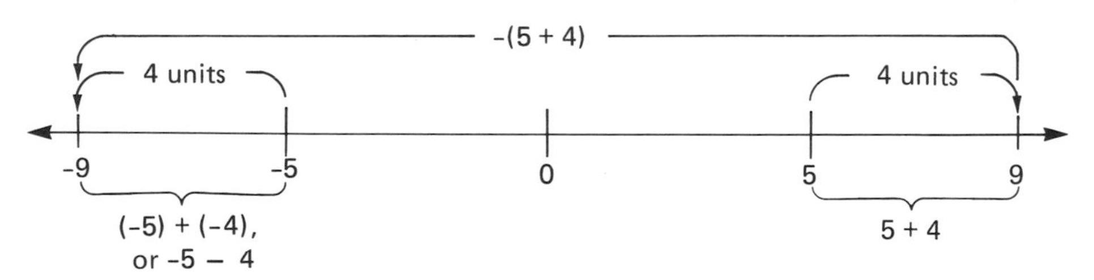

Fig. 1.30 $(-5) + (-4) = -(5 + 4) = -9$.

For any positive numbers a and b,

$$(-a) + (-b) = -(a + b)$$

Thus, *the sum of the inverses is the inverse of the sum.*

CLASS EXERCISES *Add.* **3.** $(-2) + (-1)$ **4.** $(-3) + (-5)$

ADDING NUMBERS WITH DIFFERENT SIGNS

The **sign** of a positive number is $+$ and the **sign** of a negative number is $-$.

$$\text{The sign of 8 is } +\text{; the sign of } -8 \text{ is } -$$

Note that -8 is the inverse of 8. Also,

$$|-8| = |8| = 8$$

The inverse of a is the number that is added to a to obtain 0. This is because a and $-a$ are each the same distance from the orgin, 0, but on opposite sides of it. Thus, in general, *for any number a*,

$$\boxed{a + (-a) = 0}$$

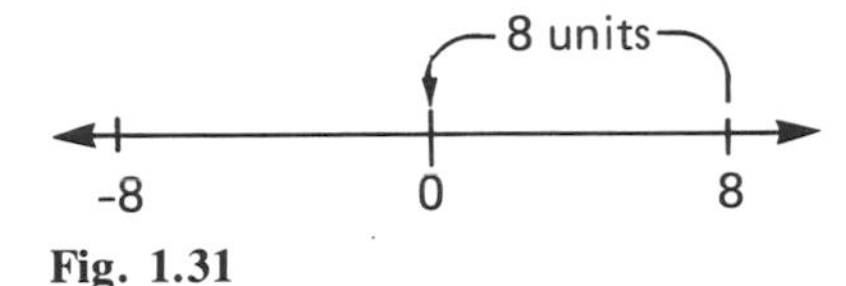

Fig. 1.31

For example, as illustrated in Figure 1.31, $8 + (-8) = 0$

Now, consider numbers with *different signs* and *different absolute values.*

Example 3 a.

$$8 - 5 = 3$$

Note that $8 - 5$ lies 5 units to the *left* of 8, as shown in Figure 1.32(a). Also,

$$8 + (-5) = 8 - 5 = 3$$

In fact, $8 > 5$. Thus, 8 is larger than -5 in absolute value, and is therefore further from the origin. When adding 8 and -5, the positive sign of 8 prevails.

b.

$$5 - 8 = -3$$

Here, $5 - 8$ lies 8 units to the *left* of 5, as shown in Figure 1.32(b). Also,

$$5 + (-8) = 5 - 8 = -3$$

Note that -8 is larger than 5 in absolute value, and is therefore further from the origin. When adding 5 and -8, the negative sign of -8 prevails.

Fig. 1.32(a) $8 - 5 > 0$ because $8 > 5$.
(b) $5 - 8 < 0$ because $5 < 8$.

Suppose that a and b are both positive. If $a > b$, then $a - b > 0$ and therefore,

$$a + (-b) > 0$$

Thus,

$$\underbrace{8 - 5}_{3} = 8 + (-5) > 0$$

If $a < b$, then $a - b < 0$ and therefore,

$$a + (-b) < 0$$

Thus,

$$\underbrace{5 - 8}_{-3} = 5 + (-8) < 0$$

> To add numbers with *different* signs and *different* absolute values:
> 1. *Subtract* the smaller absolute value from the larger absolute value.
> 2. The sign of the number with larger absolute value prevails.

Thus, in part **a** of Example 3, $|8| > |-5|$

Therefore, $8 + (-5) = +(8 - 5) = 3$ (the + is the sign of 8)

But, in part **b** of Example 3, $|-8| > |5|$

Thus, $-8 + 5 = -(8 - 5) = -3$ (the − is the sign of −8)

Example 4
Add. $-12 + 7$

Solution

$$|-12| > |7|$$

Therefore,

$$-12 + 7 = -(12 - 7) = -5$$

(the − is the sign of −12)

Example 5
Add. $(-192) + 281$

Solution

$$|281| > |-192|$$

Subtract.

$$\begin{array}{r} 281 \\ -\ 192 \\ \hline 89 \end{array}$$

281: *positive sign*; 192: *negative sign*

Thus,

$$(-192) + 281 = +(281 - 192) = 89$$

(the + is the sign of 281)

CLASS EXERCISES *Add.* **5.** $(-12) + 12$ **6.** $-53 + 89$ **7.** $292 + (-395)$

SUBTRACTING A NEGATIVE NUMBER

Recall that $6 - 2$ lies 2 units to the *left* of 6. [Figure 1.33(a)]
We now define $6 - (-2)$ so that

$6 - (-2)$ lies 2 units to the *right* of 6.

Observe that $$6 - (-2) = 6 + 2 = 8.$$ [Figure 1.33(b)]

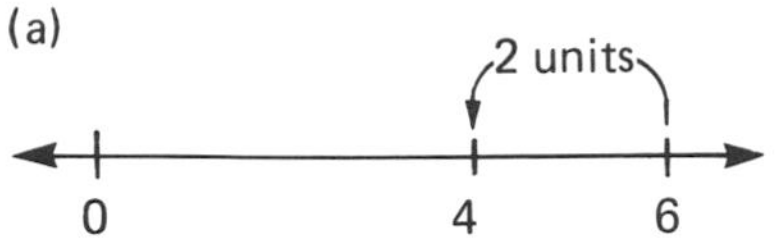

Fig. 1.33 (a) $6 - 2 = 4$. **(b)** $6 - (-2) = 6 + 2 = 8$.

DEFINITION

Let a be any number and let b be a positive number. Then,

$$a - (-b) = a + b$$

Example 6 **a.** $3 - (-10) = 3 + 10 = 13$ **b.** $(-5) - (-7) = -5 + 7 = 2$

CLASS EXERCISES *Subtract.* **8.** $8 - (-4)$ **9.** $(-9) - (-5)$

COMMUTATIVE AND ASSOCIATIVE LAWS

The Commutative Law,

$$a + b = b + a$$

and the Associative Law,

$$(a + b) + c = a + (b + c)$$

apply to the addition of numbers—positive, negative, or 0. These laws enable you to rearrange numbers when adding them. For example,

$$\begin{aligned} -5 + 6 + (-2) &= [-5 + 6] + (-2) \\ &= [6 + (-5)] + (-2) && \textit{by the Commutative Law} \\ &= 6 + [(-5) + (-2)] && \textit{by the Associative Law} \\ &= 6 + [-7] \\ &= -1 \end{aligned}$$

> If the numbers are not all of the same sign:
> 1. First, add the positive numbers to obtain a, where a is positive.
> 2. Then, add the negative numbers to obtain $-b$, where b is positive (and $-b$ is negative).
> 3. The sum of all the numbers is $a + (-b)$, or $a - b$, which can be positive, negative, or 0.

Example 7

Add.

$$\begin{array}{r} 12 \\ -8 \\ 19 \\ 20 \\ -9 \\ \underline{-17} \end{array}$$

Solution

First, add the positive and negative numbers separately:

$$\begin{array}{rr} 12 & -8 \\ 19 & -9 \\ \underline{20} & \underline{-17} \\ 51 & -34 \end{array}$$

The sum of all the numbers is 51 − 34, or 17.

CLASS EXERCISES *Add*

10. $\begin{array}{r} 57 \\ -38 \\ -59 \\ -96 \\ 27 \\ \underline{92} \end{array}$

11. $49 + (-17) + (-18) + 25 + (-33)$

PARENTHESES

When addition and subtraction are combined in the same example, parentheses must often be used to clarify the intention. *First, combine the numbers within each pair of parentheses.*

Example 8

$$8 + \underbrace{(5 - 3)}_{2} - \underbrace{(3 - 4)}_{-1} = 8 + 2 - (-1) = 8 + 2 + 1 = 11$$

There may be pairs of parentheses within a pair of brackets. *Combine within the inner pairs first*, as in Example 9.

Example 9

$$10 - [\underbrace{(7 - 2)}_{5} - \underbrace{(8 - 3)}_{5}] = 10 - \underbrace{[5 - 5]}_{0} = 10 - 0 = 10$$

Here, two pairs of parentheses lie within a pair of brackets. *Combine within both inner pairs first.*

If no parentheses are given, combine in the order written, from left to right.

Example 10

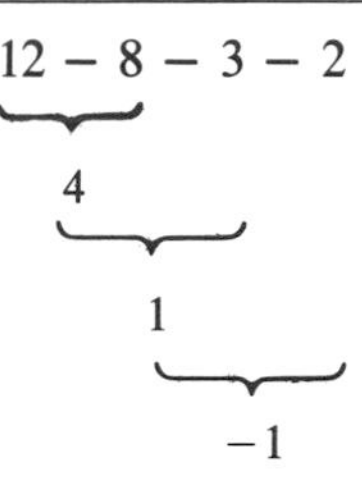

The value is −1. You could also use

$$12 - 8 - 3 - 2 = 12 + \underbrace{(-8) + (-3) + (-2)}_{-13} = -1.$$

CLASS EXERCISES *Find each value.*

12. $20 - (8 + 5 + 6) - (-7 - 4)$ **13.** $9 - [4 - (1 - 3)]$
14. $-12 - [(5 - 4) - (10 - 14)]$ **15.** $10 - 4 - 9 + 8 + 2 - 7 + 3$

SOLUTIONS TO CLASS EXERCISES

1. 1 **2.** −6 **3.** −3 **4.** −8 **5.** 0 **6.** 36 **7.** −103 **8.** $8 - (-4) = 8 + 4 = 12$
9. $-9 - (-5) = -9 + 5 = -4$

10.

57	− 38	−193
27	− 59	176
92	− 96	− 17
176	−193	

11.

49	−17	74
25	−18	−68
74	−33	6
	−68	

12. $20 - \underbrace{(8 + 5 + 6)}_{19} - \underbrace{(-7 - 4)}_{-11} = 20 - 19 - (-11) = 20 - 19 + 11 = 12$

13. $9 - [4 - \underbrace{(1 - 3)}_{-2}] = 9 - \underbrace{[4 - (-2)]}_{4 + 2} = 9 - 6 = 3$

14. $-12 - [\underbrace{(5 - 4)}_{1} - \underbrace{(10 - 14)}_{-4}] = -12 - \underbrace{[1 - (-4)]}_{1 + 4} = -12 - 5 = -17$

15. $10 - 4 - 9 + 8 + 2 - 7 + 3$

$$\underbrace{\underbrace{\underbrace{\underbrace{\underbrace{\underbrace{10 - 4}_{6} - 9}_{-3} + 8}_{5} + 2}_{7} - 7}_{0} + 3}_{3}$$

The value is 3.

HOME EXERCISES

In Exercises 1–8, add, as indicated.

1. $\begin{array}{r} -3016 \\ \underline{-2854} \end{array}$ **2.** $\begin{array}{r} -271 \\ \underline{-181} \end{array}$ **3.** $\begin{array}{r} -6089 \\ \underline{-5464} \end{array}$

4. $14 + (-8)$ **5.** $-19 + 19$ **6.** $-12 + 21$ **7.** $18 + (-15)$ **8.** $(-18) + 15$

In Exercises 9–16, subtract, as indicated.

9. $9 - 4$ **10.** $9 - (-4)$ **11.** $-9 - 4$ **12.** $-9 - (-4)$ **13.** $12 - 0$ **14.** $0 - 12$ **15.** $-18 - 17$
16. $-18 - (-17)$ **17.** Subtract -12 from 19. **18.** Subtract -30 from -60.

In Exercises 19–26, add, as indicated.

19. $\begin{array}{r} -31 \\ -17 \\ 19 \\ \underline{26} \end{array}$ **20.** $\begin{array}{r} 29 \\ 32 \\ -15 \\ -46 \\ 38 \\ \underline{-53} \end{array}$ **21.** $\begin{array}{r} -32 \\ -51 \\ 63 \\ 72 \\ \underline{-80} \end{array}$ **22.** $\begin{array}{r} -92 \\ -81 \\ 27 \\ 74 \\ 76 \\ \underline{-39} \end{array}$ **23.** $\begin{array}{r} 916 \\ -53 \\ -741 \\ 808 \\ -747 \\ \underline{831} \end{array}$ **24.** $\begin{array}{r} 1082 \\ 5019 \\ -1131 \\ -2163 \\ -1442 \\ \underline{3042} \end{array}$

25. $54 + 19 + (-17) + (-36) + (-22) + 47$ **26.** $27 + (-36) + (-79) + 45 + (-55) + (-32)$

In Exercises 27–33, subtract, as indicated.

27. $6 - (5 - 3)$ **28.** $6 - 5 - 3$ **29.** $10 - (6 - 3) - (9 - 2) - (1 - 8)$ **30.** $16 - [(4 - 5) - 1] - (-7)$
31. $16 - [4 - (5 - 1)] - (-3)$ **32.** $(10 - 4) - [6 - (3 - 1)]$ **33.** $[(7 - 12) - 5] - [8 - (2 - 7)]$

In Exercises 34–37, add or subtract, as indicated.

34. $8 + 10 - 6 + 7 - 2$ **35.** $12 + (4 - 3) - (4 + 5)$ **36.** $7 + [(6 - 3 + 2) - (7 + 19)]$
37. $3 - [(8 - 5) - (4 + 7)]$

38. The temperature in Chicago at noon is 12° Fahrenheit. By midnight the temperature has dropped 20 degrees. What is the reading at midnight?

39. The temperature in Seattle rises from $-8°$ Fahrenheit to $-2°$ Fahrenheit. How many degrees does it rise?

40. A car travels 40 miles north and then 50 miles south. How far from the starting point is the car? Is it north or is it south of the starting point?

41. A gambler wins \$20, loses \$35, wins \$37, and then loses \$6. How much has he won or lost?

42. A traveling salesman drives 10 miles west, then 4 miles east, and then 6 miles west. Where is he in relation to his starting point?

43. How much change does Susan receive if she purchases \$5, \$6, and \$8 items with a \$20 dollar bill?

1.5 MULTIPLICATION AND EXPONENTIATION

BASIC RULES OF MULTIPLICATION

The **product of** two numbers, a **and** b (or a **times** b), is usually written as

$$a \cdot b, \qquad ab, \qquad a \times b, \qquad \text{or as} \qquad \begin{array}{r} a \\ \times\, b \\ \hline \end{array}$$

The numbers a and b are called the **factors** of this product, $a \cdot b$.

Multiplication by a positive integer amounts to repeated addition. Thus,

$$5 \cdot 3 = 5 + 5 + 5 = 15$$

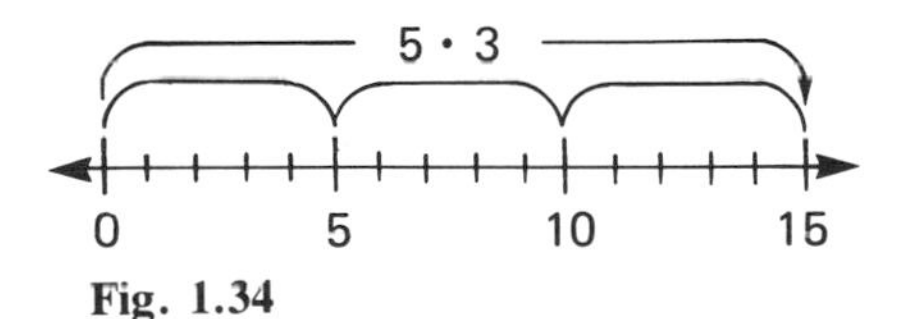

Fig. 1.34

Figure 1.34 depicts this product geometrically.

For any number a,

$$\boxed{a \cdot 1 = a \quad \text{and} \quad a \cdot 0 = 0}$$

Thus, $\quad 9 \cdot 1 = 9 \quad$ and $\quad 6 \cdot 0 = 0$

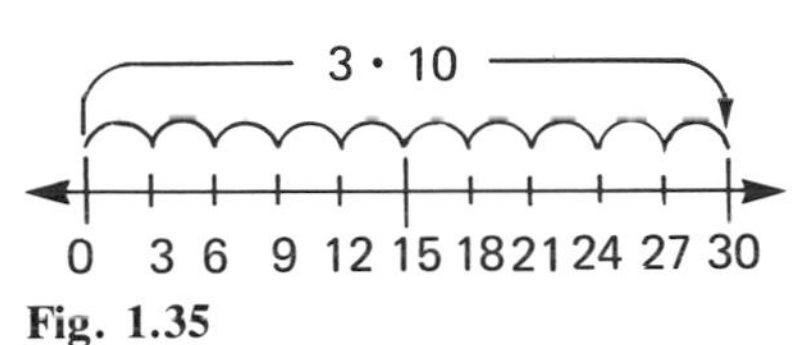

Fig. 1.35

Next, Figure 1.35 illustrates that

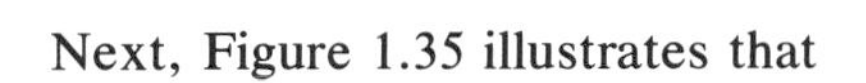

$$3 \cdot 10 = \underbrace{3 + 3 + 3 + 3 + 3 + 3 + 3 + 3 + 3 + 3}_{10 \textit{ of these}} = 30$$

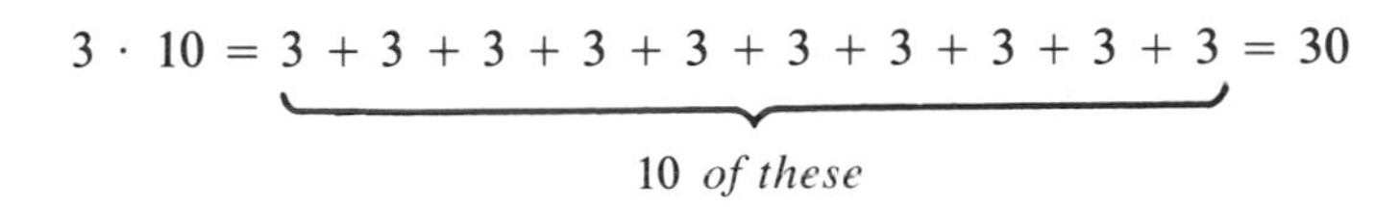

Recall that 10 is the base of our number system. Note that

$$30 = 3 \cdot 10 + 0 \cdot 1$$

$$\begin{array}{ccc} & \text{10's} & \text{1's} \\ & \downarrow & \downarrow \\ 30 = & 3 & 0 \end{array}$$

To multiply an integer by 10, which ends in *one* 0, insert *one* 0 to the right of the 1's digit of the integer. Thus,

$$3 \cdot \mathbf{10} = \mathbf{30}$$

To multiply an integer by 100, which ends in *two* 0's, insert *two* 0's to the right of the 1's digit of the integer. And **to multiply an integer by 1000**, which ends in *three* 0's, insert *three* 0's to the right of the 1's digit of the integer. How many 0's do you insert to the right of the 1's digit of the integer when multiplying by **100,000**?

Example 1 **a.** $43 \cdot 10 = 430$ **b.** $43 \cdot 100{,}000 = 4{,}300{,}000$ **c.** $43 \cdot 100 = 4300$

In general, you can multiply numbers in either order. Thus, *for any numbers a and b*,

$$\boxed{a \cdot b = b \cdot a}$$

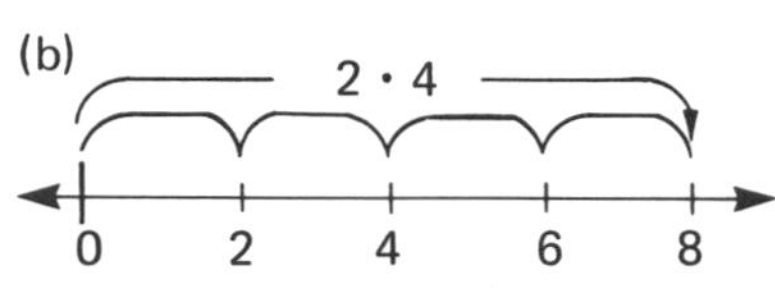

Fig. 1.36(a) $4 \cdot 2 - 8$
(b) $2 \cdot 4 = 8$.

This is the **Commutative Law of Multiplication**. For example, as illustrated in Figure 1.36,

$$4 \cdot 2 = 2 \cdot 4 = 8$$

Now, consider the product of three numbers.

Example 2 **a.** $\underbrace{(3 \cdot 2)}_{6} \cdot 5 = 6 \cdot 5 = 30$ **b.** $3 \cdot \underbrace{(2 \cdot 5)}_{10} = 3 \cdot 10 = 30$

In both cases, the product of the numbers 3, 2, and 5 is 30.

In general, *for any three numbers a, b, and c*,

$$\boxed{(a \cdot b) \cdot c = a \cdot (b \cdot c)}$$

This is the **Associative Law of Multiplication**. Here, parentheses indicate which multiplication occurs first. Because of the Associative Law, you usually write the product

$$a \cdot b \cdot c$$

without parentheses. Thus,

$$a \cdot b \cdot c = (a \cdot b) \cdot c = a \cdot (b \cdot c)$$

Next, observe that $4 \cdot \underbrace{(3 + 2)}_{5} = 4 \cdot 5 = 20$

But, you could also calculate as follows:

$$\begin{aligned} 4 \cdot (3 + 2) &= 4 \cdot 3 + 4 \cdot 2 \\ &= 12 + 8 \\ &= 20 \end{aligned}$$

The second method, though longer, has an important use in arithmetic as well as in algebra. This method illustrates the following **Distributive Laws**. *Let a, b, and c be any numbers.*

$$a \cdot (b + c) = a \cdot b + a \cdot c$$
$$(b + c) \cdot a = b \cdot a + c \cdot a$$

CLASS EXERCISES

In Exercises 1 and 2, multiply.

1. $47 \cdot 100$ **2.** $53 \cdot 1{,}000{,}000$

In Exercise 3, show that products **a** *and* **b** *are equal.*

3. a. $(6 \cdot 2) \cdot 3$ **b.** $6 \cdot (2 \cdot 3)$

4. Use the Distributive Laws to find $(6 + 2) \cdot 4$.

COLUMN MULTIPLICATION

The Distributive Laws are used when at least one factor of a product contains two or more digits.

Example 3 $72 \cdot 58 = 72 \cdot (50 + 8) = (72 \cdot 50) + (72 \cdot 8)$

$$\begin{array}{r} ^{1} \\ 72 \\ \times\ 8 \\ \hline 576 \end{array} \quad + \quad \begin{array}{r} ^{1} \\ 72 \\ \times\ 50 \\ \hline 3600 \end{array} \qquad \text{or in shortened form:} \qquad \begin{array}{r} 72 \\ \times\ 58 \\ \hline 576 \\ 3600 \\ \hline 4176 \end{array}$$

In the shortened form, the second 0 can be omitted. The other digits remain in their original position.

When one or more digits of the second factor is 0, the method can be shortened, as illustrated in the next example.

Example 4

$$\begin{array}{r} 1214 \\ \times\ 4020 \\ \hline 24280 \\ 48560 \\ \hline 4880280 \end{array} \qquad \begin{array}{l} \text{-----}1214 \cdot 20 = 24{,}280 \\ \text{-----}1214 \cdot 4000 = 4{,}856{,}000 \end{array}$$

CLASS EXERCISES *Multiply.*

5. $\begin{array}{r} 23 \\ \times\ 31 \\ \hline \end{array}$ **6.** $\begin{array}{r} 518 \\ \times\ 791 \\ \hline \end{array}$ **7.** $\begin{array}{r} 6093 \\ \times\ 5010 \\ \hline \end{array}$

NEGATIVE FACTORS

How do you multiply a *positive* integer by a *negative* integer? To see how this is done, follow the pattern in the right-hand column of Table 1.1. Refer to Figure 1.37. Then, fill in the blanks in Table 1.1.

TABLE 1.1

$10 \cdot 3 =$	30
$10 \cdot 2 =$	20
$10 \cdot 1 =$	10
$10 \cdot 0 =$	0
$10 \cdot (-1) =$	
$10 \cdot (-2) =$	
$10 \cdot (-3) =$	

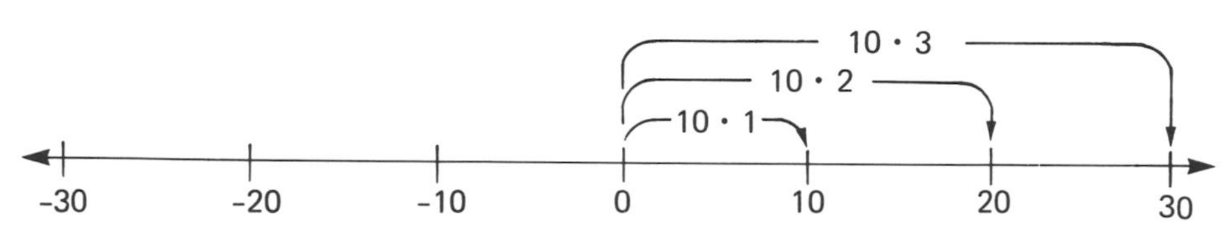

Fig. 1.37

Clearly, the pattern suggested is indicated in Table 1.2. (See Figure 1.38.)

TABLE 1.2

$10 \cdot 3 =$	30
$10 \cdot 2 =$	20
$10 \cdot 1 =$	10
$10 \cdot 0 =$	0
$10 \cdot (-1) =$	-10
$10 \cdot (-2) =$	-20
$10 \cdot (-3) =$	-30

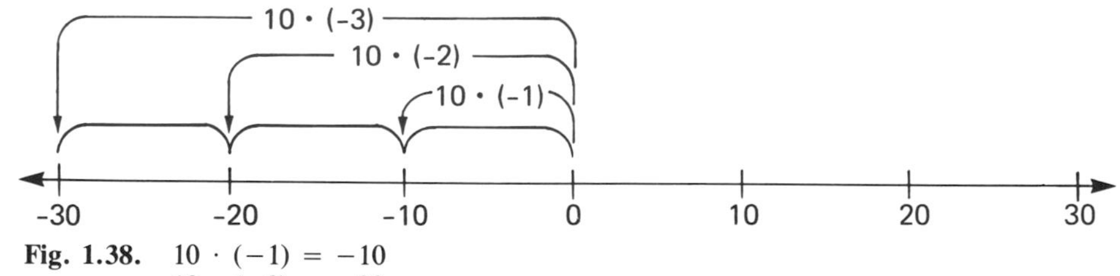

Fig. 1.38. $10 \cdot (-1) = -10$
$10 \cdot (-2) = -20$
$10 \cdot (-3) = -30$

We generally use parentheses when one or more factors of a product is negative. It is clearer to write $3(-2)$ than $3 \cdot -2$. (You might *misread* $3 \cdot -2$ as $3 - 2$.)

If a and b are positive numbers,

$$a(-b) = (-a)b = -ab \quad \text{and} \quad a(-1) = -a$$

Thus,

$$3(-2) = (-3)2 = -6$$

How do you multiply two *negative* integers? To see how this is done, follow the pattern in the right-hand column of Table 1.3. Refer to Figure 1.39. Then, fill in the blanks in Table 1.3.

TABLE 1.3

$(-10) \cdot 3 =$	-30
$(-10) \cdot 2 =$	-20
$(-10) \cdot 1 =$	-10
$(-10) \cdot 0 =$	0
$(-10) \cdot (-1) =$	
$(-10) \cdot (-2) =$	
$(-10) \cdot (-3) =$	

Fig. 1.39.

$$(-10) \cdot 3 = -(10 \cdot 3) = -30$$
$$(-10) \cdot 2 = -(10 \cdot 2) = -20$$
$$(-10) \cdot 1 = -(10 \cdot 1) = -10$$

Clearly, the pattern suggested is indicated in Table 1.4. (See Figure 1.40.)

TABLE 1.4

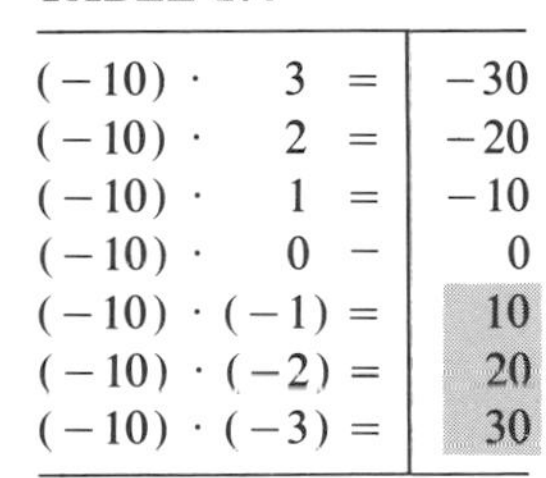

$(-10) \cdot 3 =$	-30
$(-10) \cdot 2 =$	-20
$(-10) \cdot 1 =$	-10
$(-10) \cdot 0 -$	0
$(-10) \cdot (-1) =$	10
$(-10) \cdot (-2) =$	20
$(-10) \cdot (-3) =$	30

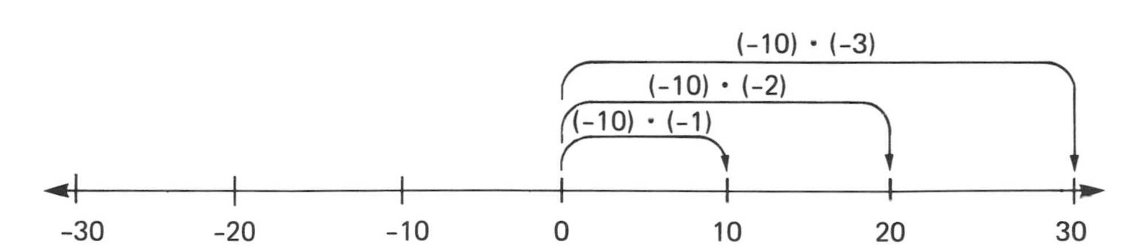

Fig. 1.40.

$$(-10) \cdot (-1) - 10$$
$$(-10) \cdot (-2) = 20$$
$$(-10) \cdot (-3) = 30$$

Let a and b be postive numbers. Then,

$$\boxed{(-a)(-b) = a \cdot b}$$

For example, $(-4)(-3) = 4 \cdot 3 = 12$

Suppose there are several negative factors.

Example 5

a. $3(-1) = -3$ **b.** $(-3)(-1) = 3$

c. $3(-1)(-2) = 6$. Here there are two negative factors, and thus, two sign changes. The product is positive.

d. $3(-1)(-2)(-2) = -12$. There are three negative factors. The product is negative.

e. $3(-1)(-2)(-2)(-2) = 24$. With four negative factors, the product is positive.

f. $3(-1)(-2)(-2)(-2)0 = 0$

The sign of the product:

1. If at least one factor is 0, the product is 0, as in part **f** of Example 5.

Now, assume none of the factors is 0. Then,

2. the product is positive if there is an *even number* (0, 2, 4, 6, . . .) of negative factors (or if all of the factors are positive), as in parts **b, c,** and **e,** of Example 5;
3. the product is negative if there is an *odd number* (1, 3, 5, . . .) of negative factors, as in parts **a** and **d** of Example 5.

CLASS EXERCISES *Multiply.* **8.** $(-3)(-2)(-2)$ **9.** $(-4)(-1)(-2)(-2)$ **10.** $(-2)(-1)(-2)(-1)(-2)$

SQUARES AND CUBES For any number a,

$$a^2 \text{ means } a \cdot a \qquad \text{and} \qquad a^3 \text{ means } a \cdot a \cdot a$$

Call a^2 the **square of *a***, or ***a* squared**, and call a^3 the **cube of *a*** or ***a* cubed**.

Example 6 **a.** $4^2 = 4 \cdot 4 = 16$ **b.** $9^2 = 9 \cdot 9 = 81$ **c.** $0^2 = 0 \cdot 0 = 0$

For negative numbers, parentheses are used to clarify the intended meaning.

Example 7 **a.** $(-2)^2 = (-2)(-2) = 4$
b. $-2^2 = -4$ because, *by convention*, $-2^2 = -(2^2) = -4$

In part **a**, square the negative number -2. In part **b**, first square 2, and then consider the inverse of 2^2.

Example 8 **a.** $3^3 = 3 \cdot 3 \cdot 3 = 27$ **b.** $1^3 = 1 \cdot 1 \cdot 1 = 1$
c. $(-2)^3 = (-2)(-2)(-2) = -(2 \cdot 2 \cdot 2) = -8$ **d.** $-2^3 = -(2^3) = -8$

CLASS EXERCISES *Find each value.*
11. 7^2 **12.** $(-4)^2$ **13.** -4^2 **14.** 10^3 **15.** $(-3)^3$ **16.** -3^3

***n*TH POWERS** Let a be any number.

$$a^1 \text{ means } a$$

Let n be a positive integer, where $n > 1$. Then,

$$a^n \text{ means } \underbrace{a \cdot a \cdot a \ldots a}_{n \text{ factors}}$$

Thus, when considering a^n, a appears as a factor n times. Call a^n the ***n*th power of *a***, or ***a* to the *n*th**.

Example 9
a. $2^4 = 2 \cdot 2 \cdot 2 \cdot 2 = 16$
b. $2^5 = 2 \cdot 2 \cdot 2 \cdot 2 \cdot 2 = 32$
c. $2^6 = 2 \cdot 2 \cdot 2 \cdot 2 \cdot 2 \cdot 2 = 64$
d. $10^4 =$ **10,000** (with **4** zeros)
e. $10^5 =$ **100,000** (with **5** zeros)
f. $10^8 =$ **100,000,000** (with **8** zeros)

Example 10
a. $1^n = 1$ for all positive integers n
b. $0^n = 0$ for all positive integers n
c. $(-1)^1 = -1$
d. $(-1)^2 = (-1)(-1) = 1$
e. $(-1)^3 = (-1)(-1)(-1) = -1$
f. $(-1)^4 = (-1)(-1)(-1)(-1) = 1$

Note the pattern in parts **c, d, e,** and **f** of Example 10.

> $(-1)^n = -1$ for *odd* (positive) integers n, that is, for $n = 1, 3, 5, 7, \ldots$
> $(-1)^n = 1$ for *even* (positive) integers n, that is, for $n = 2, 4, 6, 8, \ldots$

Example 10 can be generalized as follows. *Odd powers of negative numbers are negative. Even powers of negative numbers are positive.* To see this, count the number of negative factors, as indicated by the exponent.

Example 11
a. $(-2)^3 = (-2)(-2)(-2) = -8$
b. $(-2)^4 = (-2)(-2)(-2)(-2) = 16$
c. $(-2)^5 = (-2)(-2)(-2)(-2)(-2) = -32$
d. $(-2)^6 = (-2)(-2)(-2)(-2)(-2)(-2) = 64$

CLASS EXERCISES

Find the values of the indicated powers.

17. 2^8 **18.** 10^6 **19.** $(-1)^{25}$

20. $(-1)^{26}$ **21.** $(-10)^5$ **22.** $(-10)^6$

BASE AND EXPONENT

$$4^3 = 64$$

Here, 4 is called the *base* and 3 the *exponent*. 4^3, or 64, is called the *third power* (*or cube*) *of* 4.

DEFINITION

> BASE, EXPONENT. In the expression a^n, a is called the **base** and n the **exponent**. a^n is also called the ***n*th power of *a***.

CLASS EXERCISES *Consider the statement,* $2^5 = 32$.

23. What is the base? **24.** What is the exponent? **25.** What power of 2 is 32?

SOLUTIONS TO CLASS EXERCISES

1. 4700 **2.** 53,000,000 **3. a.** $(6 \cdot 2) \cdot 3 = 12 \cdot 3 = 36$ **b.** $6 \cdot (2 \cdot 3) = 6 \cdot 6 = 36$

4. $(6 + 2) \cdot 4 = 6 \cdot 4 + 2 \cdot 4 = 24 + 8 = 32$ **5.** 713 **6.** 409,738 **7.** 30,525,930 **8.** -12 **9.** 16

10. -8 **11.** 49 **12.** $(-4)^2 = (-4)(-4) = 16$ **13.** $-4^2 = -(4^2) = -16$ **14.** $10^3 = 10 \cdot 10 \cdot 10 = \mathbf{1000}$

15. $(-3)^3 = (-3)(-3)(-3) = -27$ **16.** $-3^3 = -(3 \cdot 3 \cdot 3) = -27$ **17.** 256

18. $10^6 = \mathbf{1{,}000{,}000}$ (with 6 zeros) **19.** -1 **20.** 1 **21.** $\mathbf{-100{,}000}$ **22.** **1,000,000** **23.** 2 **24.** 5

25. 32 is the fifth power of 2.

HOME EXERCISES

In Exercises 1–4, multiply.

1. $59 \cdot 1000$ **2.** $160 \cdot 1000$ **3.** $120 \cdot 10{,}000$ **4.** $27 \cdot 1{,}000{,}000$

*In Exercises 5 and 6, show that both products **a** and **b** are equal.*

5. a. $(4 \cdot 2) \cdot 6$ **b.** $4 \cdot (2 \cdot 6)$ **6. a.** $(10 \cdot 4) \cdot 2$ **b.** $10 \cdot (4 \cdot 2)$

In Exercises 7–10, approximate the product by rounding each factor to the nearest 100.

Sample Approximate.	**Solution** Round.
$812 \cdot 735$	812 rounds to 800 735 rounds to 700 $800 \cdot 700 = 8 \cdot 7 \cdot 100 \cdot 100$ $= 560{,}000$

7. $505 \cdot 319$ **8.** $716 \cdot 432$ **9.** $892 \cdot 194$ **10.** $1241 \cdot 919$

In Exercises 11–14, use the Distributive Laws to find each value.

11. $(3 + 5) \cdot 2$ **12.** $8 \cdot (5 + 3)$ **13.** $(7 - 2) \cdot 3$ **14.** $9 \cdot (9 + 6)$

In Exercises 15–24, multiply.

15. 23×13 **16.** 94×87 **17.** 893×79 **18.** 8039×6011 **19.** 3082×7100 **20.** 6193×5285

21. $(-8)(-2)(-2)(10)$

22. $(-5)(-4)(-3)(-2)$

23. $(-2)(-2)(-2)(-2)(-2)$

24. $(-2)(-1)(-3)(0)(-10)$

In Exercises 25–42, find each value.

25. 8^2 **26.** 11^2 **27.** $(-7)^2$ **28.** $(-10)^2$ **29.** -10^2 **30.** 20^2

31. 5^3 **32.** $(-1)^3$ **33.** -1^3 **34.** $(-5)^3$ **35.** $(-10)^3$ **36.** -20^3

37. 0^4 **38.** $(-3)^5$ **39.** $(-1)^{20}$ **40.** $(-1)^{27}$ **41.** $(-2)^7$ **42.** $(-10)^{10}$

In Exercises 43–45, consider the statement $3^4 = 81$.

43. What is the base? **44.** What is the exponent? **45.** What power of 3 is 81?

46. What power of 2 is 32? **47.** What power of (-2) is 64? **48.** What power of 10 is 100,000?

49. There are 60 minutes in an hour. An elevator operator works an eight-nour shift. How many minutes does he work?

50. A box can hold 250 sheets of paper. How many sheets can 37 boxes hold?

51. Each worker in a plant earns \$156 a week. Altogether, there are 207 workers. What is the weekly payroll?

52. 52,832 people each pay \$35 a ticket to see the heavyweight championship fight. How much money is collected for tickets?

53. Each of 5294 students at a college pays \$1760 tuition. How much money is raised in tuition fees?

54. Each of the eight buildings on a campus contains 40 classrooms. Each classroom contains 50 seats. How many seats are there altogether?

1.6 DIVISION

MULTIPLICATION AND DIVISION

$$10 = 5 \cdot 2$$

Here, 10 is the *product*. 5 and 2 are each *factors*. In division, you will write

$$\frac{10}{2} = 5$$

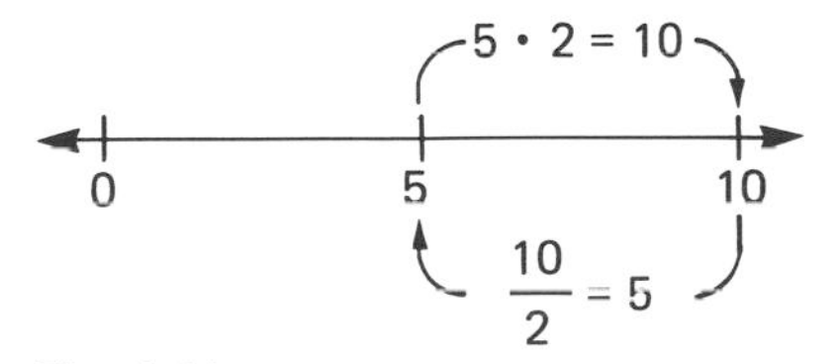

Fig. 1.41

and call 10 the *dividend*, 2 the *divisor*, and 5 the *quotient*. Figure 1.41 illustrates the preceding multiplication and division on the number line. Observe that *division undoes multiplication*, just as subtraction undoes addition.

DEFINITION

> DIVISION. Let a, b, and c be numbers, with $c \neq 0$ [*read*: c *not equal to* 0]. Let
>
> $$\frac{a}{c} = b, \quad \text{if} \quad a = b \cdot c$$
>
> In the case of division, a is called the **dividend**, c, the **divisor**, and b, the **quotient**.

Division of a by c is also written as $a \div c$ and as $c \overline{)a}$.

It is understood that whenever $\frac{a}{c} = b$, then $a = b \cdot c$. Thus, to **check division**, find the product $b \cdot c$ (or $c \cdot b$, if more convenient). *You should obtain the dividend, a.*

Example 1 *dividend* → $\frac{20}{4} = 5$ ← *quotient*, with 20 labelled *dividend* and 4 labelled *divisor*.

check $20 = 5 \cdot 4$, with 20 labelled *dividend*, 5 labelled *quotient* and 4 labelled *divisor*.

Division by 0 *is not defined*. To understand why not, first, suppose you divide a *nonzero* number, such as 6, by 0. Try to use the definition of division in terms of multiplication.

$$\frac{6}{0} = b \quad \text{would mean that} \quad 6 = b \cdot 0$$

But, by definition,

$$b \cdot 0 = 0 \text{ (instead of 6)}$$

Next, try to divide 0 by 0, according to the definition. Note that

$$0 = 1 \cdot 0, \quad 0 = 5 \cdot 0, \quad 0 = 0 \cdot 0$$

You would obtain

$$\frac{0}{0} = 1, \quad \frac{0}{0} = 5, \quad \frac{0}{0} = 0$$

There would be no *single* quotient, b, as there is when the divisor is nonzero. Therefore, *division by* 0 *is not allowed*.

Although you *cannot divide by* 0, the number 0 *can be divided by* any nonzero number. Thus,

$$\frac{0}{4} = 0 \quad \text{because} \quad 0 = 0 \cdot 4$$

For every number a other than 0,

$$\boxed{\frac{0}{a} = 0}$$

For every number a,

$$\boxed{\frac{a}{1} = a \quad \text{because} \quad a = a \cdot 1}$$

For every number a other than 0,

$$\frac{a}{a} = 1 \quad \text{because} \quad a = 1 \cdot a$$

Thus,

$$\frac{9}{1} = 9 \quad \text{and} \quad \frac{11}{11} = 1$$

Division was defined in terms of multiplication.

$$\frac{a}{c} = b, \quad \text{if} \quad a = b \cdot c$$

Recall the rules for the sign of a product.

1. The *product* a is 0 if *either* factor b or c is 0.
2. The *product* a is *positive* if *both* factors are positive or if *both* factors are negative. Thus, a is *positive* if *both* factors have the *same sign*.
3. The *product* a is *negative* if one factor is positive and the other is negative. Thus, a is *negative* if the factors have *different signs*.

Rules for the sign of the quotient. (Refer to Example 2.)
1. The *quotient* b is 0 if a is 0 and c is *not* 0.
2. The *quotient* b is *positive* if a and c have the *same sign*.
3. The *quotient* b is *negative* if a and c have *different signs*.

Example 2 **a.** $\frac{20}{5} = 4$ **b.** $\frac{-20}{-5} = 4$ **c.** $\frac{-20}{5} = -4$ **d.** $\frac{20}{-5} = -4$

CLASS EXERCISES *Divide or indicate that division is not defined.*

1. $\frac{0}{-12}$ **2.** $\frac{0}{0}$ **3.** $\frac{6}{-3}$ **4.** $\frac{-9}{-3}$

REMAINDERS When you divide 6 by 3, the quotient is 2.

$$\frac{6}{3} = 2 \quad \text{because} \quad 6 = 2 \cdot 3$$

But, if you try to divide 7 by 3, there is a "remainder."

$$7 = 2 \cdot 3 + 1$$

You say that 1 is the remainder, and you write

$$\frac{7}{3} = 2 \text{ with remainder } 1.$$

Let a and b be integers with $b > 0$. Either

$$a = b \cdot c$$

for some integer c, or else

$$a = b \cdot c + r$$

for some integer c and for some *positive* integer r *less than* b. Here, c is called the *quotient* and r the *remainder*. For example,

$$\begin{array}{ccccc} a & = & b & \cdot & c \\ \downarrow & & \downarrow & & \downarrow \\ 6 & = & 2 & \cdot & 3 \end{array} \qquad \text{and} \qquad \begin{array}{ccccccc} a & = & b & \cdot & c & + & r \\ \downarrow & & \downarrow & & \downarrow & & \downarrow \\ 7 & = & 2 & \cdot & 3 & + & 1 \end{array}$$

Note that when you divide 7 by 3, the remainder, 1, *is less than* the divisor, 3.

When the divisor is at most 12, a procedure, known as **short division**, is often employed. For example,

$$\begin{array}{rl} & 132 \leftarrow \textit{quotient} \\ \textit{divisor} \rightarrow 3 & \overline{)396} \leftarrow \textit{dividend} \end{array}$$

Example 3 Divide and find the remainder. $\quad 9\overline{)81053}$

Solution

$$\frac{8}{9} = 0 \text{ with remainder } 9$$

Carry the remainder 8 over to the next column and divide by 9.

$$\begin{array}{rl} & 9 \\ 9 & \overline{)81053} \end{array}$$

Next,

$$\frac{0}{9} = 0$$

A 0 is placed in the 100's column of the quotient.

$$9\overline{)81053}$$ with quotient so far 90

Next, $\frac{5}{9} = 0$ with remainder 5

Place a 0 in the 10's column and carry the remainder 5 over to the 1's column.

$$\frac{53}{9} = 5 \text{ with remainder } 8$$

Thus, you obtain $9\overline{)81053}$ = 9005 with remainder 8

Check for Example 3

$$\begin{array}{rl} 9005 & \leftarrow \textit{quotient} \\ \times \quad 9 & \leftarrow \textit{divisor} \\ \hline 81045 & \\ + \quad 8 & \leftarrow \textit{remainder} \\ \hline 81053 & \leftarrow \textit{dividend} \end{array}$$

To check division when there is a remainder, use the fact that

$$\text{if } \frac{a}{c} = b \text{ with remainder } r, \qquad \text{then} \qquad a = b \cdot c + r$$

Thus, *first multiply the quotient, b, by the divisor, c (in either order). Then, add the remainder, r. You should obtain the dividend, a.*

CLASS EXERCISES *a. Divide and find the remainder.* *b. Check your result.*

5. $8\overline{)4004}$ **6.** $12\overline{)9735}$

ESTIMATING THE QUOTIENT When you divide large numbers, you must first *estimate* the quotient. At this time, you need not determine the remainder.

Example 4 Estimate the quotient.

a. $\frac{84}{27}$

b. $\frac{160}{45}$

Solution

a. Because $\frac{8}{2} = 4$, *try* 4.

$$\begin{array}{r} 27 \\ \times \ 4 \\ \hline 108 > 84 \end{array}$$

Too large!

b. $\frac{160}{45}$

Because $\frac{1}{4} < 1$, *consider* $\frac{160}{45}$

Because $\frac{16}{4} = 4$, *try* 4.

Next, *try* 3.

$$\begin{array}{r} 27 \\ \times\ 3 \\ \hline 81 \end{array} < 84$$

Thus, $\frac{84}{27} = 3$ with a remainder

(Had 3 also been too large, you would have next tried 2.)

$$\begin{array}{r} 45 \\ \times\ 4 \\ \hline 180 \end{array} > 160$$

Too large!

Next, *try* 3.

$$\begin{array}{r} 45 \\ \times\ 3 \\ \hline 135 \end{array} < 160$$

Thus, $\frac{160}{45} = 3$ with a remainder

CLASS EXERCISES *Estimate the quotient.* **7.** $\frac{96}{31}$ **8.** $\frac{72}{16}$ **9.** $\frac{294}{58}$

LONG DIVISION When the divisor is greater than 12, you divide by a procedure known as **long division**.

Example 5

Divide. $21 \overline{)672}$

Solution

a. The divisor, 21, is a two-digit number. Consider the first two digits of the dividend.

$21 \overline{)672}$

Estimate the quotient. $\frac{67}{21}$

Because $\frac{6}{2} = 3$, *try* 3, and write

$$\begin{array}{r} 3 \\ 21 \overline{)672} \\ 63 \\ \hline 4 \end{array}$$

The remainder is 4.

b. Bring down the next digit of the divisor.

Example 6 $324 \overline{)69660}$

Solution

a. The divisor, 324, is a three-digit number. Consider the first three digits of the dividend.

$324 \overline{)69660}$

Estimate the quotient. $\frac{696}{324}$

Try 2.

$$\begin{array}{r} 2 \\ 324 \overline{)69660} \\ 648 \\ \hline 48 \end{array}$$

b and c

$$\begin{array}{r} 21 \\ 324 \overline{)69660} \\ 648 \\ \hline 486 \\ 324 \\ \hline 162 \end{array}$$

```
        3
21 | 672
     63
      42
```

c. Now divide 42 by the divisor, 21.

```
       32
21 | 672
     63
      42
      42
       0
```

There is no remainder.

Because all digits of the dividend, 672, have been used, the quotient is 32.

Repeat steps **b** and **c** with the remaining digit, 0, of the dividend.

```
          215
324 | 69660
      648
       486
       324
       1620
       1620
```

The quotient is 215.

Example 7

a. Divide and find the remainder. 23 | 9581

b. Check your result.

Solution

a.

```
       416
23 | 9581
     92
      38
      23
      151
      138
       13
```

All digits of the dividend, 9581, have been used. The quotient is 416. The remainder is 13.

b. **check**

```
   416   ←— quotient
 ×  23   ←— divisor
  1248
  832
  9568
 +  13   ←— remainder
  9581   ←— dividend
```

CLASS EXERCISES

Divide and find the remainder, if any. In Exercises 11 *and* 13, *check.*

10. 22 | 682 **11.** 83 | 6308 (*Check.*) **12.** 35 | 8902

13. 326 | 82151 (*Check.*)

SOLUTIONS TO CLASS EXERCISES

1. 0 **2.** not defined **3.** −2 **4.** 3

5. a. $8\overline{)4004}$ = 500 with remainder 4

b. Check.

$$\begin{array}{rl} 500 & \longleftarrow \textit{quotient} \\ \times\ 8 & \longleftarrow \textit{divisor} \\ \hline 4000 & \\ +\ 4 & \longleftarrow \textit{remainder} \\ \hline 4004 & \longleftarrow \textit{dividend} \end{array}$$

6. a. $12\overline{)9735}$ = 8 1 1 with remainder 3

b. Check.

$$\begin{array}{rl} 811 & \longleftarrow \textit{quotient} \\ \times\ 12 & \longleftarrow \textit{divisor} \\ \hline 9732 & \\ +\ 3 & \longleftarrow \textit{remainder} \\ \hline 9735 & \longleftarrow \textit{dividend} \end{array}$$

7. $\frac{96}{31}$ = 3 with a remainder

8. $\frac{72}{16}$ = 4 with a remainder

9. $\frac{294}{58}$ = 5 with a remainder

10.
$$\begin{array}{r} 31 \\ 22\overline{)682} \\ 66 \\ \hline 22 \\ 22 \\ \hline \end{array}$$

11.
$$\begin{array}{r} 76 \\ 83\overline{)6308} \\ 581 \\ \hline 498 \\ 498 \\ \hline \end{array}$$

Check.

$$\begin{array}{rl} 76 & \longleftarrow \textit{quotient} \\ \times\ 83 & \longleftarrow \textit{divisor} \\ \hline 228 & \\ 608 & \\ \hline 6308 & \longleftarrow \textit{dividend} \end{array}$$

12.
$$\begin{array}{r} 254 \\ 35\overline{)8902} \\ 70 \\ \hline 190 \\ 175 \\ \hline 152 \\ 140 \\ \hline 12 \end{array}$$

The quotient is 254. The remainder is 12.

13.
$$\begin{array}{r} 251 \\ 326\overline{)82151} \\ 652 \\ \hline 1695 \\ 1630 \\ \hline 651 \\ 326 \\ \hline 325 \end{array}$$

The quotient is 251. The remainder is 325.

Check.

$$\begin{array}{rl} 251 & \longleftarrow \textit{quotient} \\ \times\ 326 & \longleftarrow \textit{divisor} \\ \hline 1506 & \\ 502 & \\ 753 & \\ \hline 81826 & \\ +\ 325 & \longleftarrow \textit{remainder} \\ \hline 82151 & \longleftarrow \textit{dividend} \end{array}$$

HOME EXERCISES

In Exercises 1–12, divide or indicate that division is not defined.

1. $\frac{0}{4}$ 2. $\frac{4}{0}$ 3. $\frac{-50}{5}$ 4. $\frac{-36}{-12}$ 5. $\frac{-1}{0}$ 6. $\frac{0}{-1}$ 7. $\frac{9}{-1}$ 8. $\frac{-144}{-12}$ 9. $\frac{0}{0}$

10. $6\overline{)5274}$ 11. $11\overline{)9680}$ 12. $12\overline{)96084}$

13. A seven-ounce bottle of diet soda contains 63 calories. How many calories are there per ounce?

14. Which is a better buy? A nine-ounce box of soap flakes that sells for 81 cents or a 12-ounce box that sells for 96 cents?

15. A man agrees to pay $84 for a radio. He does not have to put any money down. The payments are to last over a seven-week period. How much does he pay per week?

16. The temperature change over the last eight hours is −16 degrees Celsius. Assuming a constant rate of change, how many degrees Celsius has it changed each hour?

In Exercises 17–22, divide and find the remainder.

17. $4\overline{)4813}$ 18. $6\overline{)3905}$ 19. $11\overline{)6592}$ 20. $12\overline{)592411}$ 21. $11\overline{)6421}$ 22. $12\overline{)89735}$

23. There are three feet in a yard. 56 feet = _______ yards and _______ feet.

24. There are 12 inches in a foot. 100 inches = _______ feet and _______ inches.

25. Eleven pirates find 290 gold pieces. If they share equally, **a.** how many pieces does each receive? **b.** how many pieces remain?

26. A man buys a television set for \$495. He agrees to pay \$50 a month for nine months. How much is his down payment? (*Hint*: First, consider the total monthly payments.)

In Exercises 27–32, estimate the quotient.

27. $\frac{89}{21}$ **28.** $\frac{85}{19}$ **29.** $\frac{158}{31}$ **30.** $\frac{453}{62}$ **31.** $\frac{794}{92}$ **32.** $\frac{845}{98}$

In Exercises 33–36, divide.

33. $21\overline{)462}$ **34.** $67\overline{)4824}$ **35.** $504\overline{)544320}$ **36.** $1031\overline{)2066124}$

In Exercises 37–40, divide and check.

37. $25\overline{)525}$ **38.** $55\overline{)1760}$ **39.** $48\overline{)2880}$ **40.** $343\overline{)279888}$

In Exercises 41–44, divide and find the remainder.

41. $41\overline{)892}$ **42.** $88\overline{)10000}$ **43.** $104\overline{)10485}$ **44.** $271\overline{)26953}$

In Exercises 45–48: **a.** *divide and find the remainder* **b.** *check your result.*

45. $33\overline{)996}$ **46.** $77\overline{)2943}$ **47.** $112\overline{)9694}$ **48.** $451\overline{)89264}$

49. Each of 25 players receives an equal share of a World Series earning of \$300,225. How much is each player's share?

50. There are 32 rows in an auditorium. Each row contains the same number of seats. Altogether, there are 1184 seats. How many seats are there in a row?

1.7 ORDER OF OPERATIONS

ROLE OF PARENTHESES

By convention, the expression

$$2 \cdot 3^2 \quad \text{means} \quad \text{twice the square of 3.}$$

Thus,

$$2 \cdot 3^2 = 2 \cdot 9 = 18$$

Note that the exponent 2 refers to the base 3. If you want to indicate the square of the product of 2 and 3, use parentheses. Thus,

$$(2 \cdot 3)^2 = 6^2 = 36$$

Here, the exponent applies to both factors, 2 and 3. The expression

$$3 + 5 \cdot 2 \quad \text{means} \quad \text{add the product of 5 and 2 to 3.}$$

Thus,

$$3 + 5 \cdot 2 = 3 + 10 = 13$$

If the sum of 3 and 5 is to be multiplied by 2, use parentheses. Thus,

$$(3 + 5)2 = 8 \cdot 2 = 16$$

CLASS EXERCISES *Find each value.* **1.** $5 \cdot 2^3$ **2.** $(5 \cdot 2)^3$ **3.** $4 + 2 \cdot 3$ **4.** $(4 + 2)3$

RULES FOR ORDER OF OPERATIONS

The order in which addition, subtraction, multiplication, division, and raising to a power are applied in an example is often crucial. If parentheses are given, first perform the operations within parentheses. Otherwise:

1. First, raise to a power.
2. Then, multiply or divide from left to right.
3. Then, add or subtract from left to right.

Example 1 Find the value of: $2 + 5^2 \cdot 3$

Solution First, raise to a power; then, multiply; then, add.

$$2 + 5^2 \cdot 3 = 2 + 25 \cdot 3 = 2 + 75 = 77$$

Example 2 Find the value of: $24 \div 6 + 3 \cdot 2$

Solution First, divide 24 by 6 and multiply 3 by 2. Then, add these results.

$$\underbrace{24 \div 6}_{4} + \underbrace{3 \cdot 2}_{6} = 4 + 6 = 10$$

Example 3 Find the value of: $\left(\frac{11 - 5}{3} - 1\right)^2$

Solution Think of this as $\left([(11 - 5) \div 3] - 1\right)^2$

and work from the innermost pair of parentheses outward. Thus, first find $11 - 5$, then divide $11 - 5$ by 3, then subtract 1, and finally square the result.

$$\left(\frac{11 - 5}{3} - 1\right)^2 = \left(\frac{6}{3} - 1\right)^2 = (2 - 1)^2 = 1^2 = 1$$

CLASS EXERCISES *Find each value.* **5.** $4 \cdot 2^2 + 3$ **6.** $27 \div 3 - 2 \cdot 4$ **7.** $\left(\frac{6}{3} + 1\right)^2$ **8.** $\left(\frac{8 + 4}{6} - 1\right)^3$

SOLUTIONS TO CLASS EXERCISES

1. $5 \cdot 2^3 = 5 \cdot 8 = 40$ **2.** $(5 \cdot 2)^3 = 10^3 = 1000$ **3.** $4 + 2 \cdot 3 = 4 + 6 = 10$

4. $(4 + 2)\,3 = 6 \cdot 3 = 18$ **5.** $4 \cdot 2^2 + 3 = 4 \cdot 4 + 3 = 16 + 3 = 19$

6. $\underbrace{27 \div 3}_{9} - \underbrace{2 \cdot 4}_{8} = 9 - 8 = 1$ **7.** $\left(\frac{6}{3} + 1\right)^2 = (2 + 1)^2 = 3^2 = 9$

8. $\left(\frac{8 + 4}{6} - 1\right)^3 = \left(\frac{12}{6} - 1\right)^3 = (2 - 1)^3 = 1^3 = 1$

HOME EXERCISES

In Exercises 1–36, find each value.

1. $3 + 5 \cdot 4$ **2.** $(3 + 5)4$ **3.** $4 \cdot 2^2$ **4.** $(4 \cdot 2)^2$ **5.** $\frac{4 + 2}{2}$ **6.** $\frac{4}{2} + 2$

7. $4 + \frac{2}{2} - 3 \cdot 7$ **8.** $4^2 - 2^2$ **9.** $4 - 2^2$ **10.** $(4 - 2)^2$ **11.** $(2 - 4)^2$ **12.** $2 - 4^2$

13. $(2 + 5)^2 - 1$ **14.** $2 + 5^2 - 1$ **15.** $\frac{(8 - 2)^2}{2}$ **16.** $\frac{8 - 2^2}{2}$ **17.** $5(3 - 1) + 2^2$ **18.** $[5(3 - 1) + 2]^2$

19. $\left(\frac{6 + 5}{11} - 2\right)^2$ **20.** $\frac{8 + 7}{5} - \frac{3 + 4}{-7}$ **21.** $\left[\frac{3 - (4 + 5)}{2}\right]^3$ **22.** $2\left(\frac{3 - 4 + 5}{2}\right)^3$

23. $\frac{(6 - 2)^2}{2} \quad 3(5 - 3)^3$ **24.** $\frac{(5 - 2)^2 - (-1 + 3)^2}{5}$ **25.** $6 \cdot 6 - 6 \div 6$ **26.** $6\,(6 - 6 \div 6)$

27. $6\,(6 - 6) \div 6$ **28.** $(6 \cdot 6 - 6) \div 6$ **29.** $48 \div 12 \cdot 2^2$ **30.** $(48 \div 12)^2 \cdot 2$

31. $-(2 - 3)^3$ **32.** $2 + 3 \cdot 4^2 - 1$ **33.** $(2^4 - 12) \cdot 2 + 1$ **34.** $2^5 - 12 \cdot 2 + 1$

35. $3^3 - 6 \div 2 + 5$ **36.** $8 \div 4 - 5 \cdot 3 + 2$

37. Find the value of: **a.** $4 - (3 - 2)$ **b.** $(4 - 3) - 2$ **38.** Multiply the sum of 7 and 2 by 6.

39. Add 8 to the product of 4 and 3.

40. Divide the sum of 4 and 5 by 3.

41. Subtract 4 from the cube of 2.

42. Divide the square of (-4) by the cube of (-2).

43. If each shirt costs 10 dollars, each sweater costs 12 dollars, and each tie cost 7 dollars, how much do 5 shirts, 2 sweaters, and 4 ties cost?

44. There are 16 ounces in a pound. If each pad of paper weighs 4 ounces and each book weighs 12 ounces, what is the weight in pounds of 220 pads of paper together with 8 books?

Let's Review Chapter 1

1. Consider 45,004. **a.** What is the 10,000's digit? **b.** What is the 100's digit? **c.** What is the 1's digit?

2. Express 8,506,399 in terms of 1,000,000's, 100,000's, and so on.

3. Express forty thousand, six hundred one in terms of digits.

4. In Figure 1.42, which integer is represented by each of the points Q, R, S, T, U?

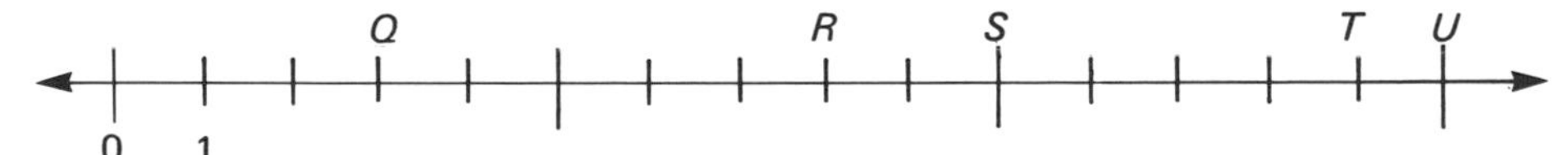

Fig. 1.42

5. Round 7,284,539 to the nearest **a.** 1,000,000 **b.** 100,000 **c.** 1000

6. In Figure 1.43, which number is represented by each of the points Q, R, S, T, U?

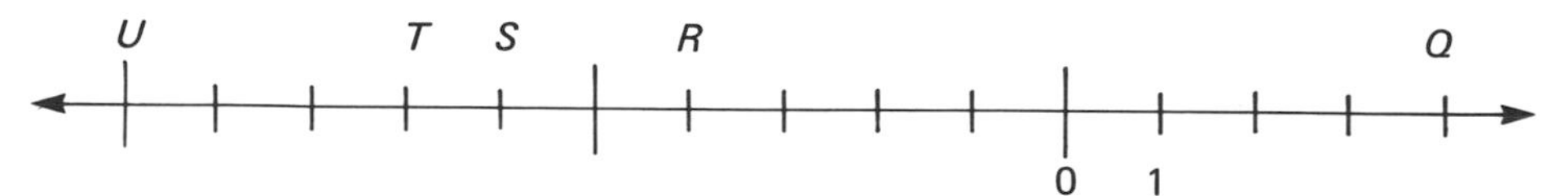

Fig. 1.43

7. Fill in "<" or ">". $-9 \square -6$

8. Fill in "left" or "right." -10 lies to the ☐ of -12 (on the number line).

9. Rearrange the integers so that you can write "<" between any two integers.

$-8, 8, 0, -4, 2, -2$

10. Find the inverse of **a.** -4 **b.** 45 **c.** 0

11. Find each absolute value. **a.** $|13|$ **b.** $|-13|$ **c.** $|0|$

12. Which numbers are 11 units from the origin?

13. Add in columns.

$$\begin{array}{r} 4215 \\ 2091 \\ 438 \\ \underline{1760} \end{array}$$

14. Approximate the following sum by first rounding each integer to the nearest 10,000 and then adding.

$$\begin{array}{r} 427\ 619 \\ 314\ 994 \\ 27\ 501 \\ \underline{8\ 455} \end{array}$$

15. It costs a wholesaler \$59 to buy a suit. A retailer buys the suit for \$95. What is the wholesaler's profit?

In Exercises 16–25, find each value.

16.
$$\begin{array}{r} 14 \\ -\ 9 \\ -\ 7 \\ 12 \\ \underline{-19} \end{array}$$

17. $-8 - (-4)$ **18.** $10 - 7 - 8$ **19.** $-20 - [(3 - 5) - 1]$ **20.** $\begin{array}{r} 4080 \\ \underline{\times\ 706} \end{array}$

21. $(-5)(-2)(-1)(-2)$ **22.** 7^2 **23.** $(-5)^3$ **24.** $(-2)^4$ **25.** $(-1)^{20}$

In Exercises 26–28, divide or indicate that division is not defined.

26. $\frac{-63}{9}$ **27.** $\frac{0}{6}$ **28.** $\frac{0}{9}$

29. a. Divide and find the remainder. **b.** Check your result. $76\overline{)9392}$

In Exercises 30–34, find each value.

30. $5(4 - 3 - 1)$ **31.** $\frac{12 - 2^2}{2}$ **32.** $\frac{(12 - 2)^2}{2}$

33. $24 \div 2(1 + 1)$ **34.** $(-3)^3 + 2^4$

2 ALGEBRAIC EXPRESSIONS

DIAGNOSTIC TEST Perhaps you are already familiar with some of the material in Chapter 2. This test will indicate which sections in Chapter 2 you need to study. The question number refers to the corresponding section in Chapter 2. If you answer *all* parts of the question correctly, you may omit the section. But *if any part of your answer is wrong, you should study the section*. When you have completed the test, turn to page A3 for the answers.

2.1 Terms

a. Find the coefficient of $-5x^2y$.
b. Are $2x^3y$ and $2x^2y$ like terms?
c. Are xyz and $-8zyx$ like terms?
d. Add. $5x + 3y + 6z + 2x - z$

2.2 Polynomials

a. Add the polynomials.

$$\begin{array}{r} 2a + b + 5c \\ 3a - b - 7c \\ \hline \end{array}$$

b. Subtract the second polynomial from the first one.

$$\begin{array}{r} 4x - 9y \\ 3x - 6y \\ \hline \end{array}$$

c. Subtract. $4a \quad b \quad [2a \quad (a \quad b)]$

2.3 Evaluating Polynomials

Evaluate each polynomial for the specified number or numbers.

a. $x^2 + 5x$; $\boxed{x = 4}$
b. $x^2y + 2xy + 1$; $\boxed{x = 2, y = -1}$
c. The area of a rectangle is given by the formula $A = lw$, where l is the length and w is the width. Find the area if $l = 14$ centimeters and $w = 12$ centimeters.

2.4 Products and Quotients

a. Multiply. $(9x^2y)(-2xy^3)$
b. Divide. $\dfrac{6a^4b}{-2ab}$
c. Multiply. $5xy(2x^2y - xy^2)$
d. Divide. $\dfrac{6ab^2 + 9ab - 18a^2b}{3ab}$

2.5 Equations

a. Check whether 7 is a root of the following equation. $2x - 1 = 13$
b. Solve. $5t - 4 = 12 + 3t$
c. Solve and check. $x - 3 = 9 - x$
d. Solve. $3(t + 1) - 2(t - 1) = 8$

2.1 TERMS

VARIABLES In arithmetic you add, subtract, multiply, and divide individual numbers. Sometimes the same arithmetic process can apply to all numbers under discussion. It is important to have symbols available that can stand for any one of several numbers. For example, if you want to express *adding 10 to a number*, you can write

$$x + 10$$

If you want to express *twice a number*, write

$$2x$$

DEFINITION

> A **variable** is a symbol that designates any one of the numbers being discussed.

Generally, the letters x, y, and z will be used as variables. Other letters will also be used.

Variables make it possible to apply arithmetic concepts to many numbers at one time. You can add, subtract, multiply, and divide these variables just as you can combine numbers. Powers of variables can also be considered. Algebraic expressions containing variables, such as

$$x + 1, \quad x - y, \quad xy, \quad x^2, \quad x^2 + 3x$$

will be considered throughout the book.

TERMS AND COEFFICIENTS

DEFINITION

> TERM. A **term** is a product of numbers and variables. A number, by itself, is also considered to be a term, as is a variable, by itself.

Example 1 Show that each of the following expressions is a term:

a. $2x$ **b.** xyz **c.** 5 **d.** t

Solution
a. $2x$ is the product of 2 and the variable x.
b. xyz is the product of the variables x, y, and z.
c. 5 is a number. As such, it is a term.
d. A variable, such as t, is also a term.

Example 2 Write each term as a product of a number and variables.

a. x^2 **b.** $3x^2y^3$ **c.** $(5x)(2y)$

Solution
a. x^2 stands for $x \cdot x$. Thus $x^2 = 1 \cdot x \cdot x$
b. $3x^2y^3 = 3 \cdot x \cdot x \cdot y \cdot y \cdot y$

c. $(5x)(2y)$ is the product of the numbers 5 and 2 and the variables x and y. The Commutative Law of Multiplication will enable you to rewrite this expression as $5 \cdot 2 \cdot x \cdot y$, or $10xy$.

DEFINITION

COEFFICIENT. The numerical factor (or the product of numerical factors) in a term is known as the **numerical coefficient**, or simply, the **coefficient**, of the term. If only variables appear, the coefficient of the term is understood to be 1 or -1. Thus, the coefficient of xy is 1, and of $-xy$ is -1.

If the term is a number, then it is its own coefficient. For example, the coefficient of the term 2 is 2, itself.

Example 3 Find the coefficient of each term. **a.** $3x$ **b.** $-4x^2y$ **c.** $(5x)(-3z)$

Solution **a.** 3 **b.** -4 **c.** -15 [Note that $-15 = 5(-3)$.]

When you multiply by 0, the product is 0. Thus, *any term*, such as $0xy$, *that has* 0 *as its coefficient reduces to* 0.

CLASS EXERCISES

In Exercises 1–3, write each term as a product of a number and variables.

1. a^3 **2.** $4y^4$ **3.** $(2a)(3b^2)$

In Exercises 4–6, find the coefficient of each term.

4. $6t$ **5.** $-xy$ **6.** $(4x)(7y^2)$

LIKE TERMS

DEFINITION

Like terms are terms that differ only in their coefficients or in the order of their variables (or in both of these).

Example 4

a. $5xy$ and $-3xy$ are like terms. They differ only in their coefficients, 5 and -3.

b. $2xz$ and $2zx$ are like terms. They differ only in the order of their variables. Note that $2xz = 2zx$.

c. $2x^2yz$ and zx^2y are like terms. They differ only in their coefficients, 2 and 1, and in the order of their variables.

d. s^2t^2 and $stst$ are like terms. Note that $s^2t^2 = stst$.

e. 3 and $2x$ are unlike terms. Only one of these contains a variable.

f. x^2 and x^3 are unlike terms. They differ in the powers of their variables.

g. x^2y and xy are unlike terms.

Any two numbers are like terms because neither contains any variables. Thus, 5 and -3 are like terms.

To add or subtract *like* terms, extend the *Distributive Laws*. For example,

$$3x + 2x = (3 + 2)x = 5x \qquad \text{and}$$

$$3x - 2x = (3 - 2)x = 1x = x$$

Let a and b be any fixed numbers, and let x be a variable. Define

$$ax + bx = (a + b)x \qquad \text{and} \qquad ax - bx = (a - b)x.$$

The Distributive Laws also apply to like terms that contain two or more variables. Thus,

$$5xy + 4xy = (5 + 4)xy = 9xy$$

As with numbers, the Distributive Laws apply to adding or subtracting more than two like terms. Thus,

$$xyz + 4xyz - 2xyz = (1 + 4 - 2)xyz = 3xyz$$

CLASS EXERCISES

In Exercises 7–9, indicate which pairs are like terms.

7. $5y$ and $9y$ **8.** $-3x$ and $3x^2$ **9.** xy^2 and $(5y)(xy)$

In Exercises 10–12, add or subtract.

10. $5a - 3a$ **11.** $2x^2y + 6x^2y$ **12.** $rst + 8rst - 7rst$

ADDING UNLIKE TERMS

Unlike terms cannot be combined in the preceding way. Simply combine unlike terms with + or − signs. For example, to add the *unlike* terms xy and $3z$, write

$$xy + 3z$$

This expression cannot be simplified. Note that $xy + 3z$ is *not* a term. (It is *not* a *product* of numbers and variables.)

Example 5 Subtract $2t$ from the sum of r and $3s$.

Solution $2t$, r, and $3s$ are unlike terms. You obtain the expression

$$(r + 3s) - 2t$$

This can be simplified to $\quad r + 3s - 2t$

To add several terms, arrange them so that like terms are grouped together. You may either group like terms in columns, as in Example 6, or you may group like terms by means of parentheses, as in Example 7. In rearranging the terms, you are extending the Commutative Law

$$a + b = b + a$$

and the Associative Law

$$(a + b) + c = a + (b + c)$$

of addition. For example,

$$\begin{aligned} 2y + (x + y) &= (x + y) + 2y && \text{by the Commutative Law} \\ &= x + (y + 2y) && \text{by the Associative Law} \\ &= x + 3y \end{aligned}$$

Example 6 Add and subtract.
$2x + 3y - z + 4x - 4y + z$

Solution Arrange like terms in columns.

$$\begin{array}{l} 2x + 3y - z \\ \underline{4x - 4y + z} \\ 6x - y \end{array}$$

Example 7 Add and subtract.
$5x + y - t + u - v + 2x - 3t$

Solution Group like terms together within parentheses. Note that

$$y - t = y + (-t)$$

Thus, you obtain

$(5x + 2x) + y + (-t - 3t)$
$+ u - v$, or
$7x + y - 4t + u - v$

CLASS EXERCISES **13.** Add $2x$ and $5y$. **14.** Subtract $3a^2$ from ab.

In Exercises 15 and 16, add or subtract.

15. $a + b + a + 3b$ **16.** $5x + y - 2z + 4x - y - z$

SOLUTIONS TO CLASS EXERCISES

1. $a^3 = 1 \cdot a \cdot a \cdot a$ **2.** $4y^4 = 4 \cdot y \cdot y \cdot y \cdot y$ **3.** $(2a)(3b^2) = 6 \cdot a \cdot b \cdot b$ **4.** 6 **5.** -1 **6.** 28
7. like **8.** unlike **9.** like **10.** $2a$ **11.** $8x^2y$ **12.** $2rst$ **13.** $2x + 5y$ **14.** $ab - 3a^2$
15. $2a + 4b$ **16.** $9x - 3z$

HOME EXERCISES

In Exercises 1–6, write each term as a product of a number and variables.

1. x^4 **2.** $5xy^2$ **3.** $10x^2yz^2$ **4.** a^3b^4 **5.** $-u$ **6.** $-a^2b^3c^5$

In Exercises 7–14, find the coefficient of each term.

7. $7x$ **8.** $-3yz$ **9.** u **10.** $-t^2$ **11.** $12xy^3z^5$ **12.** $0xyz^2$ **13.** $(3x)(2y)$ **14.** -7

In Exercises 15–26, indicate which pairs are like terms.

15. $2x$ and $4x$ **16.** $3y$ and $-3y$ **17.** $5xy$ and $5yx$ **18.** $13xy$ and $13xz$
19. $-2x^2$ and $2x^3$ **20.** 15 and 30 **21.** $2u^2v$ and $4uvu$ **22.** x^3y and xy^3
23. 1 and -1 **24.** x^2 and $0x^2$ **25.** $3x^2$ and $2x^3$ **26.** x^2y^3 and $6yxyxy$

In Exercises 27–48 add or subtract. In some cases, you will not be able to simplify the given expression.

27. $5y + 2y$ **28.** $2x - x$ **29.** $xy + 3xy$ **30.** $x^2y + 5x^2y$ **31.** $10xy - 4yx$
32. $5t - 2t + t$ **33.** $2s - s - s$ **34.** $4w + w + 2w + w$ **35.** $5r^2st^2 + 3r^2st^2$ **36.** $xy + 3x^2$
37. $6u - 3v$ **38.** $2x - y + 3x + y$ **39.** $2a + 6b + 5a + 6b - a$
40. $4a + b - c + 2a - b - 2c$ **41.** $3x^2y + 5z^2 + 3x^2y + 5z^2$ **42.** $2a + 5b - c + d + 6a - d + 7a$
43. $5x^2 + 3y^2 - z^2 + 6x^2 - z^2 - 11x^2$ **44.** $6x + y + 1 - 2x - 3 + y + 10$ **45.** $9 - 4u^2 - v^2 + 5 + 4u^2 + v^2$
46. $7 - 2u + 5v - 3 + 2u + 5v$ **47.** $9m + 2n - 6p + 2n - 4p - 9m$ **48.** $2x - 3y + 5z - 3y - 5z + 6y - 2x$
49. Add $3x$, $5x$, and $6x$. **50.** Subtract $2yz$ from the sum of $7yz$ and $-3yz$.
51. Add $2a$, $5b$, $3a$, and $-5b$. **52.** Subtract $3y$ from the sum of $4y$, $2x$, and $-y$.

2.2 POLYNOMIALS

WHAT IS A POLYNOMIAL?

DEFINITION

A *nonzero* term is also called a **monomial**. The sum (or difference) of two unlike (nonzero) terms is called a **binomial**, and of three unlike (nonzero) terms a **trinomial**.

Example 1

a. $3xy$ is a nonzero term, and hence a monomial.
b. $2x + 5y$ is the sum of two unlike terms. It is a binomial.
c. $5x - y + z$ is the sum of three unlike terms. It is a trinomial.
d. $2x + 5x + x$ is a monomial. In fact, $2x$, $5x$, and x are like terms. Thus,

$$2x + 5x + x = 8x$$

DEFINITION

A **polynomial** is either a term or a sum of terms.

Monomials, binomials, and trinomials are all polynomials.

Example 2 Each of the following is a polynomial:

$$4x - y, \quad 2x^2 + 3x - 1, \quad 2a + b - c + d,$$
$$7, \quad 0, \quad x^5 + 3x^4 - x^3 + x^2 - 2x + 1$$

CLASS EXERCISES *Classify each polynomial as (i) a monomial, (ii) a binomial, (iii) a trinomial, or (iv) none of these.*

1. x^2 **2.** $x^2 + 2x + 1$ **3.** $x^2 + y^2$ **4.** $t^3 + t^2 - t + 1$

ADDING OR SUBTRACTING POLYNOMIALS

To add or subtract polynomials, group like terms together, as in Section 2.1.

Example 3 Add the polynomials $2x + y$, $3x + y - z$, and $5y + z$.

Solution Rearrange the terms so that like terms are in the same column.

$$\begin{array}{rrr} 2x + & y & \\ 3x + & y & - z \\ & 5y & + z \\ \hline 5x + & 7y & \end{array}$$

DEFINITION The **inverse of a polynomial** P is the polynomial obtained by changing the sign of each term of P.

Let $-P$ denote the inverse of P.

Example 4 The inverse of $x - 2y + 3z$ is $-x + 2y - 3z$. Note the addition:

$$\begin{array}{rr} & x - 2y + 3z \\ + & -x + 2y - 3z \\ \hline & 0 \end{array}$$

In general, *the inverse of P is the polynomial added to P to obtain* 0.

$$\boxed{P + (-P) = 0}$$

Recall that for numbers a and b,

$$a - b = a + (-b)$$

To subtract the polynomial Q from P, add $-Q$.

$$\boxed{P - Q = P + (-Q)}$$

Example 5 *Subtract* the second polynomial from the first one.

$$\begin{array}{r} 4a - 3b + c \\ \underline{2a - 3b - c} \end{array}$$

Solution Change the sign of each term of the bottom polynomial and add.

$$\begin{array}{rrr} 4a & - 3b & + c \\ -2a & + 3b & + c \\ \hline 2a & & + 2c \end{array}$$

Example 6 Simplify.

$$10a - [4b - (3a - b)]$$

Solution Begin with the innermost parentheses and work outward, as you do when adding and subtracting real numbers.

$$\begin{aligned} 10a - [4b - (3a - b)] &= 10a - [4b - 3a + b] \\ &= 10a - 4b + 3a - b \\ &= (10a + 3a) + (-4b - b) \\ &= 13a - 5b \end{aligned}$$

CLASS EXERCISES *In Exercises 5 and 6, add the polynomials.*

5. $$\begin{array}{r} 3x + y + 1 \\ \underline{2x - y - 2} \end{array}$$

6. $$\begin{array}{rrrr} a & + 4b & + c & + d \\ a & - b & + 2c & \\ 2a & & + c & - d \\ \hline a & + 2b & & \end{array}$$

7. Find the inverse of $a - b + 4c - 3$.

8. Subtract the second polynomial from the first one.

$$\begin{array}{rrrr} 4a & - b & + c & + d \\ \hline 2a & + b & & - d \end{array}$$

9. Simplify $(5r + s) - (3r + 2s) - (r - s)$.

SOLUTIONS TO CLASS EXERCISES

1. (*i*) **2.** (*iii*) **3.** (*ii*) **4.** (*iv*) **5.** $5x - 1$ **6.** $5a + 5b + 4c$

7. $-a + b - 4c + 3$ **8.** $$\begin{array}{rrrr} 4a & - b & + c & + d \\ -2a & - b & & + d \\ \hline 2a & - 2b & + c & + 2d \end{array}$$

9. $(5r + s) - (3r + 2s) - (r - s) = 5r + s - 3r - 2s - r + s = (5r - 3r - r) + (s - 2s + s) = r$

HOME EXERCISES

In Exercises 1–10, classify each polynomial as (i) a monomial, (ii) a binomial, (iii) a trinomial, or (iv) none of these.

1. $2x$ **2.** $3x + 5y$ **3.** 5 **4.** 0 **5.** $4x^2 + 3x + 1$ **6.** $x^3 + x^2 - x + 1$ **7.** $4x + 3x + 2x + x$

8. $2x + 3y + 2x + y$ **9.** $2x - y + 2x - y$ **10.** $2x + y - 2x - y$

In Exercises 11–18, add the polynomials.

11. $\begin{array}{rr} 2x & +\ y \\ \hline x & +\ 4y \end{array}$

12. $\begin{array}{rr} 2r & -\ s \\ \hline 3r & -\ 2s \end{array}$

13. $\begin{array}{rrr} 6x & -\ 3y & +\ 7 \\ \hline 2x & +\ 3y & -\ 5 \end{array}$

14. $\begin{array}{rrr} 6r^2 & +\ s^2 & -\ 3t^2 \\ \hline r^2 & -\ s^2 & +\ t^2 \end{array}$

15. $\begin{array}{rrr} 5x & -\ 2y & +\ z \\ 6x & -\ 3y & +\ z \\ \hline x & -\ y & -\ z \end{array}$

16. $\begin{array}{rrrr} 10a & +\ 7b & -\ 3c & +\ d \\ 5a & -\ 2b & +\ 3c & +\ 9d \\ 7a & -\ 2b & +\ 5c & \\ \hline & 2b & & +\ d \end{array}$

17. $\begin{array}{rrrrr} a & +\ b & & -\ d & +\ 2e \\ & -\ b & +\ 4c & & +\ e \\ 3a & -\ b & & +\ 7d & \\ 5a & & -\ 2c & +\ d & \\ \hline -2a & & & -\ d & -\ e \end{array}$

18. $\begin{array}{rrrr} 16w & +\ 22x & -\ 17y & +\ 9z \\ 12w & -\ 17x & +\ 9y & -\ 8z \\ 32w & -\ 25x & & +\ 10z \\ 28w & +\ 7x & -\ 13y & \\ \hline 12w & -\ 5x & +\ 7y & \end{array}$

In Exercises 19–26, subtract the second polynomial from the first one.

19. $\begin{array}{rr} 4a & +\ 2b \\ \hline 2a & +\ b \end{array}$

20. $\begin{array}{rr} 9x & +\ 7y \\ \hline 8x & +\ 7y \end{array}$

21. $\begin{array}{rr} 7c & -\ 6d \\ \hline 2c & +\ 5d \end{array}$

22. $\begin{array}{rr} -7s & -\ 8t \\ \hline -7s & -\ 8t \end{array}$

23. $\begin{array}{rr} 12y & -\ 15z \\ \hline -\ 9y & +\ 18z \end{array}$

24. $\begin{array}{rrr} 6a & -\ 3b & +\ 2c \\ \hline a & & -\ c \end{array}$

25. $\begin{array}{rrrr} 2a & -\ b & +\ c & -\ d \\ \hline -\ a & -\ 2b & -\ c & +\ d \end{array}$

26. $\begin{array}{rrrr} 3r & -\ s & +\ t & -\ u \\ \hline & s & & -\ u \end{array}$

In Exercises 27–34, add or subtract.

27. $3a - (4b + 2a)$

28. $(4a - 2b) - (3a - 2b)$

29. $(5x - 2y + z) - (2x - y - z) + (4x - y)$

30. $3x - y - [2x - y - (x + y)]$

31. $6x - 2y - [-x - (y - x)]$

32. $5a - 2b + c - [(3a - b) - (2a - b + c)]$

33. $7 - 2a - 3b + [5a - (3b - c)]$

34. $2x - 3y - [3x - (z - 2y) - (x - y + z)]$

35. Add $3a + b$, $5a - 2b$, and $2a$.

36. Add $2x + y - z$, $3y - z$, and $x + z$.

37. Subtract $2x + 9y$ from $6x + 10y$.

38. Subtract $x - y$ from $x + y$.

39. Subtract $a - 2b$ from the sum of $3a + b$ and $5a - b$.

40. Subtract $2x^2$ from 0.

2.3 EVALUATING POLYNOMIALS

SUBSTITUTING A NUMBER FOR A VARIABLE

Recall that a variable designates any number under discussion. Polynomials are constructed from numbers and variables by addition and multiplication. Thus, polynomials pertain to numbers. When you are given numbers that the variables represent, you can evaluate the polynomial by substituting the numbers for the variables.

Example 1 Evaluate $3y - 1$ when $y = -2$.

Solution Replace y by -2 in the expression $3y - 1$.

$$3(-2) - 1 = -6 - 1 = -7$$

When a variable occurs more than once in a polynomial, *each time it occurs, replace it by the same number.*

Example 2 Evaluate $t^2 + 4t + 3$ when $t = 5$.

Solution Replace *each occurrence of t* by 5:

$$5^2 + 4 \cdot 5 + 3 = 25 + 20 + 3 = 48$$

CLASS EXERCISES *Evaluate each polynomial for the specified number.*

1. $5y^2 - 1$; $\boxed{y = 2}$ 2. $x^2 + x$; $\boxed{x = 5}$ 3. $2y^3 + y^2 - 4$; $\boxed{y = 2}$

SEVERAL VARIABLES

If a polynomial contains two or more variables, replace each variable by a specified number.

Example 3 Evaluate $x^2 - 2y$ when $x = 3$ and $y = -4$.

Solution Substitute 3 for x and -4 for y, and obtain

$$3^2 - 2(-4) = 9 + 8 = 17$$

Example 4 Evaluate $x^2 + 3xy + yz$ when $x = 2$, $y = 1$, $z = 4$.

Solution Replace *each occurrence of x* by 2, of **y** by **1**, and of **z** by **4**:

$$2^2 + 3 \cdot 2 \cdot 1 + 1 \cdot 4 = 4 + 6 + 4 = 14$$

CLASS EXERCISES *Evaluate each polynomial for the specified numbers.*

4. $xy + x - y$; $\boxed{x = 4, y = 3}$ 5. $s^2 + 2st + 1$; $\boxed{s = 10, t = 1}$

GEOMETRIC FORMULAS

Formulas are algebraic expressions that indicate the relationship between various quantities. Geometric formulas are often expressed in terms of polynomials.

Example 5 The area, A, of a square is given by the formula

$$A = s^2$$

where s is the length of a side. (See Figure 2.1 .) Find the area if a side is of length **a**. 10 inches, **b**. 20 inches.

Solution **a.** $A = 10^2 = 100$
The area is 100 square inches.

b. $A = 20^2 = 400$
The area is 400 square inches.

CLASS EXERCISES

6. Find the area of a square of side length 8 feet.
7. The area of a rectangle is given by the formula

$$A = lw$$

where l is the length and w the width. (See Figure 2.2 .) Find the area if $l = 9$ inches and $w = 6$ inches.

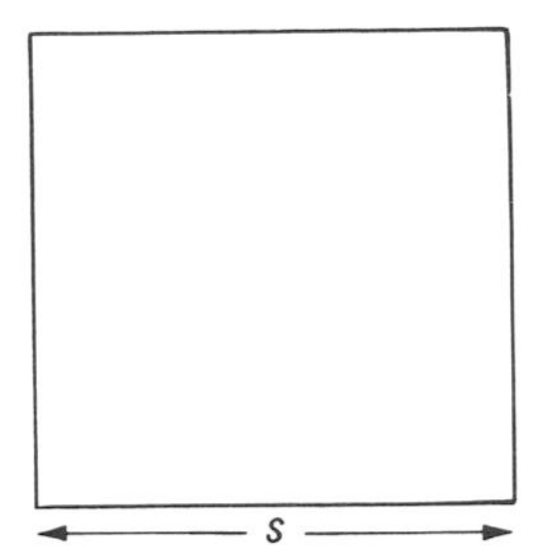

Fig. 2.1. The area of a square is s^2.

Fig. 2.2. The area of a rectangle is lw.

SOLUTIONS TO CLASS EXERCISES

1. $5 \cdot 2^2 - 1 = 5 \cdot 4 - 1 = 19$ **2.** $5^2 + 5 = 25 + 5 = 30$ **3.** $2 \cdot 2^3 + 2^2 - 4 = 2 \cdot 8 + 4 - 4 = 16$

4. $4 \cdot 3 + 4 - 3 = 12 + 4 - 3 = 13$ **5.** $10^2 + 2 \cdot 10 \cdot 1 + 1 = 100 + 20 + 1 = 121$

6. $A = 8^2 = 64$
The area is 64 square feet.

7. Substitute 9 for l and 6 for w in the formula $A = lw$.

$$A = 9 \cdot 6 = 54$$

The area is 54 square inches.

HOME EXERCISES

In Exercises 1–14, evaluate each polynomial for the specified number.

1. $3x$; $x = 4$ **2.** $t + 5$; $t = 2$ **3.** $6x + 1$; $x = 1$ **4.** $7y + 9$; $y = 0$
5. $5z - 3$; $z = -1$ **6.** z^2; $z = 6$ **7.** u^3; $u = -3$ **8.** $x^2 - 2$; $x = 4$
9. $x^2 + 4x$; $x = -1$ **10.** $x^2 + 5x + 1$; $x = -5$
11. $t^4 - t^2 + 1$; $t = 1$ **12.** $t^5 - 4t^2 - 2t + 3$; $t = -1$
13. $x^3 - 7x^2 + 14x - 3$; $x = 3$ **14.** $z^4 - z^2 + 2z + 5$; $z = 10$

In Exercises 15–26, evaluate each polynomial for the specified numbers.

15. xy; $x = 4, y = -3$ **16.** $x + 2y$; $x = -6, y = 8$
17. $s^2 - 3t^2$; $s = 10, t = 5$ **18.** $5mn + 2m - n$; $m = 8, n = -2$
19. $3a^2 - 2b^2 + c$; $a = 4, b = 3, c = 2$ **20.** $x^2 + y^2 - z^2 + xyz$; $x = 4, y = -3, z = -2$
21. $2a - 7b + 9c - 2$; $a = b = c = 3$ **22.** $a + 2b - c + 5d$; $a = 10, b = 6, c = -1, d = 2$
23. $a^3 - bc^2 + 2ab$; $a = 2, b = 5, c = -2$ **24.** $x^2y - yx^2 + x - y$; $x = 10, y = -1$
25. $x^3 + x^2 - 2y + 3xy$; $x = 100, y = -1$ **26.** $2u^2 - 3v^2 + w^2$; $u = 1, v = 2, w = 3$

In Exercises 27–34, evaluate each polynomial three separate times.

27. $x + 10$; **a.** $x = 2$, **b.** $x = -2$, **c.** $x = -10$
28. $2x - 8$; **a.** $x = 6$, **b.** $x = 0$, **c.** $x = 4$
29. $z^2 + 1$; **a.** $z = 0$, **b.** $z = 1$, **c.** $z = -1$
30. $x^2 + 5x + 2$; **a.** $x = 2$, **b.** $x = 3$, **c.** $x = 4$
31. $2x^2 - 3x + 1$; **a.** $x = 4$, **b.** $x = -4$, **c.** $x = 6$
32. $u^3 - 2u + 6$; **a.** $u = 0$, **b.** $u = -1$, **c.** $u = 2$
33. $b^4 - b^2 + 1$; **a.** $b = 1$, **b.** $b = -2$, **c.** $b = 3$
34. $m^3 - 2m^2 + m - 4$; **a.** $m = -1$, **b.** $m = 2$, **c.** $m = -3$

35. Find the area of a square if a side is of length
a. 11 inches, **b.** 15 inches, **c.** 30 inches. (See Figure 2.1 on page 61.)

36. Find the area of a rectangle of length 80 feet and width 11 feet. (See Figure 2.2 on page 61.)

37. The perimeter of (distance around) a rectangle is given by the formula $P = 2l + 2w$, where l and w are the length and width. (See Figure 2.2 on page 61.) Find the perimeter if $l = 40$ inches, $w = 22$ inches.

38. The volume of a rectangular box is given by the formula $V = lwh$, where l is the length, w the width, and h the height. (See Figure 2.3 .) Find the volume if:
a. $l = 10$ inches, $w = 6$ inches, $h = 8$ inches; **b.** $l = 20$ inches, $w = 10$ inches, $h = 12$ inches.

39. The surface area of a rectangular box is given by the formula $S = 2lw + 2lh + 2wh$, where l, w, and h are as in Exercise 38. (See Figure 2.3 .) Find the surface area if
a. $l = 5$ feet, $w = 4$ feet, $h = 3$ feet; **b.** $l = 10$ feet, $w = 6$ feet, $h = 6$ feet

40. When an object travels at a constant rate r over a period of time t, its distance d is given by the formula $d = rt$. Find d if $r = 60$ miles per hour and $t = 5$ hours.

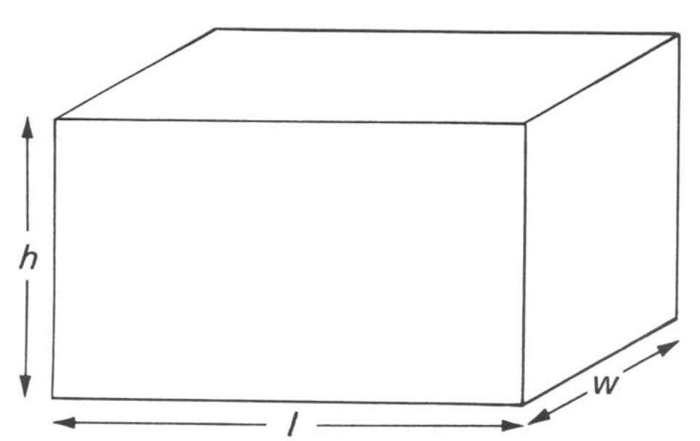

Fig. 2.3 The volume of a rectangular box is lwh. The surface area is $2lw + 2lh + 2wh$.

2.4 PRODUCTS AND QUOTIENTS

PRODUCTS OF POWERS

What happens when you multiply two powers of the same number? Observe that

$$2^2 \cdot 2^1 = 4 \cdot 2 = 8 = 2^3$$

Thus,

$$2^2 \cdot 2^1 = 2^{2+1}$$

Next, consider what happens when you multiply two powers of the same variable.

Example 1 Multiply.

$$x^3 \cdot x^2$$

Solution

$$x^3 \cdot x^2 = (x \cdot x \cdot x)(x \cdot x) = x \cdot x \cdot x \cdot x \cdot x = x^5$$

Thus,

$$x^3 \cdot x^2 = x^{3+2}$$

Let a be any number and let m and n be positive integers. To find the product

$$a^m \cdot a^n,$$

observe that

$$a^m = \underbrace{a \cdot a \cdot a \ldots a}_{m\ factors}, \qquad a^n = \underbrace{a \cdot a \cdot a \ldots a}_{n\ factors}$$

Therefore,

$$\begin{aligned} a^m \cdot a^n &= \underbrace{(a \cdot a \cdot a \ldots a)}_{m\ factors}\underbrace{(a \cdot a \cdot a \ldots a)}_{n\ factors} \\ &= \underbrace{a \cdot a \cdot a \ldots a \cdot a \cdot a \cdot a \ldots a}_{m+n\ factors} \\ &= a^{m+n} \end{aligned}$$

Thus,

$$\boxed{a^m \cdot a^n = a^{m+n}}$$

In other words, *to multiply two powers of the same number (or variable) a, write down the base a and add the exponents.* For example,

$$y^8 \cdot y^5 = y^{8+5} = y^{13}$$

Three or more powers of the same number (or variable) are multiplied by writing down the base and adding the exponents. Thus,

$$t^4 \cdot t^2 \cdot t = t^{4+2+1} = t^7$$

CLASS EXERCISES *Multiply.* **1.** $x^6 \cdot x^2$ **2.** $a^5 \cdot a$ **3.** $y^3 \cdot y^2 \cdot y^5$

ASSOCIATIVE AND COMMUTATIVE LAWS

Let a, b, and c be any numbers. According to the *Associative Law of Multiplication,*

$$(ab)c = a(bc)$$

According to the *Commutative Law of Multiplication,*

$$ab = ba$$

These laws apply to polynomials as well as to numbers. Observe how the laws are used in the following example.

Example 2 Multiply. $(2x)3$ **Solution**

$$\begin{aligned}(2x)3 &= 3(2x) && \textit{by the Commutative Law}\\ &= (3 \cdot 2)x && \textit{by the Associative Law}\\ &= 6x\end{aligned}$$

When **multiplying monomials,** use the Associative and Commutative Laws to rearrange the factors.

1. *Group all coefficients at the beginning, and multiply them.*
2. *Group powers of the same variable together and multiply them, as previously explained.*

Example 3 Multiply. $(3ab)(4ab)$

Solution $(3ab)(4ab) = \underbrace{3 \cdot 4}\,\underbrace{a\,a}\,\underbrace{b\,b} = \underbrace{12}\,\underbrace{a^2b^2}$

group all coefficients at the beginning — *group powers of the same variable together* — *multiply coefficients* *and* *powers of the same variable*

Example 4 Multiply. $(4ax^2y)(-2x^2y^3)(-3a^2bx)$

Solution

$$\begin{aligned}(4ax^2y)(-2x^2y^3)(-3a^2bx) &= [4(-2)(-3)](aa^2)\,b(x^2x^2x)(yy^3)\\ &= 24a^3bx^5y^4\end{aligned}$$

CLASS EXERCISES *Multiply.* **4.** $(5a)4$ **5.** $(-2xy)(7xy)$ **6.** $(4abc)(2ab^2)(abc^2)$

SIMPLE DIVISION Recall that $\frac{6}{2} = 3$ because $6 = 3 \cdot 2$.

Similarly, division of polynomials can be defined in terms of multiplication.

Thus, $\frac{4x}{2} = 2x$ because $4x = (2x)\,2$, and

$$\frac{x^5}{x^2} = x^3 \quad \text{because} \quad x^5 = x^3 \cdot x^2$$

Note that $\frac{x^5}{x^2} = \frac{xxxxx}{xx} = xxx = x^3.$

Thus,
$$\frac{x^5}{x^2} = x^{5-2}$$

To divide powers of the same nonzero number (or variable) *a*, *write down the base a and subtract the exponents:*

$$\boxed{\frac{a^m}{a^n} = a^{m-n}}$$

where m and n are positive integers, and m > n. For example,

$$\frac{y^8}{y^3} = y^{8-3} = y^5$$

Powers of several variables may have to be divided. Thus,

$$\frac{x^6y^2}{x^2y} = x^4y \qquad \text{because} \qquad x^6y^2 = (x^4y)(x^2y)$$

Note that
$$\frac{x^6y^2}{x^2y} = x^{6-2}y^{2-1}$$

In the next example, first divide the coefficients. Also, observe that the powers of y "cancel." In fact,

$$\frac{y^3}{y^3} = 1 \qquad \text{because} \qquad y^3 = 1 \cdot y^3$$

In Section 8.5, y^0 will be defined (for *nonzero* y) so that

$$\frac{y^3}{y^3} = y^{3-3} = \boxed{y^0 = 1}$$

Example 5

$$\frac{10x^6y^3}{5x^4y^3} = 2x^{6-4} = 2x^2$$

Note that $\dfrac{10}{2} \cdot 4 = 5 \cdot 4 = 20$ and $\dfrac{10 \cdot 4}{2} = \dfrac{40}{2} = 20$

Here, you can multiply by 4 or divide by 2 in either order. *For any numbers a, b, and c, with b* $\neq 0$,

$$\boxed{\frac{a}{b} \cdot c = \frac{a \cdot c}{b}}$$

The same applies to polynomials. Thus,

$$\frac{x}{3} \cdot 6 = \frac{x \cdot 6}{3} = \frac{6x}{3} = 2x \qquad \text{and}$$

$$\frac{x + 4}{2} \cdot 2 = \frac{2(x + 4)}{2} = x + 4$$

CLASS EXERCISES *In Exercises 7 and 8, divide.*

7. $\frac{a^5}{a^2}$ **8.** $\frac{8x^2y^7}{4xy^2}$

In Exercises 9 and 10, multiply and divide.

9. $\frac{a}{5} \cdot 15$ **10.** $\frac{b + 3}{4} \cdot 4$

DISTRIBUTIVE LAWS According to the *Distributive Laws, for any terms a, b, and c,*

$$a(b + c) = ab + ac \qquad \text{and} \qquad (b + c)a = ba + ca$$

Similarly,

$$a(b - c) = ab - ac \qquad \text{and} \qquad (b - c)a = ba - ca$$

Thus,

$$6(x + y) = 6x + 6y \qquad \text{and} \qquad 5(m - n) = 5m - 5n$$

Example 6 Use the Distributive Laws to find $7a(a + b)$.

Solution

$$\begin{aligned} 7a(a + b) &= (7a)a + (7a)b && \textit{by the Distributive Laws} \\ &= 7(aa) + 7(ab) && \textit{by the Associative Law of Multiplication} \\ &= 7a^2 + 7ab \end{aligned}$$

Example 7

$$\begin{aligned} 4xyz^2(xy - 2z + 3x^2z) &= 4xyz^2(xy) - 4xyz^2(2z) + 4xyz^2(3x^2z) \\ &= 4x^2y^2z^2 - 8xyz^3 + 12x^3yz^3 \end{aligned}$$

Finally, note that by the Distributive Laws,

$$\begin{aligned} -(b + c) &= (-1)(b + c) \\ &= (-1)b + (-1)c \\ &= -b - c \end{aligned} \qquad \text{and} \qquad \begin{aligned} -(b - c) &= (-1)(b - c) \\ &= (-1)b - (-1)c \\ &= -b + c \text{ or } c - b \end{aligned}$$

CLASS EXERCISES *Multiply.*

11. $4(a + b)$ **12.** $10s(r + s)$ **13.** $a^2(a + b)$ **14.** $-5a^4c(ab + ac^2 - 2ab^4c)$

DIVISION Observe that for $a \neq 0$,

$$\frac{ab + ac}{a} = b + c$$

because, by the Distributive Laws and the Commutative Law,

$$ab + ac = a(b + c) = (b + c)a$$

Thus, $\frac{3x + 3y}{3} = x + y$ because $3x + 3y = 3(x + y) = (x + y)3$

Note that *both* terms of the polynomial $3x + 3y$ are divided by 3. In fact,

$$\frac{3x + 3y}{3} = \frac{3x}{3} + \frac{3y}{3} = x + y$$

Also, $\frac{x^3 + 5x}{x} = x^2 + 5$ because $x^3 + 5x = (x^2 + 5)x$

Both terms of $x^3 + 5x$ are divided by x:

$$\frac{x^3 + 5x}{x} = \frac{x^3}{x} + \frac{5x}{x} = x^2 + 5$$

Example 8

$$\frac{2x + 4}{2} = \frac{2x}{2} + \frac{4}{2} = x + 2$$

Example 9

$$\frac{2x^4 - 7x^3}{x^2} = \frac{2x^4}{x^2} - \frac{7x^3}{x^2} = 2x^2 - 7x$$

Example 10

$$\frac{5x^2y + 10xy^2 - 15xy^3}{5xy} = \frac{5x^2y}{5xy} + \frac{10xy^2}{5xy} - \frac{15xy^3}{5xy} = x + 2y - 3y^2$$

CLASS EXERCISES *Divide.* **15.** $\frac{6y - 3}{3}$ **16.** $\frac{5a^4 + 8a^3}{a^2}$ **17.** $\frac{4x^2y^2 + 8xy^3 - 20xy}{4xy}$

SOLUTIONS TO CLASS EXERCISES

1. $x^6 \cdot x^2 = x^{6+2} = x^8$ **2.** $a^5 \cdot a = a^{5+1} = a^6$ **3.** $y^3 \cdot y^2 \cdot y^5 = y^{3+2+5} = y^{10}$ **4.** $(5a)4 = (5 \cdot 4)a = 20a$

5. $(-2xy)(7xy) = [(-2)7](xx)(yy) = -14x^2y^2$ **6.** $(4abc)(2ab^2)(abc^2) = (4 \cdot 2 \cdot 1)(aaa)(bb^2b)(cc^2) = 8a^3b^4c^3$

7. $\dfrac{a^5}{a^2} = a^{5-2} = a^3$ **8.** $\dfrac{8x^2y^7}{4xy^2} = \dfrac{8}{4}x^{2-1}y^{7-2} = 2xy^5$ **9.** $\dfrac{a}{5} \cdot 15 = \dfrac{15a}{5} = 3a$

10. $\dfrac{b+3}{4} \cdot 4 = \dfrac{4(b+3)}{4} = b + 3$ **11.** $4(a+b) = 4a + 4b$

12. $10s(r+s) = (10s)r + (10s)s = 10rs + 10s^2$ **13.** $a^2(a+b) = a^2 \cdot a + a^2 \cdot b = a^3 + a^2b$

14. $-5a^4c(ab + ac^2 - 2ab^4c) = -5a^4c(ab) + (-5a^4c)(ac^2) - (-5a^4c)(2ab^4c) = -5a^5bc - 5a^5c^3 + 10a^5b^4c^2$

15. $\dfrac{6y-3}{3} = \dfrac{6y}{3} - \dfrac{3}{3} = 2y - 1$ **16.** $\dfrac{5a^4 + 8a^3}{a^2} = \dfrac{5a^4}{a^2} + \dfrac{8a^3}{a^2} = 5a^2 + 8a$

17. $\dfrac{4x^2y^2 + 8xy^3 - 20xy}{4xy} = \dfrac{4x^2y^2}{4xy} + \dfrac{8xy^3}{4xy} - \dfrac{20xy}{4xy} = xy + 2y^2 - 5$

HOME EXERCISES

In Exercises 1–14, multiply.

1. $b \cdot b \cdot b$ **2.** $c^2 \cdot c$ **3.** $b^3 \cdot b^2$ **4.** $4a^2 \cdot a^4$ **5.** $(-1)\,y^2 \cdot 5 \cdot y^2$

6. $c^4 \cdot c^2 \cdot c^3$ **7.** $a^2 \cdot a \cdot a^3 \cdot a$ **8.** $(x^2y)(xy^2)$ **9.** $(2c^2d)(5cd^2)$ **10.** $(-xy)(-2x^2y^2)$

11. $(5xy)(3x^2y)(2x^2)$ **12.** $(-xy)(2xy)(3y)$ **13.** $(2xy^2z)(xyz^2)(3xz)(yz^2)$ **14.** $(-2abc)(bcd)(3abc^2)(4a^2d)$

15. Find the product of $-2a^2bc$, $-5ac$, and abc^2. **16.** Evaluate $(2ab^2)(-3a^2b)$ when $a = 1$ and $b = -1$.

In Exercises 17–26, divide.

17. $\dfrac{8x}{2}$ **18.** $\dfrac{28x^2}{-7}$ **19.** $\dfrac{a^{10}}{a^7}$ **20.** $\dfrac{a^3b^2}{ab}$ **21.** $\dfrac{m^4n^6}{m^2n^2}$

22. $\dfrac{x^4y^3z^2}{xy^2z}$ **23.** $\dfrac{4x^2}{2x}$ **24.** $\dfrac{-9a^8}{-3a^4}$ **25.** $\dfrac{64a^{12}b^9c}{16a^3b^7}$ **26.** $\dfrac{48x^3y^3z^7}{-12x^2yz^5}$

In Exercises 27–30, multiply and divide.

27. $\dfrac{x}{4} \cdot 8$ **28.** $\dfrac{5xy}{30} \cdot 6$ **29.** $\dfrac{x+2}{3} \cdot 3$ **30.** $\dfrac{1-y}{6} \cdot 6$

In Exercises 31–46, multiply.

31. $9(x^2 + y)$ **32.** $2(a - b)$ **33.** $-3(m - n)$ **34.** $-3(3x + 2y^2)$

35. $-5(2a^2 - 4b)$ **36.** $a(a + 4)$ **37.** $x(x^2y + x^3)$ **38.** $2a(a^2 + a^3)$

39. $-4y^2(y + yz)$ **40.** $4xy(xyz + y^2)$ **41.** $-mn(2m^2 - 3n^2)$ **42.** $2uv(5u^2 - 3u^2v^2)$

43. $10x^2yz^3(5x^4yz^3 + 8xy^2z^2)$ **44.** $-3(y^2 - 4y + 9)$ **45.** $4xyz^2(x^2z - 4xy + 3xz)$ **46.** $-x^2yz^2(x + 2xy - 3xyz)$

In Exercises 47–52, simplify.

Sample Simplify:	**Solution**
$4a - 7(2 - 3a)$	$4a - 7(2 - 3a) = 4a - [7(2) - 7(3a)]$ $= 4a - [14 - 21a]$ $= 4a - 14 + 21a$ $= 25a - 14$

47. $3(4 - 2a) + 2(a + 1)$
48. $4x^2(y + z^2) - 2(x - y)$
49. $a(a - b) + b(b - a)$
50. $x(x^2 - 2) + x^2(x + 2)$
51. $2(a + 2b) - 3(a - b) + 5(a + 4b)$
52. $6(x + 3y) - 4(x - y) - 2(3x + 2y)$

In Exercises 53–60, divide.

53. $\dfrac{5x + 5y}{5}$
54. $\dfrac{4m - 16n}{-4}$
55. $\dfrac{a^2 + a}{a}$
56. $\dfrac{x^3 - x^2}{x^2}$
57. $\dfrac{x^2y + x}{x}$
58. $\dfrac{a^3b^2 + ab}{ab}$
59. $\dfrac{20x^4y^3 + 30x^3y^3}{5x^2y^3}$
60. $\dfrac{16x^2y^2z^2 + 12xy^2z^2 - 4xyz}{4xyz}$

2.5 EQUATIONS

ROOTS

DEFINITION

An **equation** is a statement of equality. Thus,

$$x + 2 = 5$$

is an equation. Here, $x + 2$ is the **left side** and 5 is the **right side**.

$$3 + 2 = 5$$

is also an equation because it is a statement of equality.

DEFINITION

ROOT. A number is a **root** (or **solution**) of an equation (in one variable) if a true statement results when *you substitute the number for the variable*.

Example 1 **a.** 3 is a root of the equation

$$x + 2 = 5$$

because when you substitute *3* for x, you obtain the *true* statement

$$3 + 2 = 5$$

b. 4 is *not* a root of the equation because

$$4 + 2 = 5$$

is *false*.

First, you will **check** whether or not a given number is a root of an equation. *To check this, substitute the number for the variable of the equation, to see whether a true statement results. If the variable occurs more than once in the equation, each time it occurs, replace it by the same number.* When you check a root of an equation, write $\stackrel{?}{=}$ instead of $=$. At the end of the check, write $\stackrel{\checkmark}{=}$ if the number is a root; if not, write $\stackrel{\times}{=}$.

Example 2 Is 10 a root of the equation

$$2x - 7 = x + 3$$

Solution Replace each occurrence of x by *10*:

$$2(10) - 7 \stackrel{?}{=} 10 + 3$$

$$20 - 7 \stackrel{?}{=} 13$$

$$13 \stackrel{\checkmark}{=} 13$$

Thus, 10 is a root of the given equation.

Example 3 Is -1 a root of the equation

$$1 - 3y = 5 - y$$

Solution Replace each occurrence of the variable y by -1:

$$1 - 3(-1) \stackrel{?}{=} 5 - (-1)$$

$$1 + 3 \stackrel{?}{=} 5 + 1$$

$$4 \stackrel{\times}{=} 6$$

Thus, -1 is *not* a root of the given equation.

CLASS EXERCISES *Check whether the given number is a root of the equation.*

1. $x + 3 = 8$ $\boxed{5}$
2. $2t = 10$ $\boxed{20}$
3. $5y + 3 = y - 1$ $\boxed{-1}$

SOLVING EQUATIONS Up to now, you have *checked* whether or not a given number is the root of an equation. Now, you will learn how to find the *unknown* root (or roots) of an equation.

DEFINITION

EQUIVALENT EQUATIONS. Equations are said to be **equivalent** if they have the same root (or roots).

$$2x = 8, \qquad x + 1 = 5, \qquad \text{and} \qquad x = 4$$

are all equivalent. Each has the single root 4. Clearly, the last of the these is the simplest equation because the variable is on one side all by itself.

DEFINITION

To **solve an equation** means to find its roots.

Each equation you will be solving in this section will have a single root. You solve an equation by transforming it into simpler and simpler equivalent equations. Your goal is to obtain an equation in which one side is a variable and the other a number. For example, if after simplifying you obtain

$$x = 2$$

then 2 is the root of the given equation.

When you **add the same expression to both sides of an equation**, *you obtain an equivalent equation.* (This is called the **Addition Property**.)

Example 4 Solve and check. $x - 4 = 6$

Solution

$x - 4 = 6$ *Add 4 to both sides.*

$x - 4 + 4 = 6 + 4$

$x = 10$

The root is 10.

check Substitute *10* for x in the given equation.

$10 - 4 \stackrel{?}{=} 6$

$6 \stackrel{\checkmark}{=} 6$

Subtracting a is the same as adding $-a$. Thus, *when you* **subtract the same expression from both sides of an equation**, *you obtain an equivalent equation.*

Example 5 Solve and check. $x + 8 = 13$

Solution

$x + 8 = 13$ *Subtract 8 from both sides.*

$x + 8 - 8 = 13 - 8$

$x = 5$

check $5 + 8 \stackrel{?}{=} 13$

$13 \stackrel{\checkmark}{=} 13$

When you **multiply (or divide) both sides of an equation by the same nonzero number,** *you obtain an equivalent equation.* (This is called the **Multiplication Property.)**

Example 6 Solve. $\frac{x}{3} = 6$

Solution $\frac{x}{3} = 6$ *Multiply both sides by 3.*

$$\frac{x}{3} \cdot 3 = 6 \cdot 3$$

$$x = 18$$

Example 7 Solve. $5y = 20$

Solution $5y = 20$ *Divide both sides by 5.*

$$\frac{5y}{5} = \frac{20}{5}$$

$$y = 4$$

CLASS EXERCISES

In Exercises 4 and 5, solve each equation.

4. $6y = 42$ **5.** $y + 7 = 9$

In Exercises 6 and 7, solve and check.

6. $x - 8 = 5$ **7.** $\frac{z}{8} = -2$

USING BOTH PROPERTIES

Both the Addition and Multiplication Properties frequently apply in the same example. *In simplifying an equation, bring terms with variables to one side, numerical terms to the other side.*

Example 8 Solve. $2x - 1 = 15$

Solution $2x - 1 = 15$ *Add 1 to both sides.*

$$2x - 1 + 1 = 15 + 1$$

$2x = 16$ *Divide both sides by 2.*

$$\frac{2x}{2} = \frac{16}{2}$$

$$x = 8$$

Example 9 Solve and check. $1 - 4x = 2x - 5$

Solution Bring terms involving x to one side; bring numerical terms to the other side. Because the coefficient of x is larger on the right side, bring x's to the *right*, numerical terms to the left.

$1 - 4x = 2x - 5$ *Add $4x + 5$ to both sides.*

$$1 - 4x + 4x + 5 = 2x - 5 + 4x + 5$$

$$6 = 6x \qquad \textit{Divide both sides by 6.}$$

$$\frac{6}{\boxed{6}} = \frac{6x}{\boxed{6}}$$

$$1 = x$$

check

$$1 - 4(1) \stackrel{?}{=} 2(1) - 5$$

$$1 - 4 \stackrel{?}{=} 2 - 5$$

$$-3 \stackrel{\checkmark}{=} -3$$

Which property do you apply first when both are used in solving an equation? Consider the equation

$$\frac{x}{2} + 6 = 12$$

The *last* operation on the left side calls for *adding* 6 to $\frac{x}{2}$. To solve an equation you *undo* the operations in the *opposite order*. Thus, you first *subtract* 6 from both sides.

$$\frac{x}{2} + 6 \boxed{- 6} = 12 \boxed{- 6}$$

$$\frac{x}{2} = 6 \qquad \textit{Now, multiply both sides by 2.}$$

$$\frac{x}{2} \boxed{\cdot 2} = 6 \boxed{\cdot 2}$$

$$x = 12$$

On the other hand, consider the equation

$$\frac{x + 6}{2} = 12$$

Here, the *last* operation on the left side calls for *dividing* $x + 6$ by 2. Thus, to solve this equation, you first *multiply* both sides by 2.

$$\frac{x + 6}{2} \boxed{\cdot 2} = 12 \boxed{\cdot 2}$$

$$x + 6 = 24 \qquad \textit{Now, subtract 6 from both sides.}$$

$$x + 6 \boxed{- 6} = 24 \boxed{- 6}$$

$$x = 18$$

When you *dress*, you *put on* your socks before you *put on* your shoes. When you *undress*, you *take off* your shoes before you *take off* your socks.

CLASS EXERCISES *In Exercises 8–10, solve each equation.*

8. $5t + 3 = 4t + 5$ **9.** $\frac{x + 6}{3} = 9$ **10.** $\frac{x}{3} + 6 = 9$

EQUATIONS WITH PARENTHESES

Parentheses are necessary to clarify the meaning of arithmetic and algebraic expressions. Often, one or both sides of an equation contains parentheses.

Example 10 Solve. $\frac{x - 2(x + 3)}{3} = -1$

Solution

$$\frac{x - 2(x + 3)}{3} = -1 \quad \textit{Multiply both sides by 3.}$$

$$\left[\frac{x - 2(x + 3)}{3}\right] \cdot 3 = (-1)\,3$$

$$x - 2(x + 3) = -3 \quad \textit{Remove parentheses on the left. (Use the Distributive Laws.)}$$

$$x - 2x - 6 = -3 \quad \textit{Simplify the left side.}$$

$$-x - 6 = -3$$

$$-x - 6 + 6 = -3 + 6$$

$$-x = 3 \quad \textit{Multiply both sides by } -1.$$

$$x = -3$$

It is often easier to divide both sides of an equation by a number, rather than to use the Distributive Laws.

Example 11 Solve. $4(5x + 3) = -8$

Solution

$$4(5x + 3) = -8 \quad \textit{Divide both sides by 4.}$$

$$\frac{4(5x + 3)}{4} = \frac{-8}{4}$$

$$5x + 3 = -2$$

$$5x + 3 - 3 = -2 - 3$$

$$5x = -5$$

$$\frac{5x}{5} = \frac{-5}{5}$$
$$x = -1$$

CLASS EXERCISES *Solve each equation.*

11. $4 + 2(3x + 5) = 26$ **12.** $12(3x + 1) = -24$

SOLUTIONS TO CLASS EXERCISES

1. $\begin{aligned} 5 + 3 &\stackrel{?}{=} 8 \\ 8 &\stackrel{\checkmark}{=} 8 \end{aligned}$
5 is a root.

2. $\begin{aligned} 2 \cdot 20 &\stackrel{?}{=} 10 \\ 40 &\stackrel{\times}{=} 10 \end{aligned}$
20 is *not* a root.

3. $\begin{aligned} 5(-1) + 3 &\stackrel{?}{=} -1 - 1 \\ -5 + 3 &\stackrel{?}{=} -2 \\ -2 &\stackrel{\checkmark}{=} -2 \end{aligned}$
−1 is a root.

4. $\begin{aligned} 6y &= 42 \\ \frac{6y}{6} &= \frac{42}{6} \\ y &= 7 \end{aligned}$

5. $\begin{aligned} y + 7 &= 9 \\ y + 7 - 7 &= 9 - 7 \\ y &= 2 \end{aligned}$

6. $\begin{aligned} x - 8 &= 5 \\ x - 8 + 8 &= 5 + 8 \\ x &= 13 \end{aligned}$

check $\begin{aligned} 13 - 8 &\stackrel{?}{=} 5 \\ 5 &\stackrel{\checkmark}{=} 5 \end{aligned}$

7. $\begin{aligned} \frac{z}{8} &= -2 \\ \frac{z}{8} \cdot 8 &= (-2)8 \\ z &= -16 \end{aligned}$

check $\begin{aligned} \frac{-16}{8} &\stackrel{?}{=} -2 \\ -2 &\stackrel{\checkmark}{=} -2 \end{aligned}$

8. $\begin{aligned} 5t + 3 &= 4t + 5 \\ 5t + 3 - 4t - 3 &= 4t + 5 - 4t - 3 \\ t &= 2 \end{aligned}$

9. $\begin{aligned} \frac{x + 6}{3} &= 9 \\ \frac{x + 6}{3} \cdot 3 &= 9 \cdot 3 \\ x + 6 &= 27 \\ x + 6 - 6 &= 27 - 6 \\ x &= 21 \end{aligned}$

10. $\begin{aligned} \frac{x}{3} + 6 &= 9 \\ \frac{x}{3} + 6 - 6 &= 9 - 6 \\ \frac{x}{3} &= 3 \\ \frac{x}{3} \cdot 3 &= 3 \cdot 3 \\ x &= 9 \end{aligned}$

11. $\begin{aligned} 4 + 2(3x + 5) &= 26 \\ 4 + 6x + 10 &= 26 \\ 6x + 14 &= 26 \\ 6x + 14 - 14 &= 26 - 14 \\ 6x &= 12 \\ \frac{6x}{6} &= \frac{12}{6} \\ x &= 2 \end{aligned}$

12. $\begin{aligned} 12(3x + 1) &= -24 \\ \frac{12(3x + 1)}{12} &= \frac{-24}{12} \\ 3x + 1 &= -2 \\ 3x + 1 - 1 &= -2 - 1 \\ 3x &= -3 \\ \frac{3x}{3} &= \frac{-3}{3} \\ x &= -1 \end{aligned}$

HOME EXERCISES

In Exercises 1–10, check whether the given number is a root of the equation.

1. $x + 4 = 10;$ $\boxed{6}$ **2.** $3x = 6;$ $\boxed{2}$ **3.** $4 - t = 2;$ $\boxed{6}$ **4.** $1 - y = -1;$ $\boxed{2}$
5. $2 - 4z = 6$ $\boxed{-1}$ **6.** $-3x = 9;$ $\boxed{3}$ **7.** $2x + 7 = 1;$ $\boxed{-3}$ **8.** $4x - 2 = 3x;$ $\boxed{4}$
9. $x = 5x;$ $\boxed{0}$ **10.** $x = -x;$ $\boxed{-1}$

In Exercises 11–40, solve the equation.

11. $y - 3 = 4$ **12.** $x + 5 = 8$ **13.** $\frac{x}{2} = 7$ **14.** $7x = 21$ **15.** $-2z = 6$
16. $-t = 7$ **17.** $6 - t = -9$ **18.** $2t + 3 = 11$ **19.** $\frac{t}{2} + 1 = 7$ **20.** $\frac{t + 1}{2} = 5$

21. $6x - 3 = 5x + 2$ **22.** $2x - 5 = x + 5$ **23.** $9 - 3x = -2x + 3$ **24.** $1 + 7x = 4x + 4$

25. $x + 4 = -x + 2$ **26.** $-2x + 4 = 8$ **27.** $\frac{x}{2} + 4 = 8$ **28.** $\frac{x + 4}{2} = 8$

29. $\frac{x}{6} - 3 = 9$ **30.** $\frac{x - 3}{6} = 9$ **31.** $\frac{t + 24}{8} = 0$ **32.** $\frac{t}{8} + 24 = 0$

33. $3(x - 1) = 2x + 1$ **34.** $y - (1 + y) = 2y - 1$ **35.** $-7(t + 5) = t - 3$ **36.** $(4 + 5u) - (2u - 2) = 12$

37. $5 - (z - 3) + 6(z - 1) = 0$ **38.** $1 - [x - (1 - x)] = 0$

39. $2[4 - (3z - 5)] = 3z$ **40.** $2[y - (5 - 2y)] = -4$

In Exercises 41–50, **a.** *solve the equation and* **b.** *check the root.*

41. $3x = 12$ **42.** $2x + 1 = 3$ **43.** $7 - 3x = 1 + x$ **44.** $4x + 2 = x - 10$

45. $2x + 7 = 7x + 17$ **46.** $2 - 3x = 8 - x$ **47.** $-3(y + 2) = y - 10$ **48.** $2x + 3(x - 2) = 19$

49. $\frac{6x - (x - 5)}{5} = 8$ **50.** $\frac{2 - 3(7 - 2x)}{7} = -1$

51. What number must be added to 2 to obtain 6? **52.** What number must be subtracted from 2 to obtain 6?

53. What number must be multiplied by 2 to obtain 6? **54.** What number must be divided by 2 to obtain 6?

Let's Review Chapter 2

1. Write each term as a product of a number and variables. **a.** $5xy^3$ **b.** $\frac{uv^2w^3}{2}$

2. Find the coefficient of $-12x^2y$. **3.** Are $3x^2y$ and $6xyx$ like terms?

4. Are $10xy^3$ and $10x^3y$ like terms?

In Exercises 5–9, add or subtract.

5. $3m + 5n - m + 10n$ **6.** $4x + 3y - z + x - y - 2z$

7.
$$\begin{array}{l} w + x + 2y - z \\ w - x + 4y \\ \quad\;\; x - y + z \\ \underline{2w \qquad - y - z} \end{array}$$

8. $(5x + y - 2z) - (3x - z) + (x - y + 4z)$ **9.** $15a + b - [4a - (2a - b)]$

10. Subtract the bottom polynomial from the top one:
$$\begin{array}{l} 3a + b - c \\ \underline{2a \qquad -2c} \end{array}$$

11. Evaluate $x^2 + 3x + 10$ when $x = 5$.

12. Evaluate $x^2 - y^2$ when $x = 9$, $y = 6$.

13. Evaluate $x^3 - 2x^2 + 4$ when a. $x = 2$ b. $x = -2$

14. Find the area of a square if a side is of length 20 inches.

In Exercises 15–18, multiply.

15. $(3x^2y)(2xy^3)$ **16.** $(4m^2n)(-2mn^2)(-mn)$

17. $x^2(x - 5y)$ **18.** $4a^2bc(2a^2b + 3abc)$

In Exercises 19–22, divide.

19. $\frac{10x^3y^2}{5xy}$ **20.** $\frac{-32a^4b^3}{-4ab^3}$ **21.** $\frac{6a + 6b}{3}$ **22.** $\frac{20x^2y - 15xy^2}{5xy}$

In Exercises 23–25, check whether the given number is a root of the equation.

23. $3x + 2 = 11;\ \boxed{3}$ **24.** $5x - 2 = 2x + 3;\ \boxed{2}$
25. $5 - 2y = y + 1;\ \boxed{-2}$

In Exercises 26–29, solve each equation.

26. $\dfrac{x - 3}{4} = 9$ **27.** $3x + 9 = 15 - 3x$ **28.** $x - (8 - x) = 16$

29. $4(x - 3) - 3(x - 4) = 6$

30. *Solve and check.* $2y + 3 = 15 - 4y$

3 FACTORING

DIAGNOSTIC TEST

Perhaps you are already familiar with some of the material in Chapter 3. This test will indicate which sections in Chapter 3 you need to study. The question number refers to the corresponding section in Chapter 3. If you answer *all* parts of the question correctly, you may omit the section. But *if any part of your answer is wrong, you should study the section*. When you have completed the test, turn to page A5 for the answers.

3.1 Common Factors

a. Express 120 as the product of primes. b. Find $gcd\,(42, 56)$.
c. Isolate the common factors of the polynomial $8x^2y + 20xy^2 - 12xy$.

3.2 Products of Binomials

Combine.

a. $(a + 4)(a + 2)$ b. $(x - 7)(x + 5)$
c. $(m^2 + 4n)(m - 3n)$ d. $(a + 2)^2 \quad (a \quad 1)(a - 2)$

3.3 Difference of Squares

Factor each binomial.

a. $x^2 - 64$ b. $a^4 - 9b^2$ c. $5x^2 - 20$

3.4 Factoring Trinomials $x^2 + Mx + N$

Factor each trinomial.

a. $x^2 + 5x + 4$ b. $a^2 - a - 42$ c. $m^3 + 6m^2 + 9m$

3.5 Factoring Trinomials $Lx^2 + Mx + N$

Factor each trinomial.

a. $4x^2 + 5x + 1$ b. $8a^2 - 2a - 1$ c. $6x^2 - 7xy - 3y^2$

3.1 COMMON FACTORS

It is important to know how to write an integer as a product of other integers. This process is known as **factoring**; *it reverses multiplication*. You will also learn how to factor a polynomial, that is, how to write it as a product of other polynomials.

PRIME FACTORS

Observe that

$$15 = 3 \cdot 5$$

Thus, 15 can be expressed as the product of two integers, 3 and 5, which are called *factors* of 15.

DEFINITION

FACTOR. Let a and b be integers, where $b \neq 0$. Then, b is called a **factor of a** if

$$a = b \cdot c$$

for some integer c. In this case, if $c \neq 0$, then c is also a factor of a, and **a is divisible by b and by c**. For any integers a, b, and c, if

$$a = b \cdot c$$

then a is said to be a **multiple of b and of c**.

Example 1 Find all positive factors of 10.

Solution

$$10 = 10 \cdot 1 \quad \text{and} \quad 10 = 5 \cdot 2$$

The *positive* factors of 10 are 1, 2, 5, and 10.

Also, 10 is divisible by each of these numbers and is a multiple of each.

Note that the only way you can express 5 as the product of positive integers is

$$5 = 5 \cdot 1 = 1 \cdot 5$$

Thus, the only *positive* factors of 5 are 5 and 1.

DEFINITION

PRIME; COMPOSITE. Let p be an integer, where $p > 1$. Then, p is said to be a **prime** if the only positive factors of p are p itself and 1. An integer n, $n > 1$, that is not a prime is called a **composite**.

Thus, 2, 3, 5, and 7 are each primes. 6 and 10 are each composites.

$$6 = 2 \cdot 3 \quad \text{and} \quad 10 = 2 \cdot 5$$

4 is also a composite because $4 = 2 \cdot 2$

0, 1, and negative integers are not classified as either primes or composites.

Example 2 Which of the following are primes? Which are composites?

a. 11 **b.** 15 **c.** 23 **d.** 33

Solution **a.** 11 is a prime. **b.** 15 is a composite because $15 = 3 \cdot 5$.
c. 23 is a prime. No smaller positive integer other than 1 is a factor of 23.
d. 33 is a composite because $33 = 3 \cdot 11$.

Clearly, $$12 = 4 \cdot 3$$

But, the factor 4 can be further simplified because $4 = 2 \cdot 2$. Thus,

$$12 = 2 \cdot 2 \cdot 3$$

Every composite can be expressed as the product of primes. Thus,

$$\begin{aligned} 40 &= 8 \cdot 5 \\ &= (2 \cdot 2 \cdot 2)5 \\ &= 2^3 \cdot 5 \end{aligned}$$

It is convenient to use exponents to express repeated prime factors. *Except for the order of the factors, there is only one way of expressing a composite as the product of primes.* No matter how you begin factoring, the *prime* factors are the same. For example,

$$\begin{aligned} 40 &= 4 \cdot 10 \\ &= (2 \cdot 2)(2 \cdot 5) \\ &= 2^3 \cdot 5 \end{aligned}$$

Also, you can write $-40 = -(2^3 \cdot 5)$

Example 3 Express each integer as the product of primes.

a. 30 **b.** 50 **c.** 72

Solution **a.** $30 = 6 \cdot 5 = 2 \cdot 3 \cdot 5$ **b.** $50 = 5 \cdot 10 = 5 \cdot 2 \cdot 5 = 2 \cdot 5^2$
c. $72 = 8 \cdot 9 = 2^3 \cdot 3^2$

CLASS EXERCISES **1.** Find all *positive* factors of 14.

In Exercises 2–5, which integers are primes and which are composites?
2. 16 **3.** 17 **4.** 19 **5.** 27

In Exercises 6–8, express each integer as the product of primes.
6. 21 **7.** 45 **8.** 120

COMMON FACTORS OF INTEGERS

Clearly, 5 is a factor of both 10 and 15.

DEFINITION

COMMON FACTOR. Let m and n be integers. Then, the positive integer c is called a **common factor of m and n** if c is a factor of both m and n. The largest common factor of m and n is called the **greatest common divisor of m and n,** and is written

$$gcd\,(m, n)$$

Example 4

a. 1 and 5 are the common factors of 10 and 15. The larger of these is 5. Thus, $gcd\,(10, 15) = 5$

b. 1 is the only common factor of 8 and 9. Thus, $gcd\,(8, 9) = 1$

c. The common factors of 6 and 12 are 1, 2, 3, and 6. Thus, $gcd\,(6, 12) = 6$

Part **c.** of Example 4 illustrates the fact that *if m is a factor of n, then* $gcd\,(m, n) = m$. As a second example, $gcd\,(10, 100) = 10$

You can also speak of common factors and greatest common divisors of three or more integers. The *gcd* notation is again used.

When the integers are relatively simple, their *gcd* can be found at sight, as in Example 4. When difficulties arise, use the prime factorizations of each integer.

Example 5 Find $gcd\,(36, 54, 72)$.

Solution

$$36 = 4 \cdot 9 = 2^2 \cdot 3^2, \qquad 54 = 2 \cdot 27 = 2 \cdot 3^3, \qquad 72 = 8 \cdot 9 = 2^3 \cdot 3^2$$

Observe that 2 divides all three given integers, but 2^2 (or 4) does not divide 54. Similarly, 3^2 (or 9) divides all three given integers, but 3^3 (or 27) divides neither 36 nor 72. Thus, for each *common* prime factor, consider the *smallest* power that occurs in *any* of the prime factorizations. *The product of these smallest powers* is then a common factor of the integers 36, 54, and 72. In fact, this product is the largest common factor of these integers, and hence, the *gcd*.

Thus, here the common prime factors are 2 and 3. The smallest power of 2 that occurs is 2 (in the factorization of 54). The smallest power of 3 is 3^2.

$$gcd\,(36, 54, 72) = 2 \cdot 3^2 = 18$$

An important application of this notion occurs in finding the *gcd* of the coefficients of a polynomial. This will be used in "factoring" polynomials.

Example 6 Find the *gcd* of the coefficients of $32x - 80$.

Solution Factor both coefficients.

$$32 = 4 \cdot 8 = 2^2 \cdot 2^3 = 2^5, \qquad -80 = -(16 \cdot 5) = -2^4 \cdot 5$$

The only common prime factor is 2, and the smallest power of 2 that occurs is 2^4. Thus,

$$gcd\,(32, -80) = 2^4 = 16$$

and the *gcd* of the coefficients of $32x - 80$ is 16.

CLASS EXERCISES

In Exercises 9 and 10, find the gcd of the given integers.

9. -9 and 15 **10.** 72, 108, 132

In Exercises 11 and 12, find the gcd of the coefficients of each polynomial.

11. $9y - 18$ **12.** $12x^2 + 18x - 24$

FACTORS OF POLYNOMIALS

Recall that a nonzero integer b is a factor of (the integer) a if

$$a = bc$$

for some integer c. The same notion applies to polynomials. You will want to call 2 and x *factors* of the polynomial $2x$.

DEFINITION

Let P and Q ($\neq 0$) be polynomials. Then, Q is called a **factor of P** if

$$P = Q \cdot R$$

for some polynomial R.

Example 7 Show that 3 is a factor of $6x + 9$.

Solution First, note that 3 is the *gcd* of the coefficients of $6x + 9$. Also,

$$\frac{6x + 9}{3} = \frac{6x}{3} + \frac{9}{3} = 2x + 3$$

Thus, $6x + 9 = 3(2x + 3)$

Here, $P = 6x + 9$, $Q = 3$, $R = 2x + 3$

Example 8 Show that x^2 is a factor of $x^5 + 2x^2$.

Solution

$$\frac{x^5 + 2x^2}{x^2} = \frac{x^5}{x^2} + \frac{2x^2}{x^2} = x^3 + 2$$

Thus, $x^5 + 2x^2 = x^2(x^3 + 2)$

Here, $P = x^5 + 2x^2$, $Q = x^2$, $R = x^3 + 2$

Numbers, such as 3 (in Example 7), as well as powers of variables, such as x^2 (in Example 8), can be factors of a polynomial. In the next example, *the product of a number and of powers of several different variables is a factor of the given polynomial.*

Example 9 Let $P = 6x^2yz^3 + 12x^3yz$. Then,

$$6, x^2, y, \text{ and } z$$

are each factors of P. Also, their *product*,

$$6x^2yz$$

is a factor of P. In fact,

$$\frac{6x^2yz^3 + 12x^3yz}{6x^2yz} = z^2 + 2x$$

Thus, $$6x^2yz^3 + 12x^3yz = 6x^2yz(z^2 + 2x)$$

DEFINITION

Let P be a polynomial. The **(greatest) common (monomial) factor of P**, or for short, the **common factor of P,** is defined to be the product of

1. the greatest common divisor of the coefficients of P and
2. the smallest power of each variable that occurs in each term of P.

Thus, in Example 9, $6x^2yz$ is the *common factor of P.*

CLASS EXERCISES

13. Which of the following are factors of $5ax^3 + 10ax^2 + 25ax$?
a. 5 **b.** a **c.** x **d.** x^2
e. What is the common factor of $5ax^3 + 10ax^2 + 25ax$?

ISOLATING THE COMMON FACTOR

DEFINITION

To **isolate the common factor**, Q, of a polynomial, P, write

$$P = Q \cdot R$$

where Q is the common factor of P and where R is another polynomial.

Apply the *Distributive Laws* to isolate the common factor of a polynomial.

Example 10 Isolate the common factor of

$$5x^2y^2 - 10xy^3$$

Solution Let

$$P = 5x^2y^2 - 10xy^3$$

Then, the common factor of P is $\underbrace{5xy^2}_{Q}$. Isolate the common factor Q by writing

$$\underbrace{5x^2y^2 - 10xy^3}_{P} = \underbrace{5xy^2}_{Q}\underbrace{(x - 2y)}_{R}$$

Example 11 Isolate the common factor of

$$2a^2bc + 4a^2c + 5a^3b$$

Solution

$$gcd\,(2, 4, 5) = 1$$

The only variable that occurs in all three terms is a; the smallest power to which it occurs is the second power, a^2. Thus, a^2 is the common factor. Isolate the common factor by writing

$$\underbrace{2a^2bc + 4a^2c + 5a^3b}_{P}$$
$$= \underbrace{a^2}_{Q}\underbrace{(2bc + 4c + 5ab)}_{R}$$

Consider the inverse of the polynomial P of Example 11:

$$-2a^2bc - 4a^2c - 5a^3b$$

All of the coefficients are negative, and it is often more useful to consider

$$-Q = -a^2$$

and to write

$$\underbrace{-2a^2bc - 4a^2c - 5a^3b}_{P} = \underbrace{-a^2}_{-Q}\underbrace{(2bc + 4c + 5ab)}_{R}$$

Finally, note that the common factor of $2x + 3$ is 1. The present method of isolating the common factor does not apply to this polynomial.

CLASS EXERCISES *Isolate the common factor of each polynomial. If the common factor is 1, leave the given polynomial unchanged.*

14. $ax + a$ **15.** $12x^2 + 24x + 48$ **16.** $x^2 + xy + y^2$

SOLUTIONS TO CLASS EXERCISES

1. 1, 2, 7, and 14 **2.** composite ($16 = 8 \cdot 2 = 4 \cdot 4$) **3.** prime **4.** prime **5.** composite ($27 = 9 \cdot 3$)

6. $21 = 3 \cdot 7$ **7.** $45 = 9 \cdot 5 = 3^2 \cdot 5$ **8.** $120 = 12 \cdot 10 = (2 \cdot 2 \cdot 3) \cdot (2 \cdot 5) = 2^3 \cdot 3 \cdot 5$

9. 3 **10.** 12 **11.** 9 **12.** 6 **13.** a, b, c; e. $5ax$ **14.** $ax + a = a(x + 1)$

15. $12x^2 + 24x + 48 = 12(x^2 + 2x + 4)$ **16.** $x^2 + xy + y^2$

HOME EXERCISES

In Exercises 1–6, find all positive factors of each given integer.
1. 5 **2.** 8 **3.** 20 **4.** 22 **5.** 28 **6.** 30

In Exercises 7–12, which integers are primes and which are composites?
7. 13 **8.** 18 **9.** 31 **10.** 39 **11.** 41 **12.** 43

In Exercises 13–23, express each integer as the product of primes.
13. 22 **14.** 25 **15.** 36 **16.** 42 **17.** 48 **18.** 60 **19.** 80 **20.** 84 **21.** 90
22. 100 **23.** 144

24. Find the smallest positive integer that is the product of two different primes.

25. Find the smallest positive integer that is the product of three different primes.

26. Find the smallest positive integer that is the product of four different primes.

In Exercises 27–34, find the gcd of the given integers.
27. 9 and 18 **28.** 25 and 35 **29.** -24 and -36 **30.** 60 and 84 **31.** 4, 6, and 8
32. 25, 30, and 50 **33.** 12, 18, 24, and 36 **34.** 30, 42, 60, and 96

In Exercises 35–40, find the gcd of the coefficients of each polynomial.
35. $4x - 8$ **36.** $12x + 20$ **37.** $-x^2 + 5$ **38.** $48x^2 + 108x$
39. $24x^2 + 36x - 42$ **40.** $27x^3 - 18x^2 + 54x$

In Exercises 41–60, isolate the common factor of each polynomial. If the common factor is 1, leave the given polynomial unchanged.
41. $2a + 2$ **42.** $5x^2 + 10$ **43.** $3m^3 - 9$ **44.** $20x^2 + 50$ **45.** $16a + 25$ **46.** $t^2 + t$
47. $2x^2 + 3x$ **48.** $y^3 - y^2$ **49.** $a^{10} - a^7$ **50.** $12x^3 - 18x$ **51.** $32a^4 + 12a^3$ **52.** $x^2y + xy^2$
53. $2a^2bc + 4ab^2$ **54.** $25a^2b^3c^2 + 20ab^2c^3$ **55.** $x^7y^6z^5 - x^5y^6z^4 + x^3z^4$ **56.** $10x^3 - 20x^2 + 50x$
57. $36y^2z^5 + 60y^2z^3 - 24yz^2$ **58.** $18m^2n^3 + 54mn^4 - 72mn^3$
59. $96x^2y^5z^3 - 144x^2y^{10}z^2 + 72x^3y^4z^5$ **60.** $2x^5y + 12x^4y^2 + 8x^3y^3 - 6x^2y^4 + 4xy^5$

3.2 PRODUCTS OF BINOMIALS

Recall that the Distributive Laws apply to polynomials. Thus, let a, b, c, and d represent terms. Then,

$$a(b + c) = ab + ac, \qquad (b + c)a = ba + ca$$
$$a(b - c) = ab - ac, \qquad (b - c)a = ba - ca$$
$$a(b + c + d) = ab + ac + ad, \quad \text{and so forth}$$

USE OF DISTRIBUTIVE LAWS To multiply two polynomials (not both monomials), first use the Distributive Laws. *Every term appearing in the first polynomial is multiplied by every term of the second polynomial.* The product is then the sum of the resulting terms. Do you see that the product of two polynomials is again a polynomial?

Example 1 Multiply. $(x + 1)(x + 2)$

Solution According to the Distributive Laws,

$$\begin{aligned}(x + 1)(x + 2) &= (x + 1)x + (x + 1)2 \\ &= x^2 + x + 2x + 2 \\ &= x^2 + 3x + 2\end{aligned}$$

Observe that every term of the polynomial $x + 1$ is multiplied by every term of the polynomial $x + 2$. Thus, you obtain:

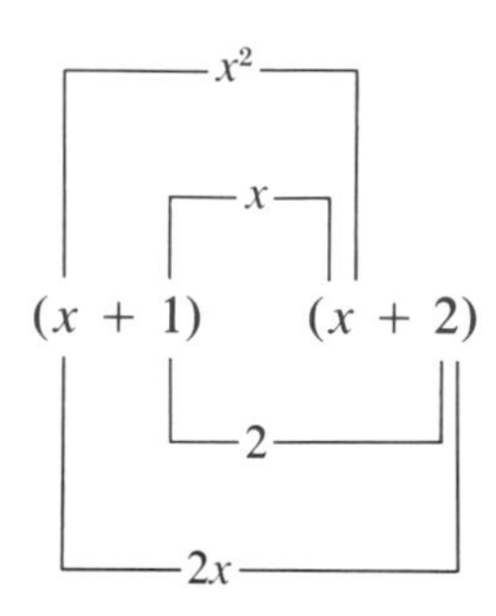

You then add these terms

$$x^2 + x + 2x + 2$$

to obtain the polynomial

$$x^2 + 3x + 2$$

Example 1 illustrates a product that is frequently encountered —namely,

$$(x + a)(x + b)$$

Of course, the letters may change. For example, you may see numbers substituted for a and b, as in

$$(x + 5)(x + 3)$$

where $a = 5$ and $b = 3$.

$$\begin{aligned}(x + a)(x + b) &= xx + ax + xb + ab \\ &= x^2 + ax + bx + ab \\ &= x^2 + (a + b)x + ab\end{aligned}$$

Therefore,

$$\boxed{(x + a)(x + b) = x^2 + (a + b)x + ab}$$

For example,

$$(x + 5)(x + 3) = x^2 + (5 + 3)x + 5 \cdot 3 = x^2 + 8x + 15$$

It is probably easier to multiply by writing the factors one above the other. Write *like terms* in the *same column.*

Example 2 Multiply. $(x - 4)(x - 2)$

Solution First, multiply by x:

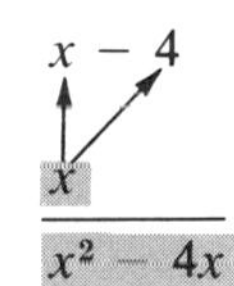

then, by -2:

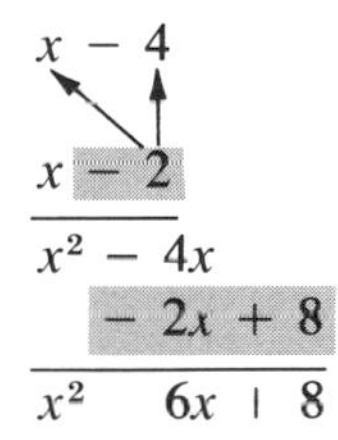

In the remaining examples, multiplication is understood.

Example 3

a.
$$\begin{array}{r} x + 6 \\ x - 3 \\ \hline x^2 + 6x \qquad \\ -\,3x - 18 \\ \hline x^2 + 3x - 18 \end{array}$$

b. $(x + 5)^2 = (x + 5)(x + 5)$

$$\begin{array}{r} x + 5 \qquad \\ x + 5 \qquad \\ \hline x^2 + \;\;5x \qquad \\ 5x + 25 \\ \hline x^2 + 10x + 25 \end{array}$$

In general,

$$\boxed{(x + a)^2 = (x + a)(x + a) = x^2 + 2ax + a^2}$$

Example 4

$$\begin{array}{rrr} x + 4 & & \\ x - 4 & & \\ \hline x^2 + 4x & & \\ -4x & - 16 & \\ \hline x^2 & - 16 & \end{array}$$

Note that the "cross-terms" $4x$ and $-4x$ are inverses. Their sum is 0.

In general,

$$(x + a)(x - a) = x^2 \underbrace{(a - a)}_{0}x - a^2$$

Thus,

$$\boxed{(x + a)(x - a) = x^2 - a^2}$$

CLASS EXERCISES

Multiply.

1. $(y + 7)(y + 5)$ **2.** $(a - 5)(a - 4)$ **3.** $(b + 8)(b - 2)$ **4.** $(x + 3)^2$
5. $(x - 3)^2$ **6.** $(y + 2)(y - 2)$

FURTHER EXAMPLES

Example 5

a.
$$\begin{array}{r} 2x + 1 \\ 2x + 3 \\ \hline 4x^2 + 2x \\ 6x + 3 \\ \hline 4x^2 + 8x + 3 \end{array}$$

b.
$$\begin{array}{r} 2a^2 + 3 \\ a + 2 \\ \hline 2a^3 + 3a \\ 4a^2 + 6 \\ \hline 2a^3 + 4a^2 + 3a + 6 \end{array}$$

c.
$$\begin{array}{r} 5a + b + 1 \\ a - b \\ \hline 5a^2 + ab + a \\ - 5ab - b^2 - b \\ \hline 5a^2 - 4ab + a - b^2 - b \end{array}$$

Example 6

$$\begin{aligned} (x + y)^2 - (x - y)^2 &= (x^2 + 2xy + y^2) - (x^2 - 2xy + y^2) \\ &= (x^2 - x^2) + (2xy + 2xy) + (y^2 - y^2) \\ &= 4xy \end{aligned}$$

CLASS EXERCISES

Combine.

7. $(2x + 5)(2x + 1)$ **8.** $(2y + 3)(y - 1)$ **9.** $(3a + b + 1)(a + 2b)$
10. $(x + 3)(x + 1) - (x + 2)(x - 1)$

SOLUTIONS TO CLASS EXERCISES

1. $y^2 + 12y + 35$ **2.** $a^2 - 9a + 20$ **3.** $b^2 + 6b - 16$ **4.** $x^2 + 6x + 9$ **5.** $x^2 - 6x + 9$

6. $y^2 - 4$ **7.** $4x^2 + 12x + 5$ **8.** $2y^2 + y - 3$ **9.** $3a^2 + 7ab + a + 2b^2 + 2b$

10.
$$\begin{aligned} (x + 3)(x + 1) - (x + 2)(x - 1) &= (x^2 + 4x + 3) - (x^2 + x - 2) \\ &= (x^2 - x^2) + (4x - x) + (3 + 2) \\ &= 3x + 5 \end{aligned}$$

HOME EXERCISES

In Exercises 1–40, multiply.

1. $(z + 8)(z + 1)$
2. $(y + 2)(y + 3)$
3. $(a + 3)(a - 2)$
4. $(b + 3)(b - 7)$
5. $(b - 2)(b - 4)$
6. $(x - 9)(x + 3)$
7. $(m - 10)(m + 3)$
8. $(n - 2)(n - 7)$
9. $5(z + 2)(z - 1)$
10. $-2(a - 9)(a + 5)$
11. $-3(b - 1)(b - 2)$
12. $-(c - 3)(c + 6)$
13. $(y + 2)^2$
14. $(a - 2)^2$
15. $(c + 6)^2$
16. $(v - 8)^2$
17. $(x + 8)(x - 8)$
18. $(y - 4)(y + 4)$
19. $(x - 12)^2$
20. $(x^2 + 4)^2$
21. $(x^2 + 3)(x^2 - 3)$
22. $(a^4 + 1)(a^4 - 1)$
23. $(u^4 + 1)(u^4 + 2)$
24. $(x^3 + 1)^2$
25. $(x^3 - 3)^2$
26. $(x^2 + 3)(x^2 + 2)$
27. $(b^2 + 5)(b^2 - 2)$
28. $(c^3 - 1)(c^3 - 4)$
29. $(c^3 - 1)(c^2 + 1)$
30. $(x + y)^2$
31. $(x + a)(x - a)$
32. $(m + 2n)(m + n)$
33. $(x + 3a)(x + 2a)$
34. $(m - 4n)(m + 2n)$
35. $(x - 2y)(x + y)$
36. $(5a - 3b)(-2a + 6b)$
37. $(2x + 2y + 1)(x - 4y)$
38. $(20m - 3n - 2)(m - 2n)$
39. $(3a + b)(2a - 2b + 3)$
40. $(x^3 + x^2 + x + 2)(x^2 - x + 4)$

In Exercises 41–44, combine and simplify.

41. $(x + 2)(x + 3) + (x + 1)(x + 4)$
42. $(x + 4)(x - 2) - (x + 3)(x + 5)$
43. $(y + 7)(y - 2) + (y + 5)^2$
44. $(y + 3)^2 - (y - 1)^2$
45. Multiply the sum of $3x + 1$ and $-2x$ by $2x + 5$.
46. Add $2x^2 + x - 1$ to the product of $x + 3$ and $x - 2$.

3.3 DIFFERENCE OF SQUARES

Recall that you factor a polynomial P by writing P as a product of other polynomials. In Section 3.1, you learned how to isolate the common factor of a polynomial. In the remainder of the chapter you will learn other techniques for factoring polynomials.

$x^2 - a^2$

One of the easiest types of factoring to recognize is known as the **difference of squares**. Recall that

$$(x + a)(x - a) = x^2 - a^2$$

The "difference of squares" method applies to a binomial, such as

$$x^2 - a^2$$

whose two terms are *squares* that are separated by a *minus* sign.

Example 1 Factor. $x^2 - 100$

Solution

$$100 = 10^2$$

Thus,

$$\begin{aligned} x^2 - 100 &= x^2 - 10^2 \\ &= (x + 10)(x - 10) \end{aligned}$$

Example 2 Factor. $4 - a^2$

Solution Although the number 4 is written first, the difference of squares method again applies.

$$4 = 2^2$$

Thus,

$$4 - a^2 = 2^2 - a^2$$
$$= (2 + a)(2 - a)$$

The above method applies to the difference of squares, but *not* to the sum of squares. For example,

$$x^2 + 9$$

cannot be factored by the present method. Notice that the cross-terms do not cancel in the following product:

$$\begin{array}{r} x + 3 \\ x + 3 \\ \hline x^2 + 3x \\ 3x + 9 \\ \hline x^2 + 6x + 9 \end{array}$$

Finally, observe that

$$s^2 - t^2 = (s + t)(s - t)$$

A common error in applying this method is to factor the individual terms, but not the polynomial as a whole. For instance, do not write

$$s^2 - t^2 = s \cdot s - t \cdot t$$

but rather,

$$s^2 - t^2 = (s + t)(s - t)$$

as above. (The first equality is correct, but is not the desired factored result.)

CLASS EXERCISES *Factor each binomial.* **1.** $y^2 - 64$ **2.** $9 - a^2$

TWO VARIABLES You may be given a binomial such as

$$x^2 - a^2b^2$$

By the Associative and Commutative Laws of Multiplication,

$$a^2b^2 = (aa)(bb)$$
$$= (ab)(ab)$$
$$= (ab)^2$$

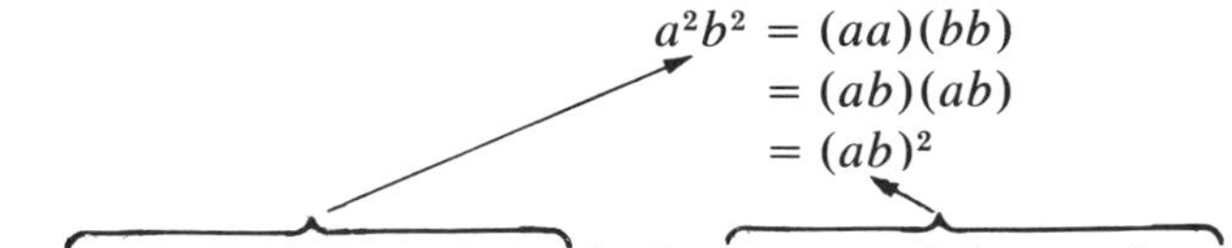

The product of the squares is the square of the product. Thus,

$$x^2 - a^2b^2 = (x + ab)(x - ab)$$

Example 3 Factor. **a.** $x^2 - 100y^2$ **b.** $16a^2 - 25b^2$

Solution **a.** $100y^2 = 10^2y^2 = (10y)^2$

Thus,

$$x^2 - 100y^2 = x^2 - (10y)^2 = (x + 10y)(x - 10y)$$

b. $16a^2 = (4a)^2$; $25b^2 = (5b)^2$

Thus,

$$16a^2 - 25b^2 = (4a)^2 - (5b)^2 = (4a + 5b)(4a - 5b)$$

CLASS EXERCISES *Factor each binomial.* **3.** $x^2 - 81a^2$ **4.** $9y^2 - 49z^2$

COMMON FACTORS You must sometimes first isolate the common factor before applying the difference of squares method.

Example 4 Factor. **a.** $5a^2 - 20b^2$ **b.** $ax^2 - a^3$

Solution **a.** $5a^2 - 20b^2 = 5(a^2 - 4b^2) = 5(a + 2b)(a - 2b)$
b. $ax^2 - a^3 = a(x^2 - a^2) = a(x + a)(x - a)$

CLASS EXERCISES *Factor each binomial.* **5.** $2x^2 - 18$ **6.** $4s^2 - 4t^2$ **7.** $xy^2 - 4x$

SOLUTIONS TO CLASS EXERCISES

1. $y^2 - 64 = (y + 8)(y - 8)$ **2.** $9 - a^2 = (3 + a)(3 - a)$ **3.** $x^2 - 81a^2 = x^2 - (9a)^2 = (x + 9a)(x - 9a)$

4. $9y^2 - 49z^2 = (3y)^2 - (7z)^2 = (3y + 7z)(3y - 7z)$ **5.** $2x^2 - 18 = 2(x^2 - 9) = 2(x + 3)(x - 3)$

6. $4s^2 - 4t^2 = 4(s^2 - t^2) = 4(s + t)(s - t)$ **7.** $xy^2 - 4x = x(y^2 - 4) = x(y + 2)(y - 2)$

HOME EXERCISES

In Exercises 1–44, factor each binomial.

1. $y^2 - 4$ **2.** $x^2 - 49$ **3.** $a^2 - 25$ **4.** $s^2 - 36$ **5.** $y^2 - 81$ **6.** $z^2 - 144$
7. $1 - x^2$ **8.** $25 - b^2$ **9.** $121 - c^2$ **10.** $x^2 - y^2$ **11.** $m^2 - n^2$ **12.** $4x^2 - 1$
13. $9a^2 - 25$ **14.** $36x^2 - 49$ **15.** $1 - 9u^2$ **16.** $4x^2 - y^2$ **17.** $9x^2 - 16a^2$ **18.** $100z^2 - 81a^2$
19. $144y^2 - 25x^2$ **20.** $400t^2 - u^2$ **21.** $5y^2 - 45$ **22.** $3s^2 - 75$ **23.** $7 - 7t^2$ **24.** $300 - 3x^2$
25. $180 - 5y^2$ **26.** $2x^2 - 32y^2$ **27.** $3y^2 - 27z^2$ **28.** $8a^2 - 200b^2$ **29.** $7x^2 - 175y^2$
30. $6m^2 - 54n^2$ **31.** $44a^2 - 99b^2$ **32.** $a^3 - a$ **33.** $x^3 - 4x$ **34.** $25y - y^3$
35. $x^3 - x^5$ **36.** $x^3 - xy^2$ **37.** $a^3 - 4ab^2$ **38.** $m^4 - 49m^2n^2$ **39.** $4s^4 - 9s^2t^2$
40. $2x^3 - 2x$ **41.** $5a^3 - 20a$ **42.** $3x^3 - 3xy^2$ **43.** $8x^3 - 50xy^2$ **44.** $12a^2b - 27b^3$

In Exercises 45–48, indicate which one binomial ***cannot*** *be factored (by the methods you have learned). Factor the others.*

45. a. $x^2 - 9$ **b.** $x^2 - 1$ **c.** $x^2 + 1$ **d.** $x^2 - a^2$
46. a. $a^2 - x^2$ **b.** $x^2 - a^2$ **c.** $x^2 + a^2$ **d.** $4x^2 + 4a^2$

47. **a.** $y^2 - 16$ **b.** $3y^2 - 48$ **c.** $16y^2 + 16$ **d.** $y^2 + 16$

48. **a.** $ax^2 + ab^2$ **b.** $a^2c + ad^2$ **c.** $25a^2 + 49b^2$ **d.** $ax^2 + a^2y$

3.4 FACTORING TRINOMIALS $x^2 + Mx + N$

THE BASIC TECHNIQUE

Example 1 Factor. $x^2 + 5x + 4$

Solution Try:

$$\begin{array}{r} x + \square \\ x + \square \\ \hline \end{array}$$

Fill in the boxes with integers whose product is 4. You could try $2 \cdot 2$ or $1 \cdot 4$.

$$\begin{array}{r} x + 2 \\ x + 2 \\ \hline x^2 + 2x \\ 2x + 4 \\ \hline x^2 + \mathbf{4}x + 4 \end{array}$$

The coefficient of the *middle* term is wrong! It should be 5, not 4.

$$\begin{array}{r} x + 1 \\ x + 4 \\ \hline x^2 + x \\ 4x + 4 \\ \hline x^2 + \mathbf{5}x + 4 \end{array}$$

This is the right coefficient of $5x$.

Thus, $x^2 + 5x + 4 = (x + 1)(x + 4)$

Here, $1 \cdot 4 = 4$

(4 is the numerical term of $x^2 + 5x + \mathbf{4}$);

and $1 + 4 = 5$

(5 is the coefficient of $5x$ in $x^2 + \mathbf{5}x + 4$.)

Example 2 Factor. $x^2 - 5x + 4$

Solution *The coefficient of* the term $-5x$ *is negative*:

$$x^2 - \mathbf{5}x + 4$$

Modify the factors of Example 1 by taking -1 and -4 (instead of 1 and 4).

$$\begin{array}{rrr} x & -\ 1 & \\ x & -\ 4 & \\ \hline x^2 & -\ x & \\ & -\ 4x & +\ 4 \\ \hline x^2 & -\ \mathbf{5}x & +\ 4 \end{array}$$

Thus, $$x^2 - 5x + 4 = (x - 1)(x - 4)$$

Example 3 Factor. $y^2 + 2y - 3$.

Solution Here, *the numerical term is negative*:

$$y^2 + 2y - \mathbf{3}$$

One of the factors of -3 must be positive, the other negative (because the product, -3, is negative). Try $1(-3)$ and $3(-1)$.

$\begin{array}{rrr} y & +\ 1 & \\ y & -\ 3 & \\ \hline y^2 & +\ y & \\ & -\ 3y & -\ 3 \\ \hline y^2 & -\ \mathbf{2}y & -\ 3 \end{array}$	$\begin{array}{rrr} y & +\ 3 & \\ y & -\ 1 & \\ \hline y^2 & +\ 3y & \\ & -\ y & -\ 3 \\ \hline y^2 & +\ \mathbf{2}y & -\ 3 \end{array}$
Wrong! The coefficient of the middle term should be 2, not -2.	Right coefficient of $2y$.

Thus, $$y^2 + 2y - 3 = (y + 3)(y - 1)$$

As you see, you may have to try more than one pair of integers. To factor

$$x^2 + Mx + N$$

try to find integers a and b whose product is N and whose sum is M. In symbols:

$$ab = N \quad \text{and} \quad a + b = M$$

Then,

$$\begin{array}{l} x + a \\ \underline{x + b} \\ x^2 + ax \\ \qquad\quad bx + ab \\ \hline x^2 + \underbrace{(a + b)}_{M}x + \underbrace{ab}_{N} = x^2 + Mx + N \end{array}$$

Table 3.1 is useful in determining the signs of a and b when factoring

$$x^2 + Mx + N, \qquad \text{that is,} \qquad x^2 + (a + b)x + ab$$

TABLE 3.1

$M(= a + b)$	$N(= ab)$	a and b
positive	positive	both positive
negative	positive	both negative
positive	negative	one positive, one negative
negative	negative	one positive, one negative

Example 4 Factor. $x^2 + 8x + 12$

Solution Here, $M = 8$ and $N = 12$. *Both M and N are positive* for the polynomial

$$x^2 + 8x + 12$$

Try *positive* integers a and b such that

$$ab = 12 \qquad \text{and} \qquad a + b = 8$$

The possibilities are $1 \cdot 12$, $2 \cdot 6$, and $3 \cdot 4$

Try these: $1 + 12 = 13$ (Wrong) $\mathbf{2 + 6 = 8}$ (Right) $3 + 4 = 7$ (Wrong)

(In practice, as soon as you find the right integral factors, immediately factor the given polynomial.)

Thus, $x^2 + 8x + 12 = (x + 2)(x + 6)$

Consider the polynomial $x^2 + x + 3$.

Note that 1 and 3, the only positive factor of 3, do not work because $1 + 3 = 4$. Thus, $x^2 + x + 3$ *cannot be factored by this method.*

In general,

$$\boxed{x^2 + 2ax + a^2 = (x + a)^2}$$

Thus, $x^2 + 10x + 25 = (x + 5)^2$ (here, $a = 5$)

and $x^2 - 10x + 25 = (x - 5)^2$ (here, $a = -5$)

CLASS EXERCISES *Factor each trinomial.*

1. $a^2 + 6a + 8$ **2.** $y^2 - 3y + 2$ **3.** $a^2 + a - 2$
4. $a^2 - a - 2$ **5.** $x^2 + 8x + 16$

COMMON FACTORS It is usually best to isolate the common factor before applying the present method.

Example 5 Factor. $3x^3 - 21x^2 + 30x$

Solution

$$3x^3 - 21x^2 + 30x = 3x(x^2 - 7x + 10)$$

(the $-$ before $7x$: negative; the 10: positive)

Try $(-1)(-10)$ and $(-2)(-5)$

$(-1) + (-10) = -11$ (Wrong) $(-2) + (-5) = -7$ (Right)

Therefore, $3x^3 - 21x^2 + 30x = 3x(x - 2)(x - 5)$

CLASS EXERCISES *Factor each trinomial.* **6.** $4x^2 + 8x - 12$ **7.** $ax^2 - 12ax + 36a$

TWO VARIABLES This method applies, with very little modification, to trinomials in two variables, as in part **b** of Example 6.

Example 6 Factor. **a.** $x^2 - 5x - 14$ **b.** $x^2 - 5xy - 14y^2$

Solution **a.** $x^2 - 5x - 14$

(the $-$ before $5x$: negative; the $-$ before 14: negative)

The possible factors of -14 are

1 and -14, -1 and 14, 2 and -7, -2 and 7

Because the coefficient of $-5x$ is negative, the negative factor must have larger absolute value than the positive factor. Thus, try only: $1(-14)$ and $2(-7)$

$1 + (-14) = -13$ (Wrong) $2 + (-7) = -5$ (Right)

Therefore, $x^2 - 5x - 14 = (x + 2)(x - 7)$

b. The considerations of part **a** lead you to try:

$$\begin{array}{r} x + 2y \\ x - 7y \\ \hline x^2 + 2xy \qquad\quad \\ - 7xy - 14y^2 \\ \hline x^2 - 5xy - 14y^2 \end{array}$$

Thus, $x^2 - 5xy - 14y^2 = (x + 2y)(x - 7y)$

CLASS EXERCISES *Factor each trinomial.* **8.** $x^2 + 5xy + 4y^2$ **9.** $m^2 - mn - 2n^2$

SOLUTIONS TO CLASS EXERCISES

1. $a^2 + 6a + 8 = (a + 4)(a + 2)$ **2.** $y^2 - 3y + 2 = (y - 2)(y - 1)$ **3.** $a^2 + a - 2 = (a + 2)(a - 1)$

4. $a^2 - a - 2 = (a - 2)(a + 1)$ **5.** $x^2 + 8x + 16 = (x + 4)^2$

6. $4x^2 + 8x - 12 = 4(x^2 + 2x - 3) = 4(x + 3)(x - 1)$ **7.** $ax^2 - 12ax + 36a = a(x^2 - 12x + 36) = a(x - 6)^2$

8. $x^2 + 5xy + 4y^2 = (x + 4y)(x + y)$ **9.** $m^2 - mn - 2n^2 = (m + n)(m - 2n)$

HOME EXERCISES

In Exercises 1–50, factor each trinomial.

1. $x^2 + 4x + 3$ **2.** $z^2 + 7z + 6$ **3.** $b^2 + 2b + 1$ **4.** $x^2 + 10x + 25$ **5.** $a^2 + a - 2$
6. $m^2 + 3m - 4$ **7.** $b^2 - b - 2$ **8.** $n^2 - 4n - 5$ **9.** $s^2 - s - 12$ **10.** $t^2 - 11t - 12$
11. $a^2 - 12a + 36$ **12.** $b^2 + 10b + 25$ **13.** $x^2 - 9x + 14$ **14.** $y^2 + 13y - 14$ **15.** $a^2 + 5a - 14$
16. $b^2 - 5b - 14$ **17.** $c^2 + 9c + 20$ **18.** $d^2 - 12d + 20$ **19.** $x^2 + 13x + 40$ **20.** $y^2 - 10y + 16$
21. $x^2 + 9x - 36$ **22.** $x^2 - 17x + 66$ **23.** $2x^2 + 6x + 4$ **24.** $2x^2 + 16x + 14$ **25.** $3x^2 + 6x - 24$
26. $4y^2 - 28y + 48$ **27.** $-y^2 + 9y - 20$ **28.** $30 + z - z^2$ **29.** $a^3 + 3a^2 - 10a$ **30.** $b^3 + 12b^2 + 32b$
31. $4m^2 + 40m + 100$ **32.** $-2n^2 + 24n - 72$ **33.** $72 + 6x - 3x^2$ **34.** $y^4 + 10y^2 + 24$
35. $u^2v^2 + 18uv^2 + 81v^2$ **36.** $uv^2 - 16uv + 64u$ **37.** $2t^3 - 24t^2 + 70t$ **38.** $63x + 12x^2 - 3x^3$
39. $x^2 + 2ax + a^2$ **40.** $y^2 - 2ay + a^2$ **41.** $x^2 + 5ax + 4a^2$ **42.** $y^2 - 3ay - 4a^2$
43. $m^2 + 4mn + 4n^2$ **44.** $m^2 + 3mn - 4n^2$ **45.** $x^2 + 6xy + 9y^2$ **46.** $u^2 - 8uv + 16v^2$
47. $2x^2 + 24xy + 72y^2$ **48.** $3x^2 + 3xy - 36y^2$ **49.** $ax^2 + 6axy + 8ay^2$ **50.** $x^3 + 6x^2y + 5xy^2$

In Exercises 51 and 52, indicate which one trinomial ***cannot*** *be factored (by the methods of this section). Factor the others.*

51. a. $x^2 + 5x + 6$ **b.** $x^2 - 5x + 6$ **c.** $x^2 - 6x + 5$ **d.** $x^2 + 6x - 5$
52. a. $y^2 + 7y + 12$ **b.** $y^2 + 8y + 12$ **c.** $y^2 + 9y + 12$ **d.** $y^2 + 13y + 12$

3.5 FACTORING TRINOMIALS $Lx^2 + Mx + N$

THE BASIC TECHNIQUE In this section you will factor trinomials, such as

$$2x^2 + 3x + 1 \quad \text{and} \quad 3x^2 + 8x + 4$$

in which the coefficient of the x^2 term is not 1. In general, these trinomials are of the form

$$Lx^2 + Mx + N$$

Example 1 Factor. $2x^2 + 3x + 1$

Solution In order to obtain $2x^2$, try:

$$\begin{array}{r} 2x + \square \\ x + \square \\ \hline \end{array}$$

Because the numerical term of

$$2x^2 + \mathbf{3}x + \mathbf{1}$$

is 1, and because the coefficient of the term $3x$ is *positive*, try $1 \cdot 1$.

$$\begin{array}{r} 2x + 1 \\ x + 1 \\ \hline 2x^2 + x \\ 2x + 1 \\ \hline 2x^2 + 3x + 1 \end{array}$$

This works. Therefore,

$$2x^2 + 3x + 1 = (2x + 1)(x + 1)$$

Example 2 Factor. $2x^2 - 5x + 2$

Solution Again, try:

$$\begin{array}{r} 2x + \square \\ x + \square \\ \hline \end{array}$$

Because of the *negative* coefficient of $-5x$ in $2x^2 - \mathbf{5}x + 2$

try $(-1)(-2)$

Now there are two combinations to try:

$\begin{array}{r} 2x - 2 \\ x - 1 \\ \hline 2x^2 - 2x \\ -2x + 2 \\ \hline 2x^2 - \mathbf{4}x + 2 \end{array}$ (Wrong)	$\begin{array}{r} 2x - 1 \\ x - 2 \\ \hline 2x^2 - x \\ -4x + 2 \\ \hline 2x^2 - \mathbf{5}x + 2 \end{array}$ (Right)

Thus, $2x^2 - 5x + 2 = (2x - 1)(x - 2)$

Example 3 Factor. $3x^2 + 8x + 4$

Solution

$$\begin{array}{r} 3x + \square \\ x + \square \\ \hline \end{array}$$

Here, try: $1 \cdot 4$ and $2 \cdot 2$

Try $1 \cdot 4$ *both* ways with $3x$ and x.

You will have to try *three* different combinations:

$$\begin{array}{r} 3x + 1 \\ x + 4 \\ \hline 3x^2 + \quad x \\ 12x + 4 \\ \hline 3x^2 + 13x + 4 \end{array}$$
(Wrong)

$$\begin{array}{r} 3x + 4 \\ x + 1 \\ \hline 3x^2 + 4x \\ 3x + 4 \\ \hline 3x^2 + 7x + 4 \end{array}$$
(Wrong)

$$\begin{array}{r} 3x + 2 \\ x + 2 \\ \hline 3x^2 + 2x \\ 6x + 4 \\ \hline 3x^2 + 8x + 4 \end{array}$$
(Right)

Therefore, $3x^2 + 8x + 4 = (3x + 2)(x + 2)$

CLASS EXERCISES *Factor each trinomial.* **1.** $2a^2 - 5a + 3$ **2.** $9x^2 + 12x + 4$

COMMON FACTORS You may have to isolate the common factor before applying the previous method.

Example 4 Factor. $-8x^2 + 6x + 2$

Solution Because you want to consider

$$Lx^2 + Mx + N, \quad \text{where } L > 0$$

isolate the common factor -2 (instead of 2):

$$-8x^2 + 6x + 2 = (-2)(4x^2 - 3x - 1)$$

Consider $4x^2 - 3x - 1$

↑ negative

Now there are two possibilities for $4x^2$:

$$\begin{array}{r} 2x + \square \\ 2x + \square \\ \hline \end{array} \quad \text{and} \quad \begin{array}{r} 4x + \square \\ x + \square \\ \hline \end{array}$$

Because the coefficient of the term $-3x$ is negative in $4x^2 - 3x - 1$, try $1(-1)$. Remember to try the integral factors *both* ways with $4x$ and x, if the first way doesn't work.

$$\begin{array}{r} 2x + 1 \\ 2x - 1 \\ \hline 4x^2 + 2x \quad\quad \\ -2x - 1 \\ \hline 4x^2 \quad\quad - 1 \\ \text{(Wrong)} \end{array} \quad\Bigg|\quad \begin{array}{r} 4x - 1 \\ x + 1 \\ \hline 4x^2 - x \quad\quad \\ 4x - 1 \\ \hline 4x^2 + 3x - 1 \\ \text{(Wrong)} \end{array} \quad\Bigg|\quad \begin{array}{r} 4x + 1 \\ x - 1 \\ \hline 4x^2 + x \quad\quad \\ -4x - 1 \\ \hline 4x^2 - 3x - 1 \\ \text{(Right)} \end{array}$$

Therefore,

$$4x^2 - 3x - 1 = (4x + 1)(x - 1) \qquad \text{and}$$

$$-8x^2 + 6x + 2 = -2(4x + 1)(x - 1)$$

CLASS EXERCISES *Factor each trinomial.* **3.** $6y^2 + 3y - 9$ **4.** $24a^2 + 28a + 8$

TWO VARIABLES Now, consider a polynomial in two variables, as in part **b.** of Example 5.

Example 5 Factor. **a.** $6x^2 + 17x + 5$ **b.** $6x^2 + 17xy + 5y^2$

Solution **a.** Try all combinations for

$$\begin{array}{r} 6x + \square \\ x + \square \\ \hline \end{array} \quad \text{and for} \quad \begin{array}{r} 3x + \square \\ 2x + \square \\ \hline \end{array}$$

along with $5 \cdot 1$.

$$\begin{array}{r} 6x + 5 \\ x + 1 \\ \hline 6x^2 + 5x \quad\quad \\ 6x + 5 \\ \hline 6x^2 + 11x + 5 \\ \text{(Wrong)} \end{array} \quad\Bigg|\quad \begin{array}{r} 6x + 1 \\ x + 5 \\ \hline 6x^2 + x \quad\quad \\ 30x + 5 \\ \hline 6x^2 + 31x + 5 \\ \text{(Wrong)} \end{array} \quad\Bigg|\quad \begin{array}{r} 3x + 5 \\ 2x + 1 \\ \hline 6x^2 + 10x \quad\quad \\ 3x + 5 \\ \hline 6x^2 + 13x + 5 \\ \text{(Wrong)} \end{array} \quad\Bigg|\quad \begin{array}{r} 3x + 1 \\ 2x + 5 \\ \hline 6x^2 + 2x \quad\quad \\ 15x + 5 \\ \hline 6x^2 + 17x + 5 \\ \text{(Right)} \end{array}$$

Thus,

$$6x^2 + 17x + 5 = (3x + 1)(2x + 5)$$

b. Modify the above factors.

$$\begin{array}{r} 3x + y \\ 2x + 5y \\ \hline 6x^2 + 2xy \\ 15xy + 5y^2 \\ \hline 6x^2 + 17xy + 5y^2 \end{array}$$

Thus,

$$6x^2 + 17xy + 5y^2 = (3x + y)(2x + 5y)$$

CLASS EXERCISES *Factor each trinomial.* **5.** $4x^2 + 8xy + 3y^2$ **6.** $4a^2 - 7ab - 15b^2$

SOLUTIONS TO CLASS EXERCISES

1. $2a^2 - 5a + 3 = (2a + 1)(a - 3)$ **2.** $9x^2 + 12x + 4 = (3x + 2)(3x + 2) = (3x + 2)^2$
3. $6y^2 + 3y - 9 = 3(2y^2 + y - 3) = 3(2y + 3)(y - 1)$
4. $24a^2 + 28a + 8 = 4(6a^2 + 7a + 2) = 4(3a + 2)(2a + 1)$
5. $4x^2 + 8xy + 3y^2 = (2x + 3y)(2x + y)$ **6.** $4a^2 - 7ab - 15b^2 = (4a + 5b)(a - 3b)$

HOME EXERCISES

Factor each trinomial.

1. $3x^2 + 4x + 1$ **2.** $7b^2 - 8b + 1$ **3.** $3x^2 + 2x - 1$ **4.** $3x^2 - 2x - 1$ **5.** $2x^2 + 7x + 3$
6. $2x^2 + 5x + 2$ **7.** $2y^2 + 5y + 3$ **8.** $2y^2 + 7y + 3$ **9.** $2a^2 + 7a + 5$ **10.** $2b^2 - 9b + 9$
11. $2m^2 - 5m + 2$ **12.** $2m^2 - 5m - 3$ **13.** $2a^2 + 7a - 4$ **14.** $2a^2 - 9a + 4$ **15.** $2a^2 + 15a + 25$
16. $4b^2 + 20b + 25$ **17.** $9y^2 - 6y + 1$ **18.** $9z^2 - 10z + 1$ **19.** $4a^2 + 4a + 1$ **20.** $4b^2 + 5b + 1$
21. $3y^2 - 4y - 7$ **22.** $3z^2 - 20z - 7$ **23.** $4m^2 + 8m + 3$ **24.** $4n^2 - 8n + 3$ **25.** $6x^2 + 5x - 6$
26. $6z^2 + 13z + 6$ **27.** $7a^2 + 9a + 2$ **28.** $7b^2 + 15b + 2$ **29.** $10x^2 + 3x - 4$ **30.** $10x^2 + 13x + 4$
31. $9x^2 - 36x - 13$ **32.** $9y^2 - 42y + 13$ **33.** $5a^2 + 17a - 12$ **34.** $5b^2 + 16b + 12$ **35.** $4x^2 + 6x + 2$
36. $6y^2 + 8y + 2$ **37.** $2b^2 + 23b - 12$ **38.** $6b^2 + 9b + 3$ **39.** $8c^2 + 20c + 12$ **40.** $5b^2 - 5b - 30$
41. $4x^3 + 10x^2 + 6x$ **42.** $10y^4 + 21y^3 + 2y^2$ **43.** $4x^2 + 4xy + y^2$ **44.** $4s^2 - 4st + t^2$
45. $2y^2 + 3yz + z^2$ **46.** $6x^2 + 5ax + a^2$ **47.** $2x^2 + 5xy + 2y^2$ **48.** $u^2 - 5uv + 4v^2$
49. $9a^2 - 24ab - 20b^2$ **50.** $16x^2 + 2xy - 3y^2$ **51.** $6x^2 + 14xy + 4y^2$ **52.** $4u^3 - 4u^2v - 3uv^2$

Let's Review Chapter 3

1. Find all positive factors of 18.

In Exercises 2–4, which integers are primes and which are composites?

2. 12 **3.** 17 **4.** 27

In Exercises 5 and 6, express each integer as the product of primes.

5. 28 **6.** 70

7. Find *gcd* (24, 60). **8.** Find the *gcd* of the coefficients of $8x^2 + 12x - 16$.

In Exercises 9–11, isolate the common factor of each polynomial.

9. $4x + 6$ **10.** $9a^6 - 15a^4$ **11.** $x^5y^4z^3 + x^3y^3z^3 + x^2y^3z$

In Exercises 12–14, multiply.

12. $(x + 1)(x + 5)$ **13.** $(z^2 + 1)(z^2 - 4)$ **14.** $(m + 2n)(m - 3n)$

In Exercises 15–17, factor each binomial.

15. $a^2 - 25$ **16.** $4y^2 - 49z^2$ **17.** $3xy^2 - 12x$

In Exercises 18–23, factor each trinomial.

18. $c^2 + 7c + 12$ **19.** $m^2 - m - 20$ **20.** $x^2 - 5x + 6$

21. $2x^2 + 7x + 5$ **22.** $2y^2 - 7y + 3$ **23.** $4x^2 + xy - 3y^2$

In Exercises 24–36, the different types of factoring are mixed up. Factor each polynomial. (One of these polynomials **cannot** *be factored by the present methods. Indicate which one.)*

24. $5m - 10n$ **25.** $36 - a^2$ **26.** $x^2 + 9x + 14$

27. $7a^2 - 28b^2$ **28.** $x^3 - 4x$ **29.** $3x^3 - 5x^2 - 2x$

30. $m^2 + 8mn + 15n^2$ **31.** $4m^2 - 16n^4$ **32.** $s^3 + 6s^2 - 7s$

33. $t^4 - 16t^2$ **34.** $a^2 - 10a + 16$ **35.** $b^2 + 6b + 4$

36. $5x^2 - 8x - 4$

4 DIVISION OF POLYNOMIALS

DIAGNOSTIC TEST Perhaps you are already familiar with some of the material in Chapter 4. This test will indicate which sections in Chapter 4 you need to study. The question number refers to the corresponding section in Chapter 4. If you answer *all* parts of the question correctly, you may omit the section. *But if any part of your answer is wrong, you should study the section.* When you have completed the test, turn to page A7 for the answers.

4.1 Rational Numbers

a. Fill in. -2 is the numerator and 5 is the denominator of the rational number $\square$

b. Each of the points Q, R, S, T, U in Figure 4A represents one of the following rational numbers:

$$\frac{1}{5}, \frac{2}{5}, \frac{7}{5}, \frac{-1}{5}, \frac{-4}{5}$$

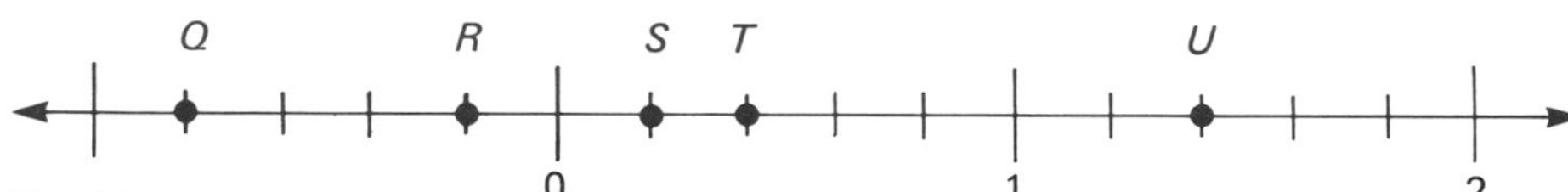

Fig. 4A

Indicate which number is represented by each point.

In c. *and* d., *simplify each fraction by dividing by the largest factor common to numerator and denominator. Express the simplified fraction with a positive denominator.*

c. $\frac{9}{12}$ d. $\frac{27}{-45}$

4.2 Rational Expressions

a. Evaluate $\frac{x^2 + 5}{3x - 2}$ when $x = 2$.

b. Evaluate $\frac{2x^2 - xy + 3y^2}{x + y + 5}$ when $x = -1$ and $y = 1$.

4.3 Monomial Divisors *Simplify.*

a. $\frac{-40x^2y^5z}{64xy^5z^4}$ b. $\frac{9(a+b)^4(a-b)}{15(a+b)(a-b)^2}$ c. $\frac{5x^4 + 4x^3}{x^4}$ d. $\frac{9abc^2}{54a^2b - 45abc}$

4.4 Factoring and Simplifying *Simplify.* a. $\frac{3a + 3b}{a^2 - b^2}$ b. $\frac{4x^2 - 9y^2}{2x^2 - xy - 3y^2}$

4.5 Polynomial Division

a. Divide $10 + 29x + 15x^2 + 2x^3$ by $5 + 2x$.
b. Find the quotient and remainder. $x + 1 \overline{)x^3 + x + 1}$

4.1 RATIONAL NUMBERS

There are times when integers do not express common notions. Other numbers, known as *fractions*, must be introduced. **Fractions** *express division of numbers*. For example, suppose a man divides a square tract of land equally among his 9 grandchildren. Then each grandchild receives $\frac{1}{9}$ (one-ninth) of the land. [See Figure 4.1(a).] If one grandson, Bob, buys his sister's share, Bob then owns $\frac{2}{9}$ (two-ninths) of the land. [See Figure 4.1(b).]

WHAT ARE RATIONAL NUMBERS

The fractions you will first consider express the quotient of two *integers*. These fractions are called *rational numbers*.

DEFINITION

A **rational number** is a number that can be written in the form $\frac{N}{D}$, where N and D are integers and $D \neq 0$. Here, N is called the **numerator** and D, the **denominator**.

Example 1

a. $\frac{3}{4}$ is a rational number with numerator 3 and denominator 4.

b. $\frac{-1}{3}$ is a rational number with numerator -1 and denominator 3.

c. $\frac{5}{2}$ is a rational number with numerator 5 and denominator 2.

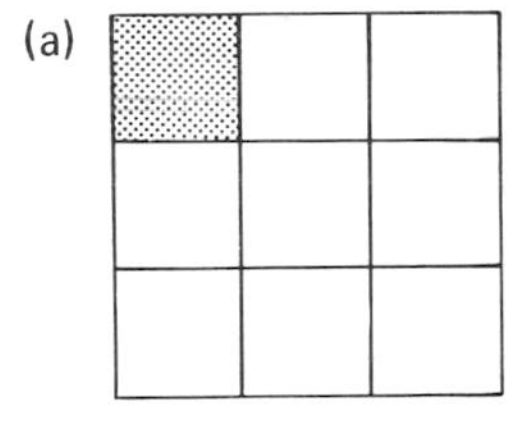

Fig. 4.1(a) $\frac{1}{9}$. **(b)** $\frac{2}{9}$

Any integer M can be expressed as the rational number $\frac{M}{1}$. For example,

$$5 = \frac{5}{1}$$

Thus, 5 can be written in the form of a rational number $\frac{N}{D}$, where N and D are integers and $D \neq 0$. Here, $N = 5$ and $D = 1$.

Example 2 Express each of the following integers in the form $\frac{N}{D}$, where N and D are integers and $D \neq 0$.

a. 17 **b.** -17 **c.** -1

Solution **a.** $17 = \frac{17}{1}$ **b.** $-17 = \frac{-17}{1}$ **c.** $-1 = \frac{-1}{1}$

CLASS EXERCISES

1. $\frac{-1}{2}$ is a rational number with numerator $\square$ and denominator $\square$.
2. 4 is the numerator and 3 the denominator of the rational number $\square$.
3. Express -6 in the form $\frac{N}{D}$, where N and D are integers and $D \neq 0$.

RATIONAL NUMBERS AND THE NUMBER LINE

Rational numbers are numbers that can be obtained by dividing the line segments between integer points into equal parts. For example (with the aid of a ruler), divide the line segment between 0 and 1 into two equal parts. The midpoint obtained represents the number $\frac{1}{2}$. (See Figure 4.2 .) Here, $\frac{1}{2}$ means that the number 1 has been divided by 2.

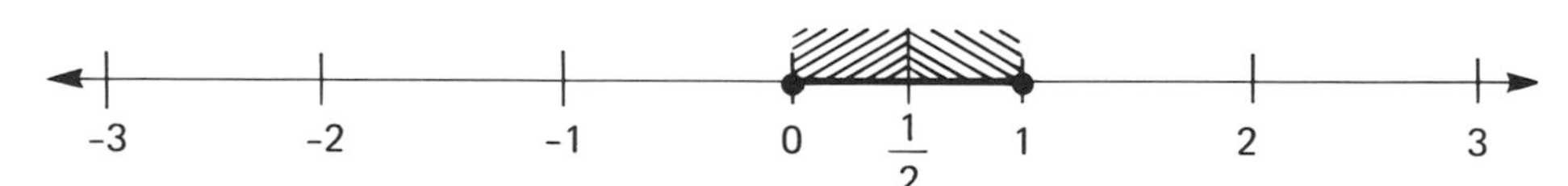

Fig. 4.2

To obtain $\frac{2}{5}$, divide the line segment between 0 and 2 into 5 equal parts. The first point of division *to the right of* 0 represents $\frac{2}{5}$. [See Figure 4.3(a).] Here, $\frac{2}{5}$ means that the number 2 has been divided by 5. It can be shown that $\frac{2}{5}$ lies twice as far from 0 as does $\frac{1}{5}$. Thus, $\frac{2}{5}$ can also be obtained by dividing the line segment between 0 and 1 into 5 equal parts, as in Figure 4.3(b). Now, the *second* point of division to the right of 0 represents $\frac{2}{5}$.

(a)

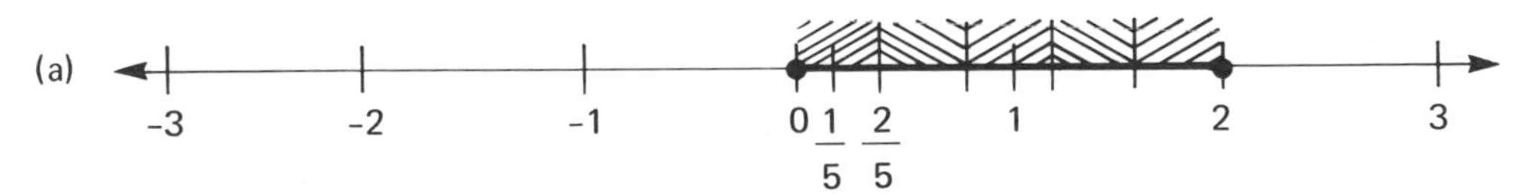

(b)

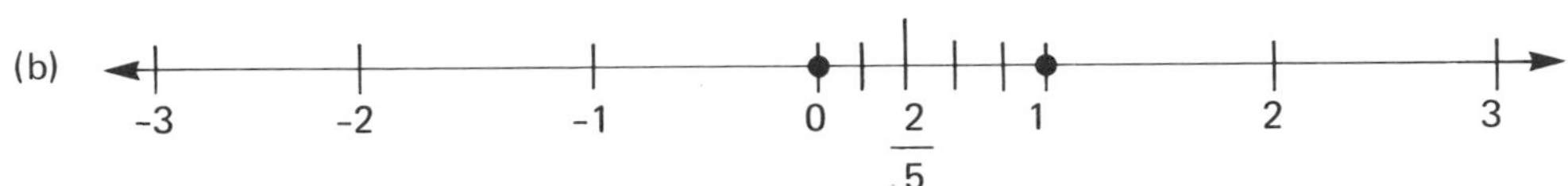

Fig. 4.3

In general, $\frac{a}{b}$ indicates that the number a has been divided by b.

To obtain $\frac{3}{2}$, divide the line segment between 0 and 3 into 2 equal parts. The point of division represents $\frac{3}{2}$. (See Figure 4.4 .) Observe that $\frac{3}{2}$ lies three times as far from 0 as does $\frac{1}{2}$.

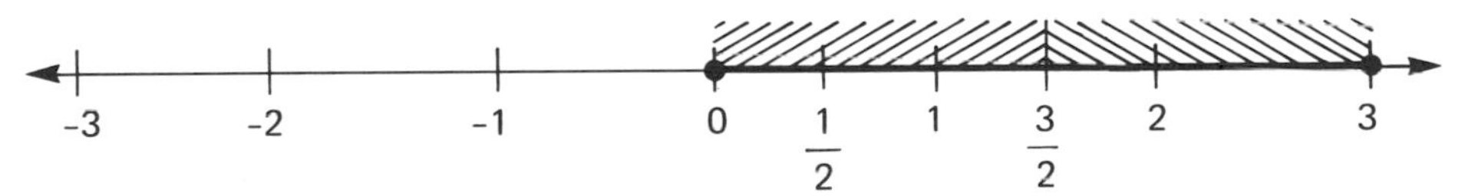

Fig. 4.4

To obtain $\frac{-1}{4}$, divide the line segment between 0 and -1 into 4 equal parts. The first point of division *to the left of* 0 represents $\frac{-1}{4}$. (See Figure 4.5 .)

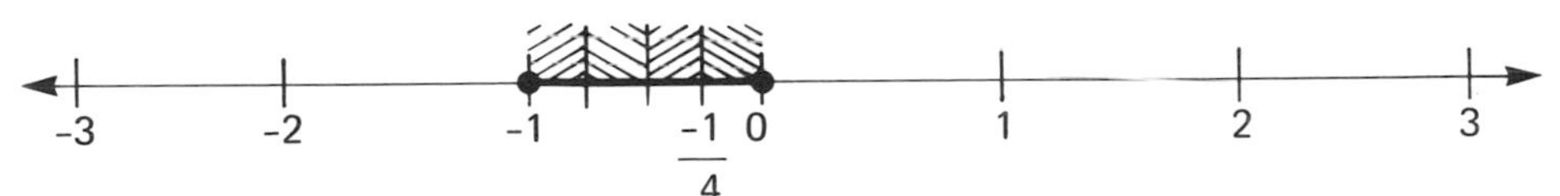

Fig. 4.5

CLASS EXERCISES *In Figure 4.6, locate the following points.*

4. $\frac{1}{5}$ **5.** $\frac{3}{5}$ **6.** $\frac{6}{5}$ **7.** $\frac{-1}{5}$

Fig. 4.6

EQUIVALENT FRACTIONS

Fractions are quotients of numbers. Rational numbers, such as $\frac{1}{4}$ and $\frac{3}{5}$, are quotients of integers and are, therefore, fractions. (In Chapter 8, you will learn about other fractions that are not rational numbers.*)

Consider the fractions

$$\frac{1}{2} \text{ and } \frac{2}{4}$$

Observe that the "cross products"

$$\frac{1}{2} \times \frac{2}{4}$$

are equal, that is,

$$1 \cdot 4 = 2 \cdot 2$$

These *fractions* are *equivalent* according to the following definition.

DEFINITION

EQUIVALENCE OF FRACTIONS. Two fractions

$$\frac{a}{b} \text{ and } \frac{c}{d}, \quad b \neq 0, d \neq 0$$

are said to be **equivalent** if their cross products

$$\frac{a}{b} \times \frac{c}{d}$$

are equal, that is, if

$$a \cdot d = b \cdot c.$$

Write

$$\frac{a}{b} = \frac{c}{d}$$

when these fractions are equivalent. Thus,

$$\frac{a}{b} = \frac{c}{d} \quad \text{if } a \cdot d = b \cdot c$$

*Although only rational fractions have been introduced thus far, much of the subsequent discussion applies to fractions in general. For this reason, many definitions and rules are stated in terms of fractions.

There is a **geometric interpretation of equivalence of fractions.** Consider the equivalent fractions $\frac{1}{2}$ and $\frac{2}{4}$. First, $\frac{1}{2}$ is obtained by dividing the line segment between 0 and 1 into 2 equal parts. The point of division represents $\frac{1}{2}$. To represent $\frac{2}{4}$, divide the line segment between 0 and 2 into 4 equal parts. The first point of division represents $\frac{2}{4}$, as you see in Figure 4.7 . But this is the same point that corresponds to $\frac{1}{2}$. Thus, the *equivalent fractions* $\frac{1}{2}$ *and* $\frac{2}{4}$ *correspond to the same point on the number line.*

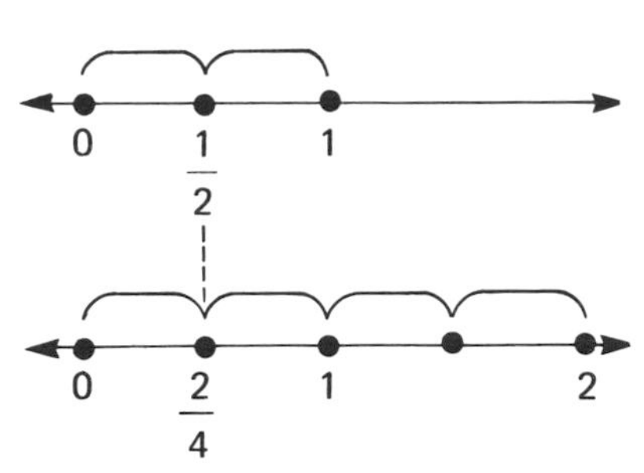

Fig. 4.7

Example 3 *Cross-multiply* to show that $\frac{15}{20} = \frac{3}{4}$.

Solution

$$\frac{15}{20} \times \frac{3}{4}$$

Does $15 \cdot 4 = 20 \cdot 3$?

Yes. $60 = 60$

Thus, $\frac{15}{20} = \frac{3}{4}$

In Example, 3, you found that $\frac{15}{20} = \frac{3}{4}$.

DEFINITION

> Let N and D be integers and let $D > 0$. The rational number $\frac{N}{D}$ is expressed in **lowest terms** when N and D have no factors in common other than 1 and -1. You **simplify a fraction** of the form $\frac{N}{D}$ by expressing it in lowest terms.

Thus, $\frac{3}{4}$ is in lowest terms because 3 and 4 have no common factors and because the denominator, 4, is positive.

In practice, how do you simplify a fraction? Observe that

$$\frac{15}{20} = \frac{3 \cdot 5}{4 \cdot 5}$$

and

$$\frac{3 \cdot 5}{4 \cdot 5} = \frac{3}{4}$$

because the cross products are equal:

$$3 \cdot 5 \cdot 4 = 4 \cdot 5 \cdot 3$$

Here, 5 is a factor common to both the numerator and the denominator of $\frac{15}{20}$. Divide 15 and 20 each by 5 to obtain

$$\frac{15}{20} = \frac{3}{4}$$

In general,

$$\frac{a \cdot k}{b \cdot k} = \frac{a}{b}, \quad b \neq 0, k \neq 0$$

Thus, *if k is a factor common to numerator and denominator, you obtain an equivalent fraction by dividing numerator and denominator by k.*

> **To simplify a fraction,** divide by the *largest* factor common to numerator and denominator. Also, if the denominator is negative, multiply numerator and denominator by -1.

Example 4 Simplify. **a.** $\frac{12}{20}$ **b.** $\frac{300}{400}$ **c.** $\frac{-25}{-40}$

Solution **a.** Divide numerator and denominator by 4.

$$\frac{12}{20} = \frac{3}{5}$$

b. Divide numerator and denominator by 100.

$$\frac{300}{400} = \frac{3}{4}$$

c. $\frac{-25}{-40} = \frac{-5}{-8} = \frac{5}{8}$

Note that if you divide numerator and denominator by -5, you simplify the fraction in one step.

Here are three equivalent ways of expressing a **negative fraction.** *Suppose* $b \neq 0$. Then,

$$\boxed{\frac{-a}{b} = \frac{a}{-b} = -\frac{a}{b}}$$

The minus sign can be in the numerator, in the denominator, or before the entire fraction. Thus,

$$\frac{-3}{5} = \frac{3}{-5} = -\frac{3}{5}$$

To see that $\frac{-3}{5} = \frac{3}{-5}$, multiply the numerator and denominator of $\frac{-3}{5}$ by -1.

CLASS EXERCISES

8. Cross-multiply to show that $\frac{-2}{10}$ and $\frac{-1}{5}$ are equivalent.

In Exercises 9–11, simplify each fraction by dividing by the largest factor common to numerator and denominator. Express the simplified fraction with a positive denominator.

9. $\frac{8}{12}$ **10.** $\frac{-5}{-15}$ **11.** $\frac{6}{4}$

12. Which of the following are equivalent to $-\frac{3}{4}$?

a. $\frac{-3}{4}$ **b.** $\frac{-3}{-4}$ **c.** $-\frac{-3}{-4}$

SOLUTIONS TO CLASS EXERCISES

1. numerator -1 and denominator 2 **2.** $\frac{4}{3}$ 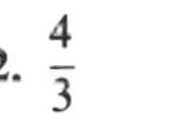 **3.** $\frac{-6}{1}$ **4.** to **7.** (See Fig. 4.8.)

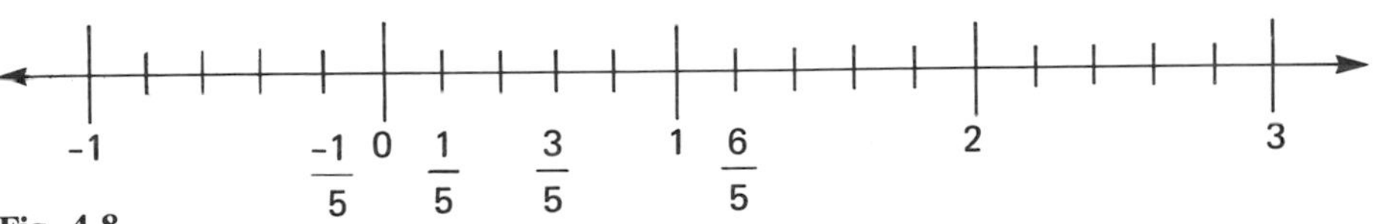

Fig. 4.8

8. Does $(-2)5 = 10(-1)$?
Yes. $-10 = -10$
Thus, $\frac{-2}{10} = \frac{-1}{5}$

9. $\frac{2}{3}$ **10.** $\frac{1}{3}$ **11.** $\frac{3}{2}$ **12.** a. and c.

HOME EXERCISES

In Exercises 1–3, fill in the integers.

1. $\frac{4}{9}$ is a rational number with numerator $\square$ and denominator $\square$.

2. $\frac{-2}{3}$ is a rational number with numerator $\square$ and denominator $\square$.

3. $\frac{-3}{2}$ is a rational number with numerator $\square$ and denominator $\square$.

In Exercises 4–6, fill in the rational number.

4. 1 is the numerator and 2 the denominator of the rational number $\square$.

5. -3 is the numerator and 5 the denominator of the rational number $\square$.

6. -5 is the numerator and 3 the denominator of the rational number $\square$.

For Exercises 7–10, draw a horizontal line and label the "integer points" as in Figure 4.9. Locate the following rational numbers on this line.

7. $\frac{1}{4}$ 8. $\frac{3}{4}$ 9. $\frac{-1}{4}$ 10. $\frac{-1}{2}$

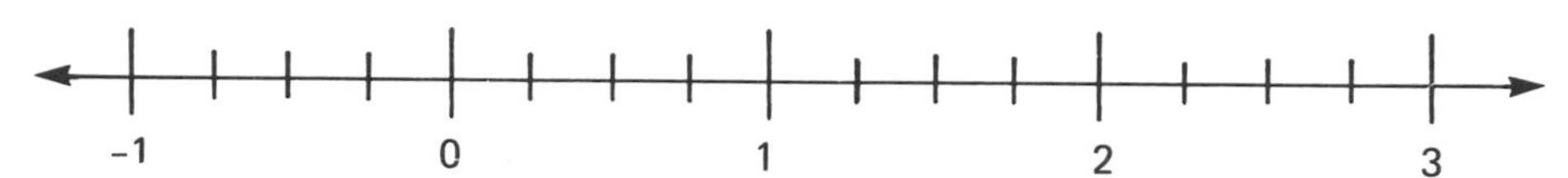

Fig. 4.9

11. Each of the points Q, R, S, T, U, V, W in Figure 4.10 represents one of the following rational numbers:

$$\frac{1}{7}, \frac{3}{7}, \frac{9}{7}, \frac{-1}{7}, \frac{-6}{7}, \frac{-8}{7}, \frac{-10}{7}$$

Indicate which number is represented by each point.

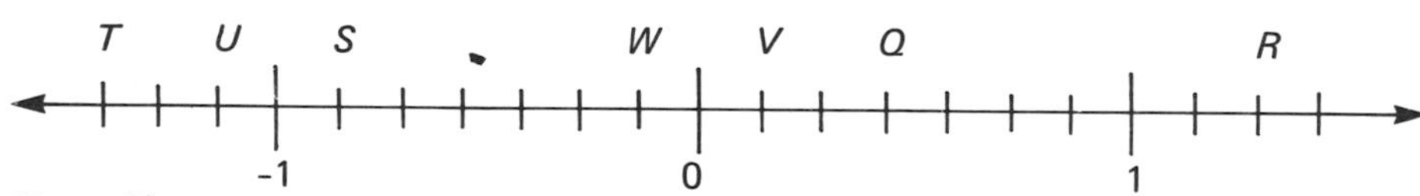

Fig. 4.10

In Exercises 12–15, express each of the following integers in the form $\frac{N}{D}$, where N and D are integers and $D \neq 0$.

12. 15 13. 1 14. -2 15. 0

In Exercises 16–20, cross-multiply to show that the given fractions are equivalent.

16. $\frac{4}{16}$ and $\frac{1}{4}$ 17. $\frac{-4}{6}$ and $\frac{-2}{3}$ 18. $\frac{-5}{8}$ and $\frac{10}{-16}$ 19. $\frac{-8}{14}$ and $\frac{4}{-7}$ 20. $\frac{100}{16}$ and $\frac{25}{4}$

In Exercises 21–44, simplify each fraction by dividing by the largest factor common to numerator and denominator. Express the simplified fraction with a positive denominator.

21. $\frac{2}{10}$ 22. $\frac{10}{20}$ 23. $\frac{3}{15}$ 24. $\frac{6}{10}$ 25. $\frac{-3}{9}$ 26. $\frac{6}{2}$ 27. $\frac{-3}{-12}$ 28. $\frac{10}{4}$

29. $\frac{7}{-14}$ 30. $\frac{-16}{20}$ 31. $\frac{32}{40}$ 32. $\frac{12}{-18}$ 33. $\frac{14}{35}$ 34. $\frac{36}{72}$ 35. $\frac{50}{300}$ 36. $\frac{18}{12}$

37. $\frac{-30}{18}$ 38. $\frac{-9}{-24}$ 39. $\frac{15}{-50}$ 40. $\frac{13}{26}$ 41. $\frac{11}{121}$ 42. $\frac{24}{-72}$ 43. $\frac{36}{42}$ 44. $\frac{25}{60}$

45. Which of the following are equivalent to $-\frac{2}{5}$?

a. $\frac{-2}{5}$ b. $\frac{-2}{-5}$ c. $\frac{2}{-5}$

46. Which of the following are equivalent to $-\frac{5}{-8}$?

a. $\frac{-5}{8}$ b. $\frac{5}{8}$ c. $\frac{-5}{-8}$

47. What is the numerical coefficient of the term $\frac{3x^2y}{4}$?

4.2 RATIONAL EXPRESSIONS

WHAT IS A RATIONAL EXPRESSION?

Rational numbers express division of integers. *Rational expressions* will now be defined so as to express division of polynomials.

DEFINITION

A **rational expression** is an algebraic expression that can be written in the form $\frac{P}{Q}$, where P and Q are polynomials and $Q \neq 0$. Here, P is called the **numerator** and Q the **denominator**.

Thus, $\frac{2x + 5}{x^2 - 10}$ is a rational expression. Its numerator is the polynomial $2x + 5$. Its denominator is the (nonzero) polynomial $x^2 - 10$. Also, $\frac{t^2 + 2t - 1}{t^3 - t^2 + t + 4}$ is a rational expression with numerator $t^2 + 2t - 1$ and denominator $t^3 - t^2 + t + 4$.

Recall that numbers are terms, and therefore (numerical) polynomials. Thus, a rational expression can have a number as its numerator or denominator.

An integer N can be written as the rational number $\frac{N}{1}$. So too, *a polynomial P can be written as the rational expression* $\frac{P}{1}$.

Example 1 Explain why each of the following is a rational expression.

a. $\frac{5}{x^2 + 1}$ b. $\frac{x + 4}{2}$ c. $\frac{-7}{3}$ d. $t^2 + 2t + 5$

Solution a. The numerator is the (numerical) polynomial 5. The denominator is the polynomial $x^2 + 1$.

b. The numerator is the polynomial $x + 4$. The denominator is the polynomial 2.

c. The numerator is the polynomial -7. The denominator is the polynomial 3.

d. $t^2 + 2t + 5$ can be written as the rational expression $\frac{t^2 + 2t + 5}{1}$, with numerator $t^2 + 2t + 5$ and denominator 1.

CLASS EXERCISES *In Exercises 1–3, fill in the polynomials.*

1. $\frac{y - 3}{y^2 + y + 1}$ is a rational expression with numerator [ ] and denominator [ ].
2. $\frac{-1}{t^2 + 1}$ is a rational expression with numerator [ ] and denominator [ ].
3. $x^2 + 5y$ can be written as the rational expression with numerator [ ] and denominator [ ].
4. 4 is the numerator and $x^2 + 3x + 2$ is the denominator of the rational expression [ ].

SUBSTITUTING NUMBERS FOR VARIABLES

You can evaluate a rational expression, just as you evaluated a polynomial, by substituting numbers for variables. Again, *substitute the same number for each occurrence of a variable.*

Example 2 Evaluate $\frac{3x}{x + 5}$ when $x = 1$.

Solution Replace x by 1 in both numerator and denominator, and obtain

$$\frac{3 \cdot 1}{1 + 5}, \quad \text{or } \frac{3}{6}, \quad \text{or finally, } \quad \frac{1}{2}$$

There may be two or more variables present in a rational expression. *For each variable, substitute the same number each time the variable occurs.*

Example 3 Evaluate $\frac{xy - 4}{x^2 + y - 1}$ when $x = 10$ and $y = 1$.

Solution Substitute *10* for each occurrence of x, and **1** for each occurrence of y.

$$\frac{10 \cdot 1 - 4}{10^2 + 1 - 1} = \frac{6}{100} = \frac{3}{50}$$

A rational expression is not defined when the denominator is 0 (because division by 0 is undefined). Thus,

$$\frac{1}{x} \text{ is undefined when } x = 0 \quad \text{and}$$

$$\frac{1}{x-1} \text{ is undefined when } x - 1 = 0 \text{, that is, when } x = 1.$$

CLASS EXERCISES

5. Evaluate $\frac{y-1}{5y+3}$ when $y = 5$.

6. Evaluate $\frac{xy+3}{x^2-y}$ when $x = 2$ and $y = -2$.

7. Evaluate $\frac{t^2+6t-1}{t+7}$ three separate times:

   a. when $t = 2$ **b.** when $t = -1$ **c.** when $t = 0$

   d. For which value of t is this rational expression undefined?

EQUIVALENT RATIONAL EXPRESSIONS

DEFINITION

Two rational expressions $\frac{P}{Q}$ and $\frac{R}{S}$, $Q \neq 0$, $S \neq 0$, are said to be **equivalent** if their cross products

$$\frac{P}{Q} \times \frac{R}{S}$$

are equal, that is, if

$$P \cdot S = Q \cdot R$$

Write

$$\frac{P}{Q} = \frac{R}{S}$$

when these rational expressions are equivalent. Thus,

$$\frac{x}{x^3+x} = \frac{1}{x^2+1} \quad \text{because} \quad \underbrace{x(x^2+1)}_{x^3+x} = (x^3+x) \cdot 1$$

Observe that the rational expression $\frac{x}{x^3 + x}$ is not defined when $x = 0$ because when 0 is substituted for x, you obtain $\frac{0}{0}$, which is undefined. The rational expression $\frac{1}{x^2 + 1}$ is defined for all x because x^2 is at least 0, and thus, the denominator, $x^2 + 1$, is at least 1 for all x. In particular, the value of $\frac{1}{x^2 + 1}$ when $x = 0$ is $\frac{1}{0^2 + 1}$, or 1. Strictly speaking, for rational expressions to be equivalent, they must be defined for the same values of their variables. However, we will not be concerned with such subtleties.

Note that

$$\frac{x}{x^3 + x} = \frac{1 \cdot x}{(x^2 + 1) \cdot x} = \frac{1}{x^2 + 1}$$

In general, for polynomials P, Q, and R, with $Q \neq 0$, $R \neq 0$,

$$\frac{P \cdot R}{Q \cdot R} = \frac{P}{Q}$$

In practice, to simplify a rational expression, you divide by factors common to numerator and denominator. The remainder of the chapter concerns dividing polynomials and simplifying rational expressions.

CLASS EXERCISES

8. Cross-multiply to show that $\frac{x^2 - 1}{3x - 3}$ and $\frac{x + 1}{3}$ are equivalent.

SOLUTIONS TO CLASS EXERCISES

1. numerator $y - 3$ and denominator $y^2 + y + 1$

2. numerator -1 and denominator $t^2 + 1$

3. numerator $x^2 + 5y$ and denominator 1

4. $\frac{4}{x^2 + 3x + 2}$

5. $\frac{5 - 1}{5 \cdot 5 + 3} = \frac{4}{28} = \frac{1}{7}$

6. $\frac{2(-2) + 3}{2^2 - (-2)} = \frac{-1}{4 + 2} = \frac{-1}{6}$

7. a. When $t = 2$: $\frac{2^2 + 6 \cdot 2 - 1}{2 + 7} = \frac{15}{9} = \frac{5}{3}$

b. When $t = -1$: $\frac{(-1)^2 + 6(-1) - 1}{-1 + 7} = \frac{-6}{6} = -1$

c. When $t = 0$: $\frac{0^2 + 6 \cdot 0 - 1}{0 + 7} = \frac{-1}{7}$

d. $t = -7$. For then, $t + 7 = -7 + 7 = 0$

8. Does $(x^2 - 1) \cdot 3 = (3x - 3)(x + 1)$?
Yes. $3x^2 - 3 = 3x^2 - 3$
Thus, $\frac{x^2 - 1}{3x - 3} = \frac{x + 1}{3}$

HOME EXERCISES

In Exercises 1–8, find **a.** *the numerator and* **b.** *the denominator of each rational expression.*

1. $\frac{x}{x+1}$ **2.** $\frac{2x+7}{x}$ **3.** $\frac{y+3}{y-4}$ **4.** $\frac{2x^2-1}{7}$ **5.** $\frac{5}{y^2+2y}$ **6.** $\frac{8}{9}$ **7.** $\frac{t^4-3t^2+1}{t^3+t+2}$ **8.** $\frac{x^2+2xy+3y}{2x+5y}$

In Exercises 9–22, evaluate each rational expression for the specified number.

9. $\frac{2x}{x+1}$ when $x = 10$
10. $\frac{7r+6}{r+8}$ when $r = 2$
11. $\frac{5}{y-3}$ when $y = 5$
12. $\frac{y-3}{y+3}$ when $y = -1$
13. $\frac{x+3}{x}$ when $x = 2$
14. $\frac{x^2+x}{x+3}$ when $x = 3$
15. $\frac{x+7}{x+6}$ when $x = 1$
16. $\frac{1}{x^3-2}$ when $x = 2$
17. $\frac{2x-1}{x+1}$ when $x = 0$
18. $\frac{t^4-t^2+1}{t+1}$ when $t = 1$
19. $\frac{z^2+1}{z^2-1}$ when $z = 4$
20. $\frac{3y^2+2y+1}{y+4}$ when $y = -1$
21. $\frac{2m-3}{m-4}$ when $m = -1$
22. $\frac{a^9-3a^4}{a+2}$ when $a = 0$

In Exercises 23–34, evaluate each rational expression for the specified numbers.

23. $\frac{2+y}{x}$ when $x = 2$ and $y = 4$
24. $\frac{x^2+x}{y^2}$ when $x = 5$ and $y = -1$
25. $\frac{2s-1}{t+3}$ when $s = 1$ and $t = 2$
26. $\frac{x+4}{xy}$ when $x = 2$ and $y = 1$
27. $\frac{ab^2}{ab+5}$ when $a = 3$ and $b = 5$
28. $\frac{x+y}{x-y}$ when $x = 6$ and $y = 5$
29. $\frac{x+y+1}{2x-3}$ when $x = y = 2$
30. $\frac{2x+3y}{y+5}$ when $x = 0$ and $y = -1$
31. $\frac{xyz}{x+y+z}$ when $x = 2$, $y = z = 1$
32. $\frac{2a-b}{c+5}$ when $a = 2$, $b = c = 1$
33. $\frac{a+b+c+d}{b+7}$ when $a = 1$, $b = 2$, $c = 3$, $d = 4$
34. $\frac{x^2+y^2+z^2}{w^2}$ when $x = 1$, $y = 2$, $z = 3$, $w = 5$

In Exercises 35–40, evaluate each rational expression three separate times.

35. $\frac{3}{2x+7}$ when **a.** $x = 1$, **b.** $x = 4$, **c.** $x = 7$
36. $\frac{1}{3t-4}$ when **a.** $t = 2$, **b.** $t = 4$, **c.** $t = 5$
37. $\frac{y+5}{y+2}$ when **a.** $y = 2$, **b.** $y = -1$, **c.** $y = 0$
38. $\frac{z+6}{z^2}$ when **a.** $z = -1$, **b.** $z = 1$, **c.** $z = 6$
39. $\frac{t^2+2t-3}{5t}$ when **a.** $t = 1$, **b.** $t = -1$, **c.** $t = 10$
40. $\frac{x^3-4}{x^2+x+7}$ when **a.** $x = 1$, **b.** $x = 3$, **c.** $x = 10$
41. For which value of x is $\frac{5}{x-2}$ undefined?
42. For which value of x is $\frac{x+1}{x+4}$ undefined?

In Exercises 43–46, cross-multiply to show that the given rational expressions are equivalent.

43. $\frac{x}{x^2+4x}$ and $\frac{1}{x+4}$
44. $\frac{3x^2}{x^3-5x^2}$ and $\frac{3}{x-5}$
45. $\frac{x^2+x}{5x+5}$ and $\frac{x}{5}$
46. $\frac{x^2-9}{2x+6}$ and $\frac{x-3}{2}$

4.3 MONOMIAL DIVISORS

DIVIDING POWERS *Assume a is any nonzero number or variable. Let m and n be positive integers. To divide a^m by a^n, divide by each factor a, common to numerator and denominator.*

$$\frac{a^m}{a^n} = \frac{\overbrace{a \cdot a \cdot a \ldots a}^{m \text{ factors}}}{\underbrace{a \cdot a \cdot a \ldots a}_{n \text{ factors}}}$$

For example,

$$\frac{a^5}{a^3} = \frac{\overset{1}{\cancel{a}} \cdot \overset{1}{\cancel{a}} \cdot \overset{1}{\cancel{a}} \cdot a \cdot a}{\underset{1}{\cancel{a}} \cdot \underset{1}{\cancel{a}} \cdot \underset{1}{\cancel{a}}} = a^2 = a^{5-3}$$

$$\frac{a^5}{a^5} = \frac{\overset{1}{\cancel{a}} \cdot \overset{1}{\cancel{a}} \cdot \overset{1}{\cancel{a}} \cdot \overset{1}{\cancel{a}} \cdot \overset{1}{\cancel{a}}}{\underset{1}{\cancel{a}} \cdot \underset{1}{\cancel{a}} \cdot \underset{1}{\cancel{a}} \cdot \underset{1}{\cancel{a}} \cdot \underset{1}{\cancel{a}}} = 1$$

$$\frac{a^3}{a^5} = \frac{\overset{1}{\cancel{a}} \cdot \overset{1}{\cancel{a}} \cdot \overset{1}{\cancel{a}}}{\underset{1}{\cancel{a}} \cdot \underset{1}{\cancel{a}} \cdot \underset{1}{\cancel{a}} \cdot a \cdot a} = \frac{1}{a^2} = \frac{1}{a^{5-3}}$$

In general,

$$\frac{a^m}{a^n} = a^{m-n}, \quad \text{if } m > n$$

$$\frac{a^m}{a^n} = \frac{a^m}{a^m} = 1, \quad \text{if } m = n$$

$$\frac{a^m}{a^n} = \frac{1}{a^{n-m}}, \quad \text{if } m < n$$

Example 1 a. $\dfrac{2^6}{2^3} = 2^{6-3} = 2^3$ or $\dfrac{64}{8} = 8$ b. $\dfrac{(-5)^3}{(-5)^3} = 1$ or $\dfrac{-125}{-125} = 1$

c. $\dfrac{3^2}{3^4} = \dfrac{1}{3^{4-2}} = \dfrac{1}{3^2}$ or $\dfrac{9}{81} = \dfrac{1}{9}$

In Example 2, powers of several variables occur. Also, it is useful to factor the numerical coefficients in the numerator and denominator.

Example 2

$$\frac{96m^8 n^{12} p^2}{-108m^{10} n^{12}} = \frac{2^5 \cdot 3}{-2^2 \cdot 3^3} \cdot \frac{m^8}{m^{10}} \cdot \frac{n^{12}}{n^{12}} \cdot p^2$$
$$= \frac{2^{5-2} p^2}{-3^{3-1} m^{10-8}}$$
$$= \frac{-8p^2}{9m^2}$$

CLASS EXERCISES *Divide.* **1.** $\frac{2^4}{2^6}$ **2.** $\frac{49a^5}{112a^5}$ **3.** $\frac{-8xyz^6}{4xy^2z^3}$

POWERS OF A POLYNOMIAL

Powers of a polynomial can appear in both numerator and denominator of a rational expression.

Example 3

$$\frac{6x(m-n)}{3x(m-n)^4} = \frac{2}{(m-n)^{4-1}} = \frac{2}{(m-n)^3}$$

CLASS EXERCISES *Simplify.* **4.** $\frac{(x+y)^5}{x+y}$ **5.** $\frac{a^4(x+2y)^3}{a^4(x+y)}$

ISOLATING COMMON FACTORS

Consider a rational expression, such as

$$\frac{5a+5b}{10c}$$

in which the denominator is a monomial but the numerator contains more than one term. You can often simplify such an expression by first isolating the common factor in the numerator.

Example 4 Simplify. $\frac{5a+5b}{10c}$

Solution Isolate the common factor in the numerator:

$$5a + 5b = 5(a+b)$$

Therefore,

$$\frac{5a+5b}{10c} = \frac{5(a+b)}{10c}$$
$$= \frac{a+b}{2c}$$

Divide numerator and denominator by 5.

Example 5

$$\frac{5a^2b + 10a^2b^2 + 25a^2b^3}{15ab^2} = \frac{\overset{a}{\cancel{5}\cancel{a^2}\cancel{b}}(1 + 2b + 5b^2)}{\underset{3\quad b}{\cancel{15}\cancel{a}\cancel{b^2}}} = \frac{a(1 + 2b + 5b^2)}{3b}$$

The same method applies when the numerator is a monomial, but the *denominator* contains more than one term. Now, whenever possible, isolate the common factor in the *denominator*.

Example 6

$$\frac{9xy^2z}{6xy^2 - 9xyz} = \frac{\overset{3\ \ y}{\cancel{9}\cancel{x}\cancel{y^2}z}}{\underset{1}{\cancel{3}\cancel{x}\cancel{y}}(2y - 3z)} = \frac{3yz}{2y - 3z}$$

CLASS EXERCISES *Simplify.*

6. $\frac{3x + 3y}{6}$ 7. $\frac{5a^2 + 4ab}{a}$ 8. $\frac{4x^2y^2 + 8xy^2}{4xy}$ 9. $\frac{8a^6b^4}{4a^3b + 8b^4 - 4a^2b^2}$

SOLUTIONS TO CLASS EXERCISES

1. $\frac{2^4}{2^6} = \frac{1}{2^{6-4}} = \frac{1}{2^2} = \frac{1}{4}$

2. $\frac{49a^5}{112a^5} = \frac{7^2}{2^4 \cdot 7} \cdot \frac{a^5}{a^5} = \frac{7}{16} \cdot 1 = \frac{7}{16}$

3. $\frac{-8xyz^6}{4xy^2z^3} = \frac{-2z^{6-3}}{y^{2-1}} = \frac{-2z^3}{y}$

4. $\frac{(x + y)^5}{x + y} = (x + y)^{5-1} = (x + y)^4$

5. $\frac{a^4(x + 2y)^3}{a^4(x + y)} = \frac{(x + 2y)^3}{x + y}$

6. $\frac{3x + 3y}{6} = \frac{3(x + y)}{6} = \frac{x + y}{2}$

7. $\frac{5a^2 + 4ab}{a} = \frac{a(5a + 4b)}{a} = 5a + 4b$

8. $\frac{4x^2y^2 + 8xy^2}{4xy} = \frac{\overset{y}{\cancel{4}\cancel{x}\cancel{y^2}}(x + 2)}{\underset{1}{\cancel{4}\cancel{x}\cancel{y}}} = y(x + 2)$

9. $\frac{8a^6b^4}{4a^3b + 8b^4 - 4a^2b^2} = \frac{\overset{2\ \ b^3}{\cancel{8}a^6\cancel{b^4}}}{\cancel{4}\cancel{b}(a^3 + 2b^3 - a^2b)} = \frac{2a^6b^3}{a^3 + 2b^3 - a^2b}$

HOME EXERCISES

Simplify each expression. (Multiply out the numbers that remain in numerator or denominator.)

1. $\frac{5^4}{5^3}$ 2. $\frac{3}{3^2}$ 3. $\frac{7^4 \cdot 11^2}{7^2 \cdot 11^3}$ 4. $\frac{a^4}{a^4}$ 5. $\frac{y^2}{y^5}$ 6. $\frac{-z^8}{z^{14}}$

7. $\frac{ab}{a^2b}$ 8. $\frac{x^4y}{xy^4}$ 9. $\frac{-m^4n^3}{-m^4n^3}$ 10. $\frac{r^2s^2t^3}{rs^2t^4}$ 11. $\frac{-pqrs}{pr^3s^2}$ 12. $\frac{-7abc}{14c^2}$

13. $\frac{20x^3y^3z}{-10x^2z^2}$ 14. $\frac{9a^2bc}{15a^4bc^2}$ 15. $\frac{-18m^{10}n^7p^8}{-12mn^8p^7}$ 16. $\frac{224x^6}{98x^3}$ 17. $\frac{250a^4b^2c}{375ab^2c}$ 18. $\frac{(x + y)^2}{(x + y)^2}$

19. $\dfrac{m+n}{(m+n)^2}$
20. $\dfrac{a-b}{(a-b)^4}$
21. $\dfrac{2^4(c-d)^5}{2(c-d)^8}$
22. $\dfrac{(x+y)^4(x-y)}{(x+y)^2(x-y)^3}$
23. $\dfrac{4(a+b)^7(a-b)^3}{8(a+b)(a-b)^3}$
24. $\dfrac{-84(x-y)(a+b)^7}{98(x-y)^6(a+b)}$
25. $\dfrac{5^2x^2(a+2)}{5^4x^3(a+2)^2}$
26. $\dfrac{108x^2y(x-y)^4}{144xy^2(x-y)}$
27. $\dfrac{3a-3b}{3}$
28. $\dfrac{ax+ay}{a}$
29. $\dfrac{2ax+2ay}{2a}$
30. $\dfrac{x^2+x^2y}{x^2}$
31. $\dfrac{a^2x-a^2y}{a^2}$
32. $\dfrac{2a+2b}{4}$
33. $\dfrac{9x+18y}{3}$
34. $\dfrac{20}{10a-5b}$
35. $\dfrac{35x^2+28y^2}{14}$
36. $\dfrac{ab+ac^2}{a^2}$
37. $\dfrac{cx^2+cy^2}{cx}$
38. $\dfrac{x^2y+x^2y^2}{x^2}$
39. $\dfrac{4x^2+8xy}{4x}$
40. $\dfrac{6x^2y+9xy^2}{6xy}$
41. $\dfrac{25a^2-25b^2}{5a}$
42. $\dfrac{15ab^2+10ab^3}{10ab^2}$
43. $\dfrac{4x^3y^2z-8x^2y^2z}{8xy^2z}$
44. $\dfrac{-24r^2s-30rst}{18rs}$
45. $\dfrac{5a-10b+5c}{25}$
46. $\dfrac{9ax+9ay-9az}{3a}$
47. $\dfrac{6r^2s+12r^2t-6r^2u}{36r^2}$
48. $\dfrac{au-av+a^2w}{a^2}$
49. $\dfrac{5a^2x-10ax^2+15a^2x^2}{20a^2x^2}$
50. $\dfrac{12ax-9bx+18cx}{6x^2}$
51. $\dfrac{20a^2b^3c^2+25a^4b^2c-30a^2b^2c^2}{25a^2b^3c}$
52. $\dfrac{40x^2yz-30x^2y^3z+20xy^2z^2}{10x^2y^2z^2}$
53. $\dfrac{2x}{4x+6x^2}$
54. $\dfrac{5xy}{10x^2y+15xy^2}$
55. $\dfrac{4xyz}{8xy^2-12xz^2}$
56. $\dfrac{25ab^2c^3}{50a^3bc^2-25ab^2c^6}$
57. $\dfrac{100a^2xy}{10a^2y-25a^2x+50a^2xy}$
58. $\dfrac{-3a^2b^3c^{10}}{6a^2b^3c+9ab^4c^4-27a^3b^3c^5}$

4.4 FACTORING AND SIMPLIFYING

Suppose both numerator and denominator of a rational expression contain more than one term. By factoring the numerator or denominator (possibly both), you can often simplify the given expression.

ISOLATING THE COMMON FACTOR

Example 1 Simplify $\dfrac{5x-5y}{5x+5y}$.

Solution Isolate the common factor in both numerator and denominator:

$$\frac{5x-5y}{5x+5y}=\frac{5(x-y)}{5(x+y)}=\frac{x-y}{x+y}$$

Note that in Example 1, you **cannot** divide the individual terms. **The following is wrong:**

$$\frac{\overset{1}{\cancel{x}}-\overset{1}{\cancel{y}}}{\underset{1}{\cancel{x}}+\underset{1}{\cancel{y}}}=\frac{0}{2}=0 \qquad \textbf{WRONG!}$$

For example, if $x = 2$, $y = 1$, then $\dfrac{x-y}{x+y}=\dfrac{2-1}{2+1}=\dfrac{1}{3}$, and *not* 0, as above.

CLASS EXERCISES *Simplify.* **1.** $\frac{4a + 4b}{4a^2 + 4}$ **2.** $\frac{5a^2 + 10a}{10a^3 + 20a^2}$

OTHER FACTORING METHODS

Example 2 Simplify $\frac{2x + 2y}{x^2 - y^2}$.

Solution Isolate the common factor in the numerator.

$$2x + 2y = 2(x + y)$$

The denominator is the difference of squares.

$$x^2 - y^2 = (x + y)(x - y)$$

Therefore, divide numerator and denominator by $x + y$.

$$\frac{2x + 2y}{x^2 - y^2} = \frac{2\overset{1}{\cancel{(x + y)}}}{\underset{1}{\cancel{(x + y)}}(x - y)} = \frac{2}{x - y}$$

Example 3 Simplify $\frac{a^2 + 3a + 2}{a^2 + 2a + 1}$.

Solution

$$a^2 + 3a + 2 = (a + 1)(a + 2)$$
$$a^2 + 2a + 1 = (a + 1)^2$$

Therefore,

$$\frac{a^2 + 3a + 2}{a^2 + 2a + 1} = \frac{\overset{1}{\cancel{(a + 1)}}(a + 2)}{\underset{a + 1}{\cancel{(a + 1)^2}}} = \frac{a + 2}{a + 1}$$

Example 4 Simplify $\frac{25ax^2 - 25ay^2}{5x^2 + 10xy + 5y^2}$.

Solution

$$25ax^2 - 25ay^2 = 25a(x^2 - y^2) = 25a(x + y)(x - y)$$
$$5x^2 + 10xy + 5y^2 = 5(x^2 + 2xy + y^2) = 5(x + y)^2$$

Therefore, $\dfrac{25ax^2 - 25ay^2}{5x^2 + 10xy + 5y^2} = \dfrac{\overset{5}{\cancel{25}}a\overset{1}{\cancel{(x + y)}}(x - y)}{\underset{1(x+y)}{\cancel{5(x + y)^2}}} = \dfrac{5a(x - y)}{x + y}$

CLASS EXERCISES *Simplify.* **3.** $\dfrac{x^2 - 1}{5x + 5}$ **4.** $\dfrac{a^2 + 3a + 2}{a^2 - 4}$ **5.** $\dfrac{ax^2 + 2axy + ay^2}{4ax + 4ay}$

SOLUTIONS TO CLASS EXERCISES

1. $\dfrac{4a + 4b}{4a^2 + 4} = \dfrac{\overset{1}{\cancel{4}}(a + b)}{\underset{1}{\cancel{4}}(a^2 + 1)} = \dfrac{a + b}{a^2 + 1}$

2. $\dfrac{5a^2 + 10a}{10a^3 + 20a^2} = \dfrac{\overset{1 \cdot 1}{\cancel{5a}} \cdot \overset{1}{\cancel{(a + 2)}}}{\underset{2a}{\cancel{10a^2}} \cdot \underset{1}{\cancel{(a + 2)}}} = \dfrac{1}{2a}$

3. $\dfrac{x^2 - 1}{5x + 5} = \dfrac{\overset{1}{\cancel{(x + 1)}}(x - 1)}{5\underset{1}{\cancel{(x + 1)}}} = \dfrac{x - 1}{5}$

4. $\dfrac{a^2 + 3a + 2}{a^2 - 4} = \dfrac{\overset{1}{\cancel{(a + 2)}}(a + 1)}{\underset{1}{\cancel{(a + 2)}}(a - 2)} = \dfrac{a + 1}{a - 2}$

5. $\dfrac{ax^2 + 2axy + ay^2}{4ax + 4ay} = \dfrac{\overset{1}{\cancel{a}}(x^2 + 2xy + y^2)}{4\underset{1}{\cancel{a}}(x + y)} = \dfrac{\overset{x + y}{\cancel{(x + y)^2}}}{4\underset{1}{\cancel{(x + y)}}} = \dfrac{x + y}{4}$

HOME EXERCISES

Simplify each rational expression.

1. $\dfrac{2a + 2b}{2x + 2y}$ **2.** $\dfrac{5a - 5b}{10a + 10b}$ **3.** $\dfrac{7x + 7y}{7x - 7y}$ **4.** $\dfrac{9x + 9y}{3x - 3y}$ **5.** $\dfrac{ax + ay}{ax - ay}$ **6.** $\dfrac{2ax + 2ay}{4ax - 4ay}$

7. $\dfrac{2x + 2y}{2x - 4y}$ **8.** $\dfrac{5m + 10n}{10m - 5n}$ **9.** $\dfrac{6x + 6y}{5x + 5y}$ **10.** $\dfrac{8a - 8b}{5a - 5b}$ **11.** $\dfrac{ax + ay}{bx + by}$ **12.** $\dfrac{2u + 2v}{4u + 4v}$

13. $\dfrac{x^2 - y^2}{3x + 3y}$ **14.** $\dfrac{5a + 5b}{a^2 - b^2}$ **15.** $\dfrac{6x - 6y}{2x^2 - 2y^2}$ **16.** $\dfrac{x^2 - 1}{2x + 2}$ **17.** $\dfrac{a^2 - 9}{4a + 12}$ **18.** $\dfrac{u^2 - 100}{2u + 20}$

19. $\dfrac{3m + 3n}{5m^2 - 5n^2}$ **20.** $\dfrac{5a^2 - 20}{3a + 6}$ **21.** $\dfrac{a^2 - 4b^2}{2a - 4b}$ **22.** $\dfrac{x^2 - 25y^2}{2x + 10y}$ **23.** $\dfrac{4x^2 - 9y^2}{12x + 18y}$ **24.** $\dfrac{100a^2 - 64b^2}{5a - 4b}$

25. $\dfrac{x^2 + 2x + 1}{x^2 + 4x + 3}$ **26.** $\dfrac{a^2 + 5a + 6}{a^2 + 4a + 4}$ **27.** $\dfrac{y^2 - 9}{y^2 + 4y + 3}$ **28.** $\dfrac{z^2 + 7z + 10}{z^2 - 25}$

29. $\dfrac{u^2 - 36}{u^2 + 7u + 6}$ **30.** $\dfrac{r^2 - 1}{2 - 2r^2}$ **31.** $\dfrac{m^2 + 7m + 12}{m^2 + 5m + 6}$ **32.** $\dfrac{a^2 - 8a + 12}{a^2 - 36}$

33. $\dfrac{x^2 + 2x - 8}{x^2 - 16}$ **34.** $\dfrac{x^2 - y^2}{x^2 + 2xy + y^2}$ **35.** $\dfrac{3a^2 - 3b^2}{6a^2 + 18ab + 12b^2}$ **36.** $\dfrac{9u^2 - 36}{3u^2 - 9u + 6}$

37. $\dfrac{5x^2 + 35x + 60}{10x^2 + 10x - 60}$ **38.** $\dfrac{a^2u^2 - a^2v^2}{au^2 + 6auv + 5av^2}$ **39.** $\dfrac{2x^2 + 11x + 15}{x^2 + 6x + 9}$ **40.** $\dfrac{3a^2 - 2ab - b^2}{4a^2 - 4b^2}$

4.5 POLYNOMIAL DIVISION

DEGREE Consider a polynomial in a single variable, such as

$$x^4 + 5x^3 - x^2 + x + 1$$

The **degree of a term** is simply the exponent of the variable of that term, if there is a variable; the **degree of a nonzero numerical term** is 0. (The degree of 0 is not defined.) For the above polynomial,

the degree of x^4 is 4, the degree of $5x^3$ is 3,
the degree of $-x^2$ is 2, the degree of x is 1,
the degree of 1, the numerical term, is 0.

The **degree of a polynomial** is the highest degree of any of its terms. Thus, the degree of $x^4 + 5x^3 - x^2 + 1$ is 4.

A *polynomial in a single variable* is said to be in **standard form** if terms of the same degree are combined and the resulting terms are arranged in order of decreasing degree. Thus,

$x^4 + 5x^3 - x^2 + x + 1$ is in standard form, but

$5x^3 + x^4 + 1 - x + x^2$ is not.

Example 1 a. Express $t^4 - t^3 + 2t^6 + t^8 - 1 + t^2 + 2t^8$ in standard form.
b. What is the degree of this polynomial?

Solution a. Combine t^8 and $2t^8$ to obtain $3t^8$. Now rearrange the terms in order of decreasing degree:

$$3t^8 + 2t^6 + t^4 - t^3 + t^2 - 1$$

is now in standard form.
b. The degree of this polynomial is 8, the highest degree of any its terms.

CLASS EXERCISES

1. Find the degree of each term.
   a. x^6 b. $12y^2$ c. $-x^9$ d. 10
2. a. Express $x^5 - x^4 + x^8 - 6x^6 + 3 + x^7 + x^5$ in standard form.
   b. What is the degree of this polynomial?

THE DIVISION PROCESS There is a method of dividing polynomials that resembles long division.

Recall that to divide 651 by 21, you proceed as follows:

$$\begin{array}{r} 31 \\ 21\,\overline{)651} \\ \underline{63} \\ 21 \\ \underline{21} \end{array}$$

For polynomials P, Q, and S, where $S \neq 0$,

$$\frac{P}{S} = Q, \quad \text{if} \quad P = Q \cdot S$$

You can also express division of polynomials by writing $S\,\overline{)P}$ with Q above.

Here, S is called the **divisor**, P the **dividend**, and Q the **quotient**.

Example 2 Divide. $x + 2\,\overline{)2x^2 + 7x + 6}$

Solution

$$\begin{array}{rl} \text{a.} \rightarrow & \overbrace{2x + 3}^{\textit{quotient}} \qquad \leftarrow \text{c.} \\ \textit{divisor} \rightarrow x + 2 & \overline{)2x^2 + 7x + 6} \leftarrow \textit{dividend} \\ \text{b.} & \underline{2x^2 + 4x} \\ & \qquad 3x + 6 \\ \text{d.} & \qquad \underline{3x + 6} \\ & \qquad\qquad 0 \end{array}$$

a. $$\frac{2x^2}{x} = 2x$$

Divide $2x^2$, the first term of the dividend, by x, the first term of the divisor, to obtain $2x$, the first term of the quotient.

b.

$$\begin{array}{rl} & 2x \\ x + 2 & \overline{)2x^2 + 7x + 6} \\ \text{Subtract.} & \underline{2x^2 + 4x} \\ & \qquad 3x + 6 \leftarrow \textit{1st difference polynomial} \end{array}$$

Now, multiply $2x$ by the divisor, $x + 2$, and *subtract* the product, $2x^2 + 4x$, from $2x^2 + 7x$, the first two terms of the divisor. The difference is

$3x$. Bring down 6, the next term of the dividend to obtain the **1st difference polynomial**, $3x + 6$.

c.

$$\frac{3x}{x} = 3$$

Next, divide $3x$, the first term of the 1st difference polynomial, by x, the first term of the divisor. The second term of the quotient is 3.

d.

$$\begin{array}{rr l}
\textit{quotient}\rightarrow & 2x + 3 & \\
\textit{divisor}\rightarrow\ x + 2 & \overline{)2x^2 + 7x + 6} & \leftarrow\textit{dividend} \\
\text{Subtract.} & \underline{2x^2 + 4x} & \\
 & 3x + 6 & \leftarrow\textit{1st difference polynomial} \\
\text{Subtract.} & \underline{3x + 6} & \\
 & 0 & \leftarrow\textit{2nd difference polynomial}
\end{array}$$

Multiply 3 by the divisor, $x + 2$, to obtain $3x + 6$. When you subtract this from the 1st difference polynomial, you obtain 0. There are no more terms to bring down. Thus, 0 is the 2nd difference polynomial, and $x + 2$ divides $2x^2 + 7x + 6$ (evenly). The quotient is $2x + 3$.

You can **check a division example** by multiplying the quotient *by the divisor* (in either order). The product should be the dividend, as you see below.

check (for Example 2)

$$\begin{array}{rr}
\textit{quotient}\rightarrow & 2x + 3 \\
\textit{divisor}\rightarrow & \underline{\times\quad x + 2} \\
 & 2x^2 + 3x \\
 & \underline{4x + 6} \\
\textit{dividend}\rightarrow & 2x^2 + 7x + 6
\end{array}$$

Example 3 Divide.

$$3x + 1\,\overline{)6x^2 + 17x + 5}$$

Solution

$$\begin{array}{rr l}
\textit{quotient}\rightarrow & 2x + 5 & \\
\textit{divisor}\rightarrow\quad 3x + 1 & \overline{)6x^2 + 17x + 5} & \leftarrow\textit{dividend} \\
\text{Subtract.} & \underline{6x^2 + 2x} & \leftarrow\textit{1st difference polynomial} \\
 & 15x + 5 & \\
\text{Subtract.} & \underline{15x + 5} & \leftarrow\textit{2nd difference polynomial} \\
 & 0 &
\end{array}$$

The quotient is $2x + 5$.

CLASS EXERCISES **3.** Divide. $x + 1\,\overline{)x^2 + 6x + 5}$ **4.** Divide and check. $t + 5\,\overline{)3t^2 + 13t - 10}$

FURTHER EXAMPLES

Example 4 Divide $6 - 7x + x^3$ by $x - 2$.

Solution Express the dividend in *standard form*. Also, add $0x^2$ for the missing term of degree 2:

$$x^3 + 0x^2 - 7x + 6$$

quotient→ $\quad x^2 + 2x - 3$

divisor→ $\quad x - 2 \overline{\big) x^2 + 0x^2 - 7x + 6}$

Subtract. $\quad \underline{x^3 - 2x^2}$

$\quad 2x^2 - 7x$ ←*1st difference polynomial*

Subtract. $\quad \underline{2x^2 - 4x}$

$\quad -3x + 6$ ←*2nd difference polynomial*

Subtract. $\quad \underline{-3x + 6}$

$\quad 0$ ←*3rd difference polynomial*

The quotient is $x^2 + 2x - 3$.

Example 5 Divide. $\quad x + 1 \overline{\big) x^3 + x^2 + x + 1}$

Solution

$\quad x^2 \quad + 1$

$x + 1 \overline{\big) x^3 + x^2 + x + 1}$

Subtract. $\quad \underline{x^3 + x^2}$

$\quad x + 1$ *The 1st difference polynomial is 0. Bring down 2 terms from the dividend because the divisor has 2 terms.*

Subtract. $\quad \underline{x + 1}$

$\quad 0$

The quotient is $x^2 + 1$.

CLASS EXERCISES **5.** Divide. $\quad x - 2 \overline{\big) x^3 - 3x - 2}$ $\qquad$ **6.** Divide $1 - x^4$ by $1 + x^2$.

REMAINDERS Observe that 2 does not divide 7 (evenly). When you divide 7 by 2, there is a "remainder."

dividend → $\dfrac{7}{2} = 3\dfrac{1}{2}$ ← *divisor* (numerator 1 ← *remainder*; 3 ← *quotient*)

divisor →

Thus, 3 is the quotient and 1 the remainder. Similarly, division of polynomials often results in a remainder. Here is a simple example.

Example 6 Divide. $2x + 1 \overline{)4x + 3}$

Solution

$$\begin{array}{r} 2 \\ 2x + 1 \overline{)4x + 3} \\ \text{Subtract.} \quad 4x + 2 \\ \hline 1 \end{array}$$

There are no more terms of the dividend to bring down. Thus, 2 is the quotient; 1 is the remainder.

$$\begin{array}{c} \textit{dividend} \rightarrow \\ \textit{divisor} \rightarrow \end{array} \frac{4x + 3}{2x + 1} = \overset{\substack{\textit{quotient} \\ \downarrow}}{2} + \frac{1}{2x + 1} \begin{array}{c} \leftarrow \textit{remainder} \\ \leftarrow \textit{divisor} \end{array}$$

Note that the remainder, 1, is a (numerical) polynomial of degree 0, whereas the divisor is a polynomial of degree 1. Thus,

$$\underbrace{\text{degree } (1)}_{0} < \underbrace{\text{degree } (2x + 1)}_{1}$$

Let P and S be polynomials, $S \neq 0$. Then, either S divides P (evenly), or else

$$\frac{P}{S} = Q + \frac{R}{S},$$

where Q and R are polynomials and degree $R <$ degree S. The polynomial Q is called the **quotient** and R the **remainder**. Thus, *if S does not divide P, add the rational expression $\frac{R}{S}$ to the quotient Q.*

When you divide polynomials, there are two possibilities for the *final difference polynomial*:

1. This difference polynomial is 0, as in Examples 2–5. In this case, there is no remainder.
2. The degree of R, the final difference polynomial, is less than the degree of the divisor, as in Example 6. Then, R is the remainder.

Continue the division process until you obtain one of these difference polynomials.

Example 7 Find the quotient and remainder. $5x + 1\overline{)10x^2 + 12x - 2}$

Solution

$$\begin{array}{lrl}
\textit{quotient} \rightarrow & & 2x + 2 + \dfrac{-4}{5x + 1} \quad \begin{array}{l}\leftarrow \textit{remainder} \\ \leftarrow \textit{divisor}\end{array} \\
\textit{divisor} \rightarrow & 5x + 1 & \overline{)10x^2 + 12x - 2} \qquad \leftarrow \textit{dividend} \\
& \text{Subtract.} & \underline{10x^2 + 2x} \\
& & \phantom{10x^2 + {}} 10x - 2 \qquad \leftarrow \textit{1st difference polynomial} \\
& \text{Subtract.} & \phantom{10x^2 + {}} \underline{10x + 2} \\
& & \phantom{10x^2 + 10x {}} -4 \qquad \leftarrow \textit{2nd difference polynomial (remainder)}
\end{array}$$

The quotient is $2x + 2$ and the remainder is -4. The result is expressed as

$$2x + 2 + \frac{-4}{5x + 1}.$$

To check a division example, where there is a remainder:

1. Multiply the quotient and divisor.
2. Add the remainder to this.

The resulting polynomial should be the dividend.

Example 8 Find the quotient and remainder. Also, check the result.

$$3x - 2\overline{)12x^3 - 11x^2 + 5x}$$

Solution

$$\begin{array}{lrl}
\textit{quotient} \rightarrow & & 4x^2 - x + 1 + \dfrac{2}{3x - 2} \quad \begin{array}{l}\leftarrow \textit{remainder} \\ \leftarrow \textit{divisor}\end{array} \\
\textit{divisor} \rightarrow & 3x - 2 & \overline{)12x^3 - 11x^2 + 5x} \qquad \leftarrow \textit{dividend} \\
& \text{Subtract.} & \underline{12x^3 - 8x^2} \\
& & - 3x^2 + 5x \qquad \leftarrow \textit{1st difference polynomial} \\
& \text{Subtract.} & \underline{- 3x^2 + 2x} \\
& & \phantom{12x^3 - 3x^2 + {}} 3x \qquad \leftarrow \textit{2nd difference polynomial} \\
& \text{Subtract.} & \phantom{12x^3 - 3x^2 + {}} \underline{3x - 2} \\
& & \phantom{12x^3 - 3x^2 + 3x - {}} 2 \qquad \leftarrow \textit{3rd difference polynomial (remainder)}
\end{array}$$

The quotient is $4x^2 - x + 1$ and the remainder is 2. The result is expressed as

$$4x^2 - x + 1 + \frac{2}{3x - 2}$$

check

$$\begin{array}{rrrrr} \textit{quotient} \rightarrow & & 4x^2 & -\ x & +\ 1 \\ \textit{divisor} \rightarrow & \times & 3x & -\ 2 & \\ \hline & 12x^3 & -\ 3x^2 & +\ 3x & \\ & & -\ 8x^2 & +\ 2x & -\ 2 \\ \hline & 12x^3 & -\ 11x^2 & +\ 5x & -\ 2 \\ \textit{remainder} \rightarrow & + & & & 2 \\ \hline \textit{dividend} \rightarrow & 12x^3 & -\ 11x^2 & +\ 5x & \end{array}$$

Example 9 Find the quotient and remainder. $x^2 \overline{)x}$

Solution 1, the degree of the dividend, x, is less than 2, the degree of the divisor, x^2. Thus, the quotient is 0 and the remainder is x.

$$\frac{x}{x^2} = 0 + \frac{x}{x^2} \leftarrow \text{remainder}$$

CLASS EXERCISES

7. **a.** What is the quotient? **b.** What is the remainder?
 c. Show that degree of remainder < degree of divisor.

$$\frac{x^3 + x^2 + 5x - 7}{x^2 + x} = x + \frac{5x - 7}{x^2 + x}$$

In Exercises 8–10, find the quotient and remainder. In Exercise 10, check the result.

8. $2x - 1 \overline{)6x^2 + x - 4}$ **9.** $x^2 + 1 \overline{)x + 1}$

10. $x + 3 \overline{)2x^3 + 7x^2 + 4x}$ (*Check.*)

SOLUTIONS TO CLASS EXERCISES

1. a. 6 **b.** 2 **c.** 9 **d.** 0

2. a. $x^8 + x^7 - 6x^6 + 2x^5 - x^4 + 3$ **b.** 8

3.

$$\begin{array}{lrl} & x + 5 & \leftarrow \textit{quotient} \\ \textit{divisor} \rightarrow x + 1 & \overline{)x^2 + 6x + 5} & \leftarrow \textit{dividend} \\ \text{Subtract.} & \underline{x^2 + x} & \\ & 5x + 5 & \\ \text{Subtract.} & \underline{5x + 5} & \\ & 0 & \end{array}$$

4.

$$\begin{array}{lrl} & 3t - 2 & \leftarrow \textit{quotient} \\ \textit{divisor} \rightarrow t + 5 & \overline{)3t^2 + 13t - 10} & \leftarrow \textit{dividend} \\ \text{Subtract.} & \underline{3t^2 + 15t} & \\ & -2t - 10 & \\ \text{Subtract.} & \underline{-2t - 10} & \\ & 0 & \end{array}$$

The quotient is $3t - 2$.

check

$$\begin{array}{rrrr} \textit{quotient} \rightarrow & 3t & -\ 2 & \\ \textit{divisor} \rightarrow & \times\ t & +\ 5 & \\ \hline & 3t^2 & -\ 2t & \\ & & 15t & -\ 10 \\ \hline \textit{dividend} \rightarrow & 3t^2 & +\ 13t & -\ 10 \end{array}$$

5.

$$\begin{array}{rl}
 & x^2 + 2x + 1 \quad \leftarrow \textit{quotient} \\
\textit{divisor} \rightarrow x - 2 & \overline{)\, x^3 + 0x^2 - 3x - 2} \quad \leftarrow \textit{dividend} \\
\text{Subtract.} & \underline{x^3 - 2x^2} \\
 & 2x^2 - 3x \\
\text{Subtract.} & \underline{2x^2 - 4x} \\
 & x - 2 \\
\text{Subtract.} & \underline{x - 2} \\
 & 0
\end{array}$$

6.

$$\begin{array}{rl}
 & -x^2 + 1 \quad \leftarrow \textit{quotient} \\
\textit{divisor} \rightarrow x^2 + 0x + 1 & \overline{)\, -x^4 + 0x^3 + 0x^2 + 0x + 1} \quad \leftarrow \textit{dividend} \\
\text{Subtract.} & \underline{-x^4 - x^2} \\
 & x^2 + 1 \\
\text{Subtract.} & \underline{x^2 + 1} \\
 & 0
\end{array}$$

7. a. x **b.** $5x - 7$ **c.** $\underbrace{\text{degree}(5x - 7)}_{1} < \underbrace{\text{degree}(x^2 + x)}_{2}$

8.

$$\begin{array}{rl}
\textit{quotient} \rightarrow & 3x + 2 + \dfrac{-2}{2x - 1} \quad \begin{array}{l}\leftarrow \textit{remainder} \\ \leftarrow \textit{divisor}\end{array} \\
\textit{divisor} \rightarrow 2x - 1 & \overline{)\, 6x^2 + x - 4} \quad \leftarrow \textit{dividend} \\
\text{Subtract.} & \underline{6x^2 - 3x} \\
 & 4x - 4 \\
\text{Subtract.} & \underline{4x - 2} \\
 & -2 \quad \leftarrow \textit{remainder}
\end{array}$$

The quotient is $3x + 2$ and the remainder is -2.

9. $\dfrac{x + 1}{x^2 + 1} = 0 + \dfrac{x + 1}{x^2 + 1}$ The quotient is 0 and the remainder is $x + 1$.

10.

$$\begin{array}{rl}
\textit{quotient} \rightarrow & 2x^2 + x + 1 + \dfrac{-3}{x + 3} \quad \begin{array}{l}\leftarrow \textit{remainder} \\ \leftarrow \textit{divisor}\end{array} \\
\textit{divisor} \rightarrow x + 3 & \overline{)\, 2x^3 + 7x^2 + 4x} \quad \leftarrow \textit{dividend} \\
\text{Subtract.} & \underline{2x^3 + 6x^2} \\
 & x^2 + 4x \\
\text{Subtract.} & \underline{x^2 + 3x} \\
 & x \\
\text{Subtract.} & \underline{x + 3} \\
 & -3 \quad \leftarrow \textit{remainder}
\end{array}$$

check

$$\begin{array}{rr}
\textit{quotient} \rightarrow & 2x^2 + x + 1 \\
\textit{divisor} \rightarrow & \underline{\times \quad x + 3} \\
 & 2x^3 + x^2 + x \\
 & \underline{+ 6x^2 + 3x + 3} \\
 & 2x^3 + 7x^2 + 4x + 3 \\
\textit{remainder} \rightarrow & \underline{+ \qquad\qquad -3} \\
\textit{dividend} \rightarrow & 2x^3 + 7x^2 + 4x
\end{array}$$

The quotient is $2x^2 + x + 1$ and the remainder is -3.

HOME EXERCISES

In Exercises 1–6: (a) Express each polynomial in standard form. (b) What is the degree of the polynomial?

1. $x^2 + 1 + 2x$ **2.** $y^3 - 1 + 2y^2 + 4y$ **3.** $t^3 + t - t^5$ **4.** $u - u^2 + u^3 - u^4$

5. $z^7 - z^4 + z + z^{10} + z^7$ **6.** $x^9 - x^8 + 1 - x^4 - x^{12} + x^{12}$

In Exercises 7–14, divide and check.

7. $x + 1\,\overline{)\,x^2 + 5x + 4}$ **8.** $y + 2\,\overline{)\,y^2 + 8y + 12}$ **9.** $x - 1\,\overline{)\,x^2 + 5x - 6}$ **10.** $u + 4\,\overline{)\,u^2 + 12u + 32}$

11. $x - 8\,\overline{)\,x^2 - 3x - 40}$ **12.** $x + 2\,\overline{)\,2x^3 + 5x^2 + 3x + 2}$

13. $t + 4\,\overline{)\,12t^2 + 3t^3 - 4 - t}$ **14.** $y^2 + 2\,\overline{)\,y^4 + 3y^2 + 2}$

In Exercises 15–24, divide as indicated.

15. $z^2 - 1\,\overline{)\,z^4 - 1}$ **16.** $t + 3\,\overline{)\,t^3 + 27}$ **17.** Divide $9 + 3a + 3a^2 + a^3$ *by* $3 + a^2$.

18. Divide $9x^2 + x^3 + 17x + 6$ by $x^2 + 7x + 3$. **19.** Divide $15 + 8a^2 + a^4$ *by* $a^2 + 3$.

20. $2b + 1 \overline{)4b^3 + 4b^2 + 5b + 2}$ **21.** $x^2 + x - 2 \overline{)x^3 + 3x^2 - 4}$ **22.** $2y^2 + 3y + 1 \overline{)2y^3 + 9y^2 + 10y + 3}$
23. $x^2 + x + 1 \overline{)x^4 + 3x^3 + 4x^2 + 3x + 1}$ **24.** $x^2 + x - 1 \overline{)2x^4 + 3x^3 + 4x^2 + 4x - 5}$

In Exercises 25–30: (a) What is the quotient? (b) What is the remainder? (c) Show that degree of remainder < degree of divisor.

25. $\dfrac{8x + 3}{2x} = 4 + \dfrac{3}{2x}$ **26.** $\dfrac{2x^3 + x + 1}{x} = 2x^2 + 1 + \dfrac{1}{x}$ **27.** $\dfrac{x^2 + 4x + 8}{x + 3} = x + 1 + \dfrac{5}{x + 3}$

28. $\dfrac{x^3 + 3x}{x^2 + 1} = x + \dfrac{2x}{x^2 + 1}$ **29.** $\dfrac{2x^3 - 4x + 5}{x^2 - 2} = 2x + \dfrac{5}{x^2 - 2}$ **30.** $\dfrac{x^4 + 4x^3 + 6x^2 + 8x + 3}{x^2 + x + 2} = x^2 + 3x + 1 + \dfrac{x + 1}{x^2 + x + 2}$

In Exercises 31–38: (a) Find the quotient, (b) find the remainder, and (c) check the result.

31. $x + 1 \overline{)3x + 2}$ **32.** $1 + x \overline{)1 + 10x}$ **33.** $x + 2 \overline{)x^2 + 5x + 8}$ **34.** $x - 1 \overline{)x^2 - 2x}$
35. $x + 1 \overline{)x^3 + 3x^2 + 3x + 5}$ **36.** $x^2 + x + 1 \overline{)2x^3 + 7x^2 + 6x + 10}$
37. $x^2 - x \overline{)3x^3 + 2x^2 + 5x + 2}$ **38.** $x^2 + 1 \overline{)x^4 - x^2 + 3}$

In Exercises 39–50: (a) Find the quotient, (b) find the remainder, and (c) express your result in the form

$$\text{QUOTIENT} + \frac{\text{REMAINDER}}{\text{DIVISOR}}.$$

39. $x + 3 \overline{)x^2 + 8x + 20}$ **40.** $x - 1 \overline{)2x^3 - 5x}$ **41.** $x - 5 \overline{)x^2}$
42. $2x + 1 \overline{)4x^2 + 6x + 5}$ **43.** $x^2 + x + 1 \overline{)4x^3 + x^2 - 2x + 10}$ **44.** $3x + 2 \overline{)12x^2 + 11x + 11}$
45. $x^2 - 2 \overline{)x^4 + x^2}$ **46.** $x^2 + x + 1 \overline{)x^4 - x^2 + x + 2}$ **47.** $x + 1 \overline{)x^3 - 3x^2 + 2x + 1}$
48. $x^2 + 1 \overline{)2x}$ **49.** $x + 2 \overline{)x^3 - 2x^2 + x + 2}$ **50.** $x^3 - x \overline{)x^2 + x}$

Let's Review Chapter 4

1. Fill in the integers. $\frac{3}{8}$ is a rational number with numerator ☐ and denominator ☐.

2. Fill in the rational number. -2 is the numerator and 5 is the denominator of the rational number ☐.

3. Each of the points Q, R, S, T, U, V in Figure 4.11 represents one of the following rational numbers:

$$\frac{1}{5}, \quad \frac{-1}{5}, \quad \frac{3}{5}, \quad \frac{-4}{5}, \quad \frac{6}{5}, \quad \frac{9}{5}$$

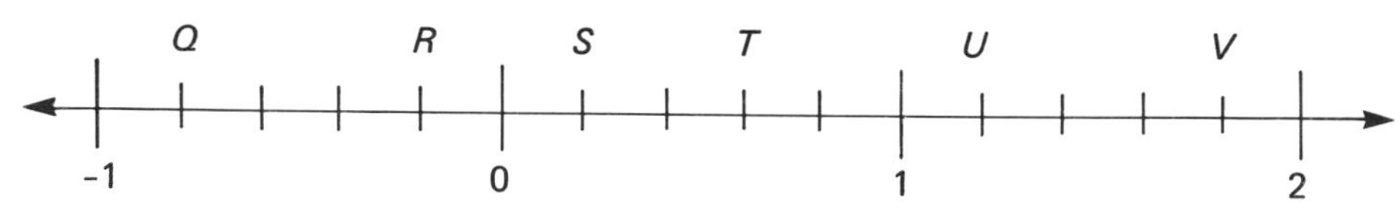

Fig. 4.11

Indicate which number is represented by each point.

4. Express -8 in the form $\frac{N}{D}$, where N and D are integers and $D \neq 0$.

5. Cross-multiply to show that $\frac{4}{6}$ and $\frac{-2}{-3}$ are equivalent.

In Exercises 6 and 7, simplify each fraction by dividing by the largest factor common to numerator and denominator. Express the simplified fraction with a positive denominator.

6. $\frac{5}{-15}$ **7.** $\frac{-24}{-40}$

8. Find **a.** the numerator and **b.** the denominator of the rational expression $\frac{2y - 1}{y^2 + 3y \nleq 5}$

9. Evaluate $\frac{3x + 2}{x - 9}$ when $x = 10$. **10.** Evaluate $\frac{xy}{x + y + 3}$ when $x = 2$ and $y = 3$.

11. Evaluate $\frac{x^2 + 1}{x^2 - 2}$ three separate times: **a.** when $x = 1$, **b.** when $x = -1$, **c.** when $x = 2$.

In Exercises 12–23, simplify each rational expression. (Multiply out the numbers that remain in numerator or denominator.)

12. $\frac{a^4 b}{a^2 b^5}$ **13.** $\frac{2 \cdot 5^4 z^5 x^2}{5^2 z^2 x^5}$ **14.** $\frac{-48a^4 bc^7}{-36abc^9}$ **15.** $\frac{3(x + y)^3 (x - y)}{3^4 (x + y)(x - y)^2}$

16. $\frac{4x - 4y}{2}$ **17.** $\frac{ax^2 + bx^2}{x^3}$ **18.** $\frac{3a^2 bc + 6abc^2}{9abc}$ **19.** $\frac{20xy^2 z}{10xy^2 z - 50xy^3 z}$

20. $\frac{5a + 5b}{25a + 50b}$ **21.** $\frac{m^2 - 9}{m^2 + 6m + 9}$ **22.** $\frac{s^3 - s^2}{s^4 - s^3}$

23. $\frac{2x^2 - 2y^2}{x^2 - 2xy + y^2}$

24. **a.** Express $-x^4 + 7x^6 + x^2 + x^2 + 5x^5$ in standard form.
b. What is the degree of this polynomial?

25. Divide and check. $y + 2 \overline{\smash{)}\, y^2 + 7y + 10}$

26. Divide. $x + 4 \overline{\smash{)}\, x^3 + 3x^2 - 3x + 4}$

27. Divide $a^4 + 4a^2 - 5$ by $a^2 + 5$.

28. **a.** What is the quotient? **b.** What is the remainder?
c. Show that the degree of the remainder is less than the degree of the divisor.

$$\frac{5x + 7}{x + 1} = 5 + \frac{2}{x + 1}$$

29. **a.** Find the quotient, **b.** find the remainder, and **c.** check the result:

$$x + 2 \overline{\smash{)}\, 4x^2 - 7x + 9}$$

30. **a.** Find the quotient, **b.** find the remainder, and **c.** express your result in the form

$$\text{QUOTIENT} + \frac{\text{REMAINDER}}{\text{DIVISOR}}$$

$$x + 4 \overline{\smash{)}\, x^3 - x^2 + 1}$$

5 ARITHMETIC OF RATIONAL EXPRESSIONS

DIAGNOSTIC TEST Perhaps you are already familiar with some of the material in Chapter 5. This test will indicate which sections in Chapter 5 you need to study. The question number refers to the corresponding section in Chapter 5. If you answer *all* parts of the question correctly, you may omit the section. But *if any part of your answer is wrong, you should study the section*. When you have completed the test, turn to page A10 for the answers.

5.1 Multiplication and Division of Rational Expressions

Multiply or divide, and simplify.

a. $\frac{3}{4} \cdot \frac{-1}{6} \cdot \frac{-2}{9}$ b. $\frac{x^2 - 1}{8} \cdot \frac{2x + 2y}{x + 1}$

c. $\frac{3a^2b}{5xy} \div \frac{9ab^2}{10x^2y}$ d. $\frac{x^2 - 9a^2}{x^2 + 3x + 2} \div \frac{4x - 12a}{2x + 2}$

5.2 Least Common Multiples

a. Find the least common multiple of the integers 12 and 20.

b. Find the least common multiple of the polynomials x^3 and $x^2 - x$.

c. Find $lcd\left(\frac{x}{x+1}, \frac{3}{x-3}\right)$. Write equivalent expressions with this *lcd* as denominator.

5.3 Addition and Subtraction of Rational Expressions

Add or subtract, and simplify.

a. $\frac{3}{10} + \frac{9}{10} - \frac{7}{10}$ b. $\frac{x}{x^2 - 4} + \frac{2}{x^2 - 4}$

c. $\frac{3}{10} + \frac{7}{20} - \frac{1}{5}$ d. $\frac{x}{x^2 - 16} + \frac{1}{x^2 + 5x + 4}$

5.4 Order of Fractions

Fill in "<" or ">".

a. $\frac{7}{10} \square \frac{3}{10}$ b. $\frac{5}{12} \square \frac{3}{7}$ c. $\frac{-9}{20} \square \frac{-7}{16}$

5.5 Complex Expressions

Simplify each expression.

a. $\dfrac{\frac{2}{9}}{\frac{4}{3}}$ b. $\dfrac{1+\frac{1}{2}}{2-\frac{1}{4}}$ c. $\dfrac{\frac{x-3}{x^2}}{\frac{x^2-9}{x^3}}$

5.6 Proportions and Equations with Rational Expressions

Solve each equation.

a. $x - \frac{3}{4} = \frac{1}{2}$ b. $\frac{3}{8} = \frac{-9}{d}$ c. $\frac{1}{x-4} + \frac{1}{x-1} = \frac{x}{x^2-5x+4}$

5.1 MULTIPLICATION AND DIVISION OF RATIONAL EXPRESSIONS

Suppose you cut a pie into quarters, as in Figure 5.1(a). You then decide to cut each piece in half. Observe that each new portion is 1/8 of the pie. [See Figure 5.1(b).] Thus,

$$\frac{1}{2} \text{ of } \frac{1}{4} \text{ equals } \frac{1}{8}$$

(Here, **the word "of" indicates multiplication.**)

$$\frac{1}{2} \cdot \frac{1}{4} = \frac{1}{8}$$

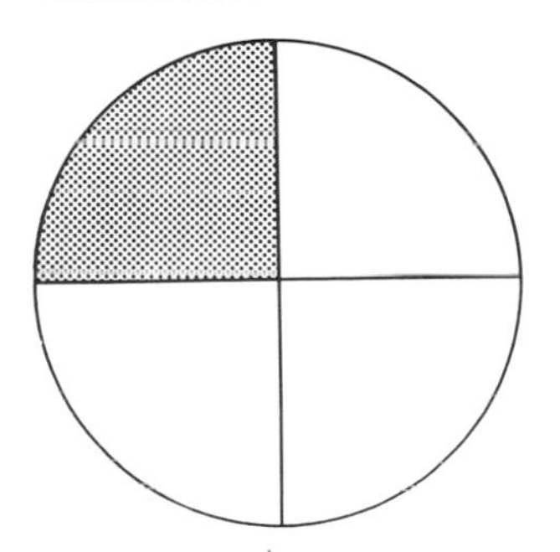

Fig. 5.1(a) $\frac{1}{4}$.

PRODUCTS OF FRACTIONS

Note that

$$\frac{1}{2} \cdot \frac{1}{4} = \frac{1 \cdot 1}{2 \cdot 4} = \frac{1}{8}$$

In Figure 5.1(c) the product $\frac{1}{2} \cdot \frac{1}{4}$ is depicted on the number line.

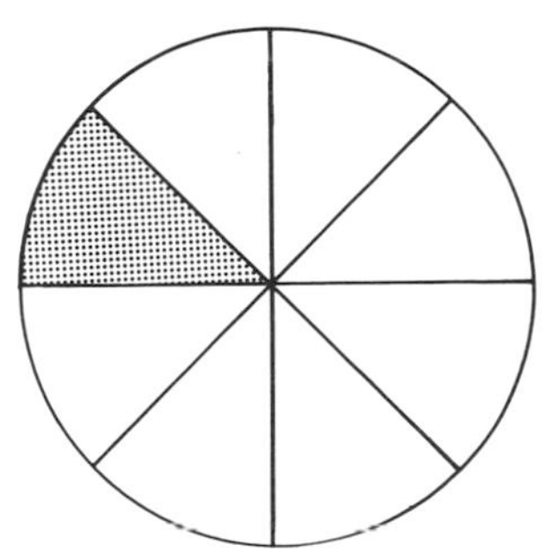

Fig. 5.1(b) $\frac{1}{2} \cdot \frac{1}{4} = \frac{1}{8}$.

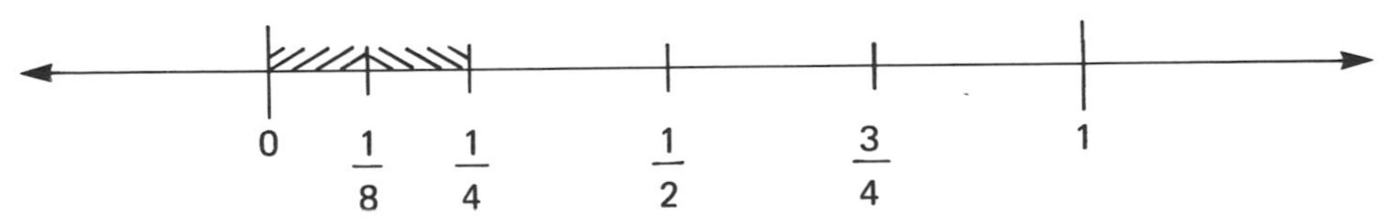

Fig. 5.1(c) $\frac{1}{2} \cdot \frac{1}{4} = \frac{1}{8}$.

To multiply two fractions:

1. Multiply their numerators.
2. Multiply their denominators.

$$\frac{a}{b} \cdot \frac{c}{d} = \frac{ac}{bd}, \qquad b \neq 0, d \neq 0$$

Example 1

$$\frac{2}{3} \cdot \frac{1}{5} = \frac{2 \cdot 1}{3 \cdot 5} = \frac{2}{15}$$

It is best to divide by factors common to the numerators and denominators before multiplying. You will then be working with smaller numbers.

Example 2 Multiply. $\frac{2}{5} \cdot \frac{15}{4}$

Solution Divide 15 and 5 by 5. Then, divide 2 and 4 by 2.

$$\frac{\overset{1}{\cancel{2}}}{\underset{1}{\cancel{5}}} \cdot \frac{\overset{3}{\cancel{15}}}{\underset{2}{\cancel{4}}} = \frac{3}{2}$$

The same procedure applies to **products of three or more fractions.**

1. Divide by factors common to numerators and denominators.
2. Multiply the resulting numerators.
3. Multiply the resulting denominators.

Example 3 Multiply. $\frac{-5}{8} \cdot \frac{12}{25} \cdot \frac{10}{-3}$

Solution Divide 12 and 8 by 4.
Divide −5 and 25 by 5.
Divide 10 and 5 by 5.
Divide 3 and −3 by 3.
Divide −1 and −1 by −1.
Divide 2 and 2 by 2.

$$\frac{\overset{\overset{1}{\cancel{-1}}}{\cancel{-5}}}{\underset{\underset{1}{\cancel{2}}}{\cancel{8}}} \cdot \frac{\overset{\overset{1}{\cancel{3}}}{\cancel{12}}}{\underset{\underset{1}{\cancel{5}}}{\cancel{25}}} \cdot \frac{\overset{\overset{1}{\cancel{2}}}{\cancel{10}}}{\underset{\underset{1}{\cancel{-1}}}{\cancel{-3}}} = 1$$

Example 4 Find $\frac{3}{4}$ of 12.

Solution Here the word "of" can be translated as "times." Thus,

$$\frac{3}{4} \cdot 12 = \frac{3}{\underset{1}{\cancel{4}}} \cdot \frac{\overset{3}{\cancel{12}}}{1} = 9$$

CLASS EXERCISES

In Exercises 1–3, multiply.

1. $\frac{3}{4} \cdot \frac{3}{5}$ **2.** $\frac{5}{6} \cdot \frac{3}{10}$ **3.** $\frac{2}{3} \cdot \frac{9}{5} \cdot \frac{-5}{12}$

4. Find $\frac{2}{3}$ of -9.

PRODUCTS OF RATIONAL EXPRESSIONS

Example 5

$$\frac{x^3}{3y} \cdot \frac{a+b}{a-b} = \frac{x^3(a+b)}{3y(a-b)}$$

Note that the resulting rational expression is left in factored form.

> **To multiply two or more rational expressions:**
>
> 1. First, divide by factors common to numerators and denominators.
> 2. Multiply the resulting numerators.
> 3. Multiply the resulting denominators.
>
> Leave the resulting expression in factored form.

Example 6 Multiply: $\frac{4x^3}{x+y} \cdot \frac{(x+y)^2}{8x}$

Solution $\frac{\overset{x^2}{\cancel{4x^3}}}{\underset{1}{\cancel{x+y}}} \cdot \frac{\overset{x+y}{\cancel{(x+y)^2}}}{\underset{2}{\cancel{8x}}} = \frac{x^2(x+y)}{2}$

CLASS EXERCISES

Multiply. Leave the resulting rational expression in factored form.

5. $\frac{-2}{a} \cdot \frac{x-1}{a^2}$ **6.** $\frac{x+y}{5x^2} \cdot \frac{15x^3}{(x+y)^3}$

FACTORING NUMERATOR OR DENOMINATOR

You may have to factor a numerator or denominator in order to divide by common factors.

Example 7

$$\frac{m+n}{m^2-3m-2} \cdot \frac{m-2}{m^2-n^2} = \frac{\overset{1}{\cancel{m+n}}}{\underset{1}{\cancel{(m-2)}}(m-1)} \cdot \frac{\overset{1}{\cancel{m-2}}}{\underset{1}{\cancel{(m+n)}}(m-n)}$$

$$= \frac{1}{(m-1)(m-n)}$$

Example 8

$$\frac{x^2-4}{x^2+5x+4}\cdot\frac{2x+8}{(x+2)^2}\cdot\frac{-1}{16x-32}=\frac{\overset{1}{\cancel{(x+2)}}\overset{1}{\cancel{(x-2)}}}{\underset{1}{\cancel{(x+4)}}(x+1)}\cdot\frac{\overset{1}{\cancel{2}}\overset{1}{\cancel{(x+4)}}}{\underset{x+2}{\cancel{(x+2)^2}}}\cdot\frac{-1}{\underset{8}{\cancel{16}}\underset{1}{\cancel{(x-2)}}}$$

$$=\frac{-1}{8(x+1)(x+2)}$$

CLASS EXERCISES *Multiply. Leave the resulting rational expression in factored form.*

7. $\dfrac{3a+3b}{x^2y^3}\cdot\dfrac{xy}{(a+b)^2}$ **8.** $\dfrac{x^2-1}{x^2+5x+6}\cdot\dfrac{x^2+4x+4}{2x-2}$

9. $\dfrac{1-x^2}{x^2+6x+9}\cdot\dfrac{x^2+2x-3}{x^2+7x+6}\cdot\dfrac{x^2-36}{x^2-2x+1}$

DIVISION OF FRACTIONS

Observe that

divisor (points to 4) *quotient* (points to 2)

$$8 \div 4 = 2$$
$$8 \div 2 = 4$$
$$8 \div 1 = 8$$

Each time the divisor is *divided by* 2, the quotient is *multiplied by* 2. According to this pattern:

$8 \div \frac{1}{2} = 16$ *Note that* $8 \cdot \frac{2}{1} = 16$.

$8 \div \frac{1}{4} = 32$ *Note that* $8 \cdot \frac{4}{1} = 32$.

Thus, to divide 8 by $\frac{1}{2}$, invert $\frac{1}{2}$ to obtain $\frac{2}{1}$, and then multiply

to divide 8 by $\frac{1}{4}$, invert $\frac{1}{4}$ to obtain $\frac{4}{1}$, and then multiply

To divide fractions or rational expressions,

$$\frac{a}{b}\div\frac{c}{d}$$

invert the divisor and multiply. Thus,

$$\boxed{\frac{a}{b}\div\frac{c}{d}=\frac{a}{b}\cdot\frac{d}{c}=\frac{ad}{bc}}$$

Example 9 Find $\frac{1}{2} \div \frac{1}{4}$.

Solution Invert the second fraction, $\frac{1}{4}$, and multiply.

$$\frac{1}{2} \div \frac{1}{4} = \frac{1}{2} \cdot \frac{4}{1} = 2$$

Example 10

$$\frac{-5}{6} \div \frac{10}{27} = \frac{-5}{6} \cdot \frac{27}{10} = \frac{-9}{4}$$

CLASS EXERCISES *Divide.* **10.** $\frac{1}{5} \div \frac{1}{3}$ **11.** $\frac{-4}{5} \div \frac{-12}{25}$

RATIONAL EXPRESSIONS The same procedure applies when you divide rational expressions. Invert the second expression and multiply.

Example 11 Find $\frac{1}{x} \div \frac{1}{x^2}$.

Solution Invert the second expression, $\frac{1}{x^2}$, and multiply.

$$\frac{1}{x} \div \frac{1}{x^2} = \frac{1}{x} \cdot \frac{x^2}{1} = x$$

Example 12 Find $\frac{x^2 + 2xy + y^2}{x^2y^3} \div \frac{x^2 - y^2}{x^4y^5}$.

Solution First, factor both numerators.

$$\frac{x^2 + 2xy + y^2}{x^2y^3} \div \frac{x^2 - y^2}{x^4y^5} = \frac{(x + y)^2}{x^2y^3} \div \frac{(x + y)(x - y)}{x^4y^5}$$

$$= \frac{(x + y)^2}{x^2y^3} \cdot \frac{x^4y^5}{(x + y)(x - y)}$$

$$= \frac{x^2y^2(x + y)}{x - y}$$

CLASS EXERCISES *Divide.* **12.** $\frac{1}{a} \div \frac{1}{b}$ **13.** $\frac{x^2y^3}{a-1} \div \frac{(a-1)^4}{xy^2}$ **14.** $\frac{a^2-4}{a^2x^2-a^2} \div \frac{a^2+3a+2}{ax+a}$

SOLUTIONS TO CLASS EXERCISES

1. $\frac{3}{4}\cdot\frac{3}{5}=\frac{3\cdot 3}{4\cdot 5}=\frac{9}{20}$

2. $\frac{\overset{1}{\cancel{5}}}{\underset{2}{\cancel{6}}}\cdot\frac{\overset{1}{\cancel{3}}}{\underset{2}{\cancel{10}}}=\frac{1}{4}$

3. $\frac{\overset{1}{\cancel{2}}}{\underset{1}{\cancel{3}}}\cdot\frac{\overset{\overset{1}{\cancel{3}}}{\cancel{9}}}{\underset{1}{\cancel{5}}}\cdot\frac{\overset{-1}{\cancel{-5}}}{\underset{\underset{2}{\cancel{6}}}{\cancel{12}}}=\frac{-1}{2}$

4. $\frac{2}{\underset{1}{\cancel{3}}}\cdot\frac{\overset{-3}{\cancel{-9}}}{1}=-6$

5. $\frac{-2}{a}\cdot\frac{x-1}{a^2}=\frac{-2(x-1)}{a^3}$, or $\frac{2(1-x)}{a^3}$

6. $\frac{\overset{1}{\cancel{x+y}}}{\underset{1\cdot 1}{\cancel{5x^2}}}\cdot\frac{\overset{3x}{\cancel{15x^3}}}{\underset{(x+y)^2}{\cancel{(x+y)^3}}}=\frac{3x}{(x+y)^2}$

7. $\frac{3a+3b}{x^2y^3}\cdot\frac{xy}{(a+b)^2}=\frac{3\overset{1}{\cancel{(a+b)}}}{\underset{xy^2}{\cancel{x^2y^3}}}\cdot\frac{\overset{1\cdot 1}{\cancel{xy}}}{\underset{a+b}{\cancel{(a+b)^2}}}=\frac{3}{xy^2(a+b)}$

8. $\frac{x^2-1}{x^2+5x+6}\cdot\frac{x^2+4x+4}{2x-2}=\frac{(x+1)\overset{1}{\cancel{(x-1)}}}{(x+3)\underset{1}{\cancel{(x+2)}}}\cdot\frac{\overset{x+2}{\cancel{(x+2)^2}}}{2\underset{1}{\cancel{(x-1)}}}=\frac{(x+1)(x+2)}{2(x+3)}$

9. $\frac{1-x^2}{x^2+6x+9}\cdot\frac{x^2+2x-3}{x^2+7x+6}\cdot\frac{x^2-36}{x^2-2x+1}=\frac{\overset{-1}{\cancel{(1-x)}}\overset{1}{\cancel{(1+x)}}}{\underset{x+3}{\cancel{(x+3)^2}}}\cdot\frac{\overset{1}{\cancel{(x+3)}}\overset{1}{\cancel{(x-1)}}}{\underset{1}{\cancel{(x+6)}}\underset{1}{\cancel{(x+1)}}}\cdot\frac{\overset{1}{\cancel{(x+6)}}(x-6)}{\underset{\underset{1}{\cancel{x-1}}}{\cancel{(x-1)^2}}}=\frac{6-x}{x+3}$

Note that $1-x=(-1)(x-1)$ and $6-x=(-1)(x-6)$.

10. $\frac{1}{5}\div\frac{1}{3}=\frac{1}{5}\cdot\frac{3}{1}=\frac{3}{5}$

11. $\frac{-4}{5}\div\frac{-12}{25}=\frac{\overset{1}{\cancel{-4}}}{\underset{1}{\cancel{5}}}\cdot\frac{\overset{5}{\cancel{25}}}{\underset{3}{\cancel{-12}}}=\frac{5}{3}$

12. $\frac{1}{a}\div\frac{1}{b}=\frac{1}{a}\cdot\frac{b}{1}=\frac{b}{a}$

13. $\frac{x^2y^3}{a-1}\div\frac{(a-1)^4}{xy^2}=\frac{x^2y^3}{a-1}\cdot\frac{xy^2}{(a-1)^4}=\frac{x^3y^5}{(a-1)^5}$

14. $\frac{a^2-4}{a^2x^2-a^2}\div\frac{a^2+3a+2}{ax+a}=\frac{(a+2)(a-2)}{a^2\underset{(x+1)(x-1)}{\cancel{(x^2-1)}}}\div\frac{(a+2)(a+1)}{a(x+1)}$

$$=\frac{\overset{1}{\cancel{(a+2)}}(a-2)}{\underset{a}{\cancel{a^2}}\underset{1}{\cancel{(x+1)}}(x-1)}\cdot\frac{\overset{1}{\cancel{a}}\overset{1}{\cancel{(x+1)}}}{\underset{1}{\cancel{(a+2)}}(a+1)}$$

$$=\frac{a-2}{a(a+1)(x-1)}$$

HOME EXERCISES

In Exercises 1–12, multiply the given fractions.

1. $\frac{1}{2} \cdot \frac{1}{3}$ **2.** $\frac{2}{3} \cdot \frac{1}{5}$ **3.** $\frac{1}{2} \cdot 3$ **4.** $\frac{-1}{3} \cdot 6$ **5.** $\frac{5}{8} \cdot \frac{2}{15}$ **6.** $\frac{3}{4} \cdot \frac{8}{9}$ **7.** $\frac{7}{10} \cdot \frac{25}{49}$ **8.** $\frac{5}{8} \cdot \frac{1}{10} \cdot \frac{2}{3}$

9. $\frac{-1}{7} \cdot \frac{-2}{5} \cdot \frac{-14}{3}$ **10.** $\frac{3}{4} \cdot \frac{49}{16} \cdot \frac{64}{21}$ **11.** $\frac{10}{7} \cdot \frac{5}{14} \cdot \frac{49}{25} \cdot \frac{2}{3}$ **12.** $\frac{7}{2} \cdot \frac{5}{8} \cdot \frac{24}{25} \cdot \frac{12}{7}$

In Exercises 13–26, multiply the given rational expressions. Leave the resulting rational expression in factored form.

13. $a \cdot \frac{x}{y}$ **14.** $\frac{a^2}{c} \cdot \frac{b}{c}$ **15.** $\frac{2a}{b} \cdot \frac{b^2}{a}$ **16.** $\frac{4xyz}{3ab} \cdot \frac{9ax}{8yz}$ **17.** $\frac{16x^2y}{5a^3b} \cdot \frac{25a^2b^2}{4xy^2}$

18. $\frac{a^2}{b} \cdot \frac{x}{a} \cdot \frac{b^2}{x}$ **19.** $\frac{2x}{3a} \cdot \frac{6a^2}{x^2} \cdot \frac{12a}{x}$ **20.** $\frac{5x+5}{x+2} \cdot \frac{1}{x+1}$

21. $\frac{5a+15b}{x^2y} \cdot \frac{xy^2}{a+3b}$ **22.** $\frac{x^2-1}{x^2+1} \cdot \frac{x^2+1}{x+1}$ **23.** $\frac{a^2-4}{3a-9} \cdot \frac{2a-6}{a^2+4a+4}$

24. $\frac{x^2-16}{2ax} \cdot \frac{4a^2}{3x+12} \cdot \frac{3x}{2x-8}$ **25.** $\frac{a^2-9}{4x^3} \cdot \frac{12x}{(a+3)^2} \cdot \frac{x^2}{a-3}$ **26.** $\frac{x^2-y^2}{4a^3} \cdot \frac{a}{y-x} \cdot \frac{16a^4}{x+y}$

27. Find $\frac{1}{2}$ of 48. **28.** Find $\frac{2}{5}$ of 60. **29.** Find $\frac{3}{4}$ of $\frac{8}{9}$. **30.** Find the product of $\frac{x}{x+1}$ and $\frac{x^2-1}{x^3}$.

In Exercises 31–40, divide the given fractions.

31. $\frac{1}{3} \div \frac{1}{6}$ **32.** $\frac{4}{5} \div \frac{8}{25}$ **33.** $\frac{7}{9} \div \frac{-7}{3}$ **34.** $\frac{2}{7} \div 4$ **35.** $5 \div \frac{1}{5}$ **36.** $\frac{15}{16} \div \frac{25}{4}$ **37.** $\frac{20}{7} \div \frac{10}{21}$

38. $\frac{121}{84} \div \frac{132}{49}$ **39.** $\frac{-55}{64} \div \frac{-125}{128}$ **40.** $\frac{81}{96} \div \frac{243}{144}$

In Exercises 41–54, divide the given rational expressions. Leave the resulting rational expression in factored form.

41. $\frac{1}{a} \div \frac{1}{b}$ **42.** $\frac{a}{b} \div a$ **43.** $\frac{x}{a^2} \div \frac{x}{a}$ **44.** $\frac{a}{b} \div \frac{b}{a}$ **45.** $\frac{ab}{c} \div \frac{a^2c}{b}$

46. $\frac{abc^3}{d} \div \frac{ab^3}{c^2d}$ **47.** $\frac{x-a}{2} \div \frac{(x-a)^2}{4}$ **48.** $\frac{a+1}{a-1} \div \frac{(a+1)^2}{(a-1)^2}$

49. $\frac{3x-3}{2} \div \frac{x-1}{4}$ **50.** $\frac{5a+5b}{a-b} \div \frac{a^2-b^2}{2}$ **51.** $\frac{x^2+3x+2}{ab^2x} \div \frac{ax+a}{bx}$

52. $\frac{x^2+7x+12}{x^2-16} \div \frac{x^2+6x+9}{2x-8}$ **53.** $\frac{a^2x^2-a^2y^2}{ax+ay} \div \frac{ax-ay}{a^2}$ **54.** $\frac{x^2-9x+14}{x^2-49} \div \frac{x^2-4}{x+7}$

55. Divide the product of $\frac{ax}{2}$ and $\frac{a^2y}{5}$ by $\frac{axy}{20}$.

56. Divide $\frac{x+y}{3ab}$ by the product of $\frac{x^2-y^2}{9a^2}$ and $\frac{x^2}{6b}$.

57. What fraction must be multiplied by $\frac{1}{3}$ to obtain $\frac{5}{6}$?

58. What fraction must be multiplied by $\frac{2}{3}$ to obtain $\frac{3}{4}$?

59. What rational expression must be multiplied by $\frac{1}{x}$ to obtain $\frac{3}{x^2}$?

60. What rational expression must be multiplied by $\frac{x}{x-a}$ to obtain $\frac{x^2y}{x^2-a^2}$?

5.2 LEAST COMMON MULTIPLES

Recall that ab is a multiple of both a and b. Thus, 12 is a multiple of both 3 and 4.

In order to add or subtract fractions or rational expressions with *different denominators* (Section 5.3), you will first have to write equivalent expressions with the *same denominator*. The new denominator will be a multiple of the given denominators.

lcm OF SEVERAL INTEGERS

DEFINITION

The **least common multiple (*lcm*) of two or more integers** is the smallest positive integer that is a multiple of each of them.

Write $lcm\ (2, 5)$ for the least common multiple of 2 and 5.

To find $lcm\ (2, 5)$, observe that $10 = 2 \cdot 5$

Thus, 10 is a multiple of both 2 and 5. None of the integers

$$1, 2, 3, 4, 5, 6, 7, 8, 9$$

is a multiple of *both* 2 and 5. For example, 4 is a multiple of 2, but not of 5.

Thus, $$lcm\ (2, 5) = 10$$

Example 1 Find $lcm\ (2, 6)$.

Solution Observe that 6 is a multiple of 2, and that 6 is a multiple of 6.

$$6 = 3 \cdot 2, \qquad 6 = 1 \cdot 6$$

By considering the integers 1, 2, 3, 4, 5, check that 6 is the *smallest* positive multiple of *both* 2 and 6. Thus,

$$lcm\ (2, 6) = 6$$

Sometimes, the least common multiple can be found almost immediately, as in Example 1. At other times, it is best to consider to prime factors of the integers.

> **To find the *lcm* of two of more integers:**
>
> 1. Express them in terms of prime factors.
> 2. The *lcm* is the product of the *highest* powers of all primes that occur.

Example 2 Find *lcm* (8, 12).

Solution Although this, too, can possibly be done at sight, it illustrates the prime factoring method.

$$8 = 2^3, \qquad 12 = 2^2 \cdot 3$$

The only primes that occur are 2 and 3. The highest power of 2 that occurs is 2^3. The only power of 3 that occurs is 3 itself. Thus,

$$lcm\ (8, 12) = 2^3 \cdot 3 = 8 \cdot 3 = 24$$

Note that the *lcm* is always a positive integer. Thus,

$$lcm\ (-8, -12) = 24$$

Example 3 Find *lcm* (40, 50, 64).

Solution

$$40 = 2^3 \cdot 5$$
$$50 = 2 \cdot 5^2$$
$$64 = 2^6$$

The highest power of 2 that occurs is 2^6. The highest power of 5 that occurs is 5^2. Thus,

$$\begin{aligned} lcm\ (40, 50, 64) &= 2^6 \cdot 5^2 \\ &= 64 \cdot 25 \\ &= 1600 \end{aligned}$$

CLASS EXERCISES *Find the lcm of the given integers.*

1. 4, 5 **2.** 3, 9 **3.** 24, 42 **4.** 25, 40, 60

lcm OF SEVERAL POLYNOMIALS

DEFINITION

> The **least common multiple of two or more polynomials** is the product of the highest powers of all "prime" factors that occur when you factor each of these polynomials.

Example 4 Find *lcm* $(2x^2y, 4xy^3)$.

Solution Write

$$4xy^3 = 2^2xy^3$$

The highest power of 2 that occurs is 2^2, or 4. The highest power of x that occurs is x^2. The highest power of y that occurs is y^3. Thus,

$$lcm\ (2x^2y, 4xy^3) = 4x^2y^3$$

Example 5 Find $lcm\ (x^2 - 1, x^2 + 2x + 1)$.

Solution Factor each polynomial.

$$x^2 - 1 = (x + 1)(x - 1)$$
$$x^2 + 2x + 1 = (x + 1)^2$$
$$lcm\ (x^2 - 1, x^2 + 2x + 1) = (x + 1)^2(x - 1)$$

Example 6 Find $lcm\ (9x + 27, x^2 + 2x - 3, 3x^2 - 6x + 3)$.

Solution Factor each polynomial.

$$9x + 27 = 9(x + 3) = 3^2(x + 3)$$
$$x^2 + 2x - 3 = (x + 3)(x - 1)$$
$$3x^2 - 6x + 3 = 3(x^2 - 2x + 1) = 3(x - 1)^2$$
$$lcm\ (9x + 27, x^2 + 2x - 3, 3x^2 - 6x + 3) = 3^2(x + 3)(x - 1)^2$$
$$= 9(x + 3)(x - 1)^2$$

CLASS EXERCISES *Find the lcm of the given polynomials.*

5. $3x^4y^2, 6xy^5$ **6.** $2x + 4, x^2 - 4$ **7.** $4x - 8, 4x^2 - 16, x^2 + 4x + 4$

LEAST COMMON DENOMINATORS

In order to add or subtract fractions or rational expressions with *different denominators*, first find equivalent expressions that have the same denominator. This denominator should be the *lcm* of the original denominators.

DEFINITION

> The **least common denominator (*lcd*) of several fractions or rational expressions** is the *lcm* of the individual denominators.

Write

$$lcd\left(\frac{1}{2}, \frac{2}{3}\right) \quad \text{for the least common denominator of } \frac{1}{2} \text{ and } \frac{2}{3}$$

Example 7 a. Find $lcd\left(\frac{1}{2}, \frac{2}{3}\right)$.

b. Write equivalent fractions that have this *lcd* as denominator.

Solution a. To find the *lcd*, determine the *lcm* of the denominators.

$$lcm\,(2, 3) = 6$$

Therefore, $$lcd\left(\frac{1}{2}, \frac{2}{3}\right) = 6$$

b. Recall that when the numerator and denominator of a fraction are each multiplied by the same number, an equivalent fraction is obtained. To obtain equivalent fractions that have denominator 6, multiply numerator and denominator of each fraction as indicated.

$$\frac{1}{2} = \frac{1 \cdot 3}{2 \cdot 3} = \frac{3}{6}$$

$$\frac{2}{3} = \frac{2 \cdot 2}{3 \cdot 2} = \frac{4}{6}$$

Example 8 a. Find $lcd\left(\frac{5}{12}, \frac{-1}{18}\right)$.

b. Write equivalent fractions that have this *lcd* as denominator.

Solution a.

$$lcd\left(\frac{5}{12}, \frac{-1}{18}\right) = lcm\,(12, 18)$$

$$12 = 2^2 \cdot 3, \qquad 18 = 2 \cdot 3^2$$

$$lcm\,(12, 18) = 2^2 \cdot 3^2 = 4 \cdot 9 = 36$$

Therefore, $$lcd\left(\frac{5}{12}, \frac{-1}{18}\right) = 36$$

b.

$$\frac{5}{12} = \frac{5 \cdot 3}{12 \cdot 3} = \frac{15}{36}$$

$$\frac{-1}{18} = \frac{-1 \cdot 2}{18 \cdot 2} = \frac{-2}{36}$$

CLASS EXERCISES a. Find the *lcd*, and b. write equivalent fractions that have this *lcd* as denominator.

8. $\frac{1}{2}$ and $\frac{1}{5}$

9. $\frac{9}{10}$ and $\frac{4}{15}$

lcd OF RATIONAL EXPRESSIONS

Example 9 a. Find $lcd\left(\frac{1}{x}, \frac{5a}{y^2}\right)$.

b. Write equivalent rational expressions that have this *lcd* as denominator.

Solution a. As in the case of fractions, find the *lcm* of the denominators x and y^2.

$$lcm\,(x, y^2) = xy^2$$

Therefore, $$lcd\left(\frac{1}{x}, \frac{5a}{y^2}\right) = xy^2$$

b. To find equivalent rational expressions that have denominator xy^2, multiply numerator and denominator of each expression as indicated.

$$\frac{1}{x} = \frac{1 \cdot y^2}{x \cdot y^2} = \frac{y^2}{xy^2}$$

$$\frac{5a}{y^2} = \frac{5a \cdot x}{y^2 \cdot x} = \frac{5ax}{xy^2}$$

Example 10 a. Find $lcd\left(\frac{x}{x^2 - a^2}, \frac{a}{x^2 + 2ax + a^2}\right)$.

b. Write equivalent expressions that have this *lcd* as denominator.

Solution a. First, factor the given denominators to find their *lcm*.

$$x^2 - a^2 = (x + a)(x - a)$$

$$x^2 + 2ax + a^2 = (x + a)^2$$

$$lcm\,(x^2 - a^2, x^2 + 2ax + a^2) = (x + a)^2(x - a)$$

Thus, $$lcd\left(\frac{x}{x^2 - a^2}, \frac{a}{x^2 + 2ax + a^2}\right) = (x + a)^2(x - a)$$

b.
$$\frac{x}{x^2 - a^2} = \frac{x}{(x + a)(x - a)} = \frac{x \cdot (x + a)}{(x + a)(x - a) \cdot (x + a)} = \frac{x^2 + ax}{(x + a)^2(x - a)}$$

$$\frac{a}{x^2 + 2ax + a^2} = \frac{a}{(x + a)^2} = \frac{a \cdot (x - a)}{(x + a)^2 \cdot (x - a)} = \frac{ax - a^2}{(x + a)^2(x - a)}$$

As you will see, *when adding rational expressions, you must usually multiply out the numerators before adding*.

CLASS EXERCISES

a. Find the *lcd*, and
b. write equivalent rational expressions that have this *lcd* as denominator.

10. $\frac{1}{a}$ and $\frac{2}{b}$ **11.** $\frac{a}{xy^2}$ and $\frac{b}{x^2y}$ **12.** $\frac{3}{x^2-1}$ and $\frac{-1}{x^2+4x+3}$

SOLUTIONS TO CLASS EXERCISES

1. 20 **2.** 9 **3.** $lcm\,(24, 42) = 2^3 \cdot 3 \cdot 7 = 168$ **4.** $lcm\,(25, 40, 60) = 2^3 \cdot 3 \cdot 5^2 = 600$ **5.** $6x^4y^5$

6. $2x + 4 = 2(x+2)$, $x^2 - 4 = (x+2)(x-2)$
$lcm\,(2x+4, x^2-4) = 2(x+2)(x-2)$

7. $4x - 8 = 4(x-2)$, $4x^2 - 16 = 4(x^2-4) = 4(x+2)(x-2)$, $x^2 + 4x + 4 = (x+2)^2$
$lcm\,(4x-8, 4x^2-16, x^2+4x+4) = 4(x+2)^2(x-2)$

8. a. 10 **b.** $\frac{1}{2} = \frac{1 \cdot 5}{2 \cdot 5} = \frac{5}{10}$, $\frac{1}{5} = \frac{1 \cdot 2}{5 \cdot 2} = \frac{2}{10}$ **9. a.** 30 **b.** $\frac{9}{10} = \frac{9 \cdot 3}{10 \cdot 3} = \frac{27}{30}$, $\frac{4}{15} = \frac{4 \cdot 2}{15 \cdot 2} = \frac{8}{30}$

10. a. ab **b.** $\frac{1}{a} = \frac{1 \cdot b}{a \cdot b} = \frac{b}{ab}$, $\frac{2}{b} = \frac{2 \cdot a}{b \cdot a} = \frac{2a}{ab}$

11. a. x^2y^2 **b.** $\frac{a}{xy^2} = \frac{a \cdot x}{xy^2 \cdot x} = \frac{ax}{x^2y^2}$, $\frac{b}{x^2y} = \frac{b \cdot y}{x^2y \cdot y} = \frac{by}{x^2y^2}$

12. a. $(x+3)(x+1)(x-1)$

b. $\frac{3}{x^2-1} = \frac{3 \cdot (x+3)}{(x+1)(x-1) \cdot (x+3)} = \frac{3x+9}{(x+3)(x+1)(x-1)}$,

$\frac{-1}{x^2+4x+3} = \frac{-1 \cdot (x-1)}{(x+3)(x+1) \cdot (x-1)} = \frac{1-x}{(x+3)(x+1)(x-1)}$

HOME EXERCISES

In Exercises 1–14, find the lcm of the given integers.

1. 2, 3 **2.** 2, 8 **3.** −3, −12 **4.** 4, 6 **5.** −10, 15 **6.** 12, 30 **7.** 12, 27
8. 50, 125 **9.** 2, 3, 5 **10.** 4, 8, 16 **11.** 2, 3, 4 **12.** 6, 9, 16 **13.** 50, 75, 100 **14.** 48, 64, 72

In Exercises 15–36, find the lcm of the given polynomials.

15. $5, x$ **16.** x^2, y^2 **17.** x, x^2 **18.** $x + 1, (x+1)^2$ **19.** $a, a + b$ **20.** a^2x, ax^2 **21.** $4a^2b, 8abc$
22. $4axy, 6a^2bx$ **23.** $12mn^2, 8m^2n$ **24.** $(x+a)^2, (x+a)(x-a)$
25. $x^2 - 4, x + 2$ **26.** $y^2 - 9, (y-3)^2$ **27.** $a^2 - 25, a^2 + 10a + 25$
28. $b^2 + 3b + 2, b^2 + 5b + 6$ **29.** $4x - 4, 8x^2 - 16x + 8$ **30.** $z^2 - 100, 7z - 70$
31. $abx - aby, a^2b^2$ **32.** $t^2 + 6t + 8, (t+2)^3$ **33.** $abx^2y^3, a^2b^7xy, ax^4yz$
34. $(x-1)^7(x+1), 4x + 4, (x-1)^9$ **35.** $a^2 - 4, 5a - 10, 25a + 50$ **36.** $a^2x^2 - a^2y^2, 4x + 4y, 6x - 6y$

In Exercises 37–56, (a) find the lcd, and (b) write equivalent fractions or rational expressions with this lcd as denominator.

37. $\frac{1}{3}$ and 1 **38.** $\frac{2}{3}$ and $\frac{-1}{5}$ **39.** $\frac{1}{2}$ and $\frac{5}{6}$ **40.** $\frac{-3}{4}$ and $\frac{-1}{12}$ **41.** $\frac{5}{4}$ and $\frac{7}{10}$ **42.** $\frac{3}{8}$ and $\frac{5}{12}$

43. $\frac{7}{20}$ and $\frac{3}{28}$ **44.** $\frac{1}{44}$ and $\frac{3}{40}$ **45.** $\frac{5}{36}$ and $\frac{1}{27}$ **46.** $\frac{1}{8}, \frac{-1}{12}$, and $\frac{2}{15}$ **47.** $\frac{1}{x}$ and $\frac{1}{y}$ **48.** $\frac{1}{a^2}$ and $\frac{a}{b}$

49. $\frac{2}{x+a}$ and -1 **50.** $\frac{a}{x^2y}$ and $\frac{b}{xy^2z}$ **51.** $\frac{x-a}{x^2y^3z^2}$ and $\frac{a}{xy^4z}$

52. $\dfrac{2}{x(x-a)}$ and $\dfrac{4}{x(x+a)}$

53. $\dfrac{1}{x^2-1}$ and $\dfrac{x}{(x-1)^2}$

54. $\dfrac{3}{x^2+6x+9}$ and $\dfrac{-1}{x^2+4x+3}$

55. $\dfrac{1}{ax-2a}$, $\dfrac{a}{x^2-4}$, and $\dfrac{-1}{ax+2a}$

56. $\dfrac{a}{x^2-25}$, $\dfrac{b}{(x+5)^2}$, and $\dfrac{c}{ax+5a}$

5.3 ADDITION AND SUBTRACTION OF RATIONAL EXPRESSIONS

Figure 5.2 illustrates how fractions that have the *same denominator* are added.

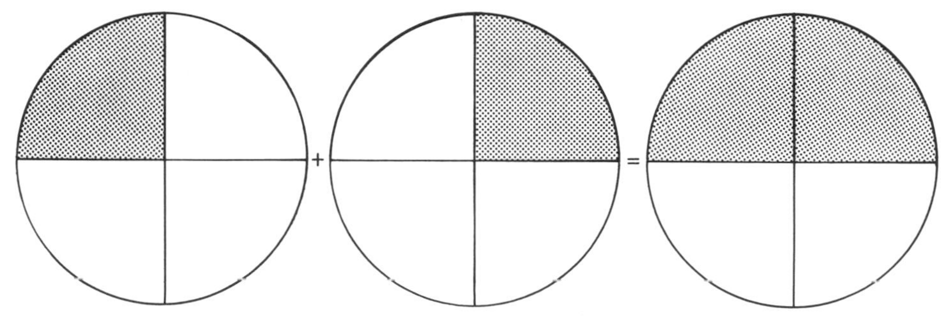

Fig. 5.2. $\frac{1}{4}+\frac{1}{4}=\frac{1+1}{4}=\frac{2}{4}=\frac{1}{2}.$

> **To add or subtract fractions or rational expressions that have the *same denominator D:***
> **1.** Add or subtract the numerators.
> **2.** The denominator is *D*.
> **3.** Simplify the resulting expression, if possible.

Fractions

Example 1

$$\frac{1}{6}+\frac{1}{6}=\frac{1+1}{6}=\frac{2}{6}$$

Simplify the sum to $\frac{1}{3}$.

Example 2

$$\frac{1}{10}+\frac{7}{10}-\frac{3}{10}=\frac{1+7-3}{10}=\frac{5}{10}=\frac{1}{2}$$

CLASS EXERCISES *Add or subtract, and simplify.*

1. $\frac{1}{9}+\frac{1}{9}$ **2.** $\frac{1}{8}+\frac{3}{8}$ **3.** $\frac{7}{12}-\left(\frac{1}{12}+\frac{5}{12}\right)$

RATIONAL EXPRESSIONS

Example 3

$$\frac{x}{yz} + \frac{1}{yz} - \frac{2}{yz} = \frac{x + 1 - 2}{yz} = \frac{x - 1}{yz}$$

CLASS EXERCISES *Add or subtract, and simplify.*

4. $\frac{1}{a} + \frac{5}{a}$ **5.** $\frac{3}{st^2} - \frac{a}{st^2} + \frac{1}{st^2}$

FACTORING AND SIMPLIFYING

Factoring the numerator or denominator of the sum or difference may simplify the result.

Example 4

$$\frac{2x}{x+1} + \frac{2}{x+1} = \frac{2x+2}{x+1} = \frac{2\overset{1}{\cancel{(x+1)}}}{\underset{1}{\cancel{x+1}}} = 2$$

Example 5

$$\frac{x}{x^2-4} - \frac{2}{x^2-4} = \frac{x-2}{x^2-4} = \frac{\overset{1}{\cancel{x-2}}}{(x+2)\underset{1}{\cancel{(x-2)}}} = \frac{1}{x+2}$$

CLASS EXERCISES *Add.* **6.** $\frac{5a}{a+2} + \frac{10}{a+2}$ **7.** $\frac{m^2}{m+4} + \frac{5m}{m+4} + \frac{4}{m+4}$

COMBINING FRACTIONS

> **To add or subtract fractions or rational expressions that have *different denominators:***
>
> 1. First, find their *lcd*.
> 2. Find equivalent expressions that have this *lcd* as denominator.
> 3. Add or subtract the numerators.
> 4. Simplify the resulting expression, if possible.

Steps (1) and (2) are what you found in Section 5.2. Steps (3) and (4) are the rules for adding expressions with the same denominator.

Example 6 Add. $\frac{3}{4} + \frac{5}{8} + \frac{1}{8}$

Solution

$$lcd\left(\frac{3}{4}, \frac{5}{8}, \frac{1}{8}\right) = lcm\,(4, 8) = 8$$

$$\frac{3}{4} = \frac{3 \cdot 2}{4 \cdot 2} = \frac{6}{8} \qquad \left(\frac{5}{8} \text{ and } \frac{1}{8} \text{ already have denominator 8.}\right)$$

$$\frac{3}{4} + \frac{5}{8} + \frac{1}{8} = \frac{6}{8} + \frac{5}{8} + \frac{1}{8} = \frac{6 + 5 + 1}{8} = \frac{12}{8} = \frac{3}{2}$$

Example 7 Find the value of: $\frac{5}{24} + \frac{1}{18} - \frac{7}{36}$

Solution

$$24 = 2^3 \cdot 3, \qquad 18 = 2 \cdot 3^2, \qquad 36 = 2^2 \cdot 3^2$$

$$lcd\left(\frac{5}{24}, \frac{1}{18}, \frac{7}{36}\right) = lcm\,(24, 18, 36) = 2^3 \cdot 3^2 = 8 \cdot 9 = 72$$

$$\frac{5}{24} = \frac{5 \cdot 3}{24 \cdot 3} = \frac{15}{72}, \qquad \frac{1}{18} = \frac{1 \cdot 4}{18 \cdot 4} = \frac{4}{72}, \qquad \frac{7}{36} = \frac{7 \cdot 2}{36 \cdot 2} = \frac{14}{72}$$

$$\frac{5}{24} + \frac{1}{18} - \frac{7}{36} = \frac{15}{72} + \frac{4}{72} - \frac{14}{72} = \frac{15 + 4 - 14}{72} = \frac{5}{72}$$

CLASS EXERCISES *Add or subtract.* **8.** $\frac{2}{5} - \frac{1}{10}$ **9.** $\frac{7}{20} + \frac{3}{28}$ **10.** $\frac{1}{4} - \left(\frac{3}{8} - \frac{5}{6}\right)$

COMBINING RATIONAL EXPRESSIONS

Example 8 Add. $\frac{1}{ax^2} + \frac{2}{a^2x}$

Solution

$$lcd\left(\frac{1}{ax^2}, \frac{2}{a^2x}\right) = lcm\,(ax^2, a^2x) = a^2x^2$$

$$\frac{1}{ax^2} = \frac{1 \cdot a}{ax^2 \cdot a} = \frac{a}{a^2x^2}$$

$$\frac{2}{a^2x} = \frac{2 \cdot x}{a^2x \cdot x} = \frac{2x}{a^2x \cdot x}$$

$$\frac{1}{ax^2} + \frac{2}{a^2x} = \frac{a}{a^2x^2} + \frac{2x}{a^2x^2} = \frac{a + 2x}{a^2x^2}$$

Example 9 Subtract. $\frac{2}{x - 4} - \frac{1}{x + 4}$

Solution

$$lcd\left(\frac{2}{x - 4}, \frac{1}{x + 4}\right) = lcm\,(x - 4, x + 4) = (x - 4)(x + 4), \quad \text{or} \quad x^2 - 16$$

$$\frac{2}{x - 4} = \frac{2(x + 4)}{(x - 4)(x + 4)} = \frac{2x + 8}{x^2 - 16}$$

$$\frac{1}{x + 4} = \frac{1(x - 4)}{(x + 4)(x - 4)} = \frac{x - 4}{x^2 - 16}$$

$$\frac{2}{x - 4} - \frac{1}{x + 4} = \frac{2x + 8}{x^2 - 16} - \frac{x - 4}{x^2 - 16}$$

$$= \frac{2x + 8 - x + 4}{x^2 - 16}$$

$$= \frac{x + 12}{x^2 - 16}$$

Example 10 Combine. $\frac{1}{ax} - \left(\frac{x}{a^2} + \frac{a^2}{x^2}\right)$

Solution

$$lcd\left(\frac{1}{ax}, \frac{x}{a^2}, \frac{a^2}{x^2}\right) = lcm\ (ax, a^2, x^2) = a^2x^2$$

$$\frac{1}{ax} - \left(\frac{x}{a^2} + \frac{a^2}{x^2}\right) = \frac{1}{ax} - \frac{x}{a^2} - \frac{a^2}{x^2}$$

$$= \frac{1 \cdot ax}{ax \cdot ax} - \frac{x \cdot x^2}{a^2 \cdot x^2} - \frac{a^2 \cdot a^2}{x^2 \cdot a^2}$$

$$= \frac{ax - x^3 - a^4}{a^2x^2}$$

CLASS EXERCISES *Combine.*

11. $\frac{a}{x^2} - \frac{b}{xy}$

12. $\frac{5}{a-1} + \frac{1}{a^2-1}$

13. $\frac{1}{x^2+5x+4} + \frac{1}{x^2-16}$

14. $\frac{1}{x+2} - \left(\frac{1}{x^2-4} - \frac{x}{x-2}\right)$

15. $\frac{x^2}{a}\left(\frac{1}{a} - \frac{1}{x}\right)$

SOLUTIONS TO CLASS EXERCISES

1. $\frac{1}{9} + \frac{1}{9} = \frac{1+1}{9} = \frac{2}{9}$

2. $\frac{1}{8} + \frac{3}{8} = \frac{1+3}{8} = \frac{4}{8} = \frac{1}{2}$

3. $\frac{7}{12} - \left(\frac{1}{12} + \frac{5}{12}\right) = \frac{7}{12} - \frac{1}{12} - \frac{5}{12} = \frac{7-1-5}{12} = \frac{1}{12}$

4. $\frac{1}{a} + \frac{5}{a} = \frac{1+5}{a} = \frac{6}{a}$

5. $\frac{3}{st^2} - \frac{a}{st^2} + \frac{1}{st^2} = \frac{3-a+1}{st^2} = \frac{4-a}{st^2}$

6. $\frac{5a}{a+2} + \frac{10}{a+2} = \frac{5a+10}{a+2} = \frac{5\overset{1}{\cancel{(a+2)}}}{\underset{1}{\cancel{a+2}}} = 5$

7. $\frac{m^2}{m+4} + \frac{5m}{m+4} + \frac{4}{m+4} = \frac{m^2+5m+4}{m+4} = \frac{\overset{1}{\cancel{(m+4)}}(m+1)}{\underset{1}{\cancel{m+4}}} = m + 1$

8. $lcd\left(\frac{2}{5}, \frac{1}{10}\right) = lcm\ (5, 10) = 10$

$\frac{2}{5} = \frac{2 \cdot 2}{5 \cdot 2} = \frac{4}{10}$

$\frac{2}{5} + \frac{1}{10} = \frac{4}{10} + \frac{1}{10} = \frac{4+1}{10} = \frac{5}{10} = \frac{1}{2}$

9. $lcd\left(\frac{7}{20}, \frac{3}{28}\right) = lcm\ (20, 28) = 2^2 \cdot 5 \cdot 7 = 140$

$\frac{7}{20} = \frac{7 \cdot 7}{20 \cdot 7} = \frac{49}{140}, \quad \frac{3}{28} = \frac{3 \cdot 5}{28 \cdot 5} = \frac{15}{140}$

$\frac{7}{20} + \frac{3}{28} = \frac{49}{140} + \frac{15}{140} = \frac{49+15}{140} = \frac{64}{140} = \frac{16}{35}$

10. $lcd\left(\frac{1}{4}, \frac{3}{8}, \frac{5}{6}\right) = lcm\ (4, 8, 6) = 2^3 \cdot 3 = 24$

$\frac{1}{4} = \frac{1 \cdot 6}{4 \cdot 6} = \frac{6}{24}, \quad \frac{3}{8} = \frac{3 \cdot 3}{8 \cdot 3} = \frac{9}{24}, \quad \frac{5}{6} = \frac{5 \cdot 4}{6 \cdot 4} = \frac{20}{24}$

$\frac{1}{4} - \left(\frac{3}{8} - \frac{5}{6}\right) = \frac{1}{4} - \frac{3}{8} + \frac{5}{6} = \frac{6}{24} - \frac{9}{24} + \frac{20}{24} = \frac{6 - 9 + 20}{24} = \frac{17}{24}$

11. $lcd\left(\frac{a}{x^2}, \frac{b}{xy}\right) = lcm\ (x^2, xy) = x^2 y$

$\frac{a}{x^2} = \frac{a \cdot y}{x^2 \cdot y} = \frac{ay}{x^2 y}, \quad \frac{b}{xy} = \frac{b \cdot x}{xy \cdot x} = \frac{bx}{x^2 y}$

$\frac{a}{x^2} - \frac{b}{xy} = \frac{ay}{x^2 y} - \frac{bx}{x^2 y} = \frac{ay - bx}{x^2 y}$

12. $lcd\left(\frac{5}{a - 1}, \frac{1}{a^2 - 1}\right) = lcm\ (a - 1, a^2 - 1) = (a + 1)(a - 1)$, or $a^2 - 1$

$\frac{5}{a - 1} = \frac{5 \quad (a + 1)}{(a - 1)(a + 1)} = \frac{5a + 5}{a^2 - 1}$

$\frac{5}{a - 1} + \frac{1}{a^2 - 1} = \frac{5a + 5}{a^2 - 1} + \frac{1}{a^2 - 1} = \frac{5a + 5 + 1}{a^2 - 1} = \frac{5a + 6}{a^2 - 1}$

13. $lcd\left(\frac{1}{x^2 + 5x + 4}, \frac{1}{x^2 - 16}\right) = lcm\ (x^2 + 5x + 4, x^2 - 16) = (x + 4)(x + 1)(x - 4)$

$\frac{1}{x^2 + 5x + 4} = \frac{1 \quad \cdot (x - 4)}{(x + 4)(x + 1) \cdot (x - 4)} = \frac{x - 4}{(x + 4)(x + 1)(x - 4)}$

$\frac{1}{x^2 - 16} = \frac{1 \quad \cdot (x + 1)}{(x + 4)(x - 4) \cdot (x + 1)} = \frac{x + 1}{(x + 4)(x + 1)(x - 4)}$

$\frac{1}{x^2 + 5x + 4} + \frac{1}{x^2 - 16} = \frac{x - 4}{(x + 4)(x + 1)(x - 4)} + \frac{x + 1}{(x + 4)(x + 1)(x - 4)}$

$= \frac{x - 4 + x + 1}{(x + 4)(x + 1)(x - 4)} = \frac{2x - 3}{(x + 4)(x + 1)(x - 4)}$

14. $lcd\left(\frac{1}{x + 2}, \frac{1}{x^2 - 4}, \frac{x}{x - 2}\right) = lcm\left(x + 2, (x + 2)(x - 2), x - 2\right) = (x + 2)(x - 2)$

$\frac{1}{x + 2} - \left(\frac{1}{x^2 - 4} - \frac{x}{x - 2}\right) = \frac{1}{x + 2} - \frac{1}{x^2 - 4} + \frac{x}{x - 2}$

$= \frac{1 \quad \cdot (x - 2)}{x + 2 \cdot (x - 2)} - \frac{1}{(x + 2)(x - 2)} + \frac{x \quad \cdot (x + 2)}{x - 2 \cdot (x + 2)}$

$= \frac{x - 2 - 1 + x^2 + 2x}{(x + 2)(x - 2)}$

$= \frac{x^2 + 3x - 3}{(x + 2)(x - 2)}$

15. First combine within parentheses.

$lcd\left(\frac{1}{a}, \frac{1}{x}\right) = lcm\ (a, x) = ax$

$\frac{x^2}{a}\left(\frac{1}{a} - \frac{1}{x}\right) = \frac{x^2}{a}\left(\frac{1 \cdot x}{a \cdot x} - \frac{1 \cdot a}{x \cdot a}\right) = \frac{\overset{x}{\cancel{x^2}}}{a}\left(\frac{x - a}{\underset{a}{a\cancel{x}}}\right) = \frac{x(x - a)}{a^2 x}$

HOME EXERCISES

In Exercises 1–58, add or subtract, and simplify.

1. $\frac{2}{7}+\frac{1}{7}$
2. $\frac{2}{3}-\frac{1}{3}$
3. $\frac{3}{4}+\frac{1}{4}$
4. $\frac{5}{8}-\frac{1}{8}$
5. $\frac{1}{6}+\frac{1}{6}+\frac{3}{6}$
6. $\frac{1}{5}+\frac{2}{5}+\frac{2}{5}$
7. $\frac{7}{9}-\frac{4}{9}+\frac{5}{9}$
8. $\frac{7}{9}-\left(\frac{4}{9}+\frac{5}{9}\right)$
9. $\frac{7}{10}-\left(\frac{3}{10}+\frac{1}{10}-\frac{9}{10}\right)$
10. $\frac{11}{20}-\left(\frac{3}{20}-\frac{1}{20}+\frac{7}{20}\right)$
11. $\frac{5}{x}+\frac{2}{x}$
12. $\frac{4}{y}-\frac{3}{y}$
13. $\frac{10}{x+1}+\frac{1}{x+1}$
14. $\frac{5x}{x+3}-\frac{2x}{x+3}$
15. $\frac{6}{a}+\frac{1}{a}+\frac{2}{a}$
16. $\frac{3}{b+1}+\frac{5}{b+1}-\frac{2}{b+1}$
17. $\frac{x}{x+2}+\frac{2}{x+2}$
18. $\frac{a}{a-4}-\frac{4}{a-4}$
19. $\frac{3a}{a+1}+\frac{3}{a+1}$
20. $\frac{6x}{x-1}-\frac{6}{x-1}$
21. $\frac{1}{x-1}-\frac{x}{x-1}$
22. $\frac{a^2}{x^2-a^2}-\frac{x^2}{x^2-a^2}$
23. $\frac{x}{x^2-4}-\frac{2}{x^2-4}$
24. $\frac{a}{a^2-9}-\frac{3}{a^2-9}$
25. $\frac{b^2}{b+1}+\frac{1}{b+1}$
26. $\frac{4m^2}{m+2}-\frac{16}{m+2}$
27. $\frac{y^2}{y+1}+\frac{2y}{y+1}+\frac{1}{y+1}$
28. $\frac{a^2}{a+2}+\frac{5a}{a+2}+\frac{6}{a+2}$
29. $\frac{1}{2}+\frac{1}{3}$
30. $\frac{1}{4}-\frac{1}{5}$
31. $\frac{1}{2}+\frac{1}{4}$
32. $\frac{3}{4}-\frac{1}{12}$
33. $\frac{7}{8}+\frac{1}{12}$
34. $\frac{7}{10}+\frac{1}{25}$
35. $\frac{3}{64}-\frac{1}{48}$
36. $\frac{13}{40}-\frac{3}{100}$
37. $\frac{1}{6}+\frac{2}{9}+1$
38. $\frac{5}{8}-\frac{1}{4}+\frac{1}{16}$
39. $\frac{1}{10}+\frac{7}{5}-\frac{2}{25}$
40. $\frac{10}{21}-\left(\frac{3}{7}+\frac{2}{9}\right)$
41. $\frac{1}{x}+\frac{1}{y}$
42. $2-\frac{1}{b}$
43. $\frac{a}{x}+\frac{a}{x^2}$
44. $\frac{a}{s^2t}-\frac{b}{st^3}$
45. $\frac{1}{x+a}+\frac{2}{(x+a)^2}$
46. $\frac{4}{y-b}-\frac{1}{y+b}$
47. $\frac{1}{at+a}-\frac{a}{t+1}$
48. $\frac{3}{x+5}+\frac{1}{x^2+10x+25}$
49. $\frac{x}{x^2-4}-\frac{1}{x^2-2x}$
50. $\frac{a}{x^2+5x+6}-\frac{a}{x^2-4}$
51. $\frac{1}{a^2-9a}+\frac{9}{a^2-81}$
52. $\frac{1}{x^2-x}+\frac{2}{x^2-x^3}$
53. $\frac{1}{x-1}+\frac{1}{x-2}+\frac{1}{x-3}$
54. $\frac{5}{a^2-25}-\left(\frac{1}{a^2+10a+25}-\frac{1}{a^2-10a+25}\right)$
55. $\frac{1}{a^2}\left(\frac{x}{a^2-4a}+\frac{1}{a^2-16}\right)$
56. $\frac{1}{x^2+6x+9}-\left(\frac{1}{x+3}+\frac{1}{x^2-9}\right)$

Hint for 57 and 58: Multiply the numerator and denominator of *one* of these rational expressions by -1.

57. $\frac{4}{x-1}+\frac{3}{1-x}$
58. $\frac{a}{a-3}-\frac{1}{3-a}$

In Exercises 59–64, combine.

59. $\frac{3}{4}\left(\frac{1}{5}+\frac{3}{5}\right)$
60. $\frac{2}{3}\left(\frac{3}{8}-\frac{3}{4}\right)$
61. $\left(\frac{3}{10}+\frac{2}{5}-\frac{1}{2}\right)\div\frac{2}{15}$
62. $\frac{x}{y}\left(\frac{1}{y^2}+\frac{6}{y^2}\right)$
63. $(x^2-a^2)\left(\frac{1}{x+a}-\frac{1}{x-a}\right)$
64. $\frac{x+1}{3x+12}\div\left(\frac{x}{x+4}+\frac{1}{(x+4)^2}\right)$

5.4 ORDER OF FRACTIONS

Figure 5.3 indicates that

$$\frac{1}{3} < \frac{2}{3} \quad \text{because} \quad \frac{1}{3} \text{ lies to the left of } \frac{2}{3}$$

SAME DENOMINATOR

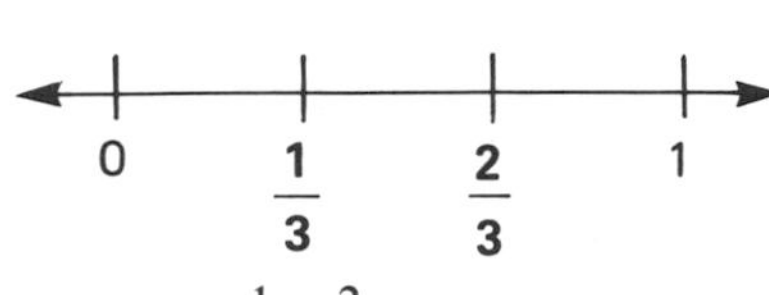

Fig. 5.3. $\frac{1}{3} < \frac{2}{3}$.

First, consider fractions whose numerators and denominators are positive. If two such fractions have the same denominator, the smaller fraction is the one with the smaller numerator. Thus, let a, A, and b be positive. Then,

$$\boxed{\frac{a}{b} < \frac{A}{b}, \text{ if } a < A}$$

Example 1

$$\frac{2}{5} < \frac{4}{5} \quad \text{because} \quad 2 < 4$$

CLASS EXERCISES *Fill in "<" or ">".*

1. $\frac{1}{6} \square \frac{5}{6}$ **2.** $\frac{8}{9} \square \frac{7}{9}$

DIFFERENT DENOMINATORS

When two fractions have different denominators, first express the fractions with the same denominator (usually, the lcd). Then, compare the new numerators.

Example 2 Fill in "<" *or* ">".

$$\frac{2}{3} \square \frac{3}{5}$$

Solution

$$lcd\left(\frac{2}{3}, \frac{3}{15}\right) = lcm\ (3, 5) = 15$$

Express both fractions with denominator 15.

$$\frac{2}{3} = \frac{2 \cdot 5}{3 \cdot 5} = \frac{10}{15}$$

$$\frac{3}{5} = \frac{3 \cdot 3}{5 \cdot 3} = \frac{9}{15}$$

Example 3 Fill in "<" or " >".

$$\frac{5}{16} \square \frac{7}{20}$$

Solution Here, instead of finding the *lcd*, it is easier to use 16 · 20, *the product of the given denominators*, as a common denominator.

$$\frac{5}{16} = \frac{5 \cdot 20}{16 \cdot 20} = \frac{100}{16 \cdot 20}$$

$$\frac{7}{20} = \frac{7 \cdot 16}{20 \cdot 16} = \frac{112}{16 \cdot 20}$$

Thus,

$$\frac{\mathbf{10}}{15} > \frac{\mathbf{9}}{15} \text{ because } 10 > 9$$

Therefore, $\frac{2}{3} > \frac{3}{5}$

$$\frac{\mathbf{100}}{16 \cdot 20} < \frac{\mathbf{112}}{16 \cdot 20}$$

because $100 < 112$

Therefore, $\frac{5}{16} < \frac{7}{20}$

CLASS EXERCISES *Fill in "<" or ">".*

3. $\frac{8}{5} \square \frac{7}{6}$ **4.** $\frac{5}{9} \square \frac{7}{12}$ **5.** $\frac{13}{4} \square 3$ (*Hint*: Express $\frac{3}{1}$ with denominator 4.)

NEGATIVE FRACTIONS To apply the previous rules to *negative* fractions, first express each fraction with a *positive denominator*. Then, express each fraction with the *lcd* as denominator.

Example 4 Fill in "<" or ">". **a.** $-\frac{1}{4} \square \frac{3}{-4}$ **b.** $\frac{-3}{10} \square \frac{-2}{7}$

Solution **a.**

$$-\frac{1}{4} = \frac{-1}{4}, \quad \frac{3}{-4} = \frac{-3}{4}$$

$$\frac{\mathbf{-1}}{4} > \frac{\mathbf{-3}}{4} \quad \text{because} \quad -1 > -3$$

-1   -3/4   -1/4   0

Fig. 5.4. $\frac{-3}{4} < \frac{-1}{4}$.

(See Figure 5.4.) Thus, $-\frac{1}{4} > \frac{3}{-4}$

Note that $\frac{-1}{4} = \frac{1}{-4}$, and $\frac{1}{-4} > \frac{3}{-4}$ even though $1 < 3$. This suggests why you should first express the fractions with *positive* denominators.

b.

$$lcd\left(\frac{-3}{10}, \frac{-2}{7}\right) = lcm(10, 7) = 70$$

$$\frac{-3}{10} = \frac{(-3) \cdot 7}{10 \cdot 7} = \frac{-21}{70}, \quad \frac{-2}{7} = \frac{(-2) \cdot 10}{7 \cdot 10} = \frac{-20}{70}$$

$$\frac{\mathbf{-21}}{70} < \frac{\mathbf{-20}}{70} \quad \text{because} \quad -21 < -20$$

Therefore, $\frac{-3}{10} < \frac{-2}{7}$

A negative number is less than any positive number. In particular, a negative fraction is less than a positive fraction. For example, $\frac{-2}{5} < \frac{1}{6}$.

CLASS EXERCISES *Fill in "<" or ">".* **6.** $-\frac{1}{10} \square \frac{3}{-10}$ **7.** $\frac{-4}{5} \square \frac{-5}{7}$ **8.** $\frac{-6}{5} \square \frac{9}{-7}$

SOLUTIONS TO CLASS EXERCISES

1. $\frac{1}{6} < \frac{5}{6}$ **2.** $\frac{8}{9} > \frac{7}{9}$ **3.** $\frac{8}{5} > \frac{7}{6}$ because $\frac{8}{5} = \frac{\mathbf{48}}{30} > \frac{\mathbf{35}}{30} = \frac{7}{6}$

4. $\frac{5}{9} < \frac{7}{12}$ because $\frac{5}{9} = \frac{\mathbf{20}}{36} < \frac{\mathbf{21}}{36} = \frac{7}{12}$, or $\frac{5}{9} = \frac{\mathbf{60}}{9 \cdot 12} < \frac{\mathbf{63}}{9 \cdot 12} = \frac{7}{12}$ **5.** $\frac{13}{4} > 3$ because $\frac{\mathbf{13}}{4} > \frac{\mathbf{12}}{4} = \frac{3}{1} = 3$

6. $-\frac{1}{10} > \frac{3}{-10}$. Note that $-\frac{1}{10} = \frac{-1}{10}$, $\frac{3}{-10} = \frac{-3}{10}$; and $\frac{\mathbf{-1}}{10} > \frac{\mathbf{-3}}{10}$ because $-1 > -3$.

7. $\frac{-4}{5} < \frac{-5}{7}$ because $\frac{-4}{5} = \frac{\mathbf{-28}}{35} < \frac{\mathbf{-25}}{35} = \frac{-5}{7}$ **8.** $\frac{-6}{5} > \frac{9}{-7} = \frac{-9}{7}$ because $\frac{-6}{5} = \frac{\mathbf{-42}}{35} > \frac{\mathbf{-45}}{35} = \frac{-9}{7}$

HOME EXERCISES

In Exercises 1–34, fill in "<" or ">".

1. $\frac{1}{5} \square \frac{2}{5}$ **2.** $\frac{5}{8} \square \frac{3}{8}$ **3.** $\frac{7}{10} \square \frac{9}{10}$ **4.** $\frac{2}{3} \square \frac{4}{3}$ **5.** $\frac{3}{4} \square \frac{1}{2}$

6. $\frac{3}{5} \square \frac{3}{4}$ **7.** $\frac{2}{7} \square \frac{3}{8}$ **8.** $\frac{4}{9} \square \frac{7}{10}$ **9.** $\frac{7}{20} \square \frac{11}{30}$ **10.** $\frac{7}{12} \square \frac{8}{15}$

11. $\frac{4}{3} \square \frac{3}{2}$ **12.** $\frac{7}{5} \square \frac{3}{2}$ **13.** $\frac{9}{4} \square \frac{5}{2}$ **14.** $\frac{21}{10} \square \frac{7}{3}$ **15.** $2 \square \frac{7}{4}$

16. $9 \square \frac{17}{2}$ **17.** $10 \square \frac{39}{4}$ **18.** $\frac{17}{20} \square \frac{15}{16}$ **19.** $\frac{61}{4} \square \frac{46}{5}$ **20.** $\frac{48}{9} \square \frac{31}{6}$

21. $\frac{-5}{9} \square \frac{-4}{9}$ **22.** $-\frac{1}{8} \square -\frac{3}{8}$ **23.** $\frac{4}{-7} \square \frac{-3}{7}$ **24.** $-\frac{7}{12} \square \frac{5}{-12}$ **25.** $\frac{-3}{4} \square \frac{-1}{2}$

26. $\frac{-5}{8} \square \frac{-1}{2}$ **27.** $\frac{-2}{3} \square \frac{-5}{6}$ **28.** $\frac{-7}{9} \square \frac{2}{-3}$ **29.** $\frac{-5}{6} \square -\frac{11}{12}$ **30.** $\frac{8}{-11} \square \frac{11}{-13}$

31. $\frac{7}{-15} \square -\frac{5}{12}$ **32.** $\frac{-1}{2} \square \frac{1}{6}$ **33.** $\frac{-9}{10} \square \frac{1}{1000}$ **34.** $\frac{-3}{8} \square 0$

35. Jerry's portion is $\frac{2}{5}$ of the pie, Jim's is $\frac{1}{4}$ of the pie. Who receives the larger portion?

36. Which weighs more: $\frac{1}{2}$ pound of feathers or $\frac{5}{12}$ pound of lead?

37. One side of a handkerchief measures $\frac{44}{3}$ inches. An adjacent side measures $\frac{57}{4}$ inches. Which side is longer?

38. A salad dressing recipe calls for $\frac{2}{3}$ cup of wine and $\frac{3}{4}$ cup of vinegar. Is there more wine or more vinegar in the dressing?

39. Which is a better buy—a 6-ounce box of soap flakes at 75¢ or an 8-ounce box at 96¢?

40. Which is a better buy—a 12-ounce can of tomatoes at 54¢ or a pound-can at 75¢? [1 pound = 16 ounces]

5.5 COMPLEX EXPRESSIONS

COMPLEX FRACTIONS

DEFINITION

A **complex fraction** is one that contains other fractions in its numerator or denominator. (There may be fractions in both numerator and denominator.)

$$\frac{\frac{1}{4}}{3}, \quad \frac{-2}{\frac{2}{3}+1}, \quad \text{and} \quad \frac{\frac{3}{8}}{\frac{7}{10}}$$

are each complex fractions. You can simplify a complex fraction by first expressing it in terms of division. For example, the numerator of $\frac{\frac{1}{4}}{3}$ is $\frac{1}{4}$; the denominator is 3. Thus,

$$\frac{\frac{1}{4}}{3} = \frac{1}{4} \div 3$$

Similarly, the numerator of $\frac{-2}{\frac{2}{3}+1}$ is -2; the denominator is $\frac{2}{3}+1$. Thus,

$$\frac{-2}{\frac{2}{3}+1} = -2 \div \left(\frac{2}{3}+1\right)$$

And the numerator of $\frac{\frac{3}{8}}{\frac{7}{10}}$ is $\frac{3}{8}$; the denominator is $\frac{7}{10}$. Therefore,

$$\frac{\frac{3}{8}}{\frac{7}{10}} = \frac{3}{8} \div \frac{7}{10}$$

Example 1 Simplify. $\dfrac{\frac{1}{2}}{5}$

Example 2 $\dfrac{\frac{2}{3}}{\frac{4}{9}} = \dfrac{2}{3} \div \dfrac{4}{9} = \dfrac{\overset{1}{\cancel{2}}}{\underset{1}{\cancel{3}}} \cdot \dfrac{\overset{3}{\cancel{9}}}{\underset{2}{\cancel{4}}} = \dfrac{3}{2}$

Solution $\dfrac{\frac{1}{2}}{5} = \dfrac{1}{2} \div 5 = \dfrac{1}{2} \div \dfrac{5}{1}$

(*Invert* $\frac{5}{1}$, *and multiply.*)

$= \dfrac{1}{2} \cdot \dfrac{1}{5} = \dfrac{1}{10}$

CLASS EXERCISES *Simplify.* **1.** $\dfrac{\frac{1}{4}}{3}$ **2.** $\dfrac{1}{\frac{4}{3}}$ **3.** $\dfrac{\frac{3}{8}}{\frac{7}{10}}$

SUMS IN NUMERATOR OR DENOMINATOR

Example 3 $\dfrac{1 + \frac{1}{2}}{\frac{1}{4}} = \left(1 + \dfrac{1}{2}\right) \div \dfrac{1}{4} = \left(\dfrac{1 \cdot 2}{1 \cdot 2} + \dfrac{1}{2}\right) \div \dfrac{1}{4}$

$= \dfrac{2 + 1}{2} \div \dfrac{1}{4} = \dfrac{3}{\underset{1}{\cancel{2}}} \cdot \dfrac{\overset{2}{\cancel{4}}}{1} = 6$

Note that the *lcd* of the given fractions is 4. You could also have multiplied the numerator and denominator of the complex fraction by 4.

$$\dfrac{\left(1 + \frac{1}{2}\right) \cdot 4}{\frac{1}{4} \cdot 4} = \dfrac{4 + 2}{1} = 6$$

CLASS EXERCISES *Simplify.* **4.** $\dfrac{3 + \frac{1}{3}}{\frac{5}{6}}$ **5.** $\dfrac{\frac{3}{4} + \frac{1}{2}}{\frac{5}{8} + \frac{1}{4}}$

COMPLEX RATIONAL EXPRESSIONS

DEFINITION

> A **complex rational expression** is one that contains other rational expressions in numerator or denominator (possibly in both).

As with fractions, you simplify a complex rational expression by first rewriting it in terms of division.

Example 4

$$\frac{\frac{1}{x}}{\frac{1}{y}} = \frac{1}{x} \div \frac{1}{y} = \frac{1}{x} \cdot \frac{y}{1} = \frac{y}{x}$$

Example 5

$$\frac{\frac{1}{x-4}}{\frac{x}{x^2-16}} = \frac{1}{x-4} \div \frac{x}{x^2-16}$$

$$= \frac{1}{\cancel{x-4}} \cdot \frac{\overset{(x+4)\cancel{(x-4)}}{\cancel{x^2-16}}}{x}$$

$$= \frac{x+4}{x}$$

Example 6

$$\frac{\frac{2}{a-2} - \frac{1}{a+2}}{\frac{a}{a^2-4}} = \left(\frac{2}{a-2} - \frac{1}{a+2}\right) \div \frac{a}{a^2-4}$$

$$= \frac{2(a+2) - 1(a-2)}{\cancel{(a-2)(a+2)}} \cdot \frac{\overset{\cancel{(a-2)(a+2)}}{\cancel{a^2-4}}}{a}$$

$$= \frac{2a + 4 - a + 2}{a}$$

$$= \frac{a+6}{a}$$

CLASS EXERCISES *Simplify.*

6. $\dfrac{\frac{x^2y}{4st}}{\frac{xy^3}{8s^2t^2}}$

7. $\dfrac{\frac{a-1}{x^2-4}}{\frac{a^2-1}{x+2}}$

8. $\dfrac{\frac{1}{x+1} + \frac{1}{x+3}}{\frac{x+2}{x^2+4x+3}}$

SOLUTIONS TO CLASS EXERCISES

1. $\dfrac{\frac{1}{4}}{3} = \dfrac{1}{4} \div 3 = \dfrac{1}{4} \cdot \dfrac{1}{3} = \dfrac{1}{12}$

2. $\dfrac{1}{\frac{4}{3}} = 1 \div \dfrac{4}{3} = 1 \cdot \dfrac{3}{4} = \dfrac{3}{4}$

3. $\dfrac{\frac{3}{8}}{\frac{7}{10}} = \dfrac{3}{8} \div \dfrac{7}{10} = \dfrac{3}{\underset{4}{\cancel{8}}} \cdot \dfrac{\overset{5}{\cancel{10}}}{7} = \dfrac{15}{28}$

4. $\dfrac{3 + \frac{1}{3}}{\frac{5}{6}} = \left(3 + \dfrac{1}{3}\right) \div \dfrac{5}{6} = \dfrac{\overset{2}{\cancel{10}}}{\underset{1}{\cancel{3}}} \cdot \dfrac{\overset{2}{\cancel{6}}}{\underset{1}{\cancel{5}}} = 4$

5. $\dfrac{\frac{3}{4} + \frac{1}{2}}{\frac{5}{8} + \frac{1}{4}} = \left(\dfrac{3}{4} + \dfrac{1}{2}\right) \div \left(\dfrac{5}{8} + \dfrac{1}{4}\right)$

$$= \frac{3 + 2}{4} \div \frac{5 + 2}{8}$$

$$= \frac{5}{4} \div \frac{7}{8}$$

$$= \frac{5}{\underset{1}{\cancel{4}}} \cdot \frac{\overset{2}{\cancel{8}}}{7}$$

$$= \frac{10}{7}$$

6. $\dfrac{\frac{x^2y}{4st}}{\frac{xy^3}{8s^2t^2}} = \dfrac{x^2y}{4st} \div \dfrac{xy^3}{8s^2t^2}$

$$= \frac{\overset{x}{\cancel{x^2}\cancel{y}}}{\underset{1}{\cancel{4st}}} \cdot \frac{\overset{2st}{\cancel{8s^2t^2}}}{\underset{y^2}{\cancel{xy^3}}}$$

$$= \frac{2stx}{y^2}$$

7. $\dfrac{\frac{a - 1}{x^2 - 4}}{\frac{a^2 - 1}{x + 2}} = \dfrac{a - 1}{x^2 - 4} \div \dfrac{a^2 - 1}{x + 2}$

$$= \frac{\overset{1}{\cancel{a - 1}}}{\underset{1}{\cancel{(x + 2)}}(x - 2)} \cdot \frac{\overset{1}{\cancel{x + 2}}}{(a + 1)\underset{1}{\cancel{(a - 1)}}}$$

$$= \frac{1}{(x - 2)(a + 1)}$$

8. $\dfrac{\frac{1}{x + 1} + \frac{1}{x + 3}}{\frac{x + 2}{x^2 + 4x + 3}} = \left(\dfrac{1}{x + 1} + \dfrac{1}{x + 3}\right) \div \dfrac{x + 2}{x^2 + 4x + 3}$

$$= \frac{(x + 3) + (x + 1)}{(x + 1)(x + 3)} \cdot \frac{x^2 + 4x + 3}{x + 2}$$

$$= \frac{\overset{\overset{1}{2\cancel{(x + 2)}}}{2x + 4}}{\underset{1}{\cancel{(x + 1)}}\underset{1}{\cancel{(x + 3)}}} \cdot \frac{\overset{1}{\cancel{(x + 1)}}\overset{1}{\cancel{(x + 3)}}}{\underset{1}{\cancel{x + 2}}}$$

$$= 2$$

HOME EXERCISES

Simplify.

1. $\dfrac{\frac{1}{2}}{2}$

2. $\dfrac{5}{\frac{1}{3}}$

3. $\dfrac{\frac{3}{4}}{-3}$

4. $\dfrac{-7}{\frac{1}{3}}$

5. $\dfrac{\frac{1}{2}}{\frac{1}{6}}$

6. $\dfrac{\frac{2}{5}}{\frac{1}{10}}$ 7. $\dfrac{\frac{-1}{8}}{\frac{3}{4}}$ 8. $\dfrac{\frac{-2}{9}}{\frac{-1}{3}}$ 9. $\dfrac{\frac{4}{3}}{\frac{16}{9}}$ 10. $\dfrac{\frac{2}{7}}{\frac{1}{14}}$

11. $\dfrac{1+\frac{1}{2}}{\frac{1}{2}}$ 12. $\dfrac{2+\frac{1}{3}}{\frac{5}{3}}$ 13. $\dfrac{1-\frac{1}{4}}{\frac{1}{8}}$ 14. $\dfrac{6-\frac{3}{4}}{\frac{-1}{2}}$ 15. $\dfrac{\frac{1}{2}+\frac{1}{3}}{\frac{1}{6}}$

16. $\dfrac{1-\frac{1}{2}}{1+\frac{1}{4}}$ 17. $\dfrac{\frac{3}{4}+\frac{1}{12}}{\frac{5}{6}+\frac{1}{8}}$ 18. $\dfrac{\frac{1}{9}-\frac{1}{6}}{\frac{2}{3}+\frac{1}{9}}$ 19. $\dfrac{\frac{3}{5}-\frac{1}{10}}{\frac{2}{25}-\frac{1}{50}}$ 20. $\dfrac{\frac{1}{12}-\frac{1}{144}}{\frac{1}{3}-\frac{2}{9}}$

21. $\dfrac{\frac{a}{x}}{y}$ 22. $\dfrac{a}{\frac{x}{y}}$ 23. $\dfrac{\frac{-1}{x}}{\frac{y}{2}}$ 24. $\dfrac{\frac{2}{x}}{\frac{4}{y}}$ 25. $\dfrac{\frac{a}{b^2}}{\frac{b}{a^2}}$

26. $\dfrac{\frac{xy}{z}}{\frac{y}{x^2}}$ 27. $\dfrac{\frac{-ab}{c^2}}{\frac{a}{c^2}}$ 28. $\dfrac{\frac{r^2s}{t^2}}{\frac{t}{r^2s^3}}$ 29. $\dfrac{\frac{6x}{y^2}}{\frac{12y}{x}}$ 30. $\dfrac{\frac{8a^2}{b}}{\frac{12a}{b^2}}$

31. $\dfrac{\frac{20a^2b}{cd}}{\frac{15ad}{bc}}$ 32. $\dfrac{\frac{1}{x+1}}{\frac{x}{x-1}}$ 33. $\dfrac{\frac{x^2-1}{x^2-4}}{\frac{x+1}{x+2}}$ 34. $\dfrac{\frac{a^2-3a}{a+3}}{\frac{a^2}{a^2-9}}$ 35. $\dfrac{\frac{1}{a}+1}{a^2}$

36. $\dfrac{1-\frac{1}{x}}{\frac{1}{x^2}}$ 37. $\dfrac{\frac{1}{x+1}-\frac{1}{x-1}}{\frac{2}{x^2-1}}$ 38. $\dfrac{\frac{x}{x-a}+\frac{a}{x+a}}{\frac{ax}{x^2-a^2}}$ 39. $\dfrac{\frac{1}{a^2-b^2}+\frac{1}{a+b}}{\frac{a}{b^2-a^2}}$ 40. $\dfrac{\frac{a^3-a^2}{a^2-1}}{\frac{a^3}{a^2+2a+1}}$

5.6 PROPORTIONS AND EQUATIONS WITH RATIONAL EXPRESSIONS

Recall that an equation is a statement of equality. You solve an equation by transforming it into simpler and simpler equivalent equations. You want to bring variables to one side, numerical terms to the other side. You can add the same expression to, or subtract the same expression from, both sides of an equation. You can also multiply or divide both sides of an equation by the same nonzero number.

EQUATIONS WITH FRACTIONS

Fractions often appear on one or both sides of an equation.

Example 1 Solve and check. $x + \frac{1}{2} = 2$

Solution

$$x + \frac{1}{2} = 2$$

Subtract $\frac{1}{2}$ from both sides.

$$x + \frac{1}{2} - \frac{1}{2} = 2 - \frac{1}{2}$$

$$2 - \frac{1}{2} = \frac{4-1}{2} = \frac{3}{2}$$

$$x = \frac{3}{2}$$

check

$$\frac{3}{2} + \frac{1}{2} \stackrel{?}{=} 2$$

$$\frac{3+1}{2} \stackrel{?}{=} 2$$

$$\frac{4}{2} \stackrel{?}{=} 2$$

$$2 \stackrel{\checkmark}{=} 2$$

Example 2 Solve. $\frac{3}{4}x = 12$

Solution

$$\frac{3}{4}x = 12$$

Divide both sides by $\frac{3}{4}$, that is, multiply by $\frac{4}{3}$.

$$\frac{\overset{1}{\cancel{4}}}{\underset{1}{\cancel{3}}} \cdot \frac{\overset{1}{\cancel{3}}}{\underset{1}{\cancel{4}}} x = \frac{4}{\underset{1}{\cancel{3}}} \cdot \overset{4}{\cancel{12}}$$

$$x = 16$$

An equation with integral coefficients can have a fractional root.

Example 3 Solve. $10x = 7$

Solution $10x = 7$ *Divide both sides by* 10.

$$\frac{10x}{10} = \frac{7}{10}$$

$$x = \frac{7}{10}$$

CLASS EXERCISES

1. Solve and check. $x + \frac{2}{5} = \frac{9}{10}$

In Exercises 2 and 3, solve each equation.

2. $\frac{2}{3}x = 3$ 3. $\frac{y}{3} + 1 = \frac{1}{2}$

SOLVING PROPORTIONS

DEFINITION

A proportion is a statement that two fractions are equivalent.

Thus, the statements

$$\frac{2}{4} = \frac{1}{2} \quad \text{and} \quad \frac{9}{12} = \frac{3}{4}$$

are each proportions.

Proportions are generally written in the form

$$\frac{a}{b} = \frac{c}{d}$$

and are often read, "a is to b as c is to d."

Example 4 Find the value of b in the following proportion.

$$\frac{8}{b} = \frac{2}{3}$$

Solution $\frac{8}{b} = \frac{2}{3}$ *Cross-multiply.* $\frac{8}{b} \times \frac{2}{3}$

$8 \cdot 3 = 2b$

$24 = 2b$ *Divide both sides by* 2.

$12 = b$

Proportions occur naturally in everyday situations

Example 5 The width and length of an 8×10 photograph are enlarged proportionally. If the length of the enlargement is 15 inches, what is its width?

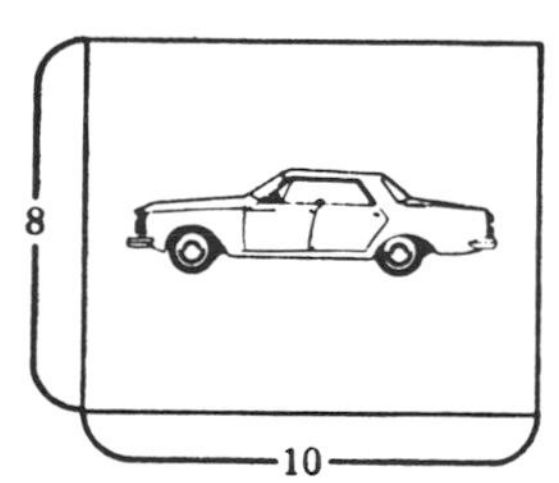

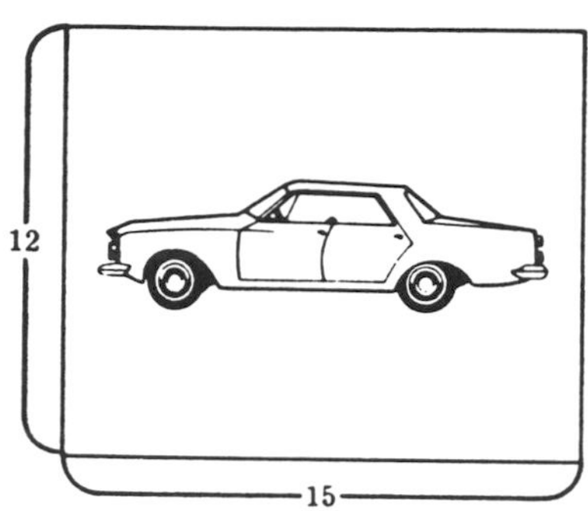

Fig. 5.5

Solution The width of the original photograph is 8 inches; the length is 10 inches. Set up the proportion:

$$\frac{8}{10} = \frac{c}{15} \qquad \textit{Cross-multiply.}$$
$$8 \cdot 15 = 10c$$
$$120 = 10c$$
$$12 = c$$

The width of the enlargement is 12 inches.

CLASS EXERCISES

4. Find the value of a in the proportion $\frac{a}{4} = \frac{1}{2}$.
5. Find the value of d in the proportion $\frac{3}{8} = \frac{24}{d}$.
6. Suppose the Phoenix Suns win 3 out of every 4 games. How many games do they win in an 84-game season?
7. Two-thirds of the entering class at the University of Iowa are men. If there are 512 men in the class, how many entering students (men and women) are there?

MULTIPLYING BY THE *lcd*

An equation, such as

$$\frac{x}{2} = \frac{x + 6}{5}$$

has the form of a proportion. To solve this type of equation, *cross-multiply*.

Example 6 Solve. $\frac{x}{2} = \frac{x + 6}{5}$

Solution

$$\frac{x}{2} = \frac{x + 6}{5} \qquad \textit{Cross-multiply.}$$
$$5x = 2(x + 6)$$
$$5x = 2x + 12$$
$$3x = 12$$
$$x = 4$$

Note that when you cross-multiplied, you actually multiplied both sides by $2 \cdot 5$. In fact, $2 \cdot 5$ is the *lcd* of $\frac{x}{2}$ and $\frac{x + 6}{5}$. Equations that involve fractions or rational expressions can often be simplified by multiplying both sides by the *lcd* of these expressions.

Example 7 Solve. $\dfrac{4}{y+3} = \dfrac{1}{y-3}$

Solution $\dfrac{4}{y+3} = \dfrac{1}{y-3}$

Multiply both sides by $(y+3)(y-3)$, the lcd [or cross-multiply].

$$4(y-3) = (y+3)1$$
$$4y - 12 = y + 3$$
$$3y = 15$$
$$y = 5$$

Example 8 Solve. $\dfrac{6}{x} + \dfrac{4}{x+1} = \dfrac{9}{x}$

Solution $lcd\left(\dfrac{6}{x}, \dfrac{4}{x+1}, \dfrac{9}{x}\right) = x(x+1)$

$$\frac{6}{x} + \frac{4}{x+1} = \frac{9}{x}$$
$$\left(\frac{6}{x} + \frac{4}{x+1}\right) x(x+1) = \frac{9}{x}\, x(x+1)$$
$$6(x+1) + 4x = 9(x+1)$$
$$6x + 6 + 4x = 9x + 9$$
$$10x + 6 = 9x + 9$$
$$x = 3$$

CLASS EXERCISES *Solve each equation.*

8. $\dfrac{x}{4} = \dfrac{x-8}{2}$ **9.** $\dfrac{1}{x} = \dfrac{2}{x+4}$ **10.** $\dfrac{4}{u+2} - \dfrac{3}{u-1} = \dfrac{-2}{u+2}$

SOLUTIONS TO CLASS EXERCISES

1.
$$x + \frac{2}{5} = \frac{9}{10}$$
$$x + \frac{2}{5} - \frac{2}{5} = \frac{9}{10} - \frac{2}{5}$$
$$x = \frac{9}{10} - \frac{4}{10}$$
$$x = \frac{1}{2}$$

Note: $\dfrac{5}{10} = \dfrac{1}{2}$

check
$$\frac{1}{2} + \frac{2}{5} \stackrel{?}{=} \frac{9}{10}$$
$$\frac{5}{10} + \frac{4}{10} \stackrel{?}{=} \frac{9}{10}$$
$$\frac{9}{10} \stackrel{\checkmark}{=} \frac{9}{10}$$

2.
$$\frac{2}{3}x = 3$$
$$\frac{3}{2} \cdot \frac{2}{3}x = \frac{3}{2} \cdot 3$$
$$x = \frac{9}{2}$$

3.
$$\frac{y}{3} + 1 = \frac{1}{2}$$
$$\frac{y}{3} + 1 - 1 = \frac{1}{2} - 1$$
$$\frac{y}{3} = \frac{-1}{2}$$
$$\frac{y}{3} \cdot 3 = \frac{-1}{2} \cdot 3$$
$$y = \frac{-3}{2}$$

4.
$$\frac{a}{4} = \frac{1}{2}$$
$$2a = 4 \cdot 1$$
$$a = 2$$

5.
$$\frac{3}{8} = \frac{24}{d}$$
$$3d = 8 \cdot 24$$
$$\frac{3d}{3} = \frac{8 \cdot \overset{8}{\cancel{24}}}{\underset{1}{\cancel{3}}}$$
$$d = 64$$

6. Let x be the number of games the Suns win.

$$\frac{3}{4} = \frac{x}{84}$$
$$3 \cdot 84 = 4x$$
$$63 = x$$

The Suns win 63 games.

7. Let x be the number of entering students.

$$\frac{2}{3} = \frac{512}{x}$$
$$2x = 3 \cdot 512$$
$$x = 768$$

8.
$$\frac{x}{4} = \frac{x-8}{2}$$
$$2x = 4(x-8)$$
$$x = 2(x-8)$$
$$x = 2x - 16$$
$$16 = x$$

9.
$$\frac{1}{x} = \frac{2}{x+4}$$
$$x + 4 = 2x$$
$$4 = x$$

10.
$$\frac{4}{u+2} - \frac{3}{u-1} = \frac{-2}{u+2}$$ *Multiply both sides by* $(u+2)(u-1)$, *the lcd.*
$$4(u-1) - 3(u+2) = -2(u-1)$$
$$4u - 4 - 3u - 6 = -2u + 2$$
$$u - 10 = -2u + 2$$
$$3u = 12$$
$$u = 4$$

HOME EXERCISES

In Exercises 1–6, solve each equation.

1. $x + \frac{1}{2} = 5$ **2.** $x - \frac{3}{4} = \frac{5}{8}$ **3.** $9x = 3$ **4.** $\frac{-1}{2}x = \frac{5}{6}$ **5.** $\frac{-2}{9}y = 18$ **6.** $\frac{3y}{4} - \frac{1}{2} = \frac{1}{3}$

In Exercises 7–10, solve and check.

7. $z - \frac{1}{4} = \frac{1}{2}$ **8.** $-3t = 8$ **9.** $1 + 5u = 3$ **10.** $\frac{4 - 3x}{7} = \frac{1}{2}$

In Exercises 11–16, find the value of a, b, c, or d in each proportion.

11. $\frac{a}{9} = \frac{2}{3}$ **12.** $\frac{1}{b} = \frac{4}{20}$ **13.** $\frac{12}{30} = \frac{c}{5}$ **14.** $\frac{7}{11} = \frac{-21}{d}$ **15.** $\frac{-15}{9} = \frac{10}{d}$ **16.** $\frac{108}{144} = \frac{c}{48}$

17. A salesman receives a commission of \$3.60 on a \$200 sale. At this rate, how much does he receive on a \$450 sale?

18. A hostess makes 40 cups of coffee for 28 guests. How many cups should she make for 42 guests?

19. A typist makes 4 errors on 10 pages. At this rate, how many errors will she make on 55 pages?

20. The price of grapefruit is three for 80¢. How much does a dozen grapefruits cost?

21. A batter gets 2 hits for every 7 times at bat. At this rate, how many hits will he have for 420 times at bat?

22. A basketball player makes 4 out of 5 free throws. At this rate, how many free throws must she attempt in order to make 16 free throws?

In Exercises 23–48, solve each equation.

23. $\frac{x}{4} = \frac{x-2}{2}$ **24.** $\frac{x+2}{5} = \frac{x}{4}$ **25.** $\frac{x+5}{2} = \frac{1-x}{4}$ **26.** $\frac{x}{3} - 1 = \frac{x}{4}$

27. $\frac{x}{5} + \frac{x}{2} = 7$ **28.** $\frac{x}{3} - \frac{x}{9} = 2$ **29.** $\frac{x+1}{2} + \frac{x-1}{4} = 4$ **30.** $\frac{x}{3} + \frac{x+2}{4} = 4$

31. $\dfrac{2x}{3} - 1 = \dfrac{x+1}{2}$

32. $\dfrac{x+5}{2} - \dfrac{x}{7} = 0$

33. $\dfrac{y+3}{5} = \dfrac{2y+3}{7}$

34. $\dfrac{z-2}{3} = \dfrac{4-z}{4}$

35. $\dfrac{u}{u-1} = 2$

36. $\dfrac{v}{v+6} = -1$

37. $\dfrac{x}{x+10} = 0$

38. $\dfrac{y+1}{y-2} = 2$

39. $\dfrac{z-1}{z+1} = 2$

40. $\dfrac{2}{u-3} = \dfrac{5}{u}$

41. $\dfrac{-1}{u+2} = \dfrac{2}{u+5}$

42. $\dfrac{3}{u-4} = \dfrac{2}{u-1}$

43. $\dfrac{1}{z} - \dfrac{1}{2} = \dfrac{1}{3}$

44. $\dfrac{7}{t} - \dfrac{3}{2} = \dfrac{5}{2t}$

45. $\dfrac{4}{x-2} - \dfrac{1}{x} = \dfrac{5}{x-2}$

46. $\dfrac{18}{v+2} - \dfrac{5}{v+1} = \dfrac{10}{v+1}$

47. $\dfrac{9}{y^2-1} - \dfrac{3}{y+1} = \dfrac{-4}{y-1}$

48. $\dfrac{2}{t+2} - \dfrac{10}{t^2-4} = \dfrac{1}{2-t}$

In Exercises 49–52, solve and check.

49. $\dfrac{y+3}{y-2} = 2$

50. $\dfrac{x}{4} + \dfrac{x}{2} = 9$

51. $\dfrac{2}{u+2} = \dfrac{5}{1-u}$

52. $\dfrac{2}{u} + \dfrac{2}{u+3} = \dfrac{6}{u+3}$

Let's Review Chapter 5

In Exercises 1–4, multiply the given fractions or rational expressions, and simplify.

1. $\dfrac{3}{5} \cdot \dfrac{5}{8}$

2. $\dfrac{-2}{3} \cdot \dfrac{-1}{9} \cdot \dfrac{6}{5}$

3. $\dfrac{a^2}{bc} \cdot \dfrac{b^3c}{a}$

4. $\dfrac{(x+3)^2}{4yz} \cdot \dfrac{8y^2}{x^2-9}$

In Exercises 5–8, divide the given fractions or rational expressions, and simplify.

5. $\dfrac{1}{3} \div \dfrac{-1}{9}$

6. $\dfrac{12}{25} \div \dfrac{3}{50}$

7. $\dfrac{4x^2y^2}{3ab} \div \dfrac{12xy}{9a^2b}$

8. $\dfrac{x^2-y^2}{x+2} \div \dfrac{x+y}{x^2+3x+2}$

In Exercises 9–12, find the lcm of the given integers or polynomials.

9. 10 and 15

10. 2, 4, and 6

11. x^2 and xy

12. $x^2 - 4$ and $x^2 - 3x + 2$

13. a. Find *lcd* $\left(\dfrac{1}{3}, \dfrac{2}{5}\right)$.
    b. Write equivalent fractions that have this *lcd* as denominator.

14. a. Find *lcd* $\left(\dfrac{1}{x-2}, \dfrac{x}{x+3}\right)$.
    b. Write equivalent fractions that have this *lcd* as denominator.

In Exercises 15–20, add or subtract, and simplify.

15. $\dfrac{1}{6} + \dfrac{1}{6}$

16. $\dfrac{2x}{x+3} + \dfrac{6}{x+3}$

17. $\dfrac{x}{x^2-9} - \dfrac{3}{x^2-9}$

18. $\dfrac{3}{4} + \dfrac{1}{2}$

19. $\dfrac{3}{x} - \dfrac{1}{x^2} + \dfrac{x+1}{x^3}$

20. $\dfrac{x}{x+a} + \dfrac{a}{x^2-a^2}$

In Exercises 21–24, fill in "<" or ">".

21. $\dfrac{1}{6} \square \dfrac{5}{6}$

22. $\dfrac{7}{12} \square \dfrac{5}{8}$

23. $\dfrac{-3}{5} \square \dfrac{-5}{7}$

24. $0 \square \dfrac{-1}{11}$

In Exercises 25–28, simplify each expression.

25. $\dfrac{\frac{3}{8}}{\frac{1}{4}}$ **26.** $\dfrac{1-\frac{1}{3}}{\frac{5}{6}}$ **27.** $\dfrac{\frac{a}{b}}{\frac{c}{b^2}}$ **28.** $\dfrac{\frac{x-a}{x^2-1}}{\frac{x^2-a^2}{x+1}}$

In Exercises 29–32, solve each equation.

29. $x - \frac{2}{3} = \frac{1}{5}$ **30.** $\frac{6}{-8} = \frac{9}{d}$ **31.** $\frac{t}{2} - \frac{t}{3} = 1$ **32.** $\frac{1}{t+1} + \frac{1}{t} = \frac{2}{t^2+t}$

6 DECIMALS AND PERCENT

DIAGNOSTIC TEST Perhaps you are already familiar with some of the material in Chapter 6. This test will indicate which sections in Chapter 6 you need to study. The question number refers to the corresponding section in Chapter 6. If you answer *all* parts of the question correctly, you may omit the section. But *if any part of your answer is wrong, you should study the section*. When you have completed the test, turn to page A13 for the answers.

6.1 Decimal Notation *Express as a decimal.* a. $\frac{7}{10}$ b. $\frac{7}{1000}$ c. $2\frac{1}{2}$

Fill in "<" or ">". d. .59 ☐ .61 e. −.203 ☐ −.231

Write in decimal notation.

f. thirty-nine thousandths

6.2 Addition and Subtraction of Decimals *Add or subtract.* a.

$$\begin{array}{r} 4.097 \\ -3.152 \\ 2.389 \\ -1.009 \\ \hline \end{array}$$

b. $-.82 - (-.53)$

6.3 Multiplication and Division of Decimals *Multiply, divide, or determine the power.*

a. 37.25×1000 b. $(-.3)^2\,(.5)$ c. $3.02 \div 1000$ d. $.036\overline{)1.44}$

6.4 Combining Operations *Find each value.* a. $100(2 - .8)$ b. $\frac{3.2 - 1.8 + .2}{.004}$ c. $\frac{3}{5} \times .835$

6.5 Equations with Decimals *Solve each equation.* a. $x - .4 = 2.1$ b. $4.4 + \frac{x}{.3} = 5$

6.6 Decimals, Fractions, and Percent

a. Express $\frac{5}{8}$ as a decimal. b. Express $\frac{2}{9}$ as an infinite repeating decimal.

c. Express .75 as a fraction in lowest terms.

d. Express 8% as a decimal and as a fraction in lowest terms.

e. Express 2.1 as a percent.

6.7 Percentage Problems

a. 75% of a number is 6. Find this number.

b. A college with 8000 students has an increase in enrollment of 25%. What is its new enrollment?

6.8 Interest

a. How much must be invested at an annual interest rate of 8% in order to earn $500 interest in one year?

b. A man invests $2500 at 6% compounded quarterly. How much interest does he earn in half a year?

6.1 DECIMAL NOTATION

The number 10 is the base of our number system. You have already seen how integers are expressed in terms of powers of 10. For example, when you write the 3-digit number

$$589$$

5 is the 100's digit, 8 is the 10's digit, and 9 is the 1's digit.

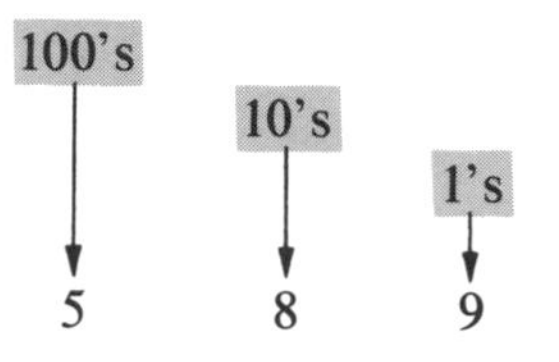

Thus, 589 stands for $5 \times 100 \quad + \quad 8 \times 10 \quad + \quad 9 \times 1$

TENTHS, HUNDREDTHS, THOUSANDTHS

Decimal notation is used to express other numbers in terms of the base 10. For example, a rational number, such as

$$\frac{7}{10}, \quad \text{or} \quad \frac{59}{100}, \quad \text{or} \quad \frac{361}{1000}$$

whose denominator is a *power of* 10, can be expressed in decimal notation as follows:

$$\frac{7}{10} = .7, \quad \frac{59}{100} = .59, \quad \frac{361}{1000} = .361$$

Digits to the *right* of the decimal point are called **decimal digits**. The *first decimal digit* indicates the number of *tenths*; the *second decimal digit* indicates the number of *hundredths*; the *third decimal digit* indicates the number of *thousandths*, etc. Thus,

$$\frac{7}{10} = .7$$

tenths

Read this decimal as "seven tenths" or as "point seven."

$$\frac{59}{100} = .59$$

Read this decimal as "fifty-nine hundredths" or as "point five, nine."

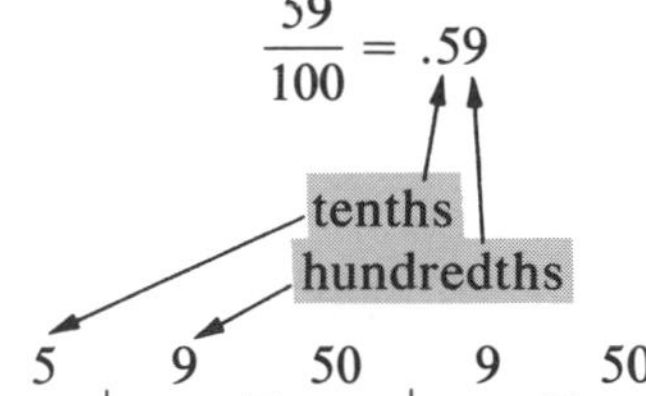

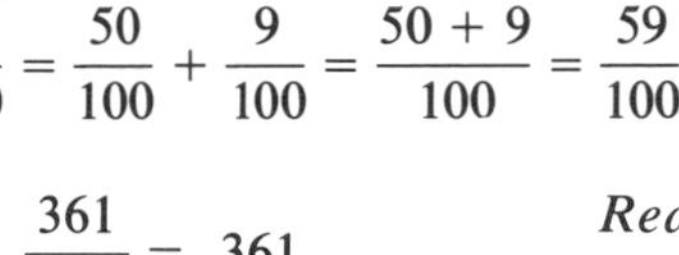

$$\frac{5}{10} + \frac{9}{100} = \frac{50}{100} + \frac{9}{100} = \frac{50+9}{100} = \frac{59}{100}$$

$$\frac{361}{1000} = .361$$

Read this decimal as "three hundred sixty-one thousandths" or as "point three, six, one."

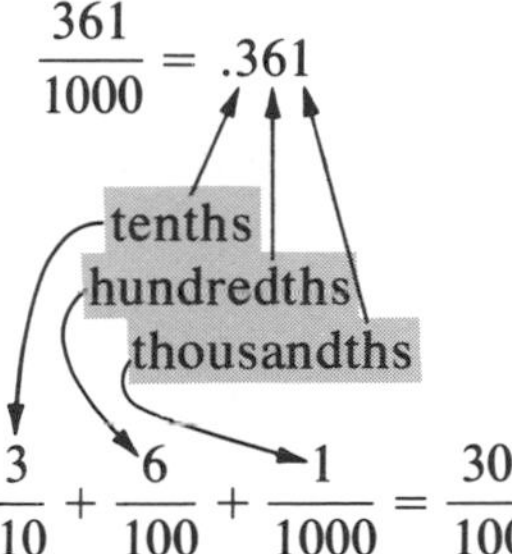

$$\frac{3}{10} + \frac{6}{100} + \frac{1}{1000} = \frac{300}{1000} + \frac{60}{1000} + \frac{1}{100} = \frac{300+60+1}{1000} = \frac{361}{1000}$$

For the decimal .834 259 1 consider the following chart for decimal digits.

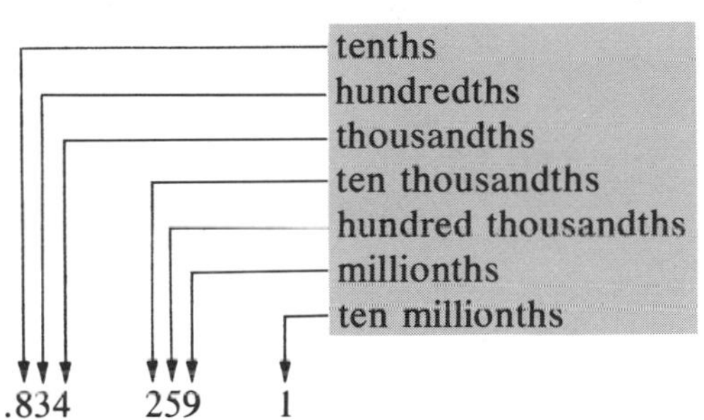

Note that when writing decimals with more than four digits, we begin at the decimal point and use spaces to separate the digits, usually, into groups of three. The right-most group can have 1, 2, or 3 digits.

Example 1 Express each fraction in decimal notation.

a. $\frac{3}{10}$ **b.** $\frac{27}{100}$

c. $\frac{889}{1000}$

Solution

a. $\frac{3}{\underbrace{10}} = \underbrace{.3}$

1 zero 1 decimal digit

b. $\frac{27}{\underbrace{100}} = \underbrace{.27}$

2 zeros 2 decimal digits

c. $\frac{889}{\underbrace{1000}} = \underbrace{.889}$

3 zeros 3 decimal digits

> When the denominator of a fraction is a power of 10, the number of zeros (0's) in the denominator equals the number of decimal digits.

CLASS EXERCISES *In Exercises 1–3, express each fraction as a decimal.*

1. $\frac{1}{10}$ **2.** $\frac{43}{100}$ **3.** $\frac{271}{1000}$

4. a. Express .39 in words.
b. Write seven hundred fifty-one thousandths in decimal notation.

0 AS THE TENTHS DIGIT How do you write $\frac{3}{100}$ in decimal notation? Recall that

$$.3 = \frac{3}{10}$$

In order to express $\frac{3}{\mathbf{100}}$ as a decimal, you must have *two* decimal digits. Write **0** as the *tenths* digit and 3 as the *hundredths* digit. Thus,

$$\frac{3}{100} = \frac{\mathbf{03}}{100} = \mathbf{.03}$$

To express $\frac{57}{1000}$ as a decimal, begin with a **0** as the *tenths* digit.

$$\frac{57}{1000} = \frac{\mathbf{057}}{1000} = \mathbf{.057}$$

To express $\frac{7}{1000}$ as a decimal, begin by writing 0's as both the *tenths* and the *hundredths* digits.

$$\frac{7}{1000} = \frac{\mathbf{007}}{1000} = \mathbf{.007}$$

Example 2 Express each fraction as a decimal.

a. $\frac{1}{10}$ **b.** $\frac{1}{100}$

c. $\frac{1}{1000}$ **d.** $\frac{11}{1000}$

Solution

a. $\frac{1}{10} = .1$ **b.** $\frac{1}{100} = .01$

c. $\frac{1}{1000} = .001$ **d.** $\frac{11}{1000} = .011$

Finally, note that .5 *five tenths*

is 10 times as large as .05 *five hundredths*

and .5 is 100 times as large as .005 *five thousandths*

Each time the 5 moves to the *right*, .5 to .05 to .005, the number *is divided by 10.*

CLASS EXERCISES *Express each fraction as a decimal.*

5. $\frac{9}{100}$ **6.** $\frac{93}{1000}$ **7.** $\frac{3}{1000}$ **8.** $\frac{71}{10,000}$

INTEGERS AND MIXED NUMBERS

An integer is expressed as a decimal simply by placing a decimal point after the 1's digit. Thus, for 475, write 475. ← *decimal point*

A mixed number, such as $41\frac{7}{10}$, can be expressed as a decimal by first writing the *integral part*, 41, *to the left of the decimal point. The fractional part*, $\frac{7}{10}$, *is then expressed to the right of the decimal point.* Thus,

$$\frac{7}{10} = .7 \quad \text{and}$$

$$\begin{aligned} 41\frac{7}{10} &= 41 + \frac{7}{10} \\ &= 41 + .7 \quad \text{Omit the ``+''.} \\ &= 41.7 \end{aligned}$$

Read: "forty-one and seven tenths" or "forty-one point seven."

Example 3 Express each mixed number as a decimal.

a. $1\frac{3}{10}$

b. $19\frac{9}{100}$

Solution

a. $1\frac{3}{10} = 1.3$

b. $19\frac{9}{100} = 19.09$

CLASS EXERCISES *Express each integer or mixed number as a decimal.*

9. 243 **10.** $39\frac{1}{100}$

EQUIVALENT DECIMALS

Observe that
$$\frac{\mathbf{30}}{\mathbf{100}} = \frac{3}{10}$$

Thus, in terms of decimals,
$$.3\mathbf{0} = .3$$

Similarly,
$$.21\mathbf{0} = .21$$

and
$$7.\mathbf{00} = 7.\mathbf{0} = 7.$$

> In general, you can add 0's to the right of the last decimal digit without changing the value. In other words, the new decimal is equivalent to the given one.

Example 4

$$.513 = .513\mathbf{0} = .513\,\mathbf{00} = .513\,\mathbf{000}$$

But
$$.513 \neq .0513$$

because $.513 = \frac{513}{1000}$, whereas $.0513 = \frac{513}{10,000}$

CLASS EXERCISES

11. Which of the following are equivalent to .2?

a. .20 **b.** .200 **c.** .02 **d.** .200 00

ORDERING DECIMALS

Clearly, $.3 < .4$ because $\frac{\mathbf{3}}{10} < \frac{\mathbf{4}}{10}$

[See Figure 6.1 .]

Next, observe that
$$\frac{\mathbf{36}}{100} < \frac{\mathbf{42}}{100}$$

Thus, to compare
$$.\mathbf{3}6 \text{ with } .\mathbf{4}2,$$

first compare their *tenths* digits. Because

$$\mathbf{3} < \mathbf{4}, \quad \text{it follows that} \quad .36 < .42$$

Fig. 6.1 .3 lies to the left of .4 . Thus, .3 < .4 .

To compare .36 with .39,

observe that both decimals have the same tenths digit, 3. Thus, compare their *hundredths* digits.

$$.36 < .39 \quad \text{because} \quad 6 < 9$$

Example 5 Fill in "$<$" or "$>$".

a. $.273 \ \square \ .259$
b. $.8567 \ \square \ .8571$

Solution

a. $.273 \boxed{>} .259$
b. Here, both decimals have the same tenths and hundredths digits. Compare their *thousandths* digits.

$$.8567 \boxed{<} .8571$$

To compare .7 with .74,

observe that $.7 = .70$ and $.70 < .74$

To compare .93 with .929,

observe that $.93 = .930$ and $.930 > .929$

To compare .45 with .4712,

use $.45 = .4500$ and $.4500 < .4712$

CLASS EXERCISES *Fill in "$<$" or "$>$".*

12. $.56 \ \square \ .65$ **13.** $.9259 \ \square \ .9291$ **14.** $.59 \ \square \ .6$ **15.** $.831 \ \square \ .85$

MIXED DECIMALS Clearly, $4.9 < 5.2$ because $4 < 5$

> Thus, to compare "mixed decimals," first compare their integral parts—then, if necessary, compare their decimal parts.

Example 6 Fill in "$<$" or "$>$".

a. $14.8 \ \square \ 11.9$
b. $2.807 \ \square \ 2.593$

Solution

a. $14.8 \boxed{>} 11.9$
b. $2.807 \boxed{>} 2.593$

CLASS EXERCISES *Fill in "$<$" or "$>$".* **16.** $7.41 \ \square \ 8.03$ **17.** $1.025 \ \square \ 1.1$

NEGATIVE DECIMALS

$$-.3 \text{ is the inverse of } .3, \text{ just as } -\frac{3}{10} \text{ is the inverse of } \frac{3}{10}$$

Similarly,

$$-1.7 \text{ is the inverse of } 1.7$$

A negative number is less than 0 or any positive number. Thus, as indicated in Figure 6.2,

$$-.4 < 0 \qquad \text{and} \qquad -.4 < .3$$

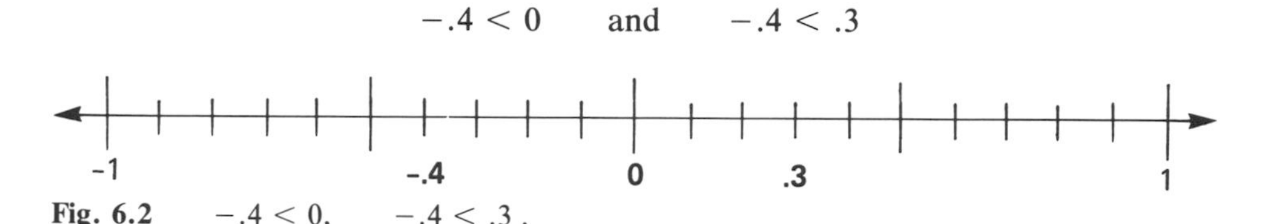

Fig. 6.2 $-.4 < 0, \quad -.4 < .3$.

Recall that $-4 < -3$ because -4 lies to the *left* of -3 on the number line. [See Figure 6.3 .] Similarly, as indicated in Figure 6.4, $-.4$ lies to the *left* of $-.3$ and therefore,

$$-.4 < -.3$$

-4 -3 -2 -1 0 1

Fig. 6.3 $-4 < -3$.

-1 -.4 -.3 0

Fig. 6.4 $-.4 < -.3$

Example 7 Fill in "<" or ">".

a. $-.54$ ☐ $-.51$
b. $-.207$ ☐ $-.199$
c. -1.3 ☐ -1.5

Solution

a. $-.54$ $\boxed{<}$ $-.51$
b. $-.207$ $\boxed{<}$ $-.199$
c. -1.3 $\boxed{>}$ -1.5

CLASS EXERCISES *Fill in "<" or ">".* **18.** $-.84$ ☐ $-.79$ **19.** -3.2 ☐ -3.09

ROUNDING

Figure 6.5 indicates that

.24 is closer to .2 than .3 whereas .26 is closer to .3 than .2

Thus, .24 rounded to the nearest tenth is .2

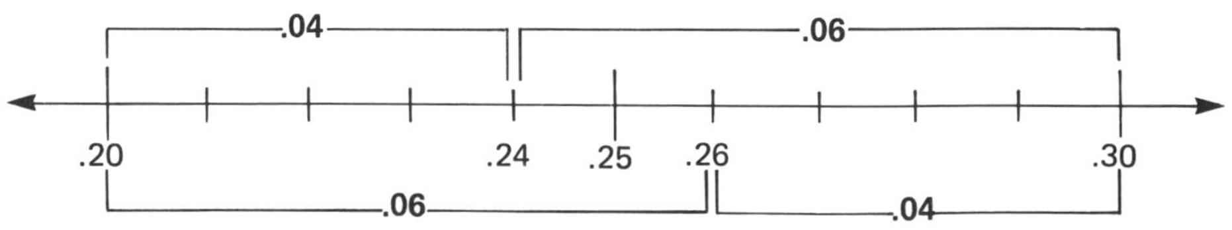

Fig. 6.5

but .26 rounded to the nearest tenth is .3

By convention, .25, which lies midway between .2 and .3, is rounded to .3 .

> **To round to the nearest *tenth*:**
> **1.** If the *hundredths* digit is *4 or less*, simply drop all digits to the right of the tenths digit.
> **2.** If the *hundredths* digit is *5 or more*, add 1 to the tenths digit and drop all digits to the right.

Example 8 Round each of the following to the nearest tenth.

a. .43 **b.** .47
c. .4509 **d.** .96

Solution **a.** .43 rounds to .4
b. .47 rounds to .5
c. .4509 rounds to .5
d. Add 1 to the tenths digit by rounding

1.0
~~.96~~ to 1.0

(Note that .10 = .1, so that .96 does *not* round to .10 .)

> **To round to the nearest *hundredth***, consider whether the *thousandths* digit is *5 or more*. **To round to the nearest *thousandth***, consider the *ten thousandths* digit.

Example 9 Round to the nearest hundredth.

a. .784 **b.** .0961

Solution **a.** .784 rounds to .78
b. **.10**
~~.0961~~ rounds to .10

> **To round to the nearest integer:**
> **1.** If the *tenths* digit is *4 or less*, simply drop all decimal digits.
> **2.** If the *tenths* digit is *5 or more*, add 1 to the 1's digit and drop all decimal digits.

Thus, to the nearest integer, 7.4 rounds to 7, 18.93 rounds to 19, and .8, which can be written as 0.8, rounds to 1.

CLASS EXERCISES

20. Round .44 to the nearest tenth.

21. Round .664 to the nearest hundredth.

22. Round 99.919 **a.** to the nearest integer **b.** to the nearest tenth **c.** to the nearest hundredth.

SOLUTIONS TO CLASS EXERCISES

1. .1 **2.** .43 **3.** .271 **4. a.** thirty-nine hundredths or point three, nine **b.** .751 **5.** .09
6. .093 **7.** .003 **8.** .0071 **9.** 243. **10.** 39.01 **11. a, b,** and **d** **12.** $.56 < .65$
13. $.9259 < .9291$ **14.** $.59 < .60$ **15.** $.831 < .850$ **16.** $7.41 < 8.03$ **17.** $1.025 < 1.100$
18. $-.84 < -.79$ **19.** $-3.20 < -3.09$ **20.** .44 rounds to .4 **21.** .664 rounds to .66
22. a. 99.919 rounds to 100 **b.** 99.919 rounds to 99.9 **c.** 99.919 rounds to 99.92

HOME EXERCISES

In Exercises 1–14, express each fraction as a decimal.

1. $\frac{5}{10}$ **2.** $\frac{19}{100}$ **3.** $\frac{127}{1000}$ **4.** $\frac{7139}{10{,}000}$ **5.** $\frac{-7}{10}$ **6.** $\frac{-109}{1000}$ **7.** $\frac{7}{100}$
8. $\frac{9}{1000}$ **9.** $\frac{9}{10{,}000}$ **10.** $\frac{41}{1000}$ **11.** $\frac{7}{100{,}000}$ **12.** $\frac{77}{100{,}000}$ **13.** $\frac{777}{100{,}000}$ **14.** $\frac{7777}{100{,}000}$

In Exercises 15–18, express each integer or mixed number as a decimal.

15. 9 **16.** $1\frac{7}{10}$ **17.** $-16\frac{3}{100}$ **18.** $102\frac{1}{1000}$

19. Which of the following are equivalent to .1?
a. .01 **b.** .10 **c.** .100 **d.** .0100

20. Which of the following are equivalent to .25?
a. .250 **b.** .025 **c.** .205 **d.** .2500

21. Which of the following are equivalent to 4?
a. 4.0 **b.** 40. **c.** 4.000 **d.** .04

22. Which of the following are equivalent to 3.2?
a. 3.02 **b.** 3.20 **c.** 30.2 **d.** 3.200

In Exercises 23–25, express each decimal in words.

23. .385 **24.** 1.54 **25.** .059

In Exercises 26–28, write each number in decimal notation.

26. point four, zero, five **27.** twenty-nine hundredths **28.** twenty-nine thousandths

In Exercises 29–50, fill in "<" or ">".

29. .1 □ .2 **30.** .83 □ .81 **31.** .06 □ .04 **32.** .136 □ .134 **33.** .19 □ .29
34. .85 □ .90 **35.** .8309 □ .8211 **36.** .2305 □ .2315 **37.** .8307 □ .8310 **38.** .080 51 □ .080 47
39. .4 □ .36 **40.** .21 □ .2 **41.** .93 □ .891 **42.** .2972 □ .3 **43.** 1.4 □ 2.1
44. 2.8 □ 2.5 **45.** −.1 □ −.2 **46.** −.68 □ −.63 **47.** −.39 □ −.41 **48.** −.103 □ −.104
49. −.4 □ −.399 **50.** − 100.01 □ −100.1

In Exercises 51–54, rearrange the numbers so that you can write "<" between any two numbers.

SAMPLE .43, .427, .39, .45, .431	**SOLUTION** $.390 < .427 < .430 < .431 < .450$

51. .2, .3, .23, .32, .302 **52.** .49, .409, .498, .489, .408 **53.** −.1, −.01, −.02, −.101, −.011

54 $-.7$, $-.679$, $-.69$, $-.697$, $-.699$

55. A steel rod is 23.25 inches long. An aluminum rod is 23.238 inches long. Which rod is longer?

56. A man lives 4.3 miles from his office and 4.25 miles from his golf course. Which is closer to his home—the office or the golf course?

In Exercises 57–60, round to the nearest a. integer b. tenth c. hundredth d. thousandth.

57. .6758 **58.** 1.1919 **59.** 3.5649 **60.** 9.9999

61. A necktie sells for $7.95. What is the price to the nearest dollar?

62. A wooden beam is 2.36 inches thick. How thick is it to the nearest tenth of an inch?

6.2 ADDITION AND SUBTRACTION OF DECIMALS

ADDITION To add or subtract decimals, line up the decimal points and add or subtract in columns. Thus,

$$\begin{array}{r} .5 \\ \underline{.6} \\ 1.1 \end{array} \quad \text{because} \quad \frac{5}{10} + \frac{6}{10} = \frac{11}{10} = 1.1$$

You may want to add 0's to the *right* of the last decimal digit, although this is not really necessary.

Example 1. Add.

$$\begin{array}{r} .409 \\ .36 \\ .5812 \\ \underline{.764} \end{array}$$

Solution

$$\begin{array}{r} .4090 \\ .3600 \\ .5812 \\ \underline{.7640} \\ 2.1142 \end{array}$$

CLASS EXERCISES *Add.*

1.
$$\begin{array}{r} .842 \\ .736 \\ \underline{.159} \end{array}$$

2.
$$\begin{array}{r} .4058 \\ .296 \\ .313 \\ \underline{.2} \end{array}$$

SUBTRACTION Throughout this section, the subtraction symbol will be printed in bold to distinguish it from the symbol for (additive) inverse. Thus, .4 **−** .2 is read: .4 *minus* .2, whereas −.5 is read: *the inverse of* .5.

Example 2 Subtract. 4.53 **−** 2.27

Solution

$$\begin{array}{r} 4.53 \\ \underline{-\ 2.27} \\ 2.26 \end{array}$$

Example 3 Subtract. $.1795 - .28$

Solution

$$.1795 - .28 = .1795 + (-.2800)$$
$$|.1795| = .1795, \ |-.2800| = .2800$$
$$.2800 > .1795$$

Subtract the smaller absolute value, .1795, from the larger absolute value, .2800

$$\begin{array}{r} -.2800 \\ .1795 \\ \hline -.1005 \end{array}$$

$-.2800$ *has a negative sign.*
$.1795$ *has a positive sign.*

The *negative* sign of $-.2800$ prevails. Thus, $.1795 - .28 = -.1005$

CLASS EXERCISES *Subtract.* **3.** $.72 - .51$ **4.** $2.4 - 1.95$ **5.** $.203 - .6999$

COMBINING SIGNED DECIMALS

Example 4 Add.

$$\begin{array}{r} .42 \\ .27 \\ -.34 \\ -.51 \\ .9 \\ \hline \end{array}$$

Solution Add the positive decimals and the negative decimals separately.

$$\begin{array}{r} .42 \\ .27 \\ .90 \\ \hline 1.59 \end{array} \qquad \begin{array}{r} -.34 \\ -.51 \\ \hline -.85 \end{array} \qquad \begin{array}{r} 1.59 \\ -\ .85 \\ \hline .74 \end{array}$$

Example 5 Subtract.

$$.6 - (-.4)$$

Solution

$$.6 - (-.4) = .6 + .4$$

Thus, change the sign of the second decimal and add.

$$\begin{array}{r} .6 \\ .4 \\ \hline 1.\not{0} \end{array}$$

CLASS EXERCISES *In Exercises 6 and 7, add.*

6.
$$\begin{array}{r} -.53 \\ -.42 \\ -.94 \\ -.49 \\ \hline \end{array}$$

7.
$$\begin{array}{r} 1.275 \\ -\ .094 \\ 3.38 \\ -1.79 \\ -\ .292 \\ \hline \end{array}$$

In Exercises 8 and 9, subtract.

8. $-.5 - (-.3)$ **9.** $-.49 - .382$

SOLUTIONS TO CLASS EXERCISES

1. 1.737 **2.** 1.2148 **3.** .21 **4.** .45 **5.** $-.4969$ **6.** -2.38

7. $\begin{array}{r} 1.275 \\ \underline{3.380} \\ 4.655 \end{array}$ $\begin{array}{r} -\ \ .094 \\ -1.790 \\ \underline{-\ \ .292} \\ -2.176 \end{array}$ $\begin{array}{r} 4.655 \\ \underline{-2.176} \\ 2.479 \end{array}$

8. $-.5 - (-.3) = -.5 + .3 = -.2$ **9.** $-.872$

HOME EXERCISES

In Exercises 1–16, add.

1. $\begin{array}{r} .9 \\ \underline{.5} \end{array}$ **2.** $\begin{array}{r} .82 \\ \underline{.63} \end{array}$ **3.** $\begin{array}{r} .921 \\ .084 \\ \underline{.911} \end{array}$ **4.** $\begin{array}{r} 1.23 \\ 2.14 \\ 3.19 \\ \underline{4.32} \end{array}$ **5.** $\begin{array}{r} .0953 \\ .3684 \\ .9291 \\ \underline{.2973} \end{array}$ **6.** $\begin{array}{r} 6.3845 \\ .009 \\ 2.12 \\ \underline{.913} \end{array}$ **7.** $\begin{array}{r} 12.0852 \\ 7.391 \\ 9.92 \\ 10.5134 \\ \underline{7.632} \end{array}$ **8.** $\begin{array}{r} .009\ 36 \\ .170\ 4 \\ .090\ 3 \\ .011\ 276 \\ \underline{.000\ 39} \end{array}$

9. $\begin{array}{r} -.4 \\ \underline{-.5} \end{array}$ **10.** $\begin{array}{r} -.47 \\ \underline{-.22} \end{array}$ **11.** $\begin{array}{r} -.462 \\ -.28 \\ -.194 \\ \underline{-.7} \end{array}$ **12.** $\begin{array}{r} .6 \\ .4 \\ -.2 \\ \underline{-.7} \end{array}$ **13.** $\begin{array}{r} .27 \\ -.39 \\ -.45 \\ \underline{-.19} \end{array}$ **14.** $\begin{array}{r} 3.3 \\ 4.7 \\ -\ .6 \\ -1.2 \\ \underline{-3.9} \end{array}$ **15.** $\begin{array}{r} 4.562 \\ -3.82 \\ -1.79 \\ -7.24 \\ 9.822 \\ \underline{6.95} \end{array}$ **16.** $\begin{array}{r} 104.836 \\ 77.09 \\ -308.7 \\ -\ 24.93 \\ 512.072 \\ \underline{9.009} \end{array}$

In Exercises 17–30, subtract.

17. $.7 - .5$ **18.** $.28 - .15$ **19.** $.738 - .116$ **20.** $.492 - .095$ **21.** $.3084 - .2196$

22. $.2 - .17$ **23.** $.34 - .285$ **24.** $2.721 - 3.58$ **25.** $.8 - (-.2)$ **26.** $.12 - (-.34)$

27. $-.7 - .5$ **28.** $-.92 - (-.29)$ **29.** $2.08 - (-1.73)$ **30.** $-6.05 - (-3.4)$

In Exercises 31–36, add or subtract.

31. $.2 + .5 - .3$ **32.** $.89 - .17 + .72$ **33.** $.86 - (.72 + .11)$ **34.** $3.9 - (1.2 - 2.5)$

35. $-.92 - [.27 - (.25 + .14)]$ **36.** $9.26 + .095 - (3.21 - 1.74) - .17$

37. A woman purchases the following items in a department store:

hair dryer	\$24.98
perfume	8.95
shoes	22.50
hat	11.45

How much does she spend?

38. One day a salesman travels distances of 17.6 miles, 31.4 miles, 28.7 miles, and 19.9 miles. How many miles does he travel that day?

39. A man buys a shirt for \$11.95. If he pays for it with a \$20 bill, what is his change?

40. A large roll of tape measures 800 inches. A medium-size roll measures 556.4 inches. How much longer is the large roll?

6.3 MULTIPLICATION AND DIVISION OF DECIMALS

MULTIPLYING BY A POWER OF 10

$.3 \times 10 = \frac{3}{10} \times \frac{10}{1} = 3.$

> To multiply a decimal by 10, move the decimal point one digit to the *right*. To multiply by 10^n, move the decimal point n digits to the *right*. If necessary, add zeros to the right of the right-most digit, as in Example 1**d**.

Example 1 Multiply.

a. .174 × 10
b. .174 × 100
c. .174 × 1000
d. .174 × 10,000

Solution

a. .174 × 10 = 1.74 → [1.74]
b. 100 = 10^2. Thus, .174 × 100 = 17.4 → [17.4]
c. 1000 = 10^3. Thus, .174 × 1000 = 174 → [174]
d. 10,000 = 10^4. Thus, .174 × 10,000 = 1740 → [1740]

CLASS EXERCISES *Multiply.* **1.** .236 × 10 **2.** .0028 × 1000 **3.** 3.621 × 10,000

DECIMAL FACTORS

When one or both factors of a product is a decimal, count the total number of decimal digits in the factors to place the decimal point in the product. Thus,

$$(.2)(.6) = \frac{2}{10} \times \frac{6}{10} = \frac{12}{100} = .12$$

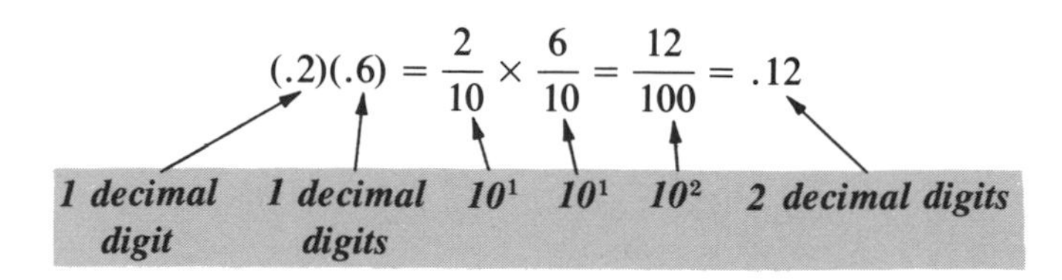

And

$$(.3)(.11) = \frac{3}{10} \times \frac{11}{100} = \frac{33}{1000} = .033$$

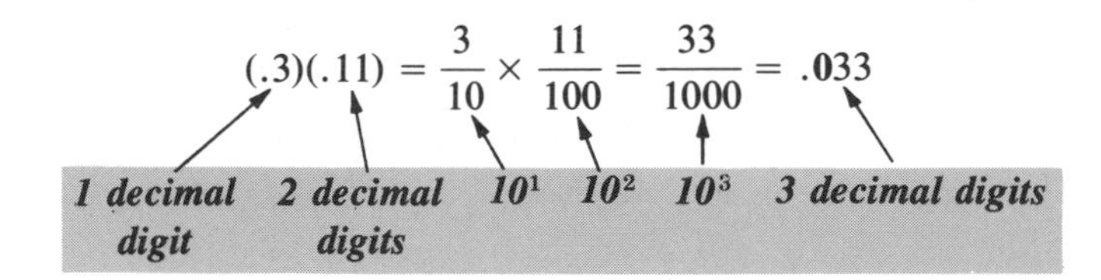

Here, *you add a 0 immediately to the right of the decimal point in order to obtain three decimal digits.*

Example 2 Multiply.

a. $4 \times .25$
b. $(.04)(.025)$

Solution **a.**

```
   .25  ←    2 decimal digits
 ×   4  ←  + 0 decimal digits
  1.00  ←    2 decimal digits
```

Thus,

$$.25 \times 4 = 1$$

b.

```
    .025   ←    3 decimal digits
  × .04    ←  + 2 decimal digits
  .00100   ←    5 decimal digits
```

Add two 0's immediately to the right of the decimal point to make five decimal digits in the product. Note that the 0's to the right of the 1 can be dropped because:

$$.001\,00 = .001$$

Thus, $(.025)(.04) = .001$

CLASS EXERCISES *Multiply.* **4.** $.8 \times .3$ **5.** $.2 \times .4$ **6.** $.025 \times .008$

NEGATIVE FACTORS

Recall that the product of nonzero factors is *positive* if there is an *even* number (0, 2, 4, 6, . . .) of negative factors and *negative* if there is an *odd* number (1, 3, 5, . . .) of negative factors.

Example 3 **a.** $(-.2)(-.4) = .08$

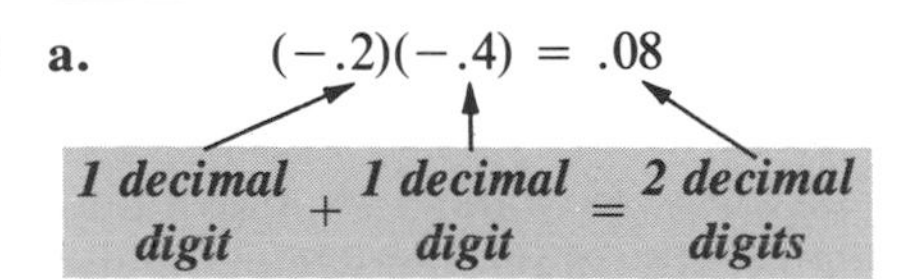

There are two negative factors.
The product is positive.

b. $(-.3)(-.5)(-.01) = -\mathbf{.0015}$

1 decimal digit + *1 decimal digit* + *2 decimal digits* = *4 decimal digits*

There are three negative factors. The product is negative.

CLASS EXERCISES *Multiply.* **7.** $(.3)(-.3)$ **8.** $(-.2)(-.7)$ **9.** $(-.1)(-.02)(-.001)(-.03)$

POWERS OF DECIMALS When finding a power of a decimal, multiply the number of decimal digits in the base by the exponent. This indicates the number of decimal digits in the resulting power.

Example 4 Determine the following power.

$$(.2)^3$$

Solution

$$2^3 = 8$$

Now, place the decimal point.

$$(.2)^3 = (.2)(.2)(.2) = \mathbf{.008}$$

1 decimal digit *3 decimal digits*

Example 5

$$(.01)^4 = (.01)(.01)(.01)(.01) = \mathbf{.000\ 000\ 01}$$

2 decimal digits *8 decimal digits*

CLASS EXERCISES *Determine each power.* **10.** $(.02)^2$ **11.** $(.1)^4$

DIVIDING BY A POWER OF 10

$$.3 \div 10 = \frac{3}{10} \times \frac{1}{10} = \frac{3}{100} = \mathbf{.03}$$

(.03)

To divide a decimal by 10, move the decimal point one digit to the *left*. To divide by 10^n, move the decimal point n digits to the *left*. If necessary, add 0's immediately to the right of the decimal point to obtain the correct number of decimal digits, as in parts **b** and **c** of Example 6.

Example 6 **a.** $3.92 \div 10 = .392$ [.3 92] **b.** $3.92 \div 100 = .0392$ [.03 92]
c. $3.92 \div 1000 = .003\ 92$ [.003 92]

CLASS EXERCISES *Divide.* **12.** $5.85 \div 10$ **13.** $72.14 \div 1000$ **14.** $.964 \div 100$

DIVIDING BY A DECIMAL If the divisor has n decimal digits, multiply both dividend and divisor by 10^n, so that the *divisor* will then be an integer.

Example 7

$$\frac{.008}{.02} = \frac{.008 \times 100}{.02 \times 100} = \frac{00.8}{2} = .4$$

3 decimal digits → .008; *2 decimal digits* → .02; 10^2; 00.8 ← *1 decimal digit*; .4 ← *1 decimal digit*

Example 8 Divide. $.027\overline{)16.2}$

Solution Note that the divisor, .027, has three decimal digits.

$$\frac{16.2}{.027} = \frac{16.2 \times 1000}{.027 \times 1000} = \frac{16\,200.}{027.}$$

```
         600.    ← line up the decimal
027. ⟌16200     ← points in the
       162       ← quotient and dividend
       ---
         00
```

The quotient is 600.

Suppose $b \neq 0$. Recall that $\frac{a}{b}$ is positive if a and b have the same sign, and $\frac{a}{b}$ is negative if a and b have different signs.

Example 9 **a.** $\frac{.015}{.5} = \frac{.015 \times 10}{.5 \times 10} = \frac{0.15}{5.} = .03$ **b.** $\frac{-.015}{.5} = -.03$

c. $\frac{.015}{-.5} = -.03$ **d.** $\frac{-.015}{-.5} = .03$

CLASS EXERCISES *Divide.* **15.** $\frac{.0016}{.04}$ **16.** $\frac{2.5}{.005}$ **17.** $.035\overline{)4.9}$ **18.** $\frac{.01}{-.2}$

ESTIMATING When calculating with decimals, you can often estimate the result quickly by first rounding to the nearest integer.

Example 10 Estimate the product

$$3.8 \times 6.2$$

by first rounding each factor to the nearest integer.

Solution 3.8 rounds to 4,
6.2 rounds to 6
$4 \times 6 = 24$

Thus, the product is approximately 24.

CLASS EXERCISES

19. Estimate the product by first rounding each factor to the nearest integer.

$$12.2 \times 9.8 \times 5.01$$

20. Estimate the quotient by first rounding dividend and divisor to the nearest integer.

$$\frac{99.9}{4.03}$$

SOLUTIONS TO CLASS EXERCISES

1. $.236 \times 10 = 2.36$ [2,36]

2. $.0028 \times 1000 = 2.8$ [002,8]

3. $3.621 \times 10{,}000 = 36{,}210$ [36210.]

4. $.8 \times .3 = .24$
1 decimal digit + 1 decimal digit = 2 decimal digits

5. $.2 \times .4 = \mathbf{.08}$
1 decimal digit + 1 decimal digit = 2 decimal digits

6. $.025 \times .008 = \mathbf{.000}\,200 = .0002$
3 decimal digits + 3 decimal digits = 6 decimal digits

7. $(.3)(-.3) = -\mathbf{.09}$
1 decimal digit + 1 decimal digit = 2 decimal digits
There is one negative factor. The product is negative.

8. $(-.2)(-.7) = .14$
1 decimal digit + 1 decimal digit = 2 decimal digits
There are two negative factors. The product is positive.

9. $(-.1)(-.02)(-.001)(-.03) = \mathbf{.000\,000\,06}$
1 decimal digit + 2 decimal digits + 3 decimal digits + 2 decimal digits = 8 decimal digits
There are four negative factors. The product is positive.

10. $(.02)^2 = (.02)(.02) = \mathbf{.0004}$
2 decimal digits + 2 decimal digits = 4 decimal digits

11. $(.1)^4 = (.1)(.1)(.1)(.1) = \mathbf{.0001}$
4 (1 decimal digit) = 4 decimal digits

12. $5.85 \div 10 = .585$ [.585]

13. $72.14 \div 1000 = .072\,14$ [.072 14]

14. $.964 \div 100 = \mathbf{.009}\,64$ [.009 64]

15. $\dfrac{.0016}{.04} = \dfrac{.0016 \times 100}{.04 \times 100} = \dfrac{00.16}{04.} = .04$

16. $\dfrac{2.5}{.005} = \dfrac{2.5 \times 1000}{.005 \times 1000} = \dfrac{2500.}{005.} = 500$

17. $035.\overline{)4900.}$ = 140.

$$\begin{array}{r} 140. \\ 035.\overline{)4900.} \\ \underline{35} \\ 140 \\ \underline{140} \\ 0 \end{array}$$

18. $\dfrac{.01}{-.2} = \dfrac{.01 \times 10}{-.2 \times 10} = \dfrac{0.10}{-2.} = -.05$

19. 12.**2** rounds to 12, 9.**8** rounds to 10, and 5.**0**1 rounds to 5.

$$12 \times 10 \times 5 = 600$$

Thus, $12.2 \times 9.8 \times 5.01$ is approximately 600.

20. 99.**9**9 rounds to 100 and 4.**0**3 rounds to 4.

$$\frac{100}{4} = 25$$

Thus, $\dfrac{99.99}{4.03}$ is approximately 25.

HOME EXERCISES

In Exercises 1–28, multiply or determine the power.

1. $.38 \times 10$ **2.** $.54 \times 100$ **3.** $.238 \times 1000$ **4.** $.7294 \times 1000$ **5.** $.0053 \times 10{,}000$
6. $.51\,386 \times 10{,}000$ **7.** 37.24×100 **8.** 37.24×1000 **9.** $.095 \times 100$ **10.** $.095 \times 1000$
11. $(.5)(.6)$ **12.** $(.3)(.2)$ **13.** $(.04)(.2)$ **14.** $(.03)(.03)$ **15.** $(.002)(.5)$
16. $(.004)(.03)$ **17.** $(.031)(.03)$ **18.** $(1.2)(.02)$ **19.** $(.005)(.02)(.01)$ **20.** $(.6)(\ \ .2)$
21. $(-.5)(-.4)$ **22.** $(-.1)(-.01)(-.2)$ **23.** $(-.4)(-.5)(-.1)(-.01)$
24. $(.2)^2$ **25.** $(.01)^2$ **26.** $(-.2)^3$ **27.** $(-.1)^3(.3)$ **28.** $(-.1)^4(.2)(.5)$

29. A book salesman sells three copies of a book priced at $10.25 . What is the total sale?

30. Each page of a 500-page book is .002 inch thick. How thick is the book? (Exclude the cover.)

31. In Exercise 30, if the front and back covers are each .2 inch thick, how thick is the book, including the covers?

32. A pair of socks sells for $1.25 . What is the price of a dozen pairs of socks?

In Exercises 33–54, divide.

33. $49.4 \div 10$ **34.** $386 \div 10$ **35.** $27.04 \div 100$ **36.** $.186 \div 100$ **37.** $.043 \div 100$ **38.** $.493 \div 1000$

39. $.012 \div 10$ **40.** $.012 \div 100$ **41.** $.012 \div 1000$ **42.** $.012 \div 10{,}000$ **43.** $\dfrac{4}{.2}$ **44.** $\dfrac{.04}{.02}$

45. $\dfrac{-.0004}{.2}$ **46.** $\dfrac{-.4}{-.0002}$ **47.** $1.5\overline{)1.35}$ **48.** $.21\overline{)147}$ **49.** $93\overline{).0186}$ **50.** $.737\overline{).0008107}$

51. $\dfrac{125}{.05}$ **52.** $\dfrac{-125}{.05}$ **53.** $\dfrac{125}{-.05}$ **54.** $\dfrac{-125}{-.05}$

55. A woman pays $572 in 10 equal monthly installments for a color television set. How much does she pay each month?

56. \$5.40 is divided equally among 10 boys. How much is each boy's share?

57. A book sells for \$4.98. One week, the total sales for that book are \$109.56. How many copies are sold?

58. A man must pay a total of \$3706.20 in 36 equal monthly installments for a motorcycle. How much is each month's installment?

In Exercises 59–62, estimate the product or quotient by first rounding each factor to the nearest integer.

59. 11.93×12.07 **60.** $8.9 \times 9.1 \times 11.9$ **61.** $\frac{19.95}{4.98}$ **62.** $\frac{7.89 \times 9.04}{12.01}$

63. A car takes 12.3 gallons of gas at 99.9 cents per gallon. Estimate the total cost by first rounding each factor to the nearest integer.

64. One kilogram is approximately 2.2 pounds. Estimate the number of pounds in 15.9 kilograms by first rounding each factor to the nearest integer.

6.4 COMBINING OPERATIONS

ORDER OF OPERATIONS

When several operations are combined in an example, recall the order in which they are performed. If parentheses are given, first perform the operations within parentheses. Otherwise:

1. First, raise to a power.
2. Then, multiply or divide from left to right.
3. Then, add or subtract from left to right.

Example 1 Find the following value.

$$1000(.6 - .42)$$

Solution First, perform operations within parentheses.

$$\begin{array}{r} .60 \\ -.42 \\ \hline .18 \end{array}$$

$$\begin{aligned} 1000(.6 - .42) &= 1000(.18) \\ &= 180. \end{aligned}$$

Example 2 Find the following value.

$$\frac{.7 + .5 + .3}{.05}$$

Solution First add, then divide.

$$\begin{aligned} \frac{.7 + .5 + .3}{.05} &= \frac{1.5}{.05} \\ &= \frac{1.5 \times 100}{.05 \times 100} \\ &= \frac{150.}{05.} \\ &= 30 \end{aligned}$$

Example 3 Find each value.

a. $100(.02)^2$

b. $(100 \times .02)^2$

Solution a. First, find the square of .02, then multiply by 100.

$$(.02)^2 = (.02)(.02) = .0004$$
$$100(.02)^2 = 100(.0004) = 00.04 = .04$$

b. First multiply, then find the square of the product.

$$(100 \times .02)^2 = (02.)^2 = 2^2 = 4$$

CLASS EXERCISES *Find each value.*

1. $(.2)(.1 + .4 + .3)$
2. $1000(.4 - .31)$
3. $\dfrac{1.2 + 3.4 + 1.8}{.08}$
4. $(1.1 + .1)^2$
5. $\dfrac{.03 + (.1)^2}{.2}$

SOLUTIONS TO CLASS EXERCISES

1. $(.2)(.1 + .4 + .3) = (.2)(.8) = .16$
2. $1000(.40 - .31) = 1000(.09) = 90$
3. $\dfrac{1.2 + 3.4 + 1.8}{.08} = \dfrac{6.4}{.08} = \dfrac{640}{8} = 80$
4. $(1.1 + .1)^2 = (1.2)^2 = 1.44$
5. $\dfrac{.03 + (.1)^2}{.2} = \dfrac{.03 + .01}{.2} = \dfrac{.04}{.2} = \dfrac{.4}{2} = .2$

HOME EXERCISES

In Exercises 1–32, find each value.

1. $10(.2 + .8)$
2. $100(2.4 + 3.6)$
3. $10(.7 + 1.5)$
4. $100(.08 + .02)$
5. $100(.04 + .32 + .05)$
6. $1000(.68 + .15 + .37 + .49)$
7. $100(.6 - .4)$
8. $100(.5 - .37)$
9. $10(.1 + .4 - .3)$
10. $100(1 - .5 + .2)$
11. $(.7 + .4)(.1)$
12. $(.396 + .115)(.01)$
13. $(1 - .1)(.43)$
14. $(1 - .01)(.001)$
15. $.2(.1 + .4 + .3)$
16. $(7.2)(.2 + .9 + .1)$
17. $(.95 - .46)(.002)$
18. $(7.42 - 1.01)(6.6 - 5.5)$
19. $(-.5)(.6 - .9)$
20. $(.7 + .3 + .2)(-.07)$
21. $5(.1)^2$
22. $(5 \times .1)^2$
23. $.6 + (.4)^2$
24. $(.6 + .4)^2$
25. $(.8 - .5)^2$
26. $.8 - (.5)^2$
27. $\dfrac{.6 + .2 + .4}{.3}$
28. $\dfrac{1.4 + 2.2 + 3.6}{.02}$
29. $\dfrac{19.1 + 5.6 + 8.3}{5}$
30. $\dfrac{2.4 - 1.2 + 3.6}{.012}$
31. $\dfrac{(.2)^3}{2}$
32. $(4 - 3.9)^4$
33. Multiply the sum of .4 and 1.6 by .01 .
34. Add 1.4 to the product of 2.2 and .003 .
35. Divide .082 by the sum of .35 and .06 .
36. Multiply the square of .5 by .3 .

6.5 EQUATIONS WITH DECIMALS

BASIC EXAMPLES

Example 1 Solve. $y + .05 = .7$

Solution

$$y + .05 = .70 \qquad \textit{Subtract .05 from both sides.}$$

$$y + .05 - .05 = .70 - .05$$

$$y = .65$$

Example 2 Solve. $\frac{x}{.5} = 2$

Solution

$$\frac{x}{.5} = 2 \qquad \textit{Multiply both sides by .5 .}$$

$$\frac{x}{.5}(.5) = 2(.5)$$

$$x = 1.0, \quad \text{or } 1$$

CLASS EXERCISES *Solve each equation.* **1.** $x - .37 = .2$ **2.** $\frac{t}{.03} = 5$ **3.** $.4x = 24$

ELIMINATING DECIMALS Frequently, it is best to solve an equation by first eliminating decimals, as in Examples 3 and 4.

Example 3 Solve. $.3t - .7 = .2$

Solution Multiply both sides by 10 to eliminate decimals.

$$(.3t - .7) \times 10 = .2 \times 10$$

On the left side, use the Distributive Laws.

$$3t - 7 = 2$$

$$3t = 9$$

$$t = 3$$

Example 4 Solve. $.4x + .05 = .15x + .3$

Solution Multiply both sides by 100 to eliminate decimals.

$$(.4x + .05) \times 100 = (.15x + .3)100$$

On both sides, use the Distributive Laws.

$$40x + 5 = 15x + 30$$

$$40x + 5 - 15x - 5 = 15x + 30 - 15x - 5$$

$$25x = 25$$

$$x = 1$$

CLASS EXERCISES

Solve each equation by first eliminating decimals.

4. $.2x - .4 = .8$ **5.** $.3y + .08 = .86y - .2$

6. $.4(t - .5) = .25(t + .1)$

SOLUTIONS TO CLASS EXERCISES

1.
$$x - .37 = .2$$
$$x - .37 + .37 = .20 + .37$$
$$x = .57$$

2.
$$\frac{t}{.03} = 5$$
$$\frac{t}{.03}(.03) = 5(.03)$$
$$t = .15$$

3.
$$.4x = 24$$
$$\frac{.4x}{.4} = \frac{24}{.4}$$
$$x = \frac{24 \times 10}{.4 \times 10}$$
$$x = \frac{\overset{6}{\cancel{24}} \times 10}{\underset{1}{\cancel{4}}}$$
$$x = 60$$

4. Multiply both sides by 10 to eliminate decimals.

$$(.2x - .4) \times 10 = .8 \times 10$$
$$2x - 4 = 8$$
$$2x = 12$$
$$x = 6$$

5. Multiply both sides by 100 to eliminate decimals.

$$(.3y + .08) \times 100 = (.86y - .2) \times 100$$
$$30y + 8 = 86y - 20$$
$$30y + 8 - 30y + 20 = 86y - 20 - 30y + 20$$
$$28 = 56y$$
$$\frac{28}{56} = y$$
$$\frac{1}{2} = y \quad [\text{or } y = .5]$$

6. First, use the Distributive Laws.

$$.4t - \underbrace{(.4)(.5)}_{.20} = .25t + \underbrace{(.25)(.1)}_{.025}$$

Now, multiply both sides by 1000 to eliminate decimals.

$$[.4t - .2] \times 1000 = [.25t + .025] \times 1000$$
$$400t - 200 = 250t + 25$$
$$400t - 200 - 250t + 200 = 250t + 25 - 250t + 200$$
$$150t = 225$$
$$t = \frac{225}{150}$$
$$t = 1.5$$

HOME EXERCISES

In Exercises 1–30, solve each equation.

1. $x - .3 = .7$ **2.** $z + .4 = 2.4$ **3.** $y + 1.02 = 3.02$ **4.** $y - .03 = 4.97$ **5.** $x - .7 = 1.44$

6. $t + .48 = .9$ **7.** $1.4 - y = 0$ **8.** $x - (.2 + .03) = 0$ **9.** $\frac{x}{.2} = 5$ **10.** $\frac{t}{.6} = 50$

11. $\frac{z}{.03} = 100$ 12. $\frac{y}{.004} = 2000$ 13. $.5x = 60$ 14. $.3t = 900$ 15. $.07y = 14$

16. $.001x = 5$ 17. $\frac{z}{.4} = 3$ 18. $\frac{y}{1.5} = 2$ 19. $\frac{x}{.25} = 8$ 20. $.2t = 2$

21. $1.4z = 28$ 22. $2.75x = 5.5$ 23. $.2(x - .1) = 3$ 24. $.4x + .05 = 9$

25. $.4y + .3 = .1 - .5y$ 26. $.75z - .25 = 1 - .5z$ 27. $.05(t + 1) = .25$

28. $t + (1 - .2t) = 2$ 29. $\frac{x - .03}{.1} = 4.1x$ 30. $\frac{y + 1}{.08} = \frac{y - 1}{.4}$

In Exercises 31–34, solve and check.

31. $x - .02 = 9.98$ 32. $x + .42 = .8$ 33. $\frac{x}{.06} = 200$ 34. $.01x = 50$

35. When .5 is added to a number, the sum is 5.5 . What is the number?

36. When .5 is subtracted from a number, the difference is 5.5 . What is the number?

37. What number must be multiplied by .5 to obtain 25? 38. What number must be divided by .5 to obtain 25?

6.6 DECIMALS, FRACTIONS, AND PERCENT

FRACTIONS TO DECIMALS

Some rational numbers, such as $\frac{3}{4}$, can be represented as "ordinary" decimals.

To express $\frac{3}{4}$ as a decimal, *divide numerator by denominator.*

$$4\overline{)3.00}\quad\text{quotient } .75$$

← ***Line up the decimal points.***

← ***Add as many 0's as necessary to the right of the decimal point.***

Thus, $\frac{3}{4} = .75$

Example 1 Express the following fractions as decimals.

a. $\frac{1}{2}$ b. $\frac{7}{8}$

Solution a. Divide numerator, 1, by denominator, 2.

$$2\overline{)1.0}\quad\text{quotient } .5$$

Thus, $\frac{1}{2} = .5$

b. Divide numerator, 7, by denominator, 8.

$$8\overline{)7.000}\quad .875$$

Thus, $\frac{7}{8} = .875$

CLASS EXERCISES *Express each fraction as a decimal.*

1. $\frac{1}{20}$ **2.** $\frac{13}{10}$

INFINITE REPEATING DECIMALS

The rational number $\frac{1}{3}$ can be expressed as an **infinite repeating decimal**. Divide numerator by denominator.

$$3\overline{)1.000000}\quad .333333$$

Thus,

$\frac{1}{3} = .333\ 333\ \ldots$ *Here, the digit 3 repeats, as indicated by the three dots.*

The rational number $\frac{2}{11}$ can be expressed as an infinite repeating decimal in which *two digits*, 1 and 8, *repeat*:

$$11\overline{)2.000000}\quad .181818$$

Thus, $\frac{2}{11} = .18\ 18\ 18\ \ldots$ *Again, three dots indicate this repetition.*

Example 2 Express the given fractions as infinite repeating decimals.

a. $\frac{2}{9}$ **b.** $\frac{1}{6}$

Solution

a. $9\overline{)2.000000}\quad .222222$

Thus,

$\frac{2}{9} = .222\ 222\ \ldots$

b. $6\overline{)1.0000000}$ with quotient $.1666666$

Here, just the 6 repeats. Thus,

$$\frac{1}{6} = .1\ 666\ 666\ldots$$

Every rational number can be written as either an ordinary decimal or as an infinite repeating decimal.

CLASS EXERCISES *Express each fraction as an infinite repeating decimal.*

3. $\frac{7}{9}$ **4.** $\frac{3}{11}$

DECIMALS TO FRACTIONS

Decimals can also be converted to fractions.

Example 3 Express each decimal as a fraction or as a mixed number.

a. .25 **b.** 1.17

Solution **a.** $.25 = \frac{25}{100} = \frac{1}{4}$

b. $1.17 = 1 + .17$

$= 1 + \frac{17}{100}$

$= 1\frac{17}{100}$ (as a mixed number)

$= \frac{117}{100}$ (as a fraction)

CLASS EXERCISES

5. Express .04 as a fraction in lowest terms.

6. Express 1.5 **a.** as a fraction in lowest terms and **b.** as a mixed number.

PRODUCTS OF FRACTIONS AND DECIMALS

Example 4 Multiply. $\frac{2}{5} \times 10.45$

Solution 1 5, the denominator of $\frac{2}{5}$, divides 10.45 evenly.

Solution 2 Convert $\frac{2}{5}$ to a decimal.

$$\frac{2}{\cancel{5}_1} \times \cancelto{2.09}{10.45} = 4.18$$

$$\frac{2}{5} = .4$$

10.45	←	2 decimal digits
× .4	←	+ 1 decimal digits
4.180	←	3 decimal digits

Example 5 Multiply. $\frac{1}{3} \times .7$

Solution Because $\frac{1}{3}$ is an infinite repeating decimal

$$\left(\frac{1}{3} = .333\ 333\ \ldots\right),$$

it is better to convert .7 to a fraction.

$$\frac{1}{3} \times .7 = \frac{1}{3} \times \frac{7}{10} = \frac{7}{30}$$

CLASS EXERCISES *Multiply*. **7.** $\frac{2}{3} \times .618$ **8.** $10\frac{3}{4} \times 72.4$ **9.** $\frac{1}{7} \times .13$

PERCENT TO DECIMALS

Percent means hundredths. The symbol

% stands for percent

Thus, 17% is read "17 percent"

It means $.17$ or $\frac{17}{100}$

Similarly, 9% means .09 or $\frac{9}{100}$

> **To convert percent to a decimal**: If there is no decimal point, insert one to the left of the percent symbol. Then, remove the percent symbol and move the decimal point two digits to the left.

Thus, 17% = 17.% = .17 and 9.% = **.09**

Note that when converting 9% to .09, you must add a **0** to the left of 9 so that you can move the decimal point two digits to the left.

Example 6 Express each percent as a decimal.

a. 50% **b.** 100% **c.** 150%

Solution **a.** $50\% = .50 = .5$
b. $100\% = 1.00 = 1.$
c. $150\% = 1.50 = 1.5$

Next, observe that

$$\frac{1}{2}\% \text{ means } \frac{1}{2} \text{ of } 1\%, \qquad \text{or } \frac{1}{2} \text{ of } .01$$

Recall that the word "of" often indicates multiplication. Thus,

$$\frac{1}{2}\% = \frac{1}{2} \times .01 = \frac{1}{2} \times .010 = .005$$

Example 7 Express each percent as a decimal.

a. $\frac{1}{4}\%$ **b.** $8\frac{1}{4}\%$

Solution **a.** $\frac{1}{4}\% = \frac{1}{4} \times .01 = \frac{1}{4} \times .0100$
$= .0025$

b. $8\frac{1}{4}\% = 8\% + \frac{1}{4}\%$
$= .08 + .0025$
$= .0825$

$$\begin{array}{r} .0800 \\ +\ .0025 \\ \hline .0825 \end{array}$$

CLASS EXERCISES *Express each percent as a decimal.*

10. 53% **11.** 250% **12.** $\frac{1}{5}\%$ **13.** $2\frac{1}{5}\%$

DECIMALS TO PERCENT

Percent means hundredths. Thus,

$$.27 = 27\%$$

(27 *hundredths* means 27 *percent*.)

> **To convert a decimal to percent**: If there are less than two decimal digits, add 0's at the right. Move the decimal point two digits to the right and insert the percent symbol at the right.

$$.27 = 27.\%, \text{ or } 27\%$$

$$.3 = .30 = 30.\%, \text{ or } 30\%$$

Example 8 Express each decimal as a percent.

a. .04 b. 1.25 c. .025
d. .5 e. .005

Solution
a. .04 = 04.% = 4%
b. 1.25 = 125.% = 125%
c. .025 = 02.5% = 2.5%
d. .5 = .50 = 50%
e. .005 = 00.5% = .5%

CLASS EXERCISES *Express each decimal as a percent.*
14. .07 **15.** 3.1 **16.** .003

PERCENT TO FRACTIONS

Percent means hundredths. Thus, $63\% = \frac{63}{100}$

> **To convert percent to fractional form**, remove the percent symbol, and divide the resulting number by 100. Simplify, if possible.

Example 9 Express each percent as a fraction in lowest terms.

a. 10% b. 12.5%

Solution
a. $10\% = \frac{10}{100} = \frac{1}{10}$
b. $12.5\% = \frac{12.5}{100} = \frac{12.5 \times 10}{100 \times 10} = \frac{125}{1000} = \frac{1}{8}$

CLASS EXERCISES *Express each percent as a fraction in lowest terms.*
17. 450% **18.** 7.5%

FRACTIONS TO PERCENT

> **To convert a fraction to a percent,** divide numerator by denominator. First, express as a decimal, then, as a percent.
> **To convert a mixed number to a percent,** first separate the integral part from the fractional part. Express as a decimal, then, as a percent.

Example 10 Express each of the following as a percent.

a. $\frac{3}{4}$ b. $2\frac{1}{2}$

Solution
a. $4\overline{)3.00}$ = .75

Thus, $\frac{3}{4} = .75 = 75\%$

b. $2\frac{1}{2} = 2 + \frac{1}{2} = 2 + .5 = 2.50$
$= 250\%$

CLASS EXERCISES *Express each fraction or mixed number as a percent.*

19. $\frac{3}{8}$ **20.** $1\frac{1}{4}$

SOLUTIONS TO CLASS EXERCISES

1. $20\overline{)1.00}$ with quotient .05. Thus, $\frac{1}{20} = .05$

2. $10\overline{)13.0}$ with quotient 1.3. Thus, $\frac{13}{10} = 1.3$

3. $9\overline{)7.000000}$ with quotient .777777. Thus, $\frac{7}{9} = .777\ 777\ldots$

4. $11\overline{)3.000000}$ with quotient .272727. Thus, $\frac{3}{11} = .27\ 27\ 27\ldots$

5. $.04 = \frac{4}{100} = \frac{1}{25}$

6. **a.** $1.5 = \frac{15}{10} = \frac{3}{2}$ **b.** $1.5 = 1\frac{5}{10} = 1\frac{1}{2}$

7. $\frac{2}{\not{3}_1} \times \not{.618}^{.206} = .412$

8. $10\frac{3}{4} = 10.75$

$$\begin{array}{r} 10.75 \\ \times\ 72.4 \\ \hline 4300 \\ 2150 \\ 7525 \\ \hline 778.3\not{0}\not{0} \end{array}$$

9. $.13 = \frac{13}{100}$

$\frac{1}{7} \times \frac{13}{100} = \frac{13}{700}$

10. $53.\% = .53$

11. $250\% = 2.50 = 2.5$

12. $\frac{1}{5}\% = \frac{1}{5} \times .010 = .002$

13. $2\frac{1}{5}\% = 2\% + \frac{1}{5}\% = .020 + .002 = .022$

14. $.07 = 07\% = 7\%$

15. $3.1 = 3.10 = 310\%$

16. $.003 = \not{0}\not{0}.3\% = .3\%$

17. $450\% = \frac{450}{100} = \frac{45}{10} = \frac{9}{2}$

18. $7.5\% = \frac{7.5}{100} = \frac{7.5 \times 10}{100 \times 10} = \frac{75}{1000} = \frac{3}{40}$

19. $8\overline{)3.000}$ with quotient .375. Thus, $\frac{3}{8} = .375 = 37.5\%$

20. $1\frac{1}{4} = \frac{5}{4}$

$4\overline{)5.00}$ with quotient 1.25. Thus, $1\frac{1}{4} = \frac{5}{4} = 1.25 = 125\%$

HOME EXERCISES

In Exercises 1–8, express each fraction as a decimal.

1. $\frac{2}{5}$ **2.** $\frac{9}{20}$ **3.** $\frac{9}{50}$ **4.** $\frac{3}{8}$ **5.** $\frac{1}{200}$ **6.** $\frac{7}{500}$ **7.** $\frac{11}{2000}$ **8.** $\frac{-1}{5000}$

In Exercises 9–12, express each fraction as an infinite repeating decimal.

9. $\frac{1}{9}$ **10.** $\frac{5}{6}$ **11.** $\frac{4}{3}$ **12.** $\frac{-7}{12}$

In Exercises 13–20, express each decimal as a fraction in lowest terms.

13. .29 **14.** .03 **15.** −.2 **16.** .06 **17.** .12 **18.** .002 **19.** .75 **20.** .075

In Exercises 21–23, express each decimal **a.** *as a fraction in lowest terms and* **b.** *as a mixed number.*

21. 1.1 **22.** 3.5 **23.** 1.25

In Exercises 24–28, multiply or divide, as indicated.

24. $\frac{1}{5} \times 4.85$ **25.** $\frac{3}{4} \times 84.96$ **26.** $.0093 \times 3\frac{1}{3}$ **27.** $.305 \div 7\frac{1}{7}$ **28.** $\frac{2}{3} \div .05$

In Exercises 29–36, express each percent as a decimal.

29. 93% **30.** 5% **31.** 300% **32.** 1000% **33.** 175% **34.** $\frac{1}{10}\%$ **35.** $\frac{3}{4}\%$ **36.** $10\frac{3}{4}\%$

In Exercises 37–44, express each decimal as a percent.

37. .43 **38.** .95 **39.** .03 **40.** 1.15 **41.** 5. **42.** .125 **43.** .0125 **44.** 1.125

In Exercises 45–50, express each percent as a fraction in lowest terms.

45. 17% **46.** 3% **47.** 20% **48.** 350% **49.** 10.5% **50.** 8.25%

In Exercises 51–58, express each fraction or mixed number as a percent.

51. $\frac{81}{100}$ **52.** $\frac{7}{10}$ **53.** $\frac{7}{100}$ **54.** $\frac{7}{20}$ **55.** $\frac{5}{8}$ **56.** $\frac{7}{12}$ **57.** $5\frac{1}{2}$ **58.** $4\frac{1}{5}$

In Exercises 59–63, fill in "<" or ">".

59. $\frac{2}{5}$ □ .2 **60.** $\frac{1}{4}$ □ .24 **61.** $-\frac{3}{10}$ □ −.33 **62.** $\frac{6}{5}$ □ 1.25 **63.** $-\frac{1}{3}$ □ −.4

6.7 PERCENTAGE PROBLEMS

TRANSLATING WORDS TO SYMBOLS

A problem involving percent is sometimes solved by setting up an equation. Your first task is to translate the problem into mathematical symbols. Then, you solve the resulting equation.

Example 1 **a.** Express in mathematical symbols.

25% of 40

Solution Here, the word "of" indicates multiplication.

a.
25% of 40
25% × 40

b. Set up an equation for the problem:

What number is 25% of 40?

c. Solve the equation of part **b**.

b. Let x represent the unknown number.

$$\underbrace{\text{What number}}_{x}\ \underset{=}{\text{is}}\ \underbrace{25\%\text{ of }40}_{25\% \times 40}$$

c. Convert percent

to a decimal:

$$x = \underbrace{.25 \times 40}_{10.00}$$
$$x = 10$$

or to a fraction:

$$x = \underbrace{\frac{1}{4} \times 40}_{10}$$
$$x = 10$$

Thus, **10** is 25% of 40.

From now on, we will use the decimal method. Note the difference in the wording of Example 2. The expressions you obtain are different from those of Example 1.

Example 2 Express the following in mathematical symbols.

a. 25% of a number

b. Set up an equation for the problem:

25% of a number is 40.

c. Solve the equation of part b.

Solution **a.** Let x represent the number.

$$\underset{25\%}{25\%}\ \underset{\times}{\text{of}}\ \underbrace{\text{a number}}_{x}$$

b.

$$\underbrace{25\% \text{ of a number}}_{25\% \cdot x}\ \underset{=}{\text{is}}\ \underset{40}{40}$$

c. Solve the equation.

$$.25 \times x = 40 \qquad \textit{Divide both sides by .25 .}$$

$$\frac{.25x}{.25} = \frac{40}{.25}$$

$$x = \frac{40 \times 100}{\underbrace{.25 \times 100}_{25}} = \frac{4000}{25} = 160$$

Therefore, 25% of **160** is 40.

CLASS EXERCISES *In Exercises 1–3, express in mathematical symbols.*

1. 10% of a number **2.** 50% of 30 **3.** What percent of 50

In Exercises 4–6, set up an equation and solve for the indicated numbers.

4. 50% of a number is 20. Find this number.

5. 100 is 40% of what number? **6.** What number is 20% of 15?

APPLICATIONS Problems involving percent arise in many practical applications. Often you must reword a problem in order to translate it into mathematical symbols.

Example 3 Five hundred seventy-six out of a total 960 graduating seniors at a high school are planning to enter college. What percentage plan to enter college?

Solution Reword the problem:

$$\underbrace{\text{What percent}}_{x\%} \ \underset{\times}{\text{of}} \ \underset{960}{960 \text{ (seniors)}} \ \underset{=}{\text{is}} \ \underset{576}{576 \text{ (college-bound seniors)?}}$$

Solve the equation:

$$x \times \underbrace{.01 \times 960}_{9.60} = 576 \qquad x\% = x \times \mathbf{.01}$$

$$\frac{x \times 9.6}{9.6} = \frac{576 \times 10}{9.6 \times 10}$$

$$x = 60$$

Thus, **60 percent** of the seniors plan to enter college.

Example 4 Oxygen constitutes 21% of the volume of air. If a room contains 3000 cubic feet of air, how much of this is oxygen ?

Solution Reword the problem:

$$\underbrace{\text{The number of cubic feet of oxygen}}_{x} \ \underset{=}{\text{is}} \ \underbrace{\underset{21\%}{21\%} \ \underset{\times}{\text{of}} \ \underset{3000}{3000}}_{630.00}$$

$$x = 630$$

Therefore, the room contains 630 cubic feet of oxygen.

CLASS EXERCISES

7. A pitcher throws 50 called strikes out of a total of 125 pitches in a game. What is his percentage of called strikes?
8. A student answers 84% of the questions on an exam correctly. If there are 50 questions on the exam, how many does she answer correctly?
9. An alloy contains 55% copper. How much copper is there in 800 tons of the alloy?

INCREASE AND DECREASE

A man earns \$10,000 a year. He receives an 8% *increase* in his salary. To determine the *amount* of his increase, find 8% of \$10,000.

$$8\% \times \$10{,}000 = .08 \times \$10{,}000 = \$800$$

His increase is \$800. To determine his *new* salary, add the increase of \$800 to his original salary of \$10,000.

$$\underbrace{\$10{,}000}_{\text{original salary}} + \underbrace{\$800}_{\text{increase}} = \underbrace{\$10{,}800}_{\text{new salary}}$$

In problems involving an *increase*,

$$\boxed{\text{Rate of increase} = \frac{\text{increase}}{\text{original amount}}}$$

Here,

$$\frac{800}{10{,}000} \begin{array}{l} \text{------ increase} \\ \text{------ original amount} \end{array}$$

Divide numerator and denominator by 100:

$$\frac{8\cancel{00}}{10{,}0\cancel{00}} = \frac{8}{100} = 8\%$$

Thus, the rate of increase (in salary) is 8%.

For convenience, let

$R = \textit{rate of increase}$, $I = \textit{increase}$, $A = \textit{original amount}$

Then,

$$\boxed{R = \frac{I}{A}}$$

or, multiplying both sides by A,

$$\boxed{R \cdot A = I}$$

Also,

$$\boxed{\text{New amount} = \text{original amount} + \text{increase}}$$

Example 5 The price of pecans, which was originally \$1.60 per pound, is increased by 5%. Find **a.** the increase and **b.** the new price.

Solution **a.** $5\% = .05$

R	$\cdot$ A	$= I$
.05	1.60	I

$I = .05 \times 1.60 = .08$
The increase is \$.08, or 8 cents per pound.

b. The new price is
\$1.60 + \$.08, or \$1.68 .

During hard times, a man who was earning \$10,000 per year is asked to take a 5% cut in his salary. The cut (or decrease) amounts to

$$5\% \times \$10{,}000, \text{ or } \$500$$

His new salary is then \$10,000 *minus* the decrease of \$500.

$$\$10{,}000 - \$500 = \$9500$$

He then earns \$9500 per year.

In general,

$$\boxed{\text{Rate of decrease} = \frac{\text{decrease}}{\text{original amount}}}$$

In symbols, let

R = *rate of decrease*, D = *decrease*, A = *original amount*

Then,

$$\boxed{R = \frac{D}{A}}$$

or, multiplying both sides by A,

$$\boxed{R \cdot A = D}$$

Also,

$$\boxed{\text{New amount} = \text{original amount} - \text{decrease}}$$

Example 6 Because of greater efficiency in management, the weekly cost of operating a plant is reduced by 3.2%. If the original operating cost was $95,000, find the new operating cost.

Solution

$$3.2\% = .032$$

R	A	$= D$
.032	95,000	D

$$D = .032 \times 95{,}000 = 3040$$

The decrease in cost amounts to $3040 per week. The new operating cost is then $91,960.

$95 000	original amount
− 3 040	decrease
$91 960	new amount

CLASS EXERCISES

10. A saleswoman who earns $12,600 per year receives a 6% increase in salary. **a.** What is the amount of her increase? **b.** What is her new salary?

11. A city with 420,000 people has a 4% increase in population. What is the new population?

12. The price of frozen orange juice increases from 35 cents per can to 42 cents. Find the rate of increase.

13. A woman who earns $15,000 a year receives a $750 cut in her salary. What is the rate of decrease?

14. The population in a resort area decreases by 60% in the winter. If there are 90,000 people there in the summer, how many are there in the winter?

SOLUTIONS TO CLASS EXERCISES

1. 10% of a number
$10\% \times x$

2. 50% of 30
$50\% \times 30$

3. What percent of 50
$x\% \times 50$

4. 50% of a number is 20.
$.50 \times x = 20$

$$x = \frac{20 \times 100}{.50 \times 100}$$

(denominator: 50)

$$x = 40$$

5. 100 is 40% of what number?

$100 = .40 \times x$

$$\frac{100 \times 100}{.40 \times 100} = x$$

$\underbrace{.40 \times 100}_{40}$

$250 = x$

6. What number is 20% of 15?

$x = .20 \times 15 = 3.00$

$= 3$

7. What percent of 125 (pitches) is 50 (called strikes)?

$x\% \times 125 = 50$

$x \times .01 \times 125 = 50$

$$\frac{x \times 1.25}{1.25} = \frac{50 \times 100}{1.25 \times 100} = 40$$

Thus, **40 percent** of the pitches are called strikes.

8. 84% of 50 (questions) is the number answered correctly.

$.84 \times 50 = x$

$42 = x$

42 questions are answered correctly.

9. 55% of 800 (tons of alloy) is the number of tons of copper.

$.55 \times 800 = x$

$440 = x$

There are **440** tons of copper in the alloy.

10. a.

R ·	A =	I
.06	12,600	I

$I = .06 \times 12{,}600 = 756$

Her increase is $756 per year.

b. Her new salary is $12,600 + $756, or $13,356 per year.

11.

R ·	A =	I
.04	420,000	I

$I = .04 \times 420{,}000 = 16{,}800$

The increase in population is 16,800.

The new population is 420,000 + 16,800, or 436,800.

12. I (increase) $= 42 - 35 = 7$

R ·	A =	I
R	35	7

$35R = 7$

$$R = \frac{1}{5} = 20\%$$

13.

R ·	A =	D
R	15,000	750

$15{,}000R = 750$

$$R = \frac{1}{20} = 5\%$$

The rate of decrease is 5%.

14.

R ·	A =	D
.60	90,000	D

$D = .60 \times 90{,}000 = 54{,}000$

The decrease in population during winter is 54,000.
The winter population is, therefore,
90,000 − 54,000, or 36,000.

HOME EXERCISES

In Exercises 1–4, express in mathematical symbols.

1. 35% of 200 **2.** 20% of a number **3.** What percent of 120 **4.** 8.5% of a number

In Exercises 5–10, set up an equation. Do not solve.

5. What number is 35% of 200? **6.** 20% of a number is 15. **7.** What percent of 120 is 30?
8. 35% of 200 is what number? **9.** What number is 40% of 80? **10.** 10.5 is 50% of what number?

In Exercises 11–18, set up an equation and solve for the indicated numbers.

11. 25% of a number is 4. Find this number. **12.** 30 is 75% of what number?

13. What percent of 80 is 40? *Check your result.* **14.** 35 is what percent of 7000?

15. Find 2.2 percent of 400. **16.** Find 300% of 15.

17. 6.5% of 200 is what number? *Check your result.* **18.** What percent of 21 is 5.25? *Check your result.*

19. Of 1250 people surveyed, 750 are Democrats. What is the percentage of Democrats?

20. A basketball player makes 78% of his foul shot attempts. If he has made 195 foul shots, how many has he attempted?

21. A student answers 75% of her exam questions correctly. If she answers 30 questions correctly, how many questions are on the exam?

22. 35% of the students at a university are foreign-born. If there are 8400 students enrolled, how many are foreign-born?

23. The leading freshman economics text has 40% of the market. If altogether 450,000 freshman economics texts are sold one year, how many copies does the leading text sell that year?

24. 36% of Americans have blood-type B. If there are 215,000,000 Americans, how many have blood-type B?

25. A family's monthly income is $2200. 25% of the money goes for rent. What is the monthly rent?

26. 16% of the people in a town are over 65. If 20,000 people there are over 65, what is the town's population?

27. A man who earns $18,000 per year receives a 4% increase in salary. **a.** How much does this increase amount to? **b.** What is his new salary?

28. The price of eggs increases from 90¢ per dozen to 99¢ per dozen. What is the rate of increase?

29. A student takes 12 credits one semester and 15 credits the next semester. By what rate does his credit load increase?

30. The population of a country increases from 7,500,000 to 9,000,000 over a period of time. What is the rate of increase during this period?

31. A woman receives a salary of $21,000 per year. Her fringe benefits amount to an additional 5.5%. What is the total yearly cost of this woman's labor to her employer?

32. A school with 9500 students has a 20% increase in enrollment. What is the new enrollment?

33. A business that makes a profit of $60,000 one year increases its profit by 50% the next year. What is its profit that next year?

34. A worker who earns $8000 has 12% deducted from her salary for taxes and union dues. What is her take-home pay?

35. A town with 18,000 people loses 3% of its population. What is its new population?

36. 8% of a 25-pound bag of potatoes spoils. How many pounds are usable?

37. Suppose the U.S. college population drops from 12,400,000 to 11,780,000. What percent decrease is this?

38. During one year a mine produces 350 million tons of ore. The next year there is a decrease in production of 7%. How much ore is produced that next year?

39. The rabbit population of a field increases at an annual rate of 50%. There are 240 rabbits in the field one year. **a.** How many will there be the following year? **b.** How many were there the preceding year?

6.8 INTEREST

ANNUAL INTEREST When money is invested, it earns *interest* over a period of time. To simplify matters, *the basic period of time will at first be one year*. The amount of money invested, called the **principal**, times the **annual**, or **yearly**, **interest rate** equals the **interest earned in one year**. Let

R *be the interest rate,* P *be the principal,* I *be the interest earned.*

Then,

$$\boxed{R \cdot P = I}$$

Thus, \$100 (principal) invested for one year at a rate of 8% earns

$$.08 \times 100 \text{ dollars (interest)}$$

or \$8 in one year. Also, \$300 invested for one year at 7% earns

$$.07 \times 300 \text{ dollars}$$

or \$21 in one year.

Example 1 Jose deposits \$850 in a bank that pays an annual interest rate of 6%. How much interest does he earn in one year?

Solution

R	P	$= I$
.06	850	I

$$I = .06 \times 850 = 51$$

He earns \$51 interest in one year.

Example 2 How much must Alice invest at $8\frac{1}{2}\%$ in order to earn \$340 interest in one year? *Check your result.*

Solution

$$I = 340$$

$$R = 8\frac{1}{2}\% = .085$$

R $\cdot$	P	$= I$
.085	P	340

$$.085P = 340$$

$$\frac{.085P}{.085} = \frac{340 \times 1000}{.085 \times 1000}$$

$$P = 4000$$

Alice must invest \$4000 at $8\frac{1}{2}\%$ to earn \$340 interest in one year.

check

$$.085 \times 4000 \stackrel{?}{=} 340$$

$$340 \stackrel{\checkmark}{=} 340$$

CLASS EXERCISES

1. Dominick deposits \$1500 in a bank that pays an annual interest rate of 7%. How much interest does he earn in one year?
2. What is the annual interest rate if \$86.40 interest is paid on a principal of \$1080 for one year?
3. How much money must be invested for a year at 6% in order to earn \$90 interest? *Check your result.*

INTEREST FOR PART OF A YEAR

When money is invested for part of a year, the above interest formula must be modified. Let

T be the time (expressed as a fraction of a year),
R be the annual interest rate, P be the principal, and
I be the interest earned.

Then,

$$\underbrace{R \cdot P}_{\text{annual interest}} \cdot \underbrace{T}_{\text{fraction of a year}} = I$$

Example 3 How much interest is earned on \$3000 invested at an annual rate of 6% for ¼ of a year?

Solution

R	$\cdot\ P$	$\cdot\ T$	$= I$
.06	3000	$\frac{1}{4}$	I

$$I = .06 \times 3000 \times \frac{1}{4} = \frac{180}{4} = 45$$

Thus, \$45 interest is earned for ¼ of a year.

CLASS EXERCISES

4. How much interest is earned on \$4200 invested at an annual rate of 8% for ½ of a year?
5. For what part of a year must \$5000 be invested at an annual rate of 6% in order to earn \$75 interest?

COMPOUND INTEREST

In practice, interest is generally *compounded.* **Compound interest** is interest paid on previously earned interest, as well as on the principal. Compounding results in greater interest, as you will see in Example 4. There, the basic period for the interest rate is one year. In this case, the interest is said to be compounded annually.

Example 4 Jerry deposits \$500 in a bank that pays an annual interest rate of 6%. How much interest does he earn if he leaves all his money, including his interest, in for two years?

Solution There are two different amounts involved—the original principal (for the *first* year) as well as the principal at the end of one year (for the *second* year). Use *subscripts* on P to distinguish between these principals. Thus, let P_1 be the principal of \$500 for the *first* year.

There are also different amounts of interest earned for each year. Let I_1 be the interest earned for the *first* year. The interest *rate* for each year is 6%, or .06 .

R	$\cdot\ P_1$	$= I_1$
.06	500	I_1

$$I_1 = .06 \times 500 = 30$$

Jerry earns \$30 interest the first year. He leaves this in the bank and, therefore, has \$530 at the end of one year. Thus, his principal, P_2, for the *second* year is \$530. Let I_2 be the interest earned during the *second* year. Again, the interest rate is .06 .

R	$\cdot\ P_2$	$= I_2$
.06	530	I_2

$$I_2 = .06 \times 530 = 31.80$$

He earns \$31.80 interest for the second year. For two years he earns

\$30.00	first year's interest
+ 31.80	second year's interest (compounded)
\$61.80	total interest for two years

If interest were paid *only on principal*, he would have earned

$$\begin{array}{rl} \$30.00 & \text{------ first year's interest} \\ +\ 30.00 & \text{------ second year's interest (\textit{not} compounded)} \\ \hline \$60.00 & \text{------ total interest for two years} \end{array}$$

Thus, he earns $1.80 more because of compounding the interest.

When the basic period for the interest rate is ¼ of a year, the interest is said to be **compounded quarterly**. When the basic period is ½ of a year, the interest is **compounded semi-annually**.

Example 5 Juan invests $1000 at 8% compounded *quarterly*. How much interest does he earn in *half* a year?

Solution

$$\frac{1}{2} = \frac{2}{4}$$

Thus, half a year is the same as two quarters (of a year). Let P_1 be the principal and I_1 the interest for the first quarter. The interest rate for each quarter is 8% or .08 .

R ·	P_1 ·	T =	I_1
.08	1000	$\frac{1}{4}$	I_1

$$I_1 = .08 \times 1000 \times \frac{1}{4} = \frac{80}{4} = 20$$

Juan earns $20 interest for the first quarter. He leaves this in the bank and has $1020 at the end of the first quarter. This is his principal, P_2, for the second quarter. Let I_2 be the interest earned for the second quarter.

R ·	P_2 ·	T =	I_2
.08	1020	$\frac{1}{4}$	I_2

$$I_2 = .08 \times 1020 \times \frac{1}{4} = \frac{81.60}{4} = 20.40$$

He earns $20.40 for the second quarter. Thus, for half a year he earns $40.40 .

$$\begin{array}{rl} \$20.00 & \text{-----}\ I_1 \\ +\ 20.40 & \text{-----}\ I_2 \\ \hline \$40.40 & \end{array}$$

CLASS EXERCISES

6. Bob deposits $1200 in a bank that pays an annual interest rate of 6%. How much interest does he earn if he leaves in all of his money, including the interest for the first year, for two years?
7. Concha invests $2500 at 8% compounded semiannually. How much interest does she earn in a year?

SOLUTIONS TO CLASS EXERCISES

1.

R ·	P =	I
.07	1500	I

$$I = .07 \times 1500 = 105$$

He earns $105 interest in one year.

2.

R ·	P =	I
R	1080	86.40

$$1080R = 86.40$$

$$108\overline{)8.64} = .08$$

8 64

The annual interest rate is 8%.

3.

R ·	P =	I
.06	P	90

$$.06P = 90$$

$$P = \frac{90 \times 100}{.06 \times 100} = 1500$$

Thus, $1500 must be invested.

check

$$.06 \times 1500 \stackrel{?}{=} 90$$

$$90.00 \stackrel{\checkmark}{=} 90$$

4.

R ·	P ·	T =	I
.08	4200	$\frac{1}{2}$	I

$$I = .08 \times 4200 \times \frac{1}{2} = \frac{336}{2} = 168$$

$168 interest is earned.

5.

R ·	P ·	T =	I
.06	5000	T	75

$$.06 \times 5000 \times T = 75$$

$$\frac{.06 \times 5000 \times T}{.06 \times 5000} = \frac{75}{.06 \times 5000}$$

$$T = \frac{75}{300} = \frac{1}{4}$$

The money must be invested for ¼ of a year.

6.

R ·	P_1 =	I_1
.06	1200	I_1

$$I = .06 \times 1200 = 72$$

He earns $72 the first year. His principal, P_2, for the second year is $1200 + $72, or $1272.

R ·	P_2 =	I_2
.06	1272	I_2

$$I_2 = .06 \times 1272 = 76.32$$

His total interest for two years is $I_1 + I_2$, or $148.32 .

7.

R	$\cdot P_1$	$\cdot T$	$= I_1$
.08	2500	$\frac{1}{2}$	I_1

$$I_1 = .08 \times 2500 \times \frac{1}{2} = \frac{200}{2} = 100$$

She earns \$100 for the first half-year. Her principal, P_2, for the second half-year, is \$2600.

R	$\cdot P_2$	$\cdot T$	$= I_2$
.08	2600	$\frac{1}{2}$	I_2

$$I_2 = .08 \times 2600 \times \frac{1}{2} = \frac{208}{2} = 104$$

Her total interest for a year is \$100 + \$104, or \$204.

HOME EXERCISES

1. How much interest is earned in one year on a principal of \$600 if the annual interest rate is 7%?
2. How much interest is earned in one year on a principal of \$400 if the annual interest rate is $7\frac{1}{2}$%?
3. What is the annual interest rate if \$45 interest is paid in a year on a principal of \$750? *Check your result.*
4. What is the annual interest rate if \$144 is paid in a year on a principal of \$1800?
5. How much money must be invested for a year at an annual interest rate of 9% in order to earn \$108 interest? *Check your result.*
6. How much money must be invested for a year at an annual interest rate of $8\frac{1}{2}$% in order to earn \$102 interest?
7. Which earns more interest? \$900 invested at 8% or \$1000 at $7\frac{1}{2}$%?
8. How much money must be invested at 8% in order to earn the same as \$800 at 7%? *Check your result.*
9. How much interest is earned on \$6000 invested at an annual rate of 7% for half a year?
10. How much interest is earned on \$4800 invested at an annual rate of 8% for ¾ of a year?
11. For what part of a year must \$1200 be invested at an annual rate of 8% in order to earn \$24 interest?
12. At what annual rate must \$8000 be invested in order to earn \$400 in half a year?
13. Which earns more interest? \$5000 at an annual rate of 7% for half a year or \$3000 at an annual rate of 8% for ¾ of a year?
14. A sum of \$250 is left as security on an apartment. The landlord must pay an interest rate of 5% annually. If all the money, including the interest, is left for two years, how much interest is earned?
15. Roberto deposits \$3200 in a bank that pays an annual interest rate of 8%. How much interest will he earn if he leaves in all his money, including the interest, in for two years?
16. Maria invests \$1000 at 6% compounded semiannually. How much interest does she earn in a year?
17. Hillel invests \$10,000 at 6% compounded quarterly. How much interest does he earn in half a year?
18. Teresa invests \$2000 at 5% compounded semiannually. How much interest does she earn in a year and a half?

Let's Review Chapter 6

In Exercises 1 and 2, express each fraction as a decimal.

1. $\frac{17}{100}$ **2.** $\frac{33}{1000}$

3. Which of the following are equivalent to .02?
a. .20 **b.** .002 **c.** .020 **d.** .020 00

In Exercises 4–6, fill in "<" or ">".

4. .46 ☐ .42 **5.** .801 ☐ .799 **6.** −.38 ☐ −.48

7. Add.

$$\begin{array}{r} .341 \\ .207 \\ .19 \\ \underline{.04} \end{array}$$

8. Subtract. .29 − .115

9. Add.

$$\begin{array}{r} -2.4 \\ 3.1 \\ -5.7 \\ 4.9 \\ \underline{-\ .8} \end{array}$$

10. Subtract. 4.5 − (1.08 − .67)

In Exercises 11–17, find each value.

11. $(.01)^2\ (10.5)$ **12.** $.198 \div 100$ **13.** $\frac{-.009}{-.3}$ **14.** $100(.4 - .07 + .35)$

15. $(.4 + .3)^2$ **16.** $\frac{3 - .02}{.04}$ **17.** $.7317 \times \frac{4}{9}$

18. Multiply the square of .4 by .03 .

19. Estimate the product by first rounding each factor to the nearest integer.
$9.31 \times 11.79 \times 2.49$

In Exercises 20–22, solve each equation.

20. $x - .02 = 1.98$ **21.** $1.2x = 8.4$ **22.** $1 + \frac{y}{.4} = 1.2$

23. Express $\frac{4}{5}$ as a decimal. **24.** Express $\frac{1}{6}$ as an infinite repeating decimal.

25. Express .48 as a fraction in lowest terms. **26.** Express $8\frac{1}{2}\%$ as a decimal.

27. Express 36% as a fraction in lowest terms. **28.** Express $\frac{3}{20}$ as a percent.

29. Express in mathematical symbols: 75% of a number.

30. 24 of the 40 children in a class are girls. What is the percentage of girls?

31. A man's salary increases from \$20,000 per year to \$24,000 per year. Find the rate of increase.

32. The population of a town drops from 50,000 to 48,000 people. What percent decrease is this?

33. How much must Donna invest at an annual interest rate of 8% in order to earn \$100 in one year?

34. Phil invests \$1200 at 6% compounded quarterly. How much does he earn in half a year?

7 EQUATIONS AND GRAPHS

DIAGNOSTIC TEST

Perhaps you are already familiar with some of the material in Chapter 7. This test will indicate which sections in Chapter 7 you need to study. The question number refers to the corresponding section in Chapter 7. If you answer *all* parts of the question correctly, you may omit the section. But *if any part of your answer is wrong, you should study the section.* When you have completed the test, turn to page A15 for the answers.

7.1 From Words to Equations

a. Find three consecutive integers whose sum is 87.

b. A man is eight years older than his wife. Twelve years ago, he was twice her age. How old is the man?

c. A 24-foot rope is cut into three pieces. One piece is four feet longer than a second piece, which in turn, is twice as long as the third piece. How long is each piece?

7.2 Distance Problems

a. A car traveling at 65 miles per hour along a straight road passes another car, headed in the same direction, and traveling at 50 miles per hour. In how many hours are they 45 miles apart?

b. Two cars leave a toll booth at the same time traveling in opposite directions along a straight road. If each car travels at 60 miles per hour, how far apart are they after four hours?

c. A motorboat goes upstream at the rate of six miles per hour and returns downstream at the rate of 10 miles per hour. If the round trip takes four hours, how far upstream does the boat go?

7.3 Literal Equations

a. Solve for a. $\quad 2a - 3b = 10$

b. Solve for y. $\quad 2xy - y = 1$

7.4 Rectangular Coordinates

For each point in Figure 7A, find a. its x-coordinate b. its y-coordinate.

c. Identify the point with its ordered pair of coordinates. In other words, express the coordinates of the point as an ordered pair.

7.5 Lines

Graph each of the following equations.

a. $y = 2x - 1$ b. $y = 5$ c. $x = -1$

7.6 Systems of Linear Equations

a. Graph to find the intersection of the lines given by:

$$L_1\colon y = 2x, \qquad L_2\colon y = 4 - 2x$$

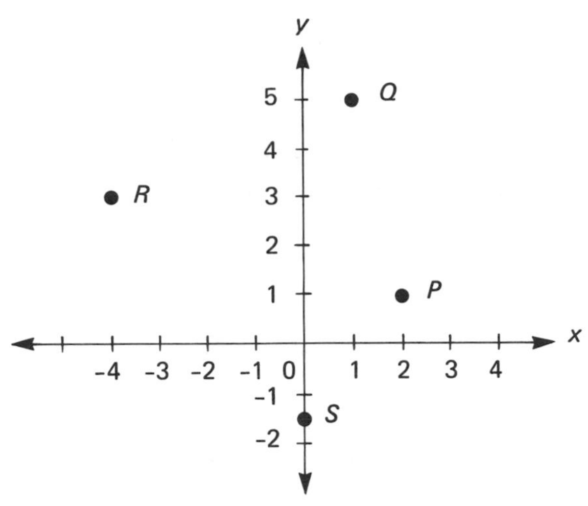

Fig. 7A

In b *and* c, *solve each system. You may use any method other than graphing.*

b. $y + 1 = x - 4$
$3y = x - 3$

c. $3x - 4y = 3$
$2x - 3y = 1$

7.7 Variation

a. Assume y varies directly as x. Suppose $y = -12$ when $x = 4$. Find y when $x = -8$.

b. Assume y varies inversely as x. Suppose $y = 3$ when $x = -2$. Find x when $y = 12$.

7.1 FROM WORDS TO EQUATIONS

TRANSLATING WORDS TO SYMBOLS

Many practical problems are solved by algebraic methods. Indeed, one of your main goals in this course is to learn how to apply mathematical methods to situations that arise in everyday life. When a problem is stated in words, you must translate the problem into mathematical symbols, just as you did with percentage problems in Chapter 6.

The problems you will now consider concern the **integers**:

$$\ldots -3, -2, -1, 0, 1, 2, 3, \ldots$$

Of course, you do not *directly* worry about integers in everyday conversation. But you constantly *use* integers in describing other concepts, such as money, time, distance, and temperature. For example, you use

100	to describe earning \$100
-100	to describe losing \$100

$32°F$ to describe the freezing point of water
$-10°F$ to describe the temperature in Minneapolis on a January morning

Furthermore, integer problems serve to develop your ability to apply mathematics to other situations.

Example 1 Let x represent an integer. Express each of the following in mathematical symbols.

a. one more than the integer (or the next *consecutive* integer)
b. two less than the integer **c.** twice the integer
d. three more than half the integer
e. half of a quantity that is three more than the integer

Solution **a.** $\underbrace{\text{one more than}}_{1\,+}\ \underbrace{\text{the integer}}_{x}$ or $x + 1$

b. $\underbrace{\text{two less than}}_{-2\,+}\ \underbrace{\text{the integer}}_{x}$ or $x - 2$

c. $\underbrace{\text{twice the integer}}_{2x}$

d. $\underbrace{\text{three more than}}_{3\,+}\ \underbrace{\text{half the integer}}_{\frac{1}{2}x}$ or $\frac{x}{2} + 3$

Here, you first halve the integer; then you add three.

e. $\underbrace{\text{half of}}_{\frac{1}{2}}\ \underbrace{\text{a quantity that is three more than an integer}}_{(3 + x)}$ or $\frac{x + 3}{2}$

Here, you first add three to the integer, and then take half.

A related type of problem concerns **age**, which is assumed to be an integer. The age may be given at various times—for example, now, two years ago, three years from now. Again, first translate the problem into mathematical symbols.

Example 2 *Five years ago*, a man was four times as old as his daughter. If x represents the daughter's *present* age, express the man's *present* age.

Solution Let x represent the daughter's age *now*. The daughter's age *five years ago* was

$$x - 5$$

The man's age *five years ago* was four times that of his daughter (*then*). His age *then* was

$$4(x - 5)$$

Now, the man is five years older. His *present* age is given by:

$$4(x - 5) + 5 = 4x - 20 + 5 = 4x - 15$$

CLASS EXERCISES

1. Express each in mathematical symbols.
   a. A certain integer is increased by 10
   b. Twice an integer is decreased by three
   c. the sum of two consecutive integers.
2. a. The sum of two integers is 20. If x represents one of these integers, express the other in terms of x.
   b. The larger of two integers is four more than the smaller, If y represents the smaller integer, express the larger one in terms of y.
3. A mother is 20 years older than her son. If x represents the son's age, express the mother's age in terms of x.
4. Let x represent a woman's age now.
   a. Express her age 10 years ago. b. Express her age in five years.
   c. If her husband is three years older than she is, express his age now.

INTEGER AND AGE PROBLEMS

In each of the following problems, translate the English expressions into mathematical symbols. You can then formulate the problems in terms of an equation. Solve the equation in order to solve the original problem.

Example 3 Find two consecutive integers whose sum is 25.

Solution You want two consecutive integers. Let x represent the smaller integer. The next consecutive integer is then $x + 1$.

Translate the problem:

$$\underbrace{\text{The sum of two consecutive integers}}_{x + (x + 1)} \ \underset{=}{\text{is}} \ \underset{25}{25.}$$

Solve the equation.

$$2x + 1 = 25$$
$$2x = 24$$
$$x = 12$$
$$x + 1 = 13$$

The two integers are 12 and 13.

Example 4 Helen is nine years older than her sister Sondra. In three years, Helen's age will be double Sondra's age. How old is each sister now?

Check your result.

Solution Let x represent Sondra's age now. Thus, Helen's age (now) is $x + 9$. In three years, Helen's age will be

$$(x + 9) + 3, \text{ or } x + 12$$

and Sondra's will be $x + 3$

Translate the problem:

$$\underbrace{\text{In three years, Helen's age}}_{x + 12} \ \underbrace{\text{will be}}_{=} \ \underbrace{\text{double Sondra's age (in three years)}}_{2(x + 3)}.$$

Solve this equation:

$$x + 12 = 2x + 6$$
$$6 = x$$
$$15 = x + 9$$

Thus, Sondra is 6 and Helen is 15.

check

$$15 + 3 \stackrel{?}{=} 2(6 + 3)$$
$$18 \stackrel{\checkmark}{=} 18$$

CLASS EXERCISES

5. Six more than an integer is four times this integer. Find the integer. *Check your result.*
6. A woman is now five times as old as her son. Three years ago, she was eight times as old as her son. How old is the son now?

OTHER PROBLEMS

Example 5 An 11-foot rope is cut into two pieces. One piece is 1 foot longer than the other. Find the length of each piece.

Solution Let x be the length of the shorter piece. Then, $x + 1$ is the length of the longer piece. Reword the first sentence of the problem as follows:

$$\underbrace{\text{The sum of the lengths of the two pieces}}_{x + (x + 1)} \ \underset{=}{\text{is}} \ \underset{11}{11 \text{ (feet).}}$$

Solve this equation.

$$\begin{aligned} 2x + 1 &= 11 \\ 2x &= 10 \\ x &= 5 \\ x + 1 &= 6 \end{aligned}$$

One piece is 5 feet long, the other, 6 feet long.

Example 6 An accountant teaches several evening courses at a college. His earnings from his accounting practice are double those from his teaching. His total annual income is \$39,000. How much does he earn from teaching?

Solution Let x be his teaching salary (in dollars). Then, $2x$ represents his earnings from his accounting practice.

$$\underbrace{\text{His total earnings}}_{x + 2x} \ \underset{=}{\text{are}} \ \underset{39{,}000}{\$39{,}000.}$$

$$\begin{aligned} 3x &= 39{,}000 \\ x &= 13{,}000 \end{aligned}$$

His annual teaching salary is \$13,000.

CLASS EXERCISES

7. A 24-foot rope is cut into three pieces. One piece is twice as long as a second piece and three times as long as the smallest piece. Find the length of each piece.
8. In a freshman class election, one of the two presidential candidates receives 20 more votes than the other. If 1048 votes are cast, how many does the winner receive?

SOLUTIONS TO CLASS EXERCISES

1. a. Let x represent the integer.

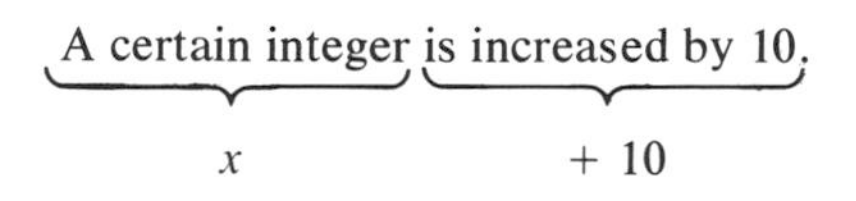

b. Let y represent the integer.

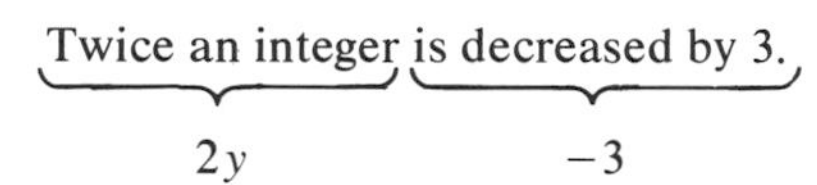

c. *Consecutive integers differ by* 1. For example, 5 and 6 are consecutive integers. If z represents an integer, the next consecutive integer is $z + 1$.

the sum of two consecutive integers

$z + (z + 1)$ or $2z + 1$

2. a. Let x represent one integer. Let $\square$ represent the other integer.

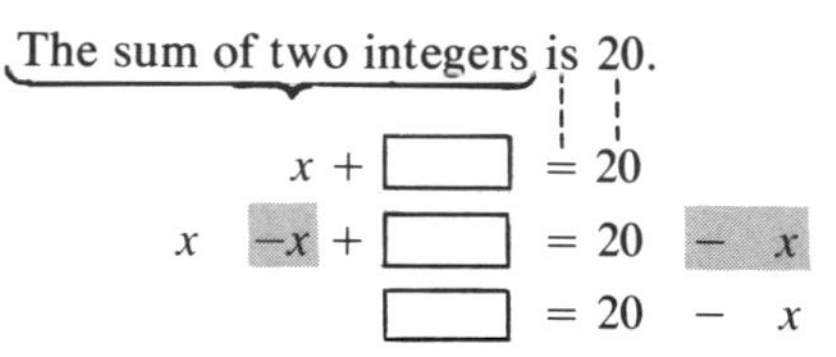

Subtract x from each side.

Thus, $20 - x$ represents the second integer.

b. Let y represent the smaller integer. You are told that the larger integer is four more than y.

$4 + y$

Thus, $y + 4$ represents the larger integer.

3. Let x be the son's age.

A mother is 20 years older than her son.

$\square = 20 + x$

The mother's age is $x + 20$.

4. Let x represent a woman's age *now*.
   a. *Ten years ago*, her age was $x - 10$.
   b. *In five years*, her age will be $x + 5$.
   c. Her husband's age *now* is $x + 3$.

5. Let x represent this integer. Translate the problem.

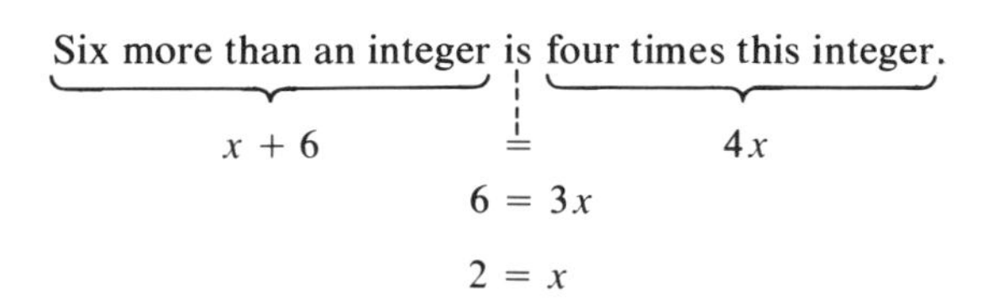

$$6 = 3x$$

$$2 = x$$

Thus, 2 is the integer you seek.

check $2 + 6 \stackrel{?}{=} 4(2)$

$$8 \stackrel{\checkmark}{=} 8$$

6. Let x represent the son's age *now*. Thus, $5x$ represents the woman's age *now*. *Three years ago*, the son's age was $x - 3$. *Three years ago*, the woman's age was $5x - 3$.

Translate the problem. In order to do this, reword the second sentence of the problem as follows:

The woman's age three years ago was eight times her son's age three years ago.

$$\underbrace{5x - 3}_{\text{The woman's age three years ago}} \stackrel{?}{=} \underbrace{8(x - 3)}_{\text{eight times her son's age three years ago}}$$

$$5x - 3 = 8x - 24$$

$$5x - 3 - 5x + 24 = 8x - 24 - 5x + 24$$

$$21 = 3x$$

$$7 = x$$

The son is now 7 years old (and his mother 35. Three years ago, he was 4 and she was 32, or $8 \cdot 4$.)

7. Let x be the length (in feet) of the smallest piece. Then, $2x$ is the length of the second piece, and $3x$ is the length of the largest piece.

The sum of the lengths of the three pieces is 24 (feet).

$$x + 2x + 3x \stackrel{?}{=} 24$$
$$6x = 24$$
$$x = 4$$
$$2x = 8$$
$$3x = 12$$

The smallest piece is 4 feet long; the second piece is 8 feet long; the largest piece is 12 feet long.

8. Let x be the winning number of votes. Then, $x - 20$ is the losing number of votes.

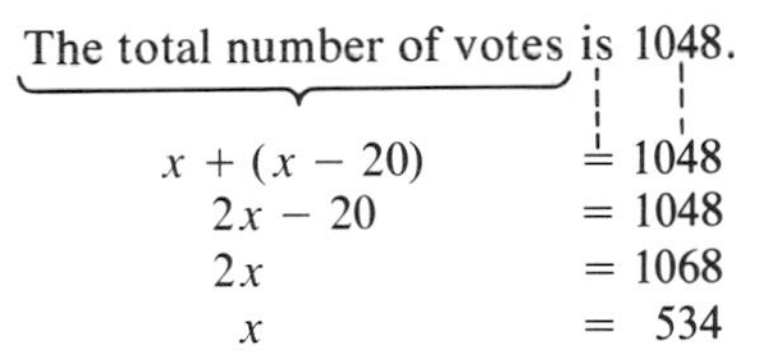

The winner receives 534 votes.

HOME EXERCISES

1. Let x represent an integer. Express each of the following in terms of x.

a. two more than the integer **b.** six less than the integer

c. triple the integer **d.** one-third of the integer

2. Let y represent an integer. Express each of the following in terms of y.
   a. twice the integer
   b. one less than twice the integer
   c. four more than twice the integer
   d. one-fourth of four more than twice the integer
3. Let x represent an integer. Express each of the following in terms of x.
   a. The integer is increased by nine.
   b. The integer is decreased by two.
   c. half of the integer
   d. the next consecutive integer
4. Suppose the sum of two integers is 50. If x represents one of these integers, express the other in terms of x.
5. The difference between two integers is 19. If x represents the larger integer, express the smaller one in terms of x.
6. The larger of two integers is eight more than the smaller. If x represents the smaller integer, express the larger one in terms of x.
7. The larger of two integers is five more than the smaller. If x represents the larger integer, express the sum of these integers in terms of x.
8. If x represents an integer, express the sum of this integer and three times the next consecutive integer.
9. Fred is six years older than Terri. If x represents Terri's age, express Fred's age.
10. A man is two years older than his wife and 25 years older than his son. If x represents the man's age, express
    a. his wife's age b. his son's age.
11. A woman is five years younger than her husband. She is twice as old as her daughter. Let x represent the woman's age. Express a. her husband's age b. her daughter's age.
12. If you are x years old now, how old will you be in 20 years?
13. Three years ago, Maria was y years old. a. How old is she now? b. How old will she be in seven years?
14. Two years ago, Dave was twice as old as his sister. If x represents his sister's present age, express Dave's present age.
15. Ten more than an integer is 23. Find this integer.
16. Three times an integer is -27. Find this integer.
17. A certain integer is decreased by seven, and the result is 25. Find this integer.
18. Half an integer is 22. What is this integer?
19. Find two consecutive integers whose sum is 31.
20. Find three consecutive integers whose sum is 27. *Check your result.*
21. One more than twice an integer is 15. Find this integer. *Check your result.*
22. One less than three times an integer is 23. Find this integer.
23. The sum of two integers is 50. The larger is six more than the smaller. Find these integers. *Check your result.*
24. Half of an integer plus one-fifth of the next consecutive integer equals 10. Find these integers.
25. Gene is four years older than Alex. Gene is 17. How old is Alex?
26. Joe is five years younger than his brother Aaron, who is 18. How old is Joe?
27. Three years ago, a mother was five times as old as her daughter. The daughter is now eight years old. How old is the mother now? *Check your result.*
28. In three years, Marco will be half as old as his brother. The brother will then be 24. How old is Marco now?
29. A man is five years older than his wife. Fifteen years ago, he was twice her age. How old is the man? *Check your result.*
30. Barbara's grandfather is five times as old as Barbara. In four years, the grandfather will be four times as old as Barbara. How old is Barbara?
31. A 31-year-old woman has a seven-year-old son. In how many years will she be double his age? *Check your result.*

32. A man has a son and daughter. The man is six times as old as his son, who is two years older than his sister. In six years, the man will be four times as old as his daughter. How old is the man?

33. A 12-foot rope is cut into two pieces. The longer piece is twice as long as the shorter one. Find the length of each piece.

34. An 18-foot rope is cut into three pieces. One piece is 3 feet longer than a second piece, which, in turn, is twice as long as the third piece. How long is each piece?

35. Carol has three dollars in nickels and dimes. She has the same number of each coin. How many nickels does she have?

36. Amy has $1.55 in nickels, dimes, and quarters. She has three more nickels than quarters and twice as many quarters as dimes. How many of each coin does she have?

37. A television channel has three minutes of programming for every minute of commercials. How long are the commercials on a one and one-half hour program?

38. In a school election 2420 votes are cast for two candidates. One candidate wins by 10 votes. How many votes does she receive?

39. Eric and Doug together have $60. Eric has $10 more than Doug. How much does Doug have?

40. Mozart wrote five more than four times the number of symphonies that Beethoven wrote. Altogether, they wrote 50 symphonies. How many did each write?

7.2 DISTANCE PROBLEMS

THE DISTANCE FORMULA

The distance that can be traveled at a *constant rate* equals the rate multiplied by the time spent traveling. For example, if a car travels at the constant rate of 50 miles per hour, in 4 hours it will travel

$$50 \cdot 4, \text{ or } 200, \text{ miles}$$

Let $r = \text{rate},\ t = \text{time},\ d = \text{distance}$

Then,

$$\boxed{r \cdot t = d}$$

This is known as the **distance formula**. It is convenient to express this formula by means of a table. Thus, for the above example, write

r ·	t =	d
50	4	200

Next, suppose an airplane flies at the constant rate of 400 miles per hour. In 6 hours, it travels 2400 miles.

r ·	t =	d
400	6	2400

The units of measurement must match. Thus, when distance is measured in miles and time in hours, then rate is given in *miles per hour*, which can be expressed as the fraction $\frac{\text{miles}}{\text{hours}}$. Corresponding to the formula $r \cdot t = d$, you have

$$\frac{\text{miles}}{\text{hours}} \cdot \text{hours} = \text{miles}$$

Throughout this section, *all rates are assumed to be constant.*

Other forms of the distance formula apply when you are given some other information. For example, suppose you are told that a man walks 12 miles at the constant rate of 3 miles per hour. To find the time he walks, divide both sides of the distance formula

$$r \cdot t = d$$

by r to obtain

$$t - \frac{d}{r}$$

Thus, $$t = \frac{12}{3} = 4$$

He walks for 4 hours.

Similarly, suppose you are told that a bus covers 180 miles of a highway in 3 hours. To find its rate, divide both sides of the distance formula by t to obtain

$$r = \frac{d}{t}$$

Then, the bus travels at the rate of $\frac{180}{3}$, or 60, miles per hour.

Example 1 A train traveling at the rate of 90 miles per hour goes 495 miles. How long does this take?

Solution Let t be the number of hours.

r	$t = \frac{d}{r}$	d
90	t	495

$$t = \frac{d}{r} = \frac{495}{90} = \frac{11}{2}$$

The trip takes $5\frac{1}{2}$ hours.

CLASS EXERCISES

1. A train travels at 70 miles an hour for five hours. How far does it travel?
2. A motorcycle travels 160 miles in four hours. At what rate does it travel?
3. How long does it take an airplane traveling at 450 miles per hour to cover 2700 miles?

SAME DIRECTION, OPPOSITE DIRECTIONS, ROUND TRIP

Suppose two cars leave from the same place at the same time along a straight road, one traveling at 60 miles per hour, the other at 50 miles per hour.

a. If they travel in *the same direction*, then *each hour* the faster car goes 10 (or 60 − 50) miles further than the slower car. (See Figure 7.1 .)

b. If they travel in opposite directions, then each hour the cars separate 110 (or 60 + 50) miles. (See Figure 7.2 .)

Fig. 7.1. If the cars travel in the same direction, the faster car goes 10 miles further than the slower car in one hour.

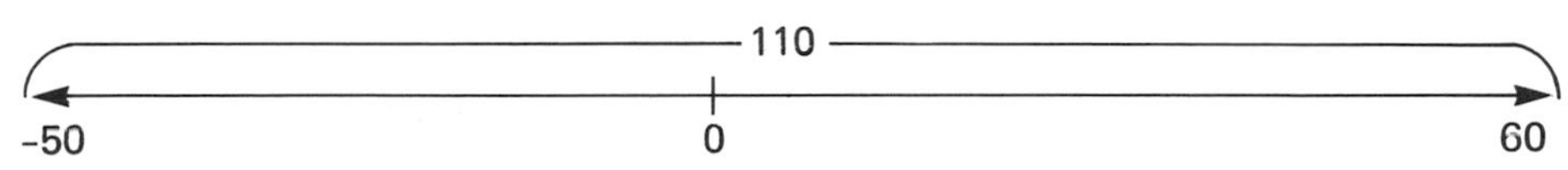

Fig. 7.2. If the cars travel in opposite directions, the cars separate 110 miles in one hour.

Example 2 Two cars leave a restaurant at the same time headed in the same direction along a straight road. One car travels at 60 miles per hour, the other at 40 miles per hour. How far apart are they after four hours?

Solution

	r ·	t =	d
faster car	60	4	240
slower car	40	4	160

The cars travel in the same direction. In 1 hour the faster car is 20 (= 60 − 40) miles ahead of the slower car. In 4 hours it is 80 (= 4 · 20) [or 240 − 160] miles ahead of the slower vehicle. (See Figure 7.3 .)

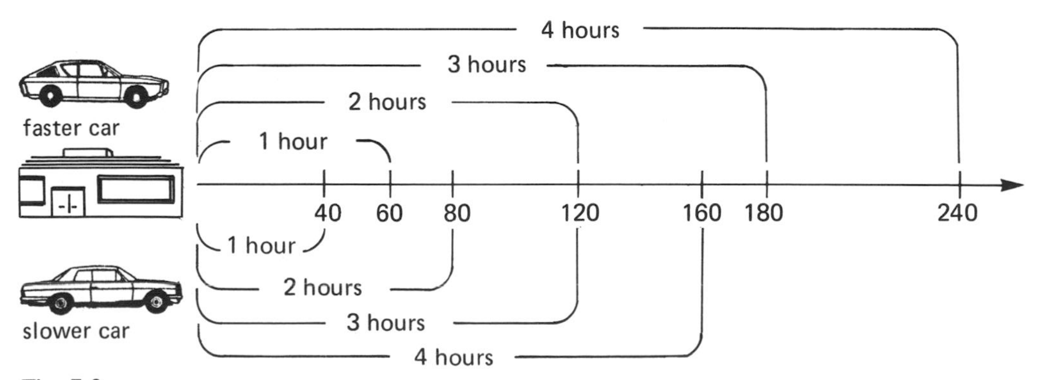

Fig. 7.3

Example 3 Two cars leave a ball park at the same time. One car travels eastward at 55 miles per hour; the other car travels westward at 70 miles per hour. How far apart are they after two hours?

Solution

	r ·	t =	d
eastward	55	2	110
westward	70	2	140

The cars travel in opposite directions. Their distance apart, d, is the sum of the distances each has traveled in 2 hours.

$$d = 110 + 140 = 250$$

The distance apart is 250 miles. Each hour the cars move apart 55 + 70 (= 125) miles. Thus, $250 = 125 \cdot 2$ (See Figure 7.4 .)

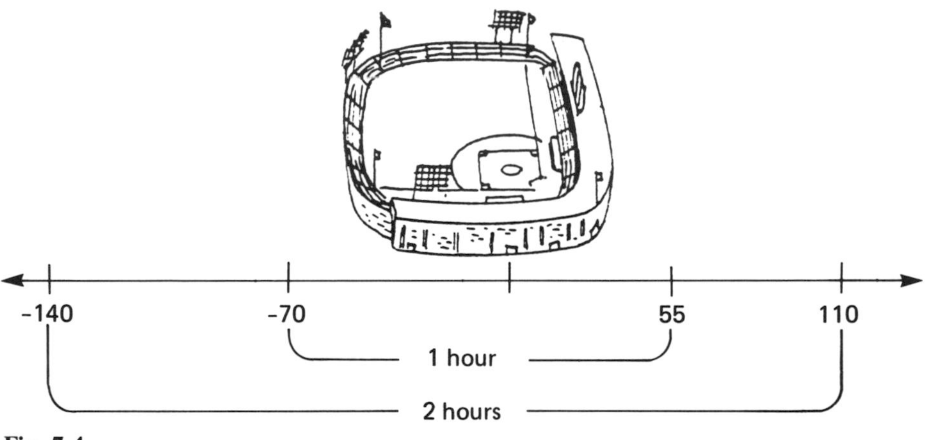

Fig. 7.4

Suppose a man runs to a mailbox and then walks back home. The round trip takes 10 minutes. If he runs for t minutes, then he walks for $(10 - t)$ minutes. For example, if he runs for 4 minutes, then he walks for $10 - 4$, or 6, minutes. Assume he takes the same route going and returning. Then, *the distance each way is the same*. (See Figure 7.5 .)

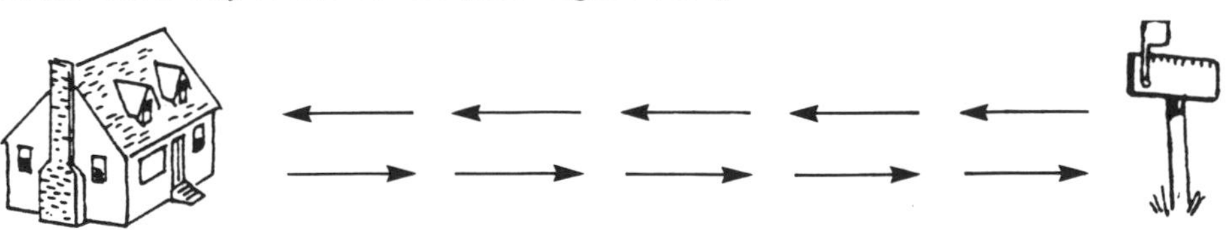

Fig. 7.5

Example 4 A canoe goes upstream at the rate of five miles per hour. It returns downstream at the rate of 10 miles per hour. The round trip takes three hours.

a. How long does the canoe travel upstream?
b. How far upstream does it go? **c.** Check your results.

Solution **a.** The canoe travels the same distance upstream as downstream. (See Figure 7.6 .) Let t be the number of hours it travels upstream. Because the round trip takes 3 hours, it travels for $(3 - t)$ hours downstream.

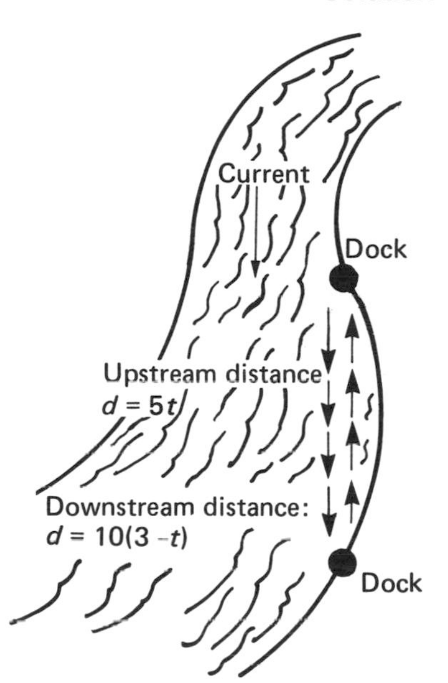

Fig. 7.6

	r ·	t =	d
upstream	5	t	$5t$
downstream	10	$3 - t$	$10(3 - t)$

The distance upstream equals the distance downstream.

$$5t = 10(3 - t)$$

Solve this equation.

$$5t = 30 - 10t$$
$$15t = 30$$
$$t = 2$$

The canoe travels for 2 hours upstream (and for 1 hour downstream).

b. To find the distance upstream, use the distance formula with this new piece of information from part **a.**

	r ·	t =	d
upstream	5	2	10

The canoe travels 10 miles upstream.

c. You can check both parts by showing that the distance downstream is also 10 (miles). Recall that the time traveling downstream is 1, or $3 - 2$, hour.

	r ·	t =	d
downstream	10	1	10

CLASS EXERCISES **4.** A car leaves a gas station traveling due south at 50 miles per hour. An hour later, a second car leaves the gas station traveling due south at 60 miles per hour. How far from the gas station are the cars when the second car passed the first?

5. Two cars leave a toll booth traveling in opposite directions, one at 45 miles per hour, the other at 65 miles per hour. In how many hours are they 275 miles apart?

6. A car drives to a convention at 60 miles per hour. The return trip at 80 miles per hour takes one hour less. How many hours does it take to drive to the convention? *Check your result.*

SOLUTIONS TO CLASS EXERCISES

1.

r	$\cdot\ t$	$= d$
70	5	d

$$d = r \cdot t = 70 \cdot 5 = 350$$

The train travels 350 miles.

2.

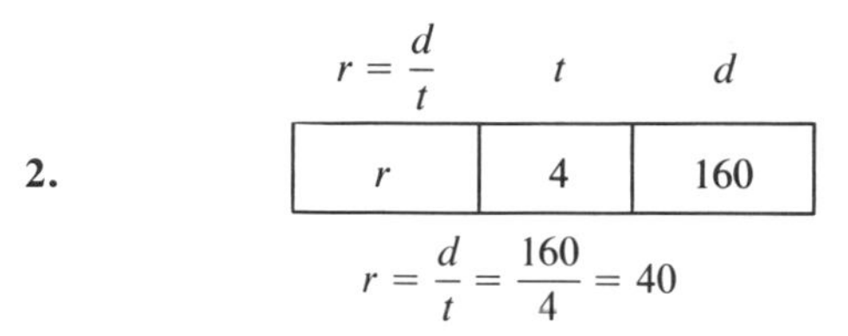

$r = \frac{d}{t}$	t	d
r	4	160

$$r = \frac{d}{t} = \frac{160}{4} = 40$$

The motorcycle travels at 40 miles per hour.

3.

r	$t = \frac{d}{r}$	d
450	t	2700

$$t = \frac{d}{r} = \frac{2700}{450} = 6$$

The trip takes 6 hours.

4. Let d be the distance from the gas station when the second car passes the first.

	r	$t = \frac{d}{r}$	d
first car	50	$\frac{d}{50}$	d
second car	60	$\frac{d}{60}$	d

$$\underbrace{\text{The first car's time}}_{\frac{d}{50}}\ \underbrace{\text{is}}_{=}\ \underbrace{\text{one hour more than the second car's time}}_{\frac{d}{60}+1}$$

$$\frac{d}{50} = \frac{d}{60} + 1 \qquad \textit{Multiply both sides by 300, the lcd.}$$

$$6d = 5d + 300$$

$$d = 300$$

The second car passes the first 300 miles from the gas station.

5.

	r ·	t =	d
first car	45	t	$45t$
second car	65	t	$65t$

Their distance apart equals 275 miles.

$$45t + 65t \stackrel{\downarrow}{=} 275$$
$$110t = 275$$
$$t = \frac{275}{110} = \frac{5}{2}$$

In $2\frac{1}{2}$ hours the cars are 275 miles apart.

6.

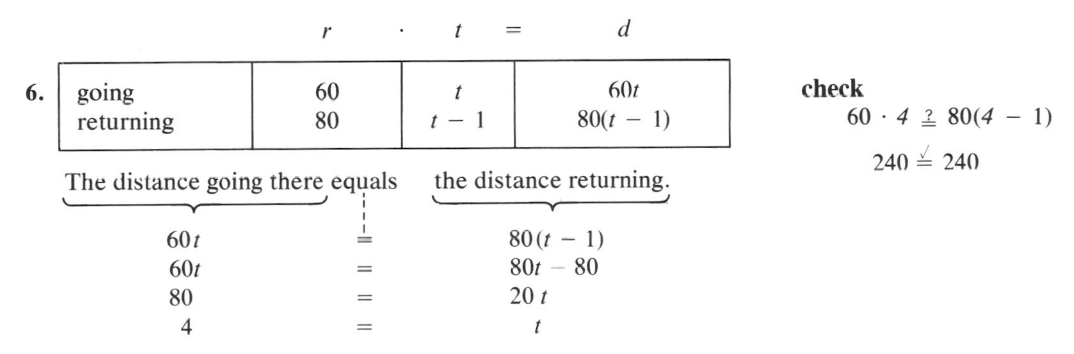

	r ·	t =	d
going	60	t	$60t$
returning	80	$t - 1$	$80(t - 1)$

The distance going there equals the distance returning.

$$60t \stackrel{\downarrow}{=} 80(t - 1)$$
$$60t = 80t - 80$$
$$80 = 20t$$
$$4 = t$$

check

$$60 \cdot 4 \stackrel{?}{=} 80(4 - 1)$$
$$240 \stackrel{\checkmark}{=} 240$$

The drive to the convention takes 4 hours.

---

*HOME EXERCISES**

1. A car travels at 30 miles per hour for three hours. How far does it travel?
2. How long does it take Bob to walk home from his office if he walks four miles per hour and lives six miles from his office?
3. An automobile travels for two hours at 60 miles per hour. It then slows down and travels at 40 miles per hour in the same direction for the next hour and a half. How far does it travel?
4. A train travels 280 miles in four hours. At what rate does it travel?
5. An airplane travels 1050 miles in three-and-one-half hours. At what rate does it travel?
6. A car travels at 45 miles per hour for four hours. On the return trip it travels at 60 miles per hour. How long does the return trip take? *Check your result.*
7. A car traveling at 70 miles per hour along a straight road passes another car cruising at 45 miles per hour. The two cars are headed in the same direction.
   a. How far apart are they after one hour? **b.** How far apart are they after three hours?
8. Two cars leave a toll booth at the same time. They travel in opposite directions along a straight road, one at 65 miles per hour, the other at 60 miles per hour.
   a. How far apart are they after one hour? **b.** How far apart are they after four hours?

---

*All rates are constant.

9. Two cars leave a town traveling in opposite direction along a straight road. One car leaves at noon and goes at 50 miles per hour. The other leaves at 1 o'clock that afternoon and goes at 65 miles per hour. How far apart are they at 3 o'clock that afternoon?

10. Two trains approach one another along (straight) parallel tracks, starting from stations 120 miles apart. One train travels at 100 miles per hour, the second train at 80 miles per hour. How long after they start do they pass each other?

11. Two joggers leave from the same place, headed in the same direction along a straight path. One goes at eight miles per hour, the other at six miles per hour. How far apart are they after 45 minutes? *Check your result.*

12. A car and a motor scooter travel along a straight highway in the same direction. The scooter travels at three-fourths the rate of the car. At the end of five hours they are 75 miles apart. How fast is the car traveling?

13. A canoe goes upstream at the rate of six miles per hour. It return downstream at the rate of nine miles per hour. If the round trip takes five hours, how far upstream does the canoe go? *Check your result.*

14. A jogger can run at the rate of eight miles per hour over level ground and at four miles per hour over hilly ground. Altogether, it takes him an hour and a half to cover 10 miles. How many of these miles are level?

7.3 LITERAL EQUATIONS

SOLVING LITERAL EQUATIONS

DEFINITION

A **literal equation** is one that contains at least two variables and that must be solved for one variable in terms of the others.

Example 1

$$x + y = 5$$

is a literal equation. You can solve for x in terms of y by subtracting y from both sides. Thus,

$$x = 5 - y$$

You can also solve for y in terms of x by subtracting x from both sides of the given equation.

$$y = 5 - x$$

To solve a literal equation, bring terms containing *the variable for which you are solving* to one side. Bring *all other terms* to the other side.

Example 2 Solve $5x - 3a + 1 = 0$ for x. *Check your result.*

Solution Leave $5x$ on the left side. Bring all other terms to the right side. Thus, add $3a - 1$ to both sides.

$$5x - 3a + 1 + 3a - 1 = 0 + 3a - 1$$

$$5x = 3a - 1$$

$$x = \frac{3a - 1}{5}$$

check Substitute $\frac{3a - 1}{5}$ for x in the given equation:

$$5\left(\frac{3a - 1}{5}\right) - 3a + 1 \stackrel{?}{=} 0$$

$$3a - 1 - 3a + 1 \stackrel{?}{=} 0$$

$$0 \stackrel{\checkmark}{=} 0$$

Example 3 Solve $4x - 3y = 12$

a. for x, **b.** for y.

Solution **a.** $4x - 3y = 12$

$$4x = 12 + 3y$$

$$x = \frac{12 + 3y}{4}$$

b. $4x - 3y = 12$

$$-3y = 12 - 4x$$

$$\frac{-3y}{-3} = \frac{12 - 4x}{-3}$$

$$y = \frac{4x - 12}{3}$$

CLASS EXERCISES

1. Solve $4a - 7b + 10 = 0$ for a. *Check your result.*
2. Solve $5x + 2y = 1$ **a.** for x, **b.** for y.
3. Solve $3x + y - 2z = 10$ for z.

SOLVING BY FACTORING

Example 4 Solve $xy + x = 1$ for x.

Solution Here, both terms containing x are on the left. In order to bring y to the right, first *factor* the polynomial on the left.

$$xy + x = x(y + 1)$$

Thus, the given equation becomes

$$x(y + 1) = 1 \qquad \textit{Divide both sides by } y + 1.$$

$$x = \frac{1}{y + 1}$$

Recall that division by 0 is undefined. In Example 4, when you divide both sides by $y + 1$, note that $y + 1 \neq 0$ (*read*: $y + 1$ does *not* equal 0), and thus, $y \neq -1$. In fact, if $y = -1$, then, from the given equation,

$$\underbrace{x(-1) + x}_{0} = 0$$
$$0 = 1$$

This cannot be. Thus, $y + 1$ cannot equal 0.

CLASS EXERCISES *Solve each equation for x.*

4. $ax - by + cx = 2$ **5.** $10x - xy + 3 = 1 - ax$

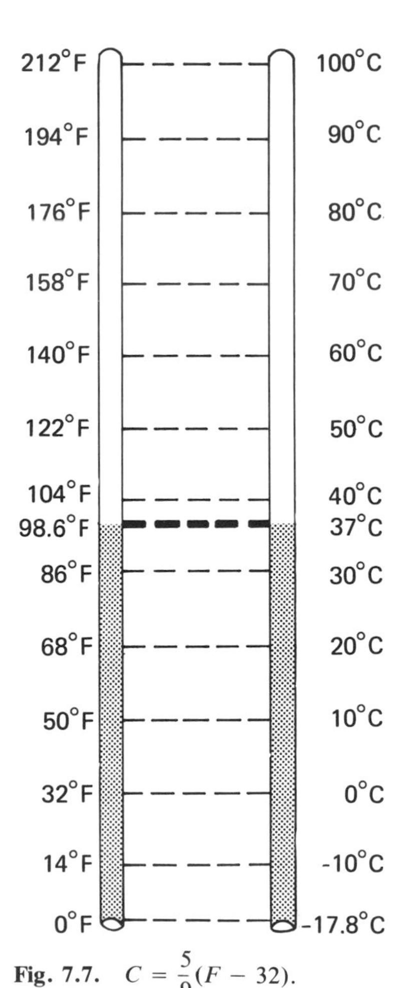

Fig. 7.7. $C = \frac{5}{9}(F - 32)$.

FORMULAS

Example 5

The formula $$C = \frac{5}{9}(F - 32)$$

relates the Celsius and Fahrenheit temperature scales. In the formula, C stands for degrees Celsius and F for degrees Fahrenheit. (See Figure 7.7 .)

a. Find C when F is 212, the boiling point of water.
b. Solve for F in terms of C. **c.** Find F when C = 37.

Solution

a. Substitute *212* for *F* in the given formula.

$$C = \frac{5}{9}(212 - 32) = \frac{5}{\cancel{9}_{1}} \cdot \cancel{180}^{20} = 100$$

b. $$C = \frac{5}{9}(F - 32) \qquad \textit{Divide both sides by } \frac{5}{9}\left(\textit{or multiply by } \frac{9}{5}\right).$$

$$\frac{9}{5}C = \frac{9}{5} \cdot \frac{5}{9}(F - 32)$$

$$\frac{9}{5}C = F - 32$$

$$\frac{9}{5}C + 32 = F$$

c. Substitute *37* for C in the formula obtained in part **b.**

$$\frac{9}{5} \cdot 37 + 32 = F$$

$$\frac{333}{5} + 32 = F$$

$$66.6 + 32 = F$$

$$98.6 = F$$

Thus, 37°C (37 degrees Celsius) corresponds to normal body temperature, 98.6°F (98.6 degrees Fahrenheit).

CLASS EXERCISES

6. Suppose that the formula $h = 10(t + 20)$ represents the height of an object in terms of time, t.
a. Find h when $t = 10$. **b.** Solve for t in terms of h.
c. Find t when $h = 400$.

SOLUTIONS TO CLASS EXERCISES

1.
$$\begin{aligned} 4a - 7b + 10 &= 0 \\ 4a - 7b + 10 + 7b - 10 &= 0 + 7b - 10 \\ 4a &= 7b - 10 \\ a &= \frac{7b - 10}{4} \end{aligned}$$

check
$$\begin{aligned} 4\left(\frac{7b - 10}{4}\right) - 7b + 10 &\stackrel{?}{=} 0 \\ 7b - 10 - 7b + 10 &\stackrel{?}{=} 0 \\ 0 &\stackrel{\checkmark}{=} 0 \end{aligned}$$

2. a.
$$\begin{aligned} 5x + 2y &= 1 \\ 5x &= 1 - 2y \\ x &= \frac{1 - 2y}{5} \end{aligned}$$

b.
$$\begin{aligned} 5x + 2y &= 1 \\ 2y &= 1 - 5x \\ y &= \frac{1 - 5x}{2} \end{aligned}$$

3.
$$\begin{aligned} 3x + y - 2z &= 10 \\ -2z &= 10 - 3x - y \\ z &= \frac{3x + y - 10}{2} \end{aligned}$$

4.
$$\begin{aligned} ax - by + cx &= 2 \\ ax + cx &= 2 + by \\ (a + c)x &= 2 + by \\ x &= \frac{2 + by}{a + c}, \quad a + c \neq 0 \end{aligned}$$

5.
$$\begin{aligned} 10x - xy + 3 &= 1 - ax \\ 10x - xy + ax &= 1 - 3 \\ (10 - y + a)x &= -2 \\ x &= \frac{-2}{10 - y + a} \end{aligned}$$

6. a. Substitute *10* for t.

$$h = 10\,(10 + 20) = 300$$

b.

$$h = 10\,(t + 20)$$
$$\frac{h}{10} = t + 20$$
$$\frac{h}{10} - 20 = t$$
$$\frac{h - 200}{10} = t$$

c. Substitute *400* for h in the formula obtained in part **b.**

$$t = \frac{400 - 200}{10} = 20$$

HOME EXERCISES

In Exercises 1–10, solve for x in terms of the other variable(s).

1. $5x = y$ **2.** $x + 2 = y$ **3.** $x - 3 = 2y$ **4.** $x + y + 4 = 0$
5. $3x - 2y + 1 = 0$ **6.** $2x - 5t = 7x + 2t$ **7.** $x + 2t - 1 = 5x + 2$
8. $x + y + z = 1$ **9.** $5x + y - z = 2x + y - 3z$ **10.** $x - 1 + y = z + 3(x - 1)$

In Exercises 11–18, solve for the indicated variable.

11. $2x + 3y = 5$ (for y) **12.** $5a - 4b = 10$ (for a) **13.** $2u - 3v = 6$ (for v)
14. $6m - \frac{n}{2} = 5$ (for n) **15.** $2a + b - 3c = 12$ (for a) **16.** $4 - 3s + 2t = 5$ (for s)
17. $xy - y = 3$ (for y) **18.** $ab + 2a = 1$ (for b)

In Exercises 19–30, solve for the indicated variables.

19. $5x - y = 20$ **a.** for x, **b.** for y
20. $3a - 4b = 7$ **a.** for a, **b.** for b
21. $5(c - 2b) = 1 + c$ **a.** for b, **b.** for c
22. $s - 3t = 2s + 3t$ **a.** for s, **b.** for t
23. $2y - z + 1 = 10z - y$ **a.** for y, **b.** for z
24. $u - 3v + 2 = 2u - v$ **a.** for u, **b.** for v
25. $xy + 4y = 1$ **a.** for x, **b.** for y
26. $3m - mn = 2n$ **a.** for m, **b.** for n
27. $5x - y + 2z = 10$ **a.** for x, **b.** for y
28. $u - 2v + 6w = 7$ **a.** for v, **b.** for w
29. $\frac{a}{b} = 1$ **a.** for a, **b.** for b
30. $\frac{2x}{y} = -3$ **a.** for x, **b.** for y

In Exercises 31–34, **a.** *solve for the indicated variable, and* **b.** *check your answer.*

31. $2x - y = 10$ (for x) **32.** $\frac{12}{x} = \frac{y}{5}$ (for y) **33.** $5a - 2b = 4$ (for b) **34.** $\frac{a}{2} + \frac{b}{3} - \frac{c}{4} = 12$ (for c)

35. The formula
$$A = lw$$
expresses the area, A, of a rectangle in terms of the length, l, and width, w. Solve for l.

36. The formula
$$C = 2\pi r$$
expresses the circumference, C, of a circle in terms of r, the length of the radius. Solve for r.

37. a. The formula
$$A = \frac{bh}{2}$$
expresses the area, A, of a triangle in terms of b, the length of the base, and h, the height. Solve for h.

b. Find h when $A = 100$ square inches and $b = 4$ inches.

38. a. The formula
$$P = 2l + 2w$$
expresses the perimeter, P, of a rectangle in terms of the length, l, and width, w. Solve for l.

b. Find l when $P = 60$ feet and $w = 10$ feet. (See Figure 7.8 .)

39. a. The formula
$$d = rt$$
expresses the distance, d, that can be traveled at a constant rate, r, over a period of time, t. Solve for r.

b. Find r when $d = 300$ miles and $t = 5$ hours.

40. a. The formula
$$V = lwh$$
expresses the volume, V, of a rectangular box in terms of the length, l, the width, w, and the height, h. Solve for h. (See Figure 7.9 .)

b. Find h when $V = 2000$ cubic inches, $l = 20$ inches, and $w = 5$ inches.

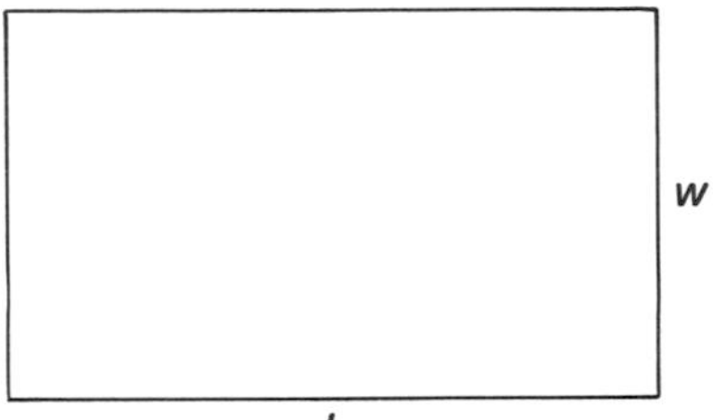

Fig. 7.8. $p = 2l + 2w$.

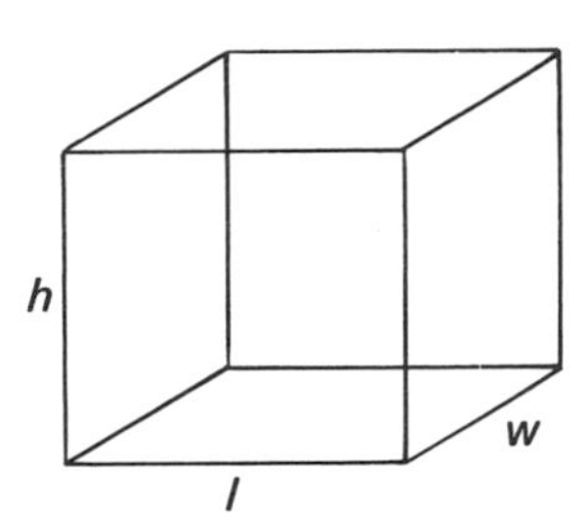

Fig. 7.9. $V = lwh$.

7.4 RECTANGULAR COORDINATES

COORDINATE AXES

Recall that numbers correspond to points on a horizontal line, the *number line*. Now, you will see how *pairs of numbers*, in a definite order, correspond to points on a plane.

Draw a vertical line through the origin, 0, on the number line. Because the number line is horizontal, the two lines are perpendicular. The horizontal line will now be called the **x-axis**, and the vertical line will be called the **y-axis**. The two lines intersect at the origin, which will represent 0 on the y-axis (as well as on the x-axis). For convenience, choose the same distance unit on the y-axis as on the x-axis. *On the y-axis, positive numbers are located upward from 0, and negative numbers downward.* (See Figure 7.10 .)

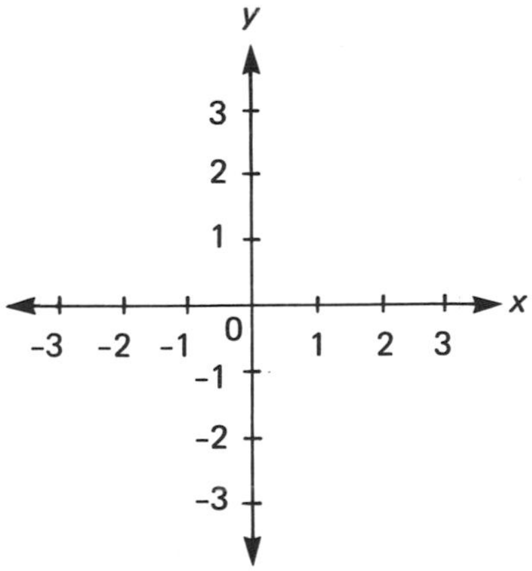

Fig. 7.10. On the y-axis, positive numbers are plotted upward from 0, and negative numbers downward.

Example 1 Locate the following points on the y-axis.

a. 3 **b.** $\frac{1}{2}$

c. -2 **d.** $\frac{-5}{2}$

Solution

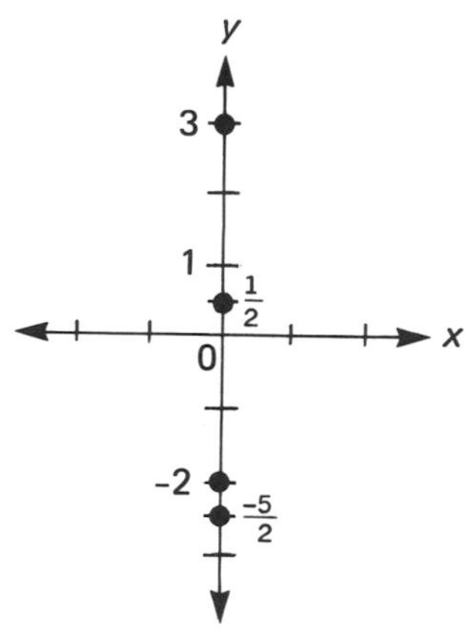

Fig. 7.11

CLASS EXERCISES

1. Locate the following points on the y-axis in Figure 7.12.
   a. 2 **b.** 5 **c.** -3 **d.** -5

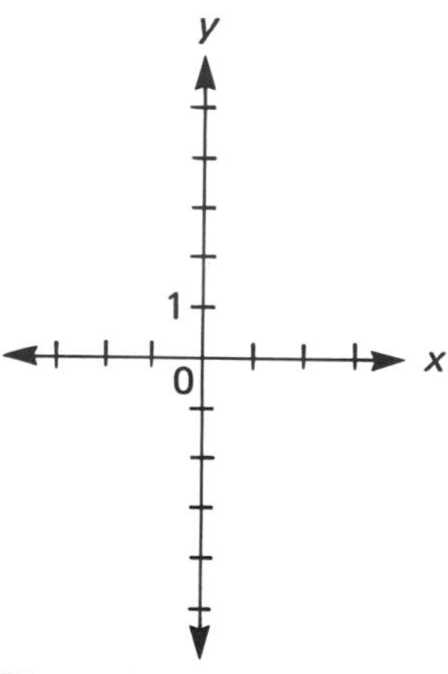

Fig. 7.12

COORDINATES

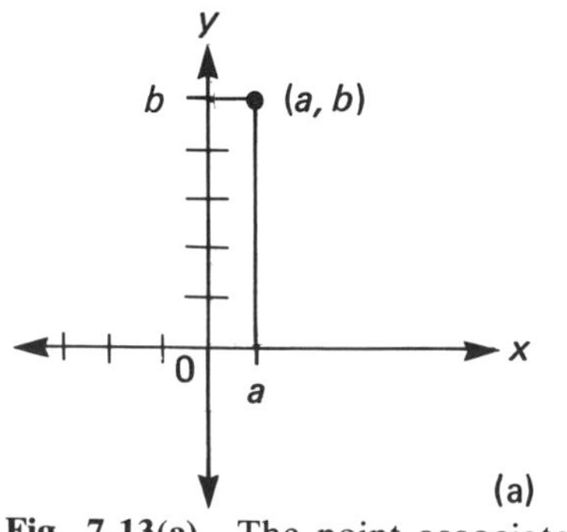

Fig. 7.13(a) The point associated with (a, b).

Every number corresponds to a point on the x-axis and to a point on the y-axis. If you consider two numbers, a and b, then a corresponds to a point on the x-axis and b to a point on the y-axis. Locate these points and draw perpendiculars to the axes through each of them, as in Figure 7.13(a). The intersection of these perpendiculars is the point P of the plane associated with a and b in this order. To indicate the order of a and b, write

$$(a, b)$$

Call (a, b) an ***ordered pair***. *The ordered pair (a, b) indicates two numbers a and b in the order written*. The number a is called **x-coordinate,** or **abscissa,** of (a, b) and b the **y-coordinate,** or **ordinate**, of (a, b).

Thus, 2 is the x-coordinate of (2, 5), and 5 is the y-coordinate. On the

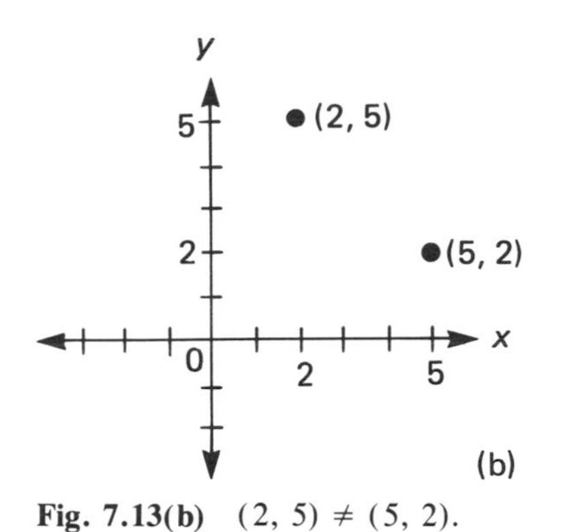

Fig. 7.13(b) $(2, 5) \neq (5, 2)$.

other hand, if you change the order of the coordinates, then 5 is the x-coordinate and 2 the y-coordinate of $(5, 2)$. Observe that

$$(2, 5) \neq (5, 2)$$

[See Figure 7.13(b).]

Every ordered pair of numbers, (a, b), corresponds to exactly one point P on the plane, as indicated. Moreover, every point P on the plane also corresponds to exactly one ordered pair of numbers, (a, b). Together, a and b are called the **rectangular coordinates of** P or, simply, the **coordinates of** P. You will identify a point P of the plane with its ordered pair of coordinates (a, b), and write

$$P = (a, b)$$

The x-axis together with the y-axis are called the **coordinate axes**.

Example 2 For each point in Figure 7.14, find

a. its x-coordinate, **b.** its y-coordinate.
c. Identify the point with its ordered pair of coordinates.

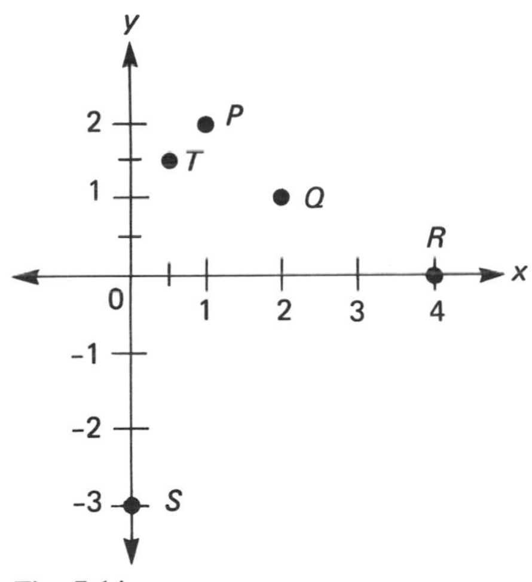

Fig. 7.14

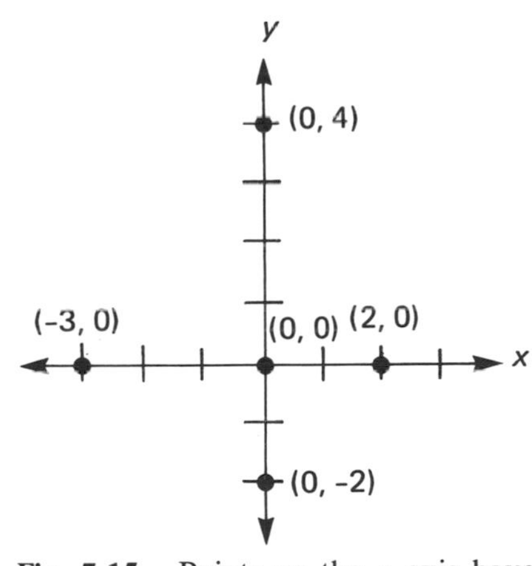

Fig. 7.15. Points on the x-axis have y-coordinate 0. Points on the y-axis have x-coordinate 0.

Solution

P:	**a.** 1,	**b.** 2,	**c.** $P = (1, 2)$
Q:	**a.** 2,	**b.** 1,	**c.** $Q = (2, 1)$
R:	**a.** 4,	**b.** 0,	**c.** $R = (4, 0)$
S:	**a.** 0,	**b.** -3,	**c.** $S = (0, -3)$
T:	**a.** $\frac{1}{2}$,	**b.** $\frac{3}{2}$,	**c.** $T = \left(\frac{1}{2}, \frac{3}{2}\right)$

Observe that points on the *x-axis* have *y-coordinate* 0 and points on the *y-axis* have *x-coordinate* 0. The origin, O, which is on both coordinate axes, has both of its coordinates equal to 0. (See Figure 7.15 .)

$$O = (0, 0)$$

CLASS EXERCISES **2.** Locate the following ordered pairs on a rectangular coordinate system.

$$(4, 1), (1, 4), (-2, -2), (0, -1), (.5, -.5)$$

SOLUTIONS TO CLASS EXERCISES

1. See Figure 7.16 . **2.** See Figure 7.17 .

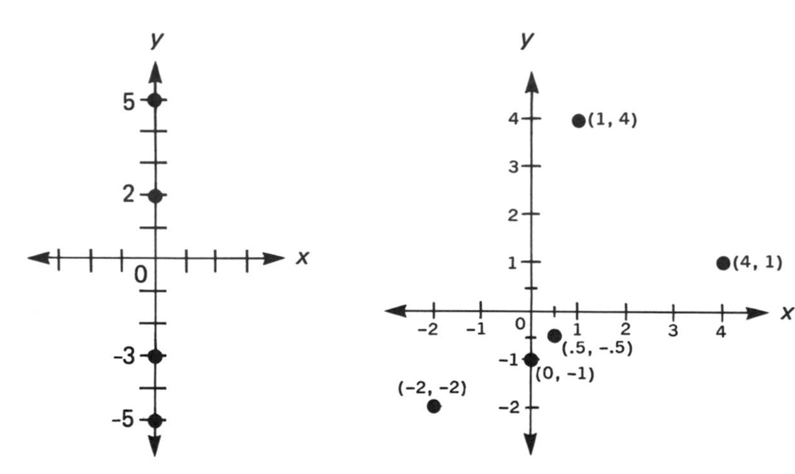

Fig. 7.16 **Fig. 7.17**

HOME EXERCISES

1. Draw a rectangular coordinate system. Locate the following points on the y-axis.

a. 4 **b.** -4 **c.** $\frac{-1}{2}$ **d.** $\frac{3}{2}$

2. In Figure 7.18, indicate which numbers are represented by the following points on the y-axis.
a. P **b.** Q **c.** R **d.** S **e.** T

3. For each point in Figure 7.19, find
a. its x-coordinate, **b.** its y-coordinate. **c.** Identify the point with its pair of coordinates.

4. For each point in Figure 7.20, find
a. its x-coordinate, **b.** its y-coordinate. **c.** Identify the point with its pair of coordinates.

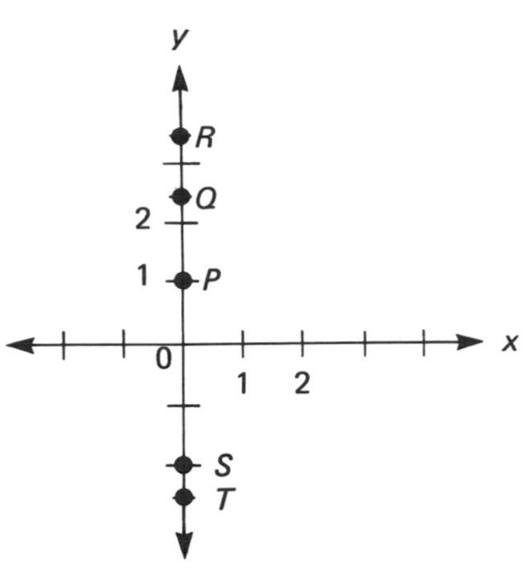

Fig. 7.18

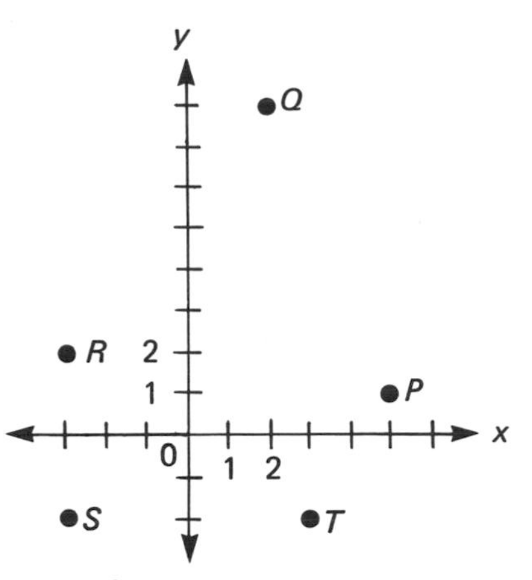

Fig. 7.19

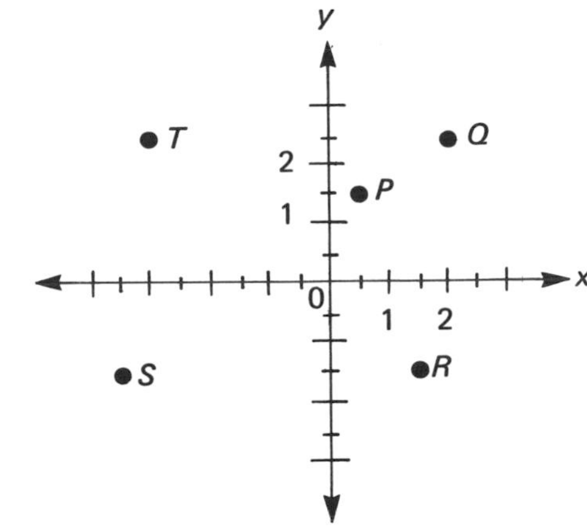

Fig. 7.20

Locate the following ordered pairs on a rectangular coordinate system. (You may wish to draw a coordinate system for the odd-numbered exercises first, and then a separate coordinate system for the even-numbered exercises.)

5. (5, 0) **6.** (0, 3) **7.** (−2, 0) **8.** (0, −3) **9.** $\left(\frac{1}{2}, 0\right)$ **10.** $\left(0, \frac{-1}{2}\right)$

11. $\left(\frac{7}{2}, 0\right)$ **12.** $\left(0, \frac{-3}{2}\right)$ **13.** (2, 2) **14.** (−1, −1) **15.** (4, 2) **16.** (2, 4)

17. (−4, 2) **18.** (−2, 4) **19.** (--2, −4) **20.** (−4, −2) **21.** $\left(6, \frac{1}{2}\right)$ **22.** $\left(\frac{-1}{2}, 4\right)$

23. (2, −5) **24.** (−6, −6) **25.** (6, −1) **26.** (1, −10) **27.** (.5, 1.5) **28.** (−.5, −2.5)

7.5 LINES

THE GRAPHING PROCEDURE

DEFINITION

> The **graph of an equation** is its pictorial representation on the plane. The graph consists of all points (x, y) for which the equation holds true.

To graph a "fairly simple" equation:

1. Consider several values of x.
2. Find the corresponding values of y.
3. Locate the points (x, y).
4. Connect these points by a straight line or curve.

The graph of an equation that can be written in the form

$$y = \square x + \square$$

is always a line. For example, the graph of each of the equations

$$y = x + 4 \quad (\text{or } y = \boxed{1}x + \boxed{4})$$
$$y = 3x - 5 \quad (\text{or } y = \boxed{3}x + \boxed{-5})$$
$$y = -2x \quad (\text{or } y = \boxed{-2}x + \boxed{0})$$
$$y = 6 \quad (\text{or } y = \boxed{0}x + \boxed{6})$$

is a line. Because *a line is determined by two points*, to obtain the line, you need only locate two points. Then, draw the line connecting them. You may wish to locate a third point to check whether a (straight) line goes through all three points.

Example 1 Graph the equation $y = x + 2$.

Solution Corresponding values of x and y are given in Table 7.1. For example, when $x = 0$, replace x by *0* in the polynomial $x + 2$ to obtain $y = 0 + 2 = 2$. The graph of the equation is drawn in Figure 7.21.

TABLE 7.1

x	$y = x + 2$
0	2
1	3
2	4

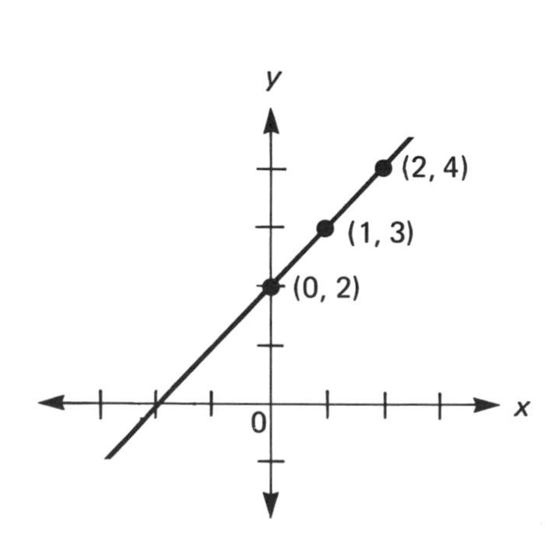

Fig. 7.21

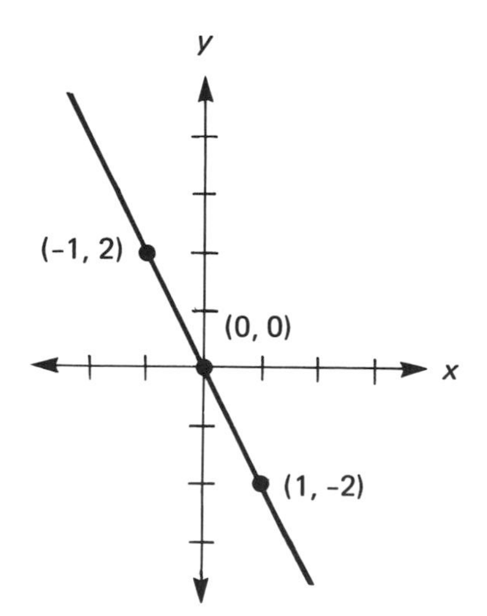

Fig. 7.22

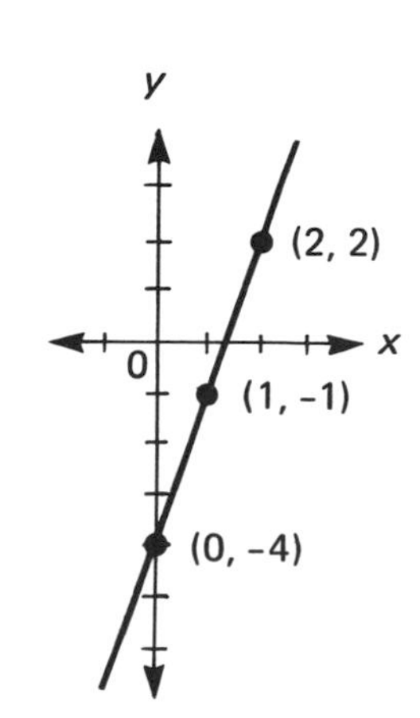

Fig. 7.23

Example 2 Graph the equation $y = -2x$.

Solution See Table 7.2 and Figure 7.22 .

TABLE 7.2

x	$y = -2x$
-1	2
0	0
1	-2

Example 3 Graph the equation $y = 3x - 4$.

Solution See Table 7.3 and Figure 7.23 .

TABLE 7.3

x	$y = 3x - 4$
0	-4
1	-1
2	2

CLASS EXERCISES *Graph each equation.*

1. $y = x - 3$ **2.** $y = 3x$ **3.** $y = 2x + 3$

HORIZONTAL AND VERTICAL LINES

Every point on a horizontal line has the same y-coordinate. For example, if one point has y-coordinate 5, every point on the line has y-coordinate 5. Therefore, $y = 5$ for every value of x. Thus, if (1, 5) is on the line, then so are (0, 5), (3, 5), and so on. (See Figure 7.24 .) The equation of the *horizontal* line that passes through (1, 5) is

$$y = 5$$

Every two points on a vertical line have the same x-coordinate. Thus, if some point on the line has x-coordinate 4, every point has x-coordinate 4. Consequently, $x = 4$ for every value of y. (See Figure 7.25, page 240.) The equation of the *vertical* line that passes through (4, 1) is

$$x = 4$$

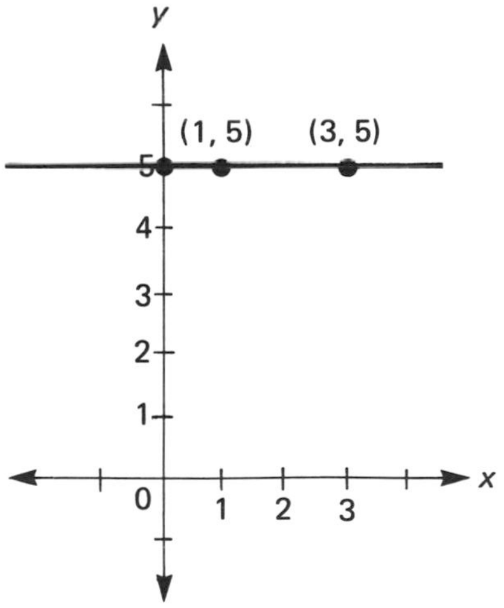

When two different points on a line have the same y-coordinate, the line is horizontal. When two different points have the same x-coordinate, the line is vertical. A horizontal line has equation

$$y = \text{constant}$$

A vertical line has equation

$$x = \text{constant}$$

Every point on the x-axis has y-coordinate 0. Thus, *the equation of the x-axis is*

$$y = 0$$

y
(1, 5) (3, 5)
5
4
3
2
1
0 1 2 3 x

Fig. 7.24. The horizontal line with equation $y = 5$.

Fig. 7.25. The vertical line with equation $x = 4$.

Every point on the y-axis has x-coordinate 0. *The equation of the y-axis is*

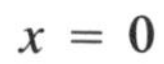

$$x = 0$$

(See Figure 7.26.)

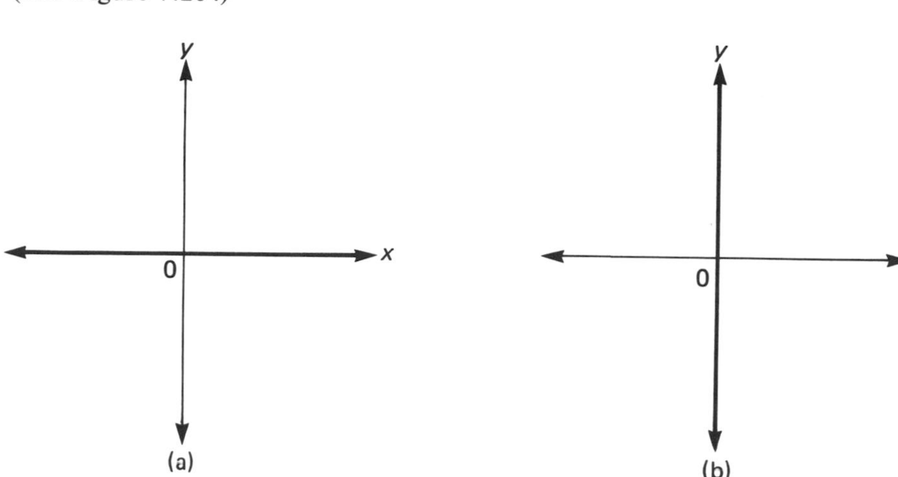

Fig. 7.26(a) The x-axis has the equation $y = 0$. **(b)** The y-axis has the equation $x = 0$.

CLASS EXERCISES *Graph each equation.* **4.** $y = 2$ **5.** $x = -2$

SOLUTIONS TO CLASS EXERCISES

1. TABLE 7.4

x	$y = x - 3$
0	-3
1	-2
2	-1

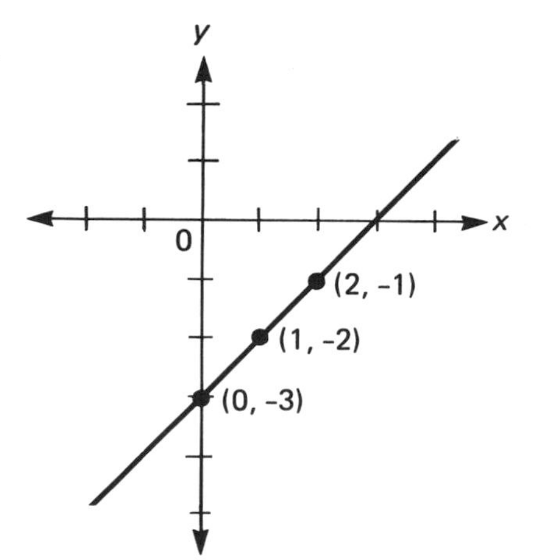

Fig. 7.27. $y = x - 3$.

2. TABLE 7.5

x	$y = 3x$
0	0
1	3
2	6

Fig. 7.28. $y = 3x$.

3. TABLE 7.6

x	$y = 2x + 3$
0	3
1	5
2	7

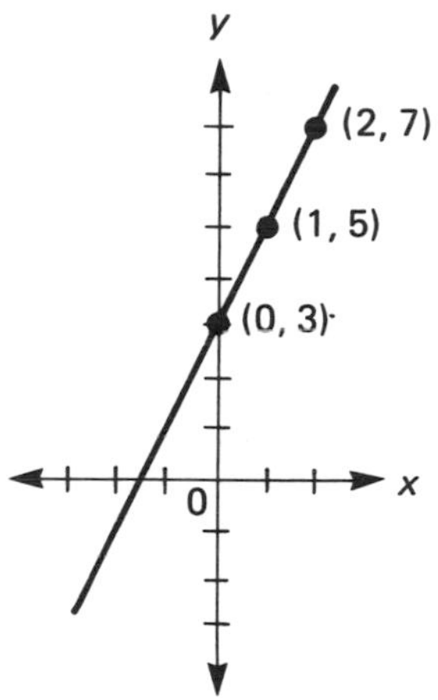

Fig. 7.29. $y = 2x + 3$.

4. TABLE 7.7

x	$y = 2$
0	2
1	2
2	2

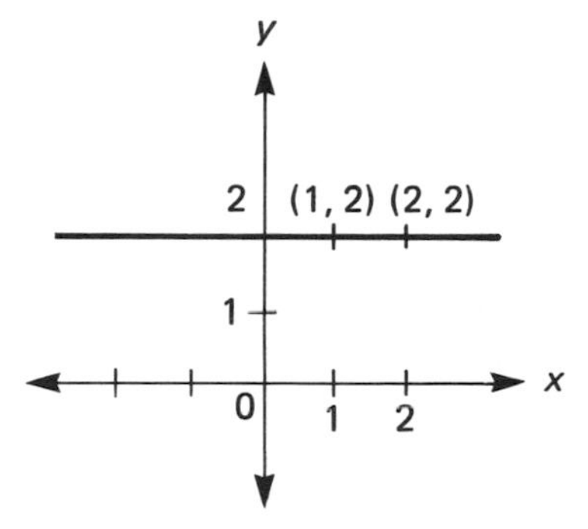

Fig. 7.30. $y = 2$.

5. TABLE 7.8

$x = -2$	y
-2	0
-2	1
-2	2

Fig. 7.31. $x = -2$.

HOME EXERCISES

In Exercises 1–20, graph each equation.

1. $y = x + 1$ **2.** $y = x + 4$ **3.** $y = x - 1$ **4.** $y = x - 2$ **5.** $y = x$

6. $y = 2x$ **7.** $y = -x$ **8.** $y = \frac{-3x}{2}$ **9.** $y = 2x + 1$ **10.** $y = 2x - 3$

11. $y = 3x - 1$ **12.** $y = 4x - 6$ **13.** $y = 1 - x$ **14.** $y = 4 - x$ **15.** $y = 6 - 2x$

16. $y = 1 - \frac{x}{2}$ **17.** $y = 3$ **18.** $x = 3$ **19.** $y = -1$ **20.** $x = -4$

In Exercises 21–24, fill in "x" or "y."

21. The □-axis has equation $y = 0$. **22.** The □-axis has equation $x = 0$.

23. All points on a horizontal line have the same □-coordinate.

24. All points on a vertical line have the same □-coordinate.

7.6 SYSTEMS OF LINEAR EQUATIONS

INTERSECTING LINES *Two different lines (on a plane) that are not "parallel" intersect (or meet) at a single point.* (See Figure 7.32 .) Now, you will find the intersection point by graphing both lines on the same coordinate system.

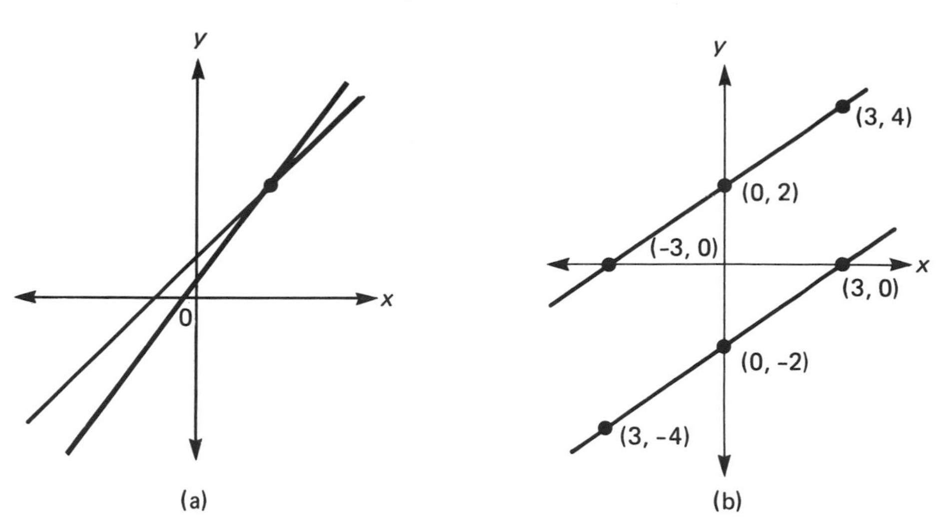

Fig. 7.32 **(a)** Two different lines (on a plane) that are not parallel intersect at a single point. **(b)** Parallel lines are lines (on the same plane) that do not intersect. For points with the same x-coordinates on the above parallel lines, the y-coordinates differ by 4.

Example 1 Find the intersection of the lines L_1, given by

$$y = x + 1$$

and L_2, given by $y = 2x - 1$

Solution

TABLE 7.9

(a) L_1:

x	y
0	1
1	2

(b) L_2:

x	y
0	-1
1	1

Graph both lines on the same coordinate system. As you see in Figure 7.33, the intersection is the point (2, 3).

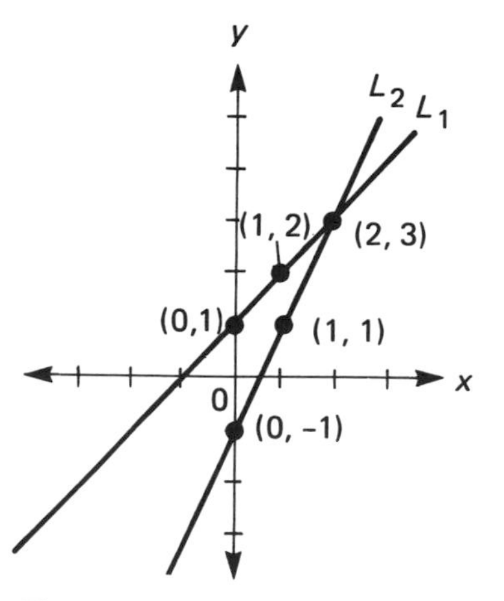

Fig. 7.33

check You can check your graphical solution by replacing x by 2 and y by **3** in each of the given equations.

$y = x + 1$	$y = 2x - 1$
$\mathbf{3} \stackrel{?}{=} 2 + 1$	$\mathbf{3} \stackrel{?}{=} 2 \cdot 2 - 1$
$3 \stackrel{\checkmark}{=} 3$	$3 \stackrel{\checkmark}{=} 3$

In Example 1, the check showed that you found the *exact* point by graphing. However, in general, the intersection point can only be *approximated* by graphing. It is often difficult to approximate closely an intersection point, such as $\left(\frac{3}{5}, \frac{4}{7}\right)$, with fractional coordinates.

Example 2 Find the intersection of the lines L_1, given by

$$y - 2 = 3(x - 5)$$

and L_2, given by

$$2y = x - 6$$

Solution To see that the first equation represents a line, L_1, add 2 to both sides of the equation.

$$\begin{aligned} y - 2 &= 3(x - 5) \\ y &= 3(x - 5) + 2 \\ \text{or} \quad y &= 3x - 13 \end{aligned}$$

[See Table 7.10(a).]

To see that the second equation represents a line, L_2, divide both sides of the equation by 2.

$$\begin{aligned} 2y &= x - 6 \\ y &= \frac{x - 6}{2} = \frac{1}{2}x - 3 \end{aligned}$$

[See Table 7.10(b).]

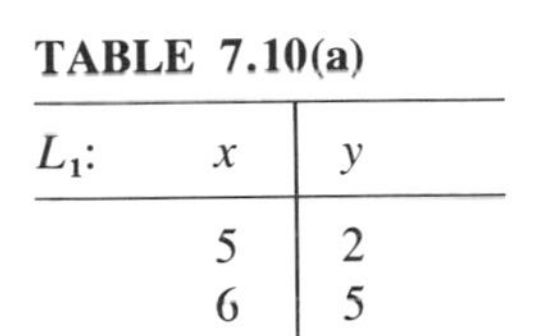

TABLE 7.10(a)

L_1: x	y
5	2
6	5

TABLE 7.10(b)

L_2: x	y
0	−3
2	−2

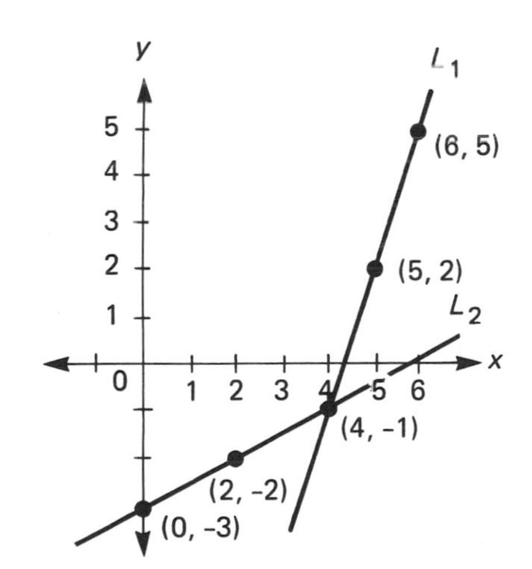

Fig. 7.34

As you see in Figure 7.34, the intersection is the point $(4, -1)$.

CLASS EXERCISES

Use the graphical method.

1. Find the intersection of the lines L_1, given by $y = x - 2$, and L_2, given by $y = 4 - x$. *Check your solution.*
2. Find the intersection of the lines L_1, given by $y = -4x$, and L_2, given by $y + 3 = 2x$.

SYSTEMS AND THEIR SOLUTIONS

The graphical method that you used to find the intersection of two lines tends to be inaccurate, particularly when the coordinates of the intersection point are not integers. There are also algebraic methods of determining this intersection.

An equation of a line is called a **linear equation**. Thus,

$$y = 2x + 3$$

is a linear equation. Also,

$$y + 1 = 3(x + 1)$$

is a linear equation because then,

$$y + 1 = 3x + 3$$

and thus,

$$y = 3x + 2$$

DEFINITION

> An ordered pair (x_1, y_1) is a **solution** of a (linear) equation in the variables x and y if a true statement results when x_1 replaces x and y_1 replaces y.

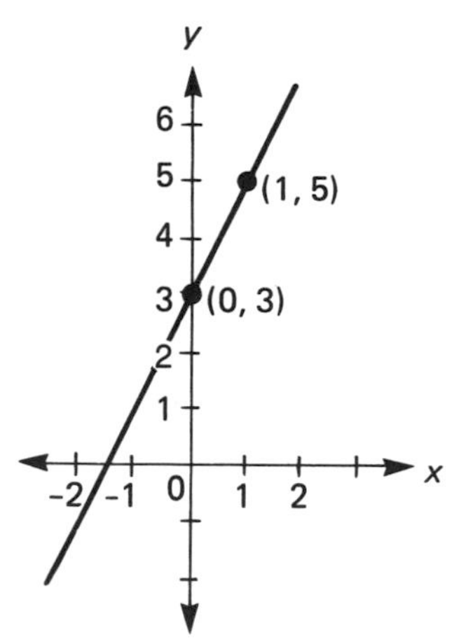

Fig. 7.35. Every point on a line corresponds to a solution of an equation of the line. Thus, (0, 3) and (1, 5) are both solutions of the equation, $y = 2x + 3$, for the above line.

Thus, (0, 3) is a solution of the equation

$$y = 2x + 3$$

because when you replace x by *0* and **y** by **3**, a true statement,

$$3 = 2 \cdot 0 + 3$$

results. Note that (1, 5) is also a solution because when you replace x by *1* and **y** by **5**, a true statement,

$$\mathbf{5} = 2 \cdot 1 + 3$$

results.

Geometrically, a solution represents a point on the line given by this equation. There are "infinitely many" solutions of a linear equation because every point on the line corresponds to a solution. (See Figure 7.35 .)

Two linear equations, taken together, form a **system of linear equations (in two variables)**. For example, the two equations

$$y = 2x + 3$$
$$y + 1 = 3(x + 1)$$

together form such a system.

DEFINITION

> An ordered pair (x_1, y_1) is a **solution of a system of linear equations** if (x_1, y_1) is a solution of both equations of the system.

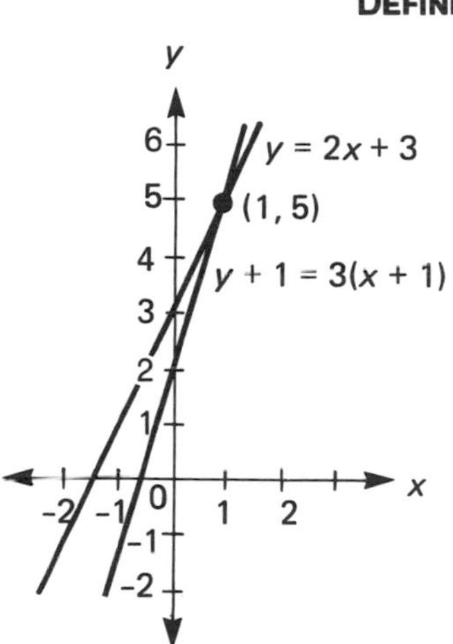

Fig. 7.36. A solution of a system of linear equations represents the intersection of the corresponding lines. Thus, (1, 5) is the solution of the system $y = 2x + 3$, $y + 1 = 3(x + 1)$.

The ordered pair (1, 5) is a solution of

$$y + 1 = 3(x + 1)$$

because $5 + 1 = 3(1 + 1)$ [or $6 = 6$]

Recall that (1, 5) is also a solution of $y = 2x + 3$. Thus, (1, 5) is a solution of the system:

$$y = 2x + 3$$
$$y + 1 = 3(x + 1)$$

Geometrically, a solution of a system of linear equations represents the intersection of the lines whose equations comprise the system. (See Figure 7.36.)

CLASS EXERCISES

3. Show that the ordered pair (2, 3) is a solution of the system

$$y = x + 1$$
$$2y = 3x$$

by replacing x by 2 and y by 3 in each equation.

Substituting You can solve a system of equations by substituting for one of the variables, as in Example 3.

Example 3 Solve the system.

$$y = 2x + 3$$
$$y + 2 = 4x + 1$$

Check the solution.

Solution Substitute

$$2x + 3 \text{ for } y$$

in the second equation, and solve for x.

$$2x + 3 + 2 = 4x + 1$$
$$2x + 5 = 4x + 1$$
$$4 = 2x$$
$$2 = x$$

Now, replace x by **2** in the first equation, and find y:

$$y = 2 \cdot 2 + 3$$
$$y = 7$$

The solution of the system is (2, 7). [See Figure 7.37 .]

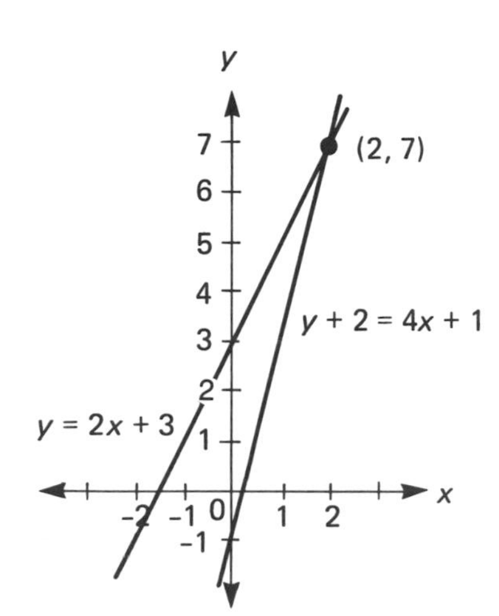

Fig. 7.37

check You can check the solution by replacing x by **2** and y by 7 in each of the given equations.

$$7 \stackrel{?}{=} 2 \cdot \mathbf{2} + 3$$
$$7 \stackrel{\checkmark}{=} 7$$

$$7 + \mathbf{2} \stackrel{?}{=} 4 \cdot \mathbf{2} + 1$$
$$9 \stackrel{\checkmark}{=} 9$$

In the **substitution method**:

1. Solve one equation for one of the variables, say y, in terms of the other, x.
2. Replace y by this expression in the other equation.
3. Solve this other equation for x.
4. To find y, replace x by this value in the equation for y in Step 1.

In Example 3, Step 1 was automatic. You were given y in terms of x in the first equation, $y = 2x + 3$.

Example 4 Solve the system.

$$y + 3 = 4(x - 1)$$
$$y - 2 = -5x$$

Solution **1.** From the second equation,

$$y = 2 - 5x$$

2. Replace y by the expression $2 - 5x$ in the first equation.

$$2 - 5x + 3 = 4(x - 1)$$

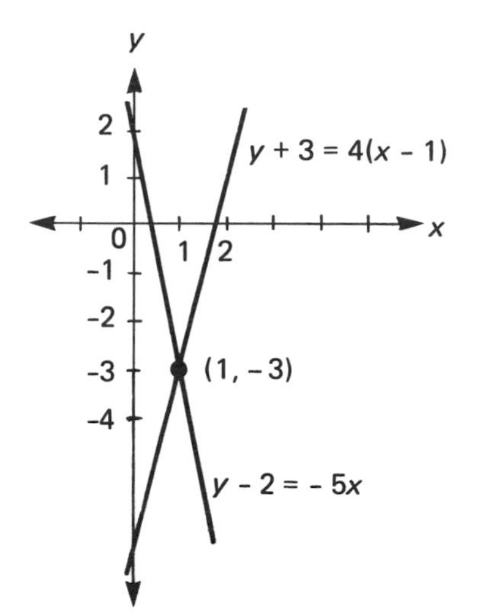

Fig. 7.38

3. Solve this equation for x.

$$5 - 5x = 4x - 4$$
$$9 = 9x$$
$$1 = x$$

4. Replace x by **1** in the equation $y = 2 - 5x$ in Step 1.

$$y = 2 - 5 \cdot 1$$
$$y = -3$$

The solution is $(1, -3)$. [See Figure 7.38 .]

A linear equation can be written in the form

$$Ax + By = C$$

where A, B, and C are numbers, and where A and B are not both 0. For example,

$$y = 2x + 5$$

becomes

$$-2x + y = 5$$

Here, $A = -2$, $B = 1$, $C = 5$.

Often the equations of a (linear) system are given in the form†

$$Ax + By = C$$
$$ax + by = c$$

Example 5 Solve the system.

$$x + 2y = 10$$
$$2x - 3y = 6$$

Solution **1.** Because x appears in the first equation with coefficient 1, solve this equation for x in terms of y.

$$x = 10 - 2y$$

---

†It turns out that the equations represent parallel or identical lines if $\frac{A}{a} = \frac{B}{b}$, $a \neq 0$ or if $A = a = 0$.

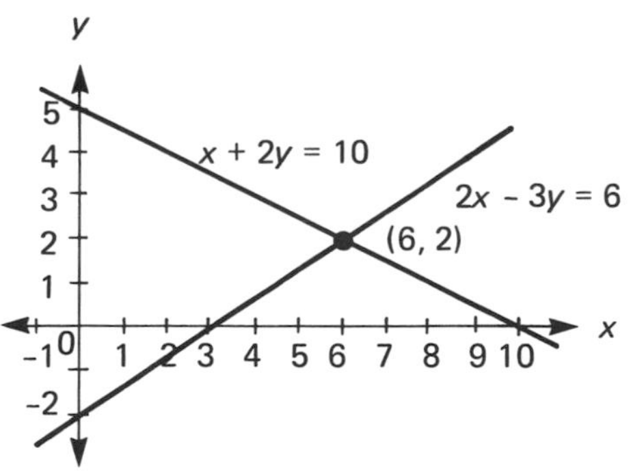

Fig. 7.39

2. Replace x by the expression $10 - 2y$ in the second equation.

$$2(10 - 2y) - 3y = 6$$

3. Solve this for y.

$$20 - 4y - 3y = 6$$
$$20 - 7y = 6$$
$$14 = 7y$$
$$2 = y$$

4. Replace y by **2** in the equation $x = 10 - 2y$ in Step 1.

$$x = 10 - 2 \cdot \mathbf{2}$$
$$x = 6$$

The solution is (6, 2). [See Figure 7.39 .]

CLASS EXERCISES *Solve by substituting. In Exercise 4, check your solution.*

4. $y = x + 4$
$y = 3x$

5. $2x - y = 3$
$3x - y = 5$

ADDING You can also solve a system of equations by first adding or subtracting the corresponding sides of the equations to eliminate one of the variables.

Example 6 Solve the system.

$$2x + 3y = 5$$
$$4x - 3y = 1$$

Solution Add the corresponding sides of the equations to eliminate y.

$$\begin{array}{r} 2x + 3y = 5 \\ \underline{4x - 3y = 1} \\ 6x \qquad = 6 \\ x \qquad = 1 \end{array}$$

Example 7 Twice a number plus a second number is ten. The sum of the two numbers is six. Set up a system of equations and solve for these numbers.

Solution Let x be the first number and y the second number.

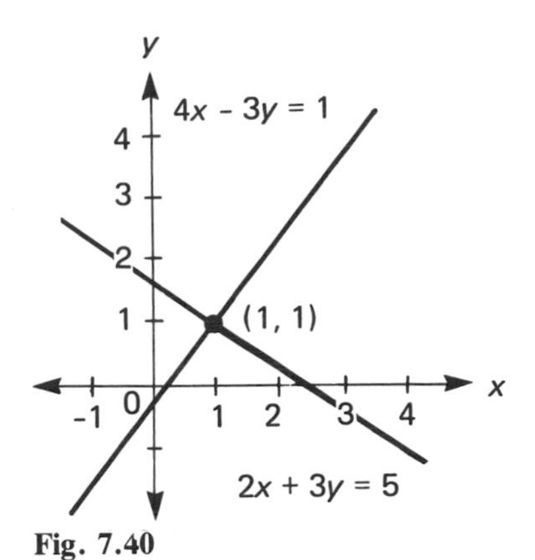

Fig. 7.40

Now, replace x by *1* in either equation. (For example, use the first equation.)

$$2 \cdot 1 + 3y = 5$$
$$3y = 3$$
$$y = 1$$

The solution is (1, 1). [See Figure 7.40 .]

Twice a number plus a second number is ten.

$$2x \quad + \quad y \quad = \quad 10$$

The sum of the two numbers is six.

$$x + y \quad = \quad 6$$

Solve the system.

$$2x + y = 10$$
$$x + y = 6$$

Subtract each side of the second equation from the corresponding side of the first equation to eliminate y.

$$\begin{aligned} 2x + y &= 10 \\ x + y &= 6 \\ \hline x \quad &= 4 \end{aligned}$$

Replace x by *4* in the second equation.

$$4 + y = 6$$
$$y = 2$$

The solution is (4, 2). Thus, the two numbers are 4 and 2.

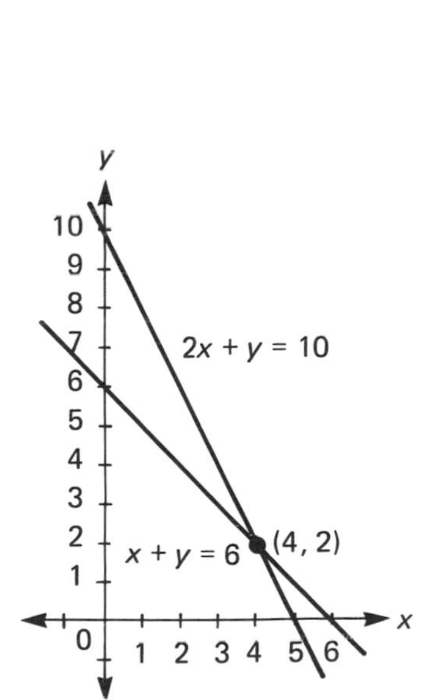

Fig. 7.41

Although Example 7 concerns numbers, the solution can be regarded as representing the intersection point of the lines with equations $2x + y = 10$ and $x + y = 6$. [See Figure 7.41 .]

These algebraic methods apply to systems in variables other than x and y. Also, in Example 8, *both* equations are first transformed, so that when the resulting equations are added, a variable is eliminated.

Example 8 Solve the system.

$$3s + 2t = 10$$
$$2s + 3t = 5$$

Solution Multiply both sides of the first equation by 2, and both sides of the second equation by -3. You will then eliminate s.

$$2(3s + 2t) = 2 \cdot 10 \quad \text{or} \quad 6s + 4t = 20$$
$$-3(2s + 3t) = -3 \cdot 5 \quad \text{or} \quad -6s - 9t = -15$$
$$-5t = 5$$
$$t = -1$$

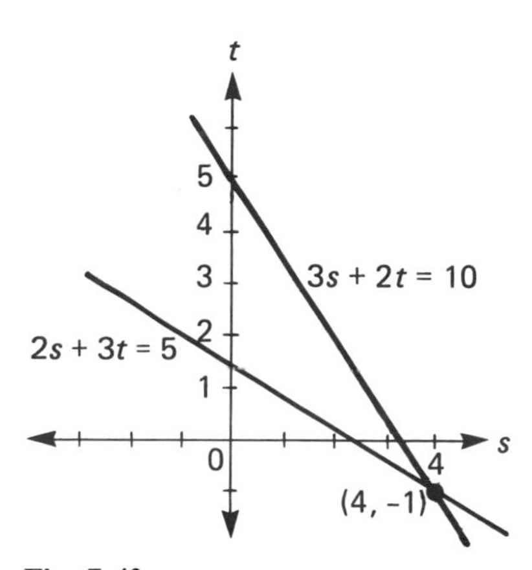

Fig. 7.42

Replace t by -1 in the (given) first equation.

$$3s + 2(-1) = 10$$
$$3s = 12$$
$$s = 4$$

The solution is $(4, -1)$.

You can graph this system by letting the horizontal axis be the s-axis and the vertical axis be the t-axis, as in Figure 7.42 .

CLASS EXERCISES *Solve by the "adding" method.*

6. $x + y = 8$, $x - y = 2$

7. $x + 2y = 6$, $2x + 3y = 10$

8. $5a + 4b = -1$, $4a + 5b = 1$

9. Four times a number is two more than five times another number. Twice the first number is two less than three times the second number. Set up a system of equations, and solve for these numbers.

SOLUTIONS TO CLASS EXERCISES

1. TABLE 7.11

a. L_1 :

x	y
0	−2
2	0

b. L_2 :

x	y
0	4
4	0

As you see in Figure 7.43, the intersection is the point (3, 1).

check

Replace x by *3* and y by **1** in each equation.

$y = x - 2$ | $y = 4 - x$

$\mathbf{1} \stackrel{?}{=} 3 - 2$ | $\mathbf{1} \stackrel{?}{=} 4 - 3$

$1 \stackrel{\checkmark}{=} 1$ | $1 \stackrel{\checkmark}{=} 1$

2. TABLE 7.12(a) **TABLE 7.12(b)**

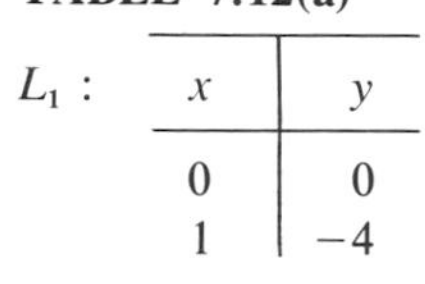

L_1 :

x	y
0	0
1	-4

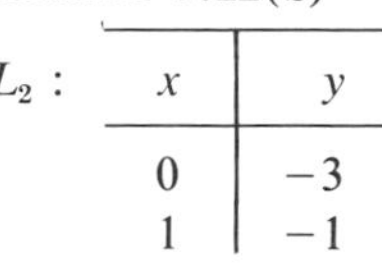

L_2 :

x	y
0	-3
1	-1

In the second equation, to obtain y in terms of x, subtract 3 from both sides.

$$y + 3 = 2x$$
$$y = 2x - 3$$

As you see in Figure 7.44, the intersection is the point $\left(\frac{1}{2}, -2\right)$.

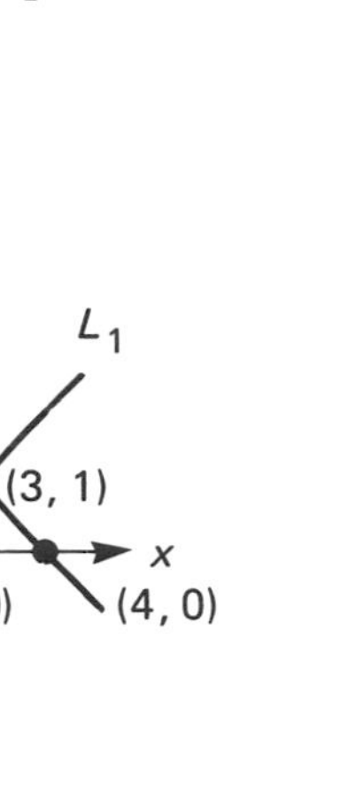

Fig. 7.43

Fig. 7.44

3. $\mathbf{3} \stackrel{?}{=} 2 + 1$ | $2 \cdot \mathbf{3} \stackrel{?}{=} 3 \cdot 2$

$3 \stackrel{\checkmark}{=} 3$ | $6 \stackrel{\checkmark}{=} 6$

4. $x + 4 = 3x$

$4 = 2x$

$2 = x$

$y = x + 4$

This is the first equation. Replace x by 2.

$y = 2 + 4$

$y = 6$

The solution is (2, 6).

check

$\mathbf{6} \stackrel{?}{=} 2 + 4$ | $\mathbf{6} \stackrel{?}{=} 3 \cdot 2$

$6 \stackrel{\checkmark}{=} 6$ | $6 \stackrel{\checkmark}{=} 6$

5. The first equation can be rewritten as

$$y = 2x - 3$$

and the second as

$$y = 3x - 5$$

Thus, solve.

$2x - 3 = 3x - 5$

$2 = x$

$y = 2x - 3$

$y = 2 \cdot 2 - 3$

$y = 1$

The solution is (2, 1).

6.
$$\begin{aligned} x + y &= 8 \\ x - y &= 2 \\ \hline 2x \quad &= 10 \\ x \quad &= 5 \\ x + y &= 8 \\ 5 + y &= 8 \\ y &= 3 \end{aligned}$$

The solution is (5, 3).

7.
$$x + 2y = 6$$
$$2x + 3y = 10$$
Multiply both sides of the first equation by -2.
$$-2x - 4y = -12$$
$$2x + 3y = 10$$
$$-y = -2$$
$$y = 2$$
$$x + 2y = 6$$
$$x + 2 \cdot 2 = 6$$
$$x + 4 = 6$$
$$x = 2$$
The solution is (2, 2).

8.
$$5a + 4b = -1$$
$$4a + 5b = 1$$
Multiply both sides of the first equation by -4 and both sides of the second equation by 5.
$$-20a - 16b = 4$$
$$20a + 25b = 5$$
$$9b = 9$$
$$b = 1$$
$$5a + 4b = -1$$
$$5a + 4 \cdot 1 = -1$$
$$5a = -5$$
$$a = -1$$
The solution is $(-1, 1)$.

9. Let x be the first number and y the second number.

Four times a number | is | two more than | five times another number.
$4x$ | $=$ | $2 +$ | $5y$

Twice the first number | is | two less than | three times the second number.
$2x$ | $=$ | $-2 +$ | $3y$

Multiply both sides of the second equation by -2, and "add" the first equation.

$$-4x = 4 - 6y$$
$$4x = 2 + 5y$$
$$0 = 6 - y$$
$$y = 6$$
$$4x = 2 + 5(6)$$
$$4x = 32$$
$$x = 8$$
The two numbers are 8 and 6.

HOME EXERCISES

In Exercises 1–10, find the intersection of the given lines, graphically. In Exercises 1–4, check your solution.

1. L_1: $y = x$
L_2: $y = 4 - x$

2. L_1: $y = 2x$
L_2: $y - 3 = \frac{x}{2}$

3. L_1: $y - 4 = x$
L_2: $y - 2 = 2(x + 1)$

4. L_1: $y + 2 = \frac{x + 4}{2}$
L_2: $y = 2x$

5. L_1: $2y = x + 3$
L_2: $y = 2x$

6. L_1: $y = 4$
L_2: $x = 5$

7. L_1: $y - 4 = 3x$
L_2: $y = x$

8. L_1: $y = 2x + 5$
L_2: $y = x + 1$

9. L_1: $y = \frac{x}{4}$
L_2: $y + 2 = 2(x - 6)$

10. L_1: $3y + 1 = 2x$
L_2: $y + 1 = x$

In Exercises 11–16, solve by substituting. In Exercises 11–14, check your solution.

11. $y = x + 1$
$y = 3x - 5$

12. $y = x - 2$
$y = 2x - 8$

13. $y = 2x - 3$
$y = 3x - 7$

14. $y = 4x - 1$
$y = 2x + 3$

15. $y = 4 - 3x$
$y - 3 = 2(x - 2)$

16. $y - 8 = 5x$
$y + 1 = 4(x + 2)$

In Exercises 17–22, solve by the "adding" method. In Exercises 17–20, check your solution.

17. $x + y = 4$
$x - y = 2$

18. $c + 5d = 2$
$5c + 5d = 6$

19. $r + 2s = 8$
$4r - s = 5$

20. $3m - n = 2$
$2m - 3n = -1$

21. $6x - 5y = 7$
$2x + 3y = 7$

22. $2x + 3y = 9$
$3x + 2y = 11$

In Exercises 23–32, solve by either substituting or adding.

23. $x + 2y = 6$
$3x - y = 4$

24. $u + v = 8$
$5v = 3u$

25. $3y = 4x$
$y - 4 = x + 2$

26. $x + y = 3$
$2x - 3y = 6$

27. $a + 2b = 5$
$4a - 6b = 6$

28. $6x - 3y = 3$
$3x + y = 9$

29. $3y = 2x$
$y - 4 = 2x$

30. $2x + 5y = 9$
$4x + 3y - 11$

31. $x - 3y = 2$
$4x - 10y = 10$

32. $2s + 3t = 10$
$3s + 2t = 10$

33. Twice a number is six less than three times another number. Four times the first number is two more than the second number. Find these numbers.

34. With $50 you can buy seven records and five books, or four records and ten books. Assuming each record is the same price and each book is the same price, how much does a record cost?

35. Diane has $2.85 in nickels and quarters. If the number of nickels and quarters were reversed, she would have $3.45 . How many quarters does she have?

36. A bookbinder has two machines. The slower machine in five minutes and the faster machine in ten minutes together bind 700 books. The slower machine in 10 minutes and the faster machine in four minutes together bind 600 books. How many books per minute can each machine bind?

7.7 VARIATION

DIRECT VARIATION A linear equation of the form

$$y = kx$$

describes many important relationships in science and everyday affairs. For example, if a car is traveling at the constant rate of 60 miles per hour, the distance, y, traveled in x hours is given by the linear equation

$$y = 60x$$

Thus, in half an hour, the car travels $60 \cdot \frac{1}{2}$, or 30, miles; in 3 hours the car travels $60 \cdot 3$, or 180, miles.

DEFINITION

> Let x and y be variables. Then y **varies directly** as x if
>
> $$y = kx$$
>
> for some nonzero number k. This number k is called the **constant of variation**.

Example 1 Suppose

$$y = 3x$$

Then, y varies directly as x. The constant of variation is 3. Some of the corresponding values of x and y are given in Table 7.13 .

The relationship between x and y is described by a line through the origin (See Figure 7.45 .)

TABLE 7.13

x	y
1	3
2	6
5	15
−1	−3
−3	−9

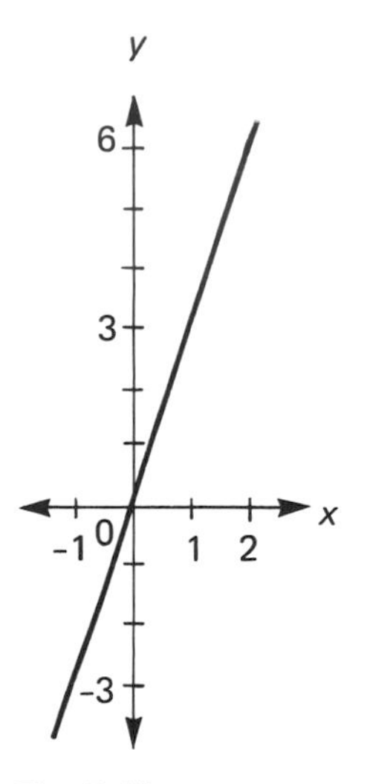

Fig. 7.45

When y varies directly as x, then

$$y = kx, \qquad k \neq 0.$$

It follows that *x also varies directly as y* because if you divide both sides of the equation by k, you obtain

$$\frac{1}{k} y = x$$

The constant of variation is now $\frac{1}{k}$. Also, if y varies directly as x, then

$$\frac{y}{x} = k, \qquad x \neq 0$$

Thus, *the quotient, $\frac{y}{x}$, remains fixed.* Let x_1 and x_2 be values of x and let y_1 and y_2 be the corresponding values of y. Then,

$$\frac{y_1}{x_1} = \frac{y_2}{x_2} = k$$

Thus, *corresponding values of y and x are proportional. When one variable increases, the other variable also increases.*

Example 2 Suppose $y = 5x$

The constant of variation is 5. Then, you can also write

$$x = \frac{1}{5}y$$

Here, the constant of variation is $\frac{1}{5}$. Also,

$$\frac{y}{x} = 5$$

Some corresponding values of x and y are given in Table 7.14.

TABLE 7.14

x	y
1	5
2	10
3	15
4	20

Example 3 Suppose y varies directly as x, and $y = 12$ when $x = 3$.

a. Find the constant of variation. **b.** Find y when $x = 5$. **c.** Find x when $y = 8$.

Solution **a.**

$$y = kx$$

Replace y by *12* and x by **3**.

$$12 = k \cdot 3$$

$$4 = k$$

Thus, the constant of variation is 4, and $y = 4x$.

b. Replace x by *5*.

$$y = 4 \cdot 5 = 20$$

c. Replace y by *8*.

$$8 = 4x$$

$$2 = x$$

CLASS EXERCISES *Suppose y varies directly as x.*

1. Find the constant of variation if $y = 8$ when $x = 2$.
2. Suppose $y = 6$ when $x - 4$. Find y when $x = 10$.
3. Suppose $y = -2$ when $x = 4$. Find x when $y = 4$.

INVERSE VARIATION

DEFINITION

Let x and y be variables. Then, y **varies inversely as** x if

$$y = \frac{k}{x}, \qquad x \neq 0$$

for some nonzero number k. Again, k is called the constant of variation.

Example 4 Let

$$y = \frac{2}{x}, \qquad x \neq 0$$

Then, y varies inversely as x. The constant of variation is 2. Some corresponding values of x and y are given in Table 7.15(a).

TABLE 7.15(a)

x	y
1	2
2	1
4	$\frac{1}{2}$
6	$\frac{1}{3}$
8	$\frac{1}{4}$

TABLE 7.15(b)

x	y
$\frac{1}{2}$	4
$\frac{1}{3}$	6
$\frac{1}{4}$	8

TABLE 7.15(c)

x	y
-1	-2
-2	-1
-4	$\frac{-1}{2}$
$\frac{-1}{2}$	-4

Recall

$$\frac{2}{\frac{1}{2}} = 2 \div \frac{1}{2} = 2 \cdot \frac{2}{1} = 4$$

and that

$$\frac{2}{\frac{1}{3}} = 2 \div \frac{1}{3} = 2 \cdot \frac{3}{1} = 6$$

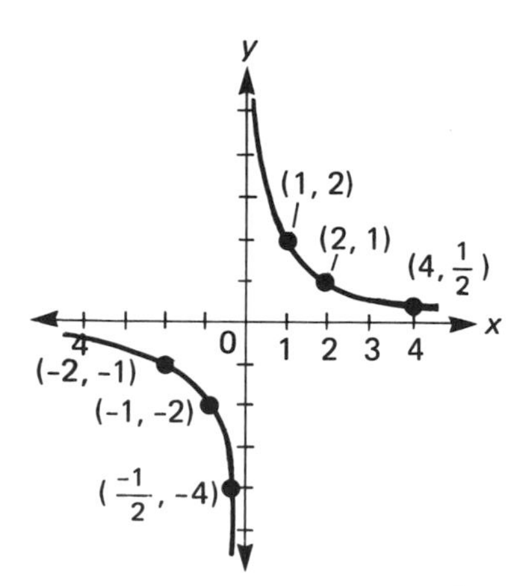

Fig. 7.46. The hyperbola $y = \frac{2}{x}$. Note the two branches.

Thus, when x takes on the fractional values $\frac{1}{2}, \frac{1}{3}, \frac{1}{4}$, the values of x and y given in Table 7.15(a) are interchanged. See Table 7.15(b). Note that x can also take on negative values. The corresponding values of y are also negative. See Table 7.15(c). However, neither x nor y can take on the value 0 in the equation

$$y = \frac{2}{x}$$

First, x cannot be 0 because division by 0 is undefined. Also, y cannot be 0 because y equals the fraction $\frac{2}{x}$, and a fraction is 0 only if the numerator is 0.

The graph is given in Figure 7.46 and is known as a **hyperbola**. Each of the curves is called a **branch** of the hyperbola.

If y varies inversely as x, then

$$y = \frac{k}{x}, \qquad k \neq 0$$

Cross-multiply, and obtain

$$x = \frac{k}{y}, \qquad y \neq 0$$

Thus, *x varies inversely as y, with the same constant of variation, k*. Also, *if y varies inversely as x*, then

$$xy = k$$

Thus, *the product of the two variables is constant. When one variable increases, the other variable decreases.*

Example 5 Suppose

$$y = \frac{-6}{x}, \qquad x \neq 0$$

Here, the constant of variation is -6. Cross-multiply to obtain

$$x = \frac{-6}{y}, \qquad y \neq 0$$

Again, the constant of variation is -6. Finally,

$$xy = -6$$

Some corresponding values of x and y are given in Table 7.16.

TABLE 7.16

x	y
1	-6
2	-3
3	-2
6	-1
12	$\frac{-1}{2}$
18	$\frac{-1}{3}$
$\frac{1}{2}$	-12
$\frac{1}{3}$	-18

Example 6 **Boyle's Law** states that the pressure, p, of a compressed gas varies inversely as the volume, v, of gas. Suppose the pressure is 40 pounds per square inch when the volume is 100 cubic inches. Find the pressure when the gas is compressed to 80 cubic inches.

Solution Because p varies inversely as v,

$$p = \frac{k}{v}, \qquad v \neq 0$$

To find k, let $p = 40$ and $v = 100$.

$$40 = \frac{k}{100}$$

$$4000 = k$$

Thus, the relationship between pressure and volume is given by the equation

$$p = \frac{4000}{v}$$

Now, replace **v** by **80** to find the corresponding value of p.

$$p = \frac{4000}{\mathbf{80}} = 50$$

CLASS EXERCISES *Suppose y varies inversely as x.*

4. Find the constant of variation if $y = 4$ when $x = 5$.
5. Suppose $y = 3$ when $x = 2$. Find y when $x = 1$.
6. Suppose $y = -5$ when $x = 5$. Find x when $y = 20$.

SOLUTIONS TO CLASS EXERCISES

1.
$$y = kx$$
$$8 = k \cdot 2$$
$$4 = k$$

2.
$$y = kx$$
$$6 = k \cdot 4$$
$$\frac{3}{2} = k$$
$$y = \frac{3}{2}x$$
When $x = 10$,
$$y = \frac{3}{2} \cdot 10 = 15$$

3.
$$y = kx$$
$$-2 = k \cdot 4$$
$$\frac{-1}{2} = k$$
$$y = \frac{-1}{2}x$$
When $y = 4$,
$$4 = \frac{-1}{2}x$$
$$-8 = x$$

4.
$$y = \frac{k}{x}, \quad x \neq 0$$
$$4 = \frac{k}{5}$$
$$20 = k$$

5.
$$y = \frac{k}{x}, \quad x \neq 0$$
$$3 = \frac{k}{2}$$
$$6 = k$$
$$y = \frac{6}{x}, \quad x \neq 0$$
When $x = 1$,
$$y = \frac{6}{1} = 6$$

6.
$$y = \frac{k}{x}, \quad x \neq 0$$
$$-5 = \frac{k}{5}$$
$$-25 = k$$
$$y = \frac{-25}{x}, \quad x \neq 0$$
When $y = 20$,
$$20 = \frac{-25}{x}$$
$$20x = -25$$
$$x = \frac{-5}{4}$$

HOME EXERCISES

In Exercises 1–8, suppose y varies directly as x. Find the constant of variation.

1. $y = 4$ when $x = 2$ **2.** $y = 10$ when $x = 2$ **3.** $y = 9$ when $x = 3$ **4.** $y = 36$ when $x = 6$

5. $y = 4$ when $x = -2$ **6.** $y = 10$ when $x = 4$ **7.** $y = \frac{1}{2}$ when $x = \frac{1}{4}$ **8.** $y = \frac{1}{2}$ when $x = \frac{1}{3}$

In Exercises 9–16, assume y varies directly as x.

9. Suppose $y = 6$ when $x = 3$. Find y when $x = 5$.
10. Suppose $y = 12$ when $x = 2$. Find y when $x = 6$.
11. Suppose $y = 15$ when $x = 3$. Find y when $x = 2$.
12. Suppose $y = 20$ when $x = 4$. Find y when $x = -2$.
13. Suppose $y = 100$ when $x = 10$. Find x when $y = 400$.
14. Suppose $y = 49$ when $x = 7$. Find x when $y = 7$.
15. Suppose $y = -12$ when $x = 36$. Find x when $y = 48$.
16. Suppose $y = -64$ when $x = -16$. Find x when $y = 16$.

In Exercises 17–24, suppose y varies inversely as x. Find the constant of variation.

17. $y = 4$ when $x = 2$
18. $y = 3$ when $x = 4$
19. $y = -2$ when $x = -1$
20. $y = 12$ when $x = 2$
21. $y = \frac{1}{2}$ when $x = 4$
22. $y = -3$ when $x = 2$
23. $y = -8$ when $x = -4$
24. $y = \frac{1}{3}$ when $x = \frac{1}{2}$

In Exercises 25–32, assume y varies inversely as x.

25. Suppose $y = 2$ when $x = 2$. Find y when $x = 4$.
26. Suppose $y = 5$ when $x = 2$. Find y when $x = 1$.
27. Suppose $y = 10$ when $x = 5$. Find y when $x = 25$.
28. Suppose $y = 3$ when $x = 4$. Find y when $x = 3$.
29. Suppose $y = 12$ when $x = 5$. Find x when $y = -5$.
30. Suppose $y = 3$ when $x = 10$. Find x when $y = 15$.
31. Suppose $y = 7$ when $x = 4$. Find x when $y = 28$.
32. Suppose $y = 5$ when $x = 3$. Find x when $y = 10$.

In Exercises 33–38, fill in "directly" or "inversely."

33. When width is held constant, the area of a rectangle varies ______________ as the length. (Note: Area of rectangle = length · width)

34. For rectangles of a fixed area, the length of the rectangle varies ______________ as the width.

35. The cost of a sack of potatoes varies ______________ as the price per pound of potatoes.

36. If you have twenty dollars to spend on cookies, the number of pounds you can buy varies ______________ as the price per pound.

37. For cars traveling at a constant rate along the throughway from Albany to Buffalo, the time each car spends traveling varies ______________ as its rate.

38. If all seats are \$2.50, the gross of a movie theatre varies ______________ as the number of tickets sold.

39. The cost of a slice of cheese varies directly as its weight. If a 12-ounce slice costs \$1.80, how much does a 14-ounce slice cost?

40. The tension in a spring varies directly as the distance it is stretched. If the tension is 36 pounds when the distance stretched is 9 inches, what is the tension when the distance stretched is 1 foot?

41. The boll weevil population of a cotton field varies inversely as the amount of insecticide used. If 5000 boll weevils are in a field when 60 pounds of insecticide are used, how many boll weevils will be left when 300 pounds are used?

42. The weight of a body varies inversely as its distance from the center of earth. The earth's radius is (approximately) 4000 miles. How much does a 200-pound man weigh 16,000 miles above the surface of the earth?

Let's Review Chapter 7

1. Six more than twice an integer is 20. Find the integer.

2. Find three consecutive integers whose sum is 90. *Check your answer.*

3. Two sisters are five years apart in age. In two years, the younger sister will be 19. How old is the older sister now?

4. A man is twice as old as his daughter. Twenty years ago, he was six times her age. How old is the man now?

5. How long does it take to drive 175 miles at 50 miles per hour?

6. Two cars on a straight highway pass each other, traveling in opposite directions. Each car is traveling at 65 miles per hour. In how many hours will they be 260 miles apart?

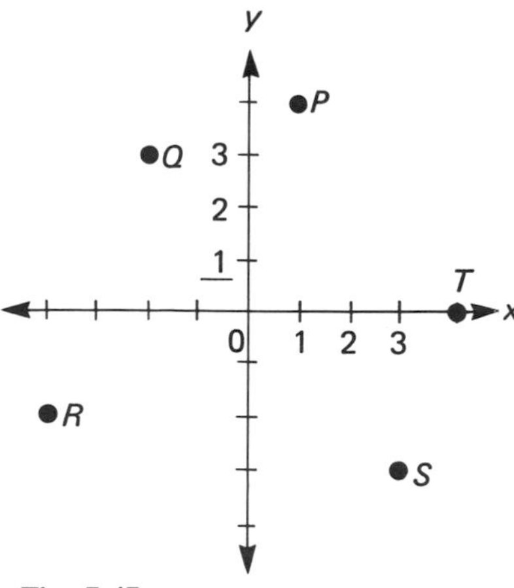

Fig. 7.47

7. A canoe goes upstream at the rate of four miles per hour. It returns downstream at the rate of eight miles per hour. If the round trip takes three hours, how far upstream does the canoe go?
8. Solve for y. $3 - 2x + 7y = 6$
9. Solve $\frac{3a}{b}$ for -2 **a.** for a, **b.** for b.
10. **a.** Solve $7u - 3v = uv$ for v. **b.** *Check your answer.*
11. For each point in Figure 7.47, find
    a. its x-coordinate,
    b. its y-coordinate.
    c. Identify the point with its ordered pair of coordinates.

In Exercises 12–16, locate the following points on a rectangular coordinate system.

12. $(4, 0)$ **13.** $(0, -2)$ **14.** $(5, 3)$ **15.** $(-3, 4)$ **16.** $\left(\frac{-1}{2}, \frac{-3}{2}\right)$

Graph the equations in Exercises 17–20.

17. $y = x + 2$ **18.** $y = 2x$ **19.** $y = 2 - x$ **20.** $y = \frac{x}{2}$

21. Graph the intersection of the lines given by

$$L_1: y = 2x + 4, \qquad L_2: y = 7 - x$$

22. Solve by substituting.

$$y = 2x + 1$$
$$y = 6 - x$$

23. Solve by the "adding" method.

$$6a - 5b = 7$$
$$3a - 2b = 4$$

24. Solve by either substituting or adding.

$$3x + 4y = 2$$
$$2x + 3y = -1$$

25. Suppose y varies directly as x. Find the constant of variation if $y = 4$ when $x = -2$.
26. Assume y varies directly as x. Suppose $y = 10$ when $x = 4$. Find y when $x = 6$.
27. Assume y varies inversely as x. Suppose $y = 8$ when $x = 2$. Find x when $y = \frac{1}{2}$.
28. Fill in "directly" or "inversely." If y varies inversely as x, then x varies ______ as y.

8 ROOTS

DIAGNOSTIC TEST

Perhaps you are already familiar with some of the material in Chapter 8. This test will indicate which sections in Chapter 8 you need to study. The question number refers to the corresponding section in Chapter 8. If you answer *all* parts of the question correctly, you may omit the section. But *if any part of your answer is wrong, you should study the section*. When you have completed the test, turn to page A23 for the answers.

8.1 Square Roots

a. Find $\sqrt{121}$. b. Find $\sqrt{(ab)^2}$, if $a > 0$, $b < 0$.

c. Find consecutive integers N and $N + 1$ such that $N < \sqrt{111} < N + 1$.

d. Use $\sqrt{2} \approx 1.41$ to approximate $\frac{\sqrt{2}}{5}$ to the nearest tenth.

In the remaining problems, assume $a > 0$, $x > 0$, $b \neq 0$.

8.2 Roots of Products and Quotients

a. Find $\sqrt{490{,}000}$. b. Find $\sqrt{\frac{36a^2}{b^4}}$. c. Simplify $\sqrt{28a^3}$.

8.3 Addition and Subtraction of Roots

Simplify each expression.

a. $\sqrt{25x} - \sqrt{9x}$ b. $\frac{2\sqrt{3}}{5} - \frac{\sqrt{3}}{2}$

8.4 Multiplication and Division of Roots

In parts a–c, multiply or divide, and simplify.

a. $\sqrt{2} \cdot \sqrt{3} \cdot \sqrt{6}$ b. $\frac{\sqrt{20}}{3} \cdot \frac{\sqrt{27}}{2}$ c. $\frac{\sqrt{75}}{\sqrt{50}} \div \frac{\sqrt{48}}{\sqrt{32}}$

d. Factor. $3 + \sqrt{18}$ e. Use part d to simplify $\frac{3 + \sqrt{18}}{9}$.

8.5 Zero and Negative Exponents

Find each value. a. 6^{-2} b. $(-5)^0$ c. $\left(\frac{3}{4}\right)^{-3}$

8.6 Scientific Notation

a. Express 1,960,000 in scientific notation.

b. Convert 3.89×10^{-6} back into the usual decimal notation.

c. Find the value of $\frac{54{,}000{,}000 \times .008}{.000\,018 \times 4000}$. Express your result in scientific notation.

8.1 SQUARE ROOTS

ROOTS OF NUMBERS

You know that

$$2^2 = 2 \cdot 2 = 4$$

Suppose you are asked,

What *positive* number times itself is 4?

In other words, you are asked to find a *positive* number such that

$$(\text{the number}) \cdot (\text{the number}) = 4$$

Clearly, 2 satisfies this condition. 2 is called the "positive square root of 4." Note that the *negative* number -2 also works because

$$(-2)(-2) = 4$$

-2 is called the "negative square root of 4."

DEFINITION

SQUARE ROOT. Let a be positive. Then, b is called the **positive**, or **principal, square root of *a*** if b is positive and if

$$b^2 = a$$

In this case, $-b$ is called the **negative square root of *a***. Also, the **square root of 0** is 0.

Write

$$b = \sqrt{a} \qquad [\textit{Read: b equals the positive square root of a.}]$$

if b is the *positive* square root of a. Here, the symbol $\sqrt{}$ is known as a **radical sign.** Thus,

$\sqrt{4} = 2$ *The positive square root of 4 is 2.*

whereas $-\sqrt{4} = -2$ *The negative square root of 4 is -2.*

Example 1

a. $\sqrt{9} = 3$ because 3 is positive and $3^2 = 9$.

b. $-\sqrt{9} = -3$ **c.** $\sqrt{100} = 10$ **d.** $\sqrt{144} = 12$

Example 2 Find:

a. $\sqrt{4^2}$

Solution a.

$$4^2 = 16$$

Thus, $\sqrt{4^2} = \sqrt{16} = 4$

b. $\sqrt{(-4)^2}$

b. $(-4)^2 = (-4)(-4) = 16$

Thus, $\sqrt{(-4)^2} = \sqrt{16} = 4$

c. $-\sqrt{4^2}$

c. $-\sqrt{4^2} = -\sqrt{16} = -4$

Negative numbers do not have square roots (within the so-called "real number system"). Thus,

$$\sqrt{-4} \text{ is not defined (as a real number)}$$

-2 is *not* the square root of -4 because

$$(-2)(-2) = 4, \text{ rather than } -4$$

Note the difference between $\sqrt{-4}$, which is not defined, and $-\sqrt{4}$, the negative square root of 4, which equals -2.

CLASS EXERCISES *Find each square root.*

1. a. $\sqrt{25}$ b. $-\sqrt{25}$
2. a. $\sqrt{2^2}$ b. $-\sqrt{2^2}$ c. $-\sqrt{(-2)^2}$

ROOTS OF ALGEBRAIC EXPRESSIONS

Just as you have considered squares such as

$$a^2, (3x)^2, \text{ and } (st)^2$$

so too, you will consider square roots of algebraic expressions. *When working with algebraic expressions, you must consider the signs of the variables.*

Example 3 Assume
$a > 0, x > 0, s > 0, t > 0.$
Find:

a. $\sqrt{a^2}$

b. $\sqrt{(3x)^2}$

c. $\sqrt{(st)^2}$

Solution a. $\sqrt{a^2} = a$

because
$a > 0$ and $a^2 = a \cdot a$

b. Note that $3x \cdot 3x = (3x)^2$. Thus,

$$\sqrt{(3x)^2} = 3x$$

c. $\sqrt{(st)^2} = st$

The product of positive numbers is positive. Thus, in **b** and **c**, $3x$ and st are each positive.

Next, recall that

$$|-3| = |3| = 3$$

(You may want to review the notion of absolute value in Section 1.2.) Observe that

$$\sqrt{(-3)^2} = \sqrt{9} = 3$$

Thus, $$\sqrt{(-3)^2} = |-3|$$

In part **a** of Example 4, this is generalized.

Example 4 Let $p > 0$ and $n < 0$. Find:

a. $\sqrt{n^2}$

b. $\sqrt{(pn)^2}$

Solution **a.** $n < 0$
Therefore, $\sqrt{n^2} = |n|$

Observe that because n is negative, it follows that $-n$ is positive and therefore,

$$|n| = -n$$

Thus, you could also say that

$$\sqrt{n^2} = -n$$

b. $p > 0$, $n < 0$
Therefore, $pn < 0$

Thus, $\sqrt{(pn)^2} = |pn|$

Note that if p is positive, then

$$\sqrt{p^2} = p = |p|$$

Also, $$\sqrt{0^2} = 0 = |0|$$

Therefore, *for any number a, positive, negative, or 0,*

$$\boxed{\sqrt{a^2} = |a|}$$

CLASS EXERCISES *Assume $s > 0$ and $t < 0$. Find each square root.*

3. $\sqrt{s^2}$ **4.** $\sqrt{t^2}$ **5.** $\sqrt{(st)^2}$

ROOTS OF EVEN POWERS

Observe that

$$10{,}000 = 100 \cdot 100 \text{ and } 1{,}000{,}000 = 1000 \cdot 1000$$

Thus,
$$\sqrt{10^4} = \sqrt{10{,}000} = 100 = 10^2$$

and
$$\sqrt{10^6} = \sqrt{1{,}000{,}000} = 1000 = 10^3$$

Therefore,
$$\sqrt{10^{2 \cdot 2}} = 10^2$$

and
$$\sqrt{10^{2 \cdot 3}} = 10^3$$

You can also consider square roots of even powers of variables.

Example 5 Let $b > 0$. Find:

a. $\sqrt{a^4}$

b. $\sqrt{b^6}$

c. $\sqrt{c^{20}}$

Solution **a.**
$$\sqrt{a^4} = a^2$$

because $a^4 = a^2 \cdot a^2$

and a^2 is nonnegative (positive or 0).

b.
$$\sqrt{b^6} = b^3$$

Note that $b^6 = b^3 \cdot b^3$.

Also, $b^3 > 0$ because $b > 0$

c.
$$\sqrt{c^{20}} = c^{10}$$

In general, *if k is a positive integer, then for every number a,*

$$\sqrt{a^{2k}} = a^k, \text{ for } k \text{ even}$$
$$\sqrt{a^{2k}} = |a^k|, \text{ for } k \text{ odd}$$

Thus,
$$\sqrt{(-1)^8} = (-1)^4 = 1$$

but
$$\sqrt{(-1)^{10}} = |(-1)^5| = |-1| = 1$$

CLASS EXERCISES *Let a, b, c be any numbers. Find each square root.*

6. $\sqrt{a^{12}}$ **7.** $\sqrt{(bc)^4}$

RATIONAL AND IRRATIONAL NUMBERS

Recall that a *rational number* is a number that can be written in the form $\frac{N}{D}$, where N and D are integers and $D \neq 0$. For example, $\frac{3}{4}$, $\frac{-1}{2}$, and $\frac{5}{3}$ are rational numbers. Also, an integer such as 6 is a rational number because $6 = \frac{6}{1}$, and a decimal such as .3 is a rational number because $.3 = \frac{3}{10}$. An **irrational number** is a number that cannot be expressed as the quotient of integers, as previously stated. To obtain an example of an irrational number, consider the following.

The integers

$$0, 1, 4, 9, 16, 25, \ldots$$

are squares of other integers. Their square roots are also integers. Thus,

$$\sqrt{0} = 0, \quad \sqrt{1} = 1, \quad \sqrt{4} = 2, \quad \sqrt{9} = 3, \quad \text{and so on}$$

These square roots are rational numbers, but the square roots of integers such as

$$2, 3, 5, 6, 7, \ldots$$

are irrational. Thus,

$$\sqrt{2}, \sqrt{3}, \sqrt{5}, \sqrt{6}, \sqrt{7}, \ldots$$

are irrational.

Every rational number can be expressed as either a (terminating) decimal or as an infinite repeating decimal (in which one or more digits repeat). Thus,

$$\frac{1}{4} = .25, \quad \frac{2}{3} = .666\,666\ldots, \quad \frac{4}{11} = .\underbrace{36}\underbrace{36}\underbrace{36}\ldots$$

Every irrational number can be expressed as an infinite nonrepeating decimal. The digits do not repeat in any regular pattern. The irrational number $\sqrt{2}$ can be *approximated* by an ordinary decimal to as many decimal digits as desired. Observe how this can be done.

$$\underbrace{\sqrt{1}}_{1} < \sqrt{2} < \underbrace{\sqrt{4}}_{2} \qquad [\textit{Read: } \sqrt{1} < \sqrt{2} \textit{ and } \sqrt{2} < \sqrt{4}]$$

Thus,

$$1 < \sqrt{2} < 2$$

Try to find a decimal between 1 and 2 whose square is close to 2. By considering $(1.1)^2$, $(1.2)^2$, and so on, you find

$$\begin{array}{r} 1.4 \\ \times\ 1.4 \\ \hline 5\,6 \\ 1\,4 \\ \hline 1.9\,6 \end{array} \qquad \begin{array}{r} 1.5 \\ \times\ 1.5 \\ \hline 7\,5 \\ 1\,5 \\ \hline 2.2\,5 \end{array}$$

$(1.4)^2 = 1.96$ and $(1.5)^2 = 2.25$. Thus,

$$(1.4)^2 < 2 < (1.5)^2$$

Therefore, it seems reasonable that

$$1.4 < \sqrt{2} < 1.5$$

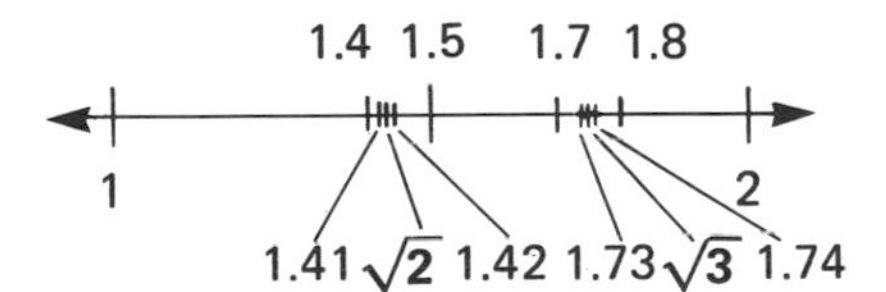

Fig. 8.1. The irrational number $\sqrt{2}$ lies between the rational numbers 1.41 and 1.42. The irrational number $\sqrt{3}$ lies between the rational numbers 1.73 and 1.74.

$$\begin{array}{r} 1.41 \\ \times 1.41 \\ \hline 141 \\ 564 \\ 141 \\ \hline 1.9881 \end{array} \qquad \begin{array}{r} 1.42 \\ \times 1.42 \\ \hline 284 \\ 568 \\ 142 \\ \hline 2.0164 \end{array}$$

Next, $(1.41)^2 = 1.9881$ and $(1.42)^2 = 2.0164$

Thus,

$$(1.41)^2 < 2 < (1.42)^2$$

Therefore,

$$1.41 < \sqrt{2} < 1.42$$

(See Figure 8.1.) To the nearest thousandth, it can be shown that

$$\sqrt{2} \approx 1.414 \qquad [\textit{Read: } \sqrt{2} \textit{ is approximately equal to } 1.414.]$$

Similarly, it can be shown that

$$\sqrt{3} \approx 1.732$$

(See Figure 8.1.)

In Table A on page 386 $\sqrt{n}$ is given for integers n between 1 and 100. When $\sqrt{n}$ is irrational, the Table gives an approximation to the nearest thousandth. For example, when $n = 44$, then $\sqrt{n} \approx 6.633$.

Example 6 Use $\sqrt{2} \approx 1.41$ to approximate

a. $5\sqrt{2}$, **b.** $5 + \sqrt{2}$,

c. $\dfrac{\sqrt{2}}{2}$

to the nearest tenth.

Solution **a.** $5\sqrt{2} \approx 5(1.41) = 7.05$

When the last digit is 5 or more, round by increasing the next to last digit by 1. Thus, to the nearest tenth,

$$5\sqrt{2} \approx 7.1$$

b. $5 + \sqrt{2} \approx 5 + 1.41$

Thus,

$$5 + \sqrt{2} \approx 6.41$$

Because the last digit is less than 5, to the nearest tenth,

$$5 + \sqrt{2} \approx 6.4$$

c. $\dfrac{\sqrt{2}}{2} \approx \dfrac{1.410}{2} = .705$

To the nearest tenth,

$$\frac{\sqrt{2}}{2} \approx .7$$

Example 7 Find consecutive integers N and $N + 1$ such that

$$N < \sqrt{92} < N + 1$$

$$(N < \sqrt{92} \text{ and } \sqrt{92} < N + 1).$$

Solution 81 and 100 are squares and

$$81 < 92 < 100$$

Thus, $\sqrt{81} < \sqrt{92} < \sqrt{100}$

or $9 < \sqrt{92} < 10$

Therefore, $N = 9$ and $N + 1 - 10$. (See Figure 8.2.)

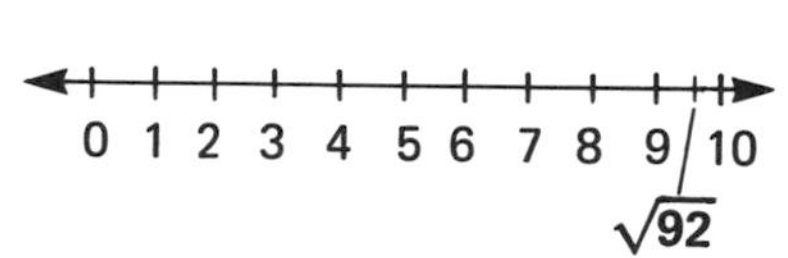

Fig. 8.2

Table A on page 386 can be employed to estimate the square root of a large number. Use

$$\sqrt{n^2} = n \text{ for } n \geqslant 0$$

To estimate $\sqrt{650}$, for example, find the nearest numbers to 650 in the n^2 column. Observe that by taking $n^2 = 625$ and then $n^2 = 676$, you obtain

$$\sqrt{625} = 25 \text{ and } \sqrt{676} = 26$$

Thus,

$$25 < \sqrt{650} < 26$$

Finally, what is the geometric significance of $\sqrt{2}$? The area of a square is given by s^2, where s is the length of a side. Consider the squares in Figure 8.3 . It is easy enough to draw the first and third squares, whose side lengths are rational. Whether or not you can draw the middle square, doesn't it seem reasonable that there should be a square whose area is 2 square inches? The length of a side is then $\sqrt{2}$ inches.

1.96 square in	2 square in	2.25 square in
1.4 in		1.5 in

Fig. 8.3

CLASS EXERCISES *In Exercises 8 and 9, use* $\sqrt{3} \approx 1.73$ *to approximate each of the following to the nearest tenth.*

8. $3\sqrt{3}$ **9.** $\frac{\sqrt{3}}{3}$

10. Find consecutive integers N and $N + 1$ such that $N < \sqrt{70} < N + 1$.

SOLUTIONS TO CLASS EXERCISES

1. a. 5 **b.** -5 **2. a.** 2 **b.** -2 **c.** -2 **3.** s **4.** $|t|$, or $-t$ **5.** $|st|$, or $-st$ **6.** a^6

7. $(bc)^2$, or b^2c^2 **8.** 5.2 **9.** .6 **10.** $8 < \sqrt{70} < 9$

HOME EXERCISES

In Exercises 1–18, find each square root.

1. $\sqrt{16}$ **2.** $\sqrt{36}$ **3.** $\sqrt{49}$ **4.** $\sqrt{64}$ **5.** $\sqrt{81}$ **6.** $\sqrt{121}$
7. $-\sqrt{144}$ **8.** $\sqrt{900}$ **9.** $\sqrt{2500}$ **10.** $-\sqrt{1}$ **11.** $\sqrt{0}$ **12.** $\sqrt{3^2}$
13. $\sqrt{(-3)^2}$ **14.** $-\sqrt{3^2}$ **15.** $-\sqrt{(-3)^2}$ **16.** $\sqrt{6^2}$ **17.** $\sqrt{(-6)^2}$ **18.** $\sqrt{(-9)^2}$

In Exercises 19–28, find each square root. Assume a, b, c, x, y, and z are each positive or 0.

19. $\sqrt{x^2}$ **20.** $-\sqrt{z^2}$ **21.** $\sqrt{(ab)^2}$ **22.** $\sqrt{c^4}$ **23.** $\sqrt{x^8}$
24. $\sqrt{y^{10}}$ **25.** $\sqrt{a^{12}}$ **26.** $\sqrt{(xy)^4}$ **27.** $\sqrt{x^{18}}$ **28.** $\sqrt{z^{64}}$

In Exercises 29–38, find each square root. Assume $m > 0$ *and* $n < 0$.

29. $\sqrt{m^2}$ **30.** $\sqrt{(-m)^2}$ **31.** $-\sqrt{m^2}$ **32.** $\sqrt{n^2}$ **33.** $\sqrt{(-n)^2}$
34. $-\sqrt{n^2}$ **35.** $\sqrt{(mn)^2}$ **36.** $\sqrt{(-mn)^2}$ **37.** $\sqrt{n^4}$ **38.** $\sqrt{n^6}$

In Exercises 39–46, find consecutive integers N and N + 1 such that for the indicated square root $\sqrt{M}$, $N < \sqrt{M} < N + 1$. *In Exercises 44–46, use Table A on page 386.*

39. $\sqrt{14}$ **40.** $\sqrt{32}$ **41.** $\sqrt{65}$ **42.** $\sqrt{99}$ **43.** $\sqrt{145}$ **44.** $\sqrt{999}$ **45.** $\sqrt{1670}$ **46.** $\sqrt{9745}$

In Exercises 47–54, use $\sqrt{2} \approx 1.41$, $\sqrt{3} \approx 1.73$ *to approximate each of the following to the nearest tenth.*

47. $\sqrt{2} - 1$ **48.** $7 - \sqrt{2}$ **49.** $2\sqrt{3}$ **50.** $\dfrac{\sqrt{3}}{2}$

51. $\sqrt{2} + \sqrt{3}$ **52.** $\sqrt{2} + 3\sqrt{3}$ **53.** $\dfrac{\sqrt{3}}{5} + 10\sqrt{2}$ **54.** $\sqrt{2}\sqrt{3}$

8.2 ROOTS OF PRODUCTS AND QUOTIENTS

ROOTS OF PRODUCTS

Observe that $\sqrt{9 \cdot 4} = \sqrt{36} = 6$

Furthermore, $\sqrt{9} \cdot \sqrt{4} = 3 \cdot 2 = 6$

It follows that $\sqrt{9 \cdot 4} = \sqrt{9} \cdot \sqrt{4}$

This suggests that (with suitable restrictions) *the square root of a product equals the product of the square roots*. More precisely,

$$\boxed{\sqrt{ab} = \sqrt{a}\sqrt{b}, \text{ where } a > 0,\ b > 0}$$

Example 1 **a.**
$$\begin{aligned}\sqrt{2500} &= \sqrt{25 \cdot 100} \\ &= \sqrt{25} \cdot \sqrt{100} \\ &= 5 \cdot 10 \\ &= 50\end{aligned}$$

b.
$$\begin{aligned}\sqrt{250{,}000} &= \sqrt{25\ 00\ 00} \\ &= \sqrt{25 \cdot 100 \cdot 100} \\ &= \sqrt{25} \cdot \sqrt{100} \cdot \sqrt{100} \\ &= 5 \cdot 10 \cdot 10 \\ &= 500\end{aligned}$$

c. Let $x > 0$. Then, $\sqrt{9x^2} = \sqrt{9}\sqrt{x^2} = 3x$

The preceding product rule applies to three or more factors. For example,

$$\boxed{\sqrt{abc} = \sqrt{a}\sqrt{b}\sqrt{c}, \text{ where } a > 0,\ b > 0,\ c > 0}$$

Example 2 Let $x > 0$, $y > 0$. Then, $\sqrt{25x^6y^{10}} = \sqrt{25}\sqrt{x^6}\sqrt{y^{10}} = 5x^3y^5$

CLASS EXERCISES *Assume* $a > 0$, $b > 0$. *Find each square root.*

1. $\sqrt{3600}$ **2.** $\sqrt{1{,}000{,}000}$ **3.** $\sqrt{a^2b^6}$ **4.** $\sqrt{100b^8c^{16}}$

SIMPLIFYING BY THE PRODUCT RULE

The product rule enables you to simplify some irrational square roots, and often, to approximate them quickly, as in Example 3. You can also simplify square roots of odd powers of a variable, as in Example 4.

Example 3 **a.** Simplify. $\sqrt{8}$ **b.** Estimate $\sqrt{8}$, roughly.

Solution **a.**

$$8 = 4 \cdot 2$$
$$\sqrt{8} = \sqrt{4}\sqrt{2} = 2\sqrt{2}$$

The equivalent form $2\sqrt{2}$ is simpler than $\sqrt{8}$ in that the radical sign applies to a *smaller* positive integer. *Leave the answer in radical form.* Note that $2\sqrt{2}$ is the *exact* answer. In part b you will *approximate* $\sqrt{8}$.

b.

$$\sqrt{2} \approx 1.41$$

Thus,

$$\sqrt{8} = 2\sqrt{2} \approx 2(1.41) \approx 2.8$$

Example 4 Let $x > 0$. Simplify. $\sqrt{x^9}$

Solution The "largest square factor" that can be removed from the radical symbol is x^8. Use $x^9 = x^8 \cdot x$

Then,

$$\sqrt{x^9} = \sqrt{x^8 \cdot x} = \sqrt{x^8}\sqrt{x} = x^4\sqrt{x}$$

CLASS EXERCISES

5. a. Simplify. $\sqrt{200}$ **b.** Estimate $\sqrt{200}$ to the nearest tenth, using $\sqrt{2} \approx 1.414$.

6. Simplify. $\sqrt{a^{11}}$, $a > 0$

ROOTS OF QUOTIENTS

Observe that

$$\sqrt{\frac{100}{25}} = \sqrt{4} = 2$$

Also,

$$\frac{\sqrt{100}}{\sqrt{25}} = \frac{10}{5} = 2$$

It follows that

$$\sqrt{\frac{100}{25}} = \frac{\sqrt{100}}{\sqrt{25}}$$

The square root of a quotient equals the quotient of the square roots. More precisely,

$$\boxed{\sqrt{\frac{a}{b}} = \frac{\sqrt{a}}{\sqrt{b}}, \text{ where } a > 0,\ b > 0}$$

Thus,

$$\sqrt{\frac{4}{9}} = \frac{\sqrt{4}}{\sqrt{9}} = \frac{2}{3}$$

Example 5 **a.** Let $b \neq 0$. Then,

$$\sqrt{\frac{a^4}{b^8}} = \frac{\sqrt{a^4}}{\sqrt{b^8}} = \frac{a^2}{b^4}$$

b. Let $x > 0$ and $z \neq 0$. Then,

$$\sqrt{\frac{25x^2y^4}{81z^{12}}} = \frac{\sqrt{25x^2y^4}}{\sqrt{81z^{12}}} = \frac{\sqrt{25}\sqrt{x^2}\sqrt{y^4}}{\sqrt{81}\sqrt{z^{12}}} = \frac{5xy^2}{9z^6}$$

A fraction, such as

$$\frac{1}{\sqrt{5}}$$

is often easier to work with when the denominator is rational. **Rationalize the denominator** of this fraction as follows.

$$\frac{1}{\sqrt{5}} = \frac{1 \cdot \sqrt{5}}{\sqrt{5} \cdot \sqrt{5}} = \frac{\sqrt{5}}{5}$$

Similarly,

$$\frac{4}{\sqrt{2}} = \frac{4 \cdot \sqrt{2}}{\sqrt{2} \cdot \sqrt{2}} = \frac{4\sqrt{2}}{2} = 2\sqrt{2}$$

Example 6 Let $c > 0$. Simplify. $\sqrt{\frac{c^5}{32}}$

Solution

$$\sqrt{\frac{c^5}{32}} = \frac{\sqrt{c^5}}{\sqrt{32}} = \frac{\sqrt{c^4}\sqrt{c}}{\sqrt{16}\sqrt{2}} = \frac{c^2\sqrt{c} \cdot \sqrt{2}}{4\sqrt{2} \cdot \sqrt{2}} = \frac{c^2\sqrt{2c}}{8}$$

CLASS EXERCISES *In Exercises 7 and 8, assume $y > 0$, $z > 0$. Find each square root.*

7. $\sqrt{\dfrac{25}{4}}$ **8.** $\sqrt{\dfrac{x^4y^{14}}{9z^2}}$

9. Simplify $\sqrt{\dfrac{x^5}{27}}$, $x > 0$. Rationalize the resulting denominator.

ROOTS OF SUMS AND DIFFERENCES

Let $a > 0$, $b > 0$. Although

$$\sqrt{ab} = \sqrt{a}\sqrt{b} \quad \text{and} \quad \sqrt{\frac{a}{b}} = \frac{\sqrt{a}}{\sqrt{b}}$$

make sure that you realize that, in general,

$$\sqrt{a + b} \neq \sqrt{a} + \sqrt{b} \quad \text{and} \quad \sqrt{a - b} \neq \sqrt{a} - \sqrt{b}$$

For example, if $a = 25$ and $b = 4$, then

$$\sqrt{25 + 4} = \sqrt{29} < \sqrt{36} = 6$$

whereas

$$\sqrt{25} + \sqrt{4} = 5 + 2 = 7$$

Also,

$$\sqrt{25 - 4} = \sqrt{21} > \sqrt{16} = 4$$

whereas,

$$\sqrt{25} - \sqrt{4} = 5 - 2 = 3$$

CLASS EXERCISES

10. Which of these statements are true?

a. $\sqrt{36 + 9} = \sqrt{36} + \sqrt{9}$ **b.** $\sqrt{36 \cdot 9} = \sqrt{36} \cdot \sqrt{9}$

c. $\sqrt{36 - 9} = \sqrt{36} - \sqrt{9}$ **d.** $\sqrt{\dfrac{36}{9}} = \dfrac{\sqrt{36}}{\sqrt{9}}$

ROOTS OF DECIMALS

Observe that

$$\sqrt{\frac{1}{100}} = \frac{\sqrt{1}}{\sqrt{100}} = \frac{1}{10} \quad \text{and that} \quad \sqrt{\frac{1}{10{,}000}} = \frac{\sqrt{1}}{\sqrt{10{,}000}} = \frac{1}{100}$$

Thus, in decimal notation,

$$\sqrt{.01} = .1 \quad \text{and} \quad \sqrt{.0001} = .01$$

Example 7 a. $\sqrt{.04} = \sqrt{4(.01)} = \sqrt{4}\sqrt{.01} = 2(.1) = .2$

b. $\sqrt{.0025} = \sqrt{25(.0001)} = \sqrt{25}\sqrt{.0001} = 5(.01) = .05$

c. For any x, $\sqrt{.36x^4} = \sqrt{36(.01)x^4} = \sqrt{36}\sqrt{.01}\sqrt{x^4} = 6(.1)x^2 = .6x^2$

CLASS EXERCISES *Let a be any number. Find each square root.*

11. $\sqrt{.01}$ **12.** $\sqrt{.0001}$ **13.** $\sqrt{.49a^8}$

SOLUTIONS TO CLASS EXERCISES

1. $\sqrt{3600} = \sqrt{36 \cdot 100} = \sqrt{36} \cdot \sqrt{100} = 6 \cdot 10 = 60$

2. $\sqrt{1{,}000{,}000} = \sqrt{100 \cdot 100 \cdot 100} = \sqrt{100} \cdot \sqrt{100} \cdot \sqrt{100} = 10 \cdot 10 \cdot 10 - 1000$

3. $\sqrt{a^2b^6} = \sqrt{a^2}\sqrt{b^6} = ab^3$ **4.** $\sqrt{100b^8c^{16}} = \sqrt{100}\sqrt{b^8}\sqrt{c^{16}} = 10b^4c^8$

5. **a.** $\sqrt{200} = \sqrt{100 \cdot 2} = \sqrt{100}\sqrt{2} = 10\sqrt{2}$ **b.** $\sqrt{200} = 10\sqrt{2} \approx 10(1.414)$. To the nearest tenth, $\sqrt{200} \approx 14.1$

6. $\sqrt{a^{11}} = \sqrt{a^{10} \cdot a} = \sqrt{a^{10}}\sqrt{a} = a^5\sqrt{a}$

7. $\sqrt{\frac{25}{4}} = \frac{\sqrt{25}}{\sqrt{4}} = \frac{5}{2}$ **8.** $\sqrt{\frac{x^4y^{14}}{9z^2}} = \frac{\sqrt{x^4}\sqrt{y^{14}}}{\sqrt{9}\sqrt{z^2}} = \frac{x^2y^7}{3z}$ **9.** $\sqrt{\frac{x^5}{27}} = \frac{\sqrt{x^5}}{\sqrt{27}} = \frac{\sqrt{x^4}\sqrt{x}}{\sqrt{9}\sqrt{3}} = \frac{x^2\sqrt{x} \cdot \sqrt{3}}{3\sqrt{3} \cdot \sqrt{3}} = \frac{x^2\sqrt{3x}}{9}$

10. b and d. **11.** .1 **12.** .01 **13.** $.7a^4$

HOME EXERCISES

In Exercises 1–32, find each square root. Assume $a > 0$, $b > 0$, $c > 0$, $d \neq 0$, $x > 0$, $y > 0$, $z > 0$.

1. $\sqrt{4900}$ **2.** $\sqrt{8100}$ **3.** $\sqrt{12{,}100}$ **4.** $\sqrt{40{,}000}$ **5.** $\sqrt{250{,}000}$ **6.** $\sqrt{36{,}000{,}000}$

7. $\sqrt{100{,}000{,}000}$ **8.** $\sqrt{4x^2}$ **9.** $\sqrt{16a^4}$ **10.** $\sqrt{25b^8}$ **11.** $\sqrt{100a^{18}}$ **12.** $\sqrt{a^2b^2}$ **13.** $\sqrt{x^2y^4}$

14. $\sqrt{x^4y^4}$ **15.** $\sqrt{a^8x^{10}}$ **16.** $\sqrt{x^2y^2z^2}$ **17.** $\sqrt{a^4b^2c^6}$ **18.** $\sqrt{36a^4x^2}$ **19.** $\sqrt{121y^2z^{12}}$

20. $\sqrt{\frac{1}{4}}$ **21.** $\sqrt{\frac{9}{25}}$ **22.** $\sqrt{\frac{36}{49}}$ **23.** $\sqrt{\frac{81}{4}}$ **24.** $\sqrt{\frac{25}{144}}$ **25.** $\sqrt{\frac{100}{121}}$ **26.** $\sqrt{\frac{a^2}{c^2}}$

27. $\sqrt{\frac{x^2}{y^8}}$ **28.** $\sqrt{\frac{x^2}{4}}$ **29.** $\sqrt{\frac{25}{d^8}}$ **30.** $\sqrt{\frac{a^2b^6}{9}}$ **31.** $\sqrt{\frac{144x^4}{y^2z^{10}}}$ **32.** $\sqrt{\frac{25x^4y^8}{4z^{10}}}$

In Exercises 33–38: **a.** *Simplify each expression. (Leave your answer in radical form.)* **b.** *Estimate the square root to the nearest tenth using* $\sqrt{2} \approx 1.414$ *and* $\sqrt{3} \approx 1.732$.

33. $\sqrt{18}$ **34.** $\sqrt{50}$ **35.** $\sqrt{72}$ **36.** $\sqrt{48}$ **37.** $\sqrt{75}$ **38.** $\sqrt{300}$

In Exercises 39–48, simplify each expression. Leave answers in radical form and with rational denominator.

39. $\sqrt{a^3}$ **40.** $\sqrt{c^7}$ **41.** $\sqrt{\dfrac{a^5}{b^6}}$ **42.** $\sqrt{4b^9}$ **43.** $\sqrt{27b^9}$

44. $\sqrt{a^3b^5}$ **45.** $\sqrt{\dfrac{9a^9}{2b^2}}$ **46.** $\sqrt{98a^4b^3}$ **47.** $\sqrt{\dfrac{50a^3}{98b^8}}$ **48.** $\sqrt{\dfrac{175c^5}{108a^2b^4}}$

In Exercises 49–56, find each square root.

49. $\sqrt{.09}$ **50.** $\sqrt{.36}$ **51.** $\sqrt{.0004}$ **52.** $\sqrt{.0064}$

53. $\sqrt{.000\,001}$ **54.** $\sqrt{.0144x^4}$ **55.** $\sqrt{.81x^8y^4}$ **56.** $\sqrt{.0121x^{12}y^{16}}$

8.3 ADDITION AND SUBTRACTION OF ROOTS

SIMPLIFYING EXPRESSIONS WITH RADICALS

You can add or subtract irrational square roots in the same way you combine terms involving variables. Thus,

$$2\sqrt{3} + 3\sqrt{3} = 5\sqrt{3}$$

just as

$$2x + 3x = 5x$$

But

$$2\sqrt{3} + 3\sqrt{5}$$

cannot be further simplified, just as $2x + 3y$ cannot be.

Example 1 Simplify.

$$\sqrt{7} + 5\sqrt{7} - 3\sqrt{7}$$

Solution Note that

$$\sqrt{7} = 1\sqrt{7}$$

Thus,

$$\sqrt{7} + 5\sqrt{7} - 3\sqrt{7} = (1 + 5 - 3)\sqrt{7} = 3\sqrt{7}$$

Example 2 Simplify.

$$\sqrt{8} + \sqrt{18}$$

Solution

$$\sqrt{8} = \sqrt{4}\sqrt{2} = 2\sqrt{2}$$
$$\sqrt{18} = \sqrt{9}\sqrt{2} = 3\sqrt{2}$$

Therefore,

$$\sqrt{8} + \sqrt{18} = 2\sqrt{2} + 3\sqrt{2} = 5\sqrt{2}$$

Example 3 Let $a > 0$. Simplify. $\sqrt{4a^5} + \sqrt{25a^5}$

Solution

$$\sqrt{4a^5} = \sqrt{4}\sqrt{a^4}\sqrt{a} = 2a^2\sqrt{a}$$
$$\sqrt{25a^5} = \sqrt{25}\sqrt{a^4}\sqrt{a} = 5a^2\sqrt{a}$$

Thus, $\sqrt{4a^5} + \sqrt{25a^5} = 2a^2\sqrt{a} + 5a^2\sqrt{a} = 7a^2\sqrt{a}$

CLASS EXERCISES *Simplify each expression. Assume* $a > 0$.

1. $9\sqrt{2} - 3\sqrt{2}$ **2.** $8\sqrt{3} + \sqrt{5} + 2\sqrt{3} - 3\sqrt{5}$

3. $\sqrt{20} + \sqrt{45}$ **4.** $\sqrt{50a^3} + \sqrt{98a^3}$

COMBINING FRACTIONS Fractions can be irrational as well as rational. The rules of Sections 5.3 for adding or subtracting fractions apply here.

Example 4 $\dfrac{\sqrt{2}}{3} + \dfrac{2\sqrt{3}}{3} = \dfrac{\sqrt{2} + 2\sqrt{3}}{3}$

Example 5 Subtract. $\dfrac{2}{3} - \dfrac{\sqrt{2}}{5}$

Solution The *lcd* is 15. Thus,

$$\frac{2}{3} - \frac{\sqrt{2}}{5} = \frac{5 \cdot 2 - 3\sqrt{2}}{15} = \frac{10 - 3\sqrt{2}}{15}$$

CLASS EXERCISES *Simplify.* **5.** $\dfrac{\sqrt{5}}{2} + \dfrac{\sqrt{5}}{2}$ **6.** $\dfrac{\sqrt{2}}{2} + \dfrac{\sqrt{2}}{5}$ **7.** $\dfrac{4\sqrt{7}}{3} - \dfrac{2\sqrt{7}}{5} + \dfrac{\sqrt{7}}{15}$

SOLUTIONS TO CLASS EXERCISES

1. $9\sqrt{2} - 3\sqrt{2} = (9 - 3)\sqrt{2} = 6\sqrt{2}$ **2.** $8\sqrt{3} + \sqrt{5} + 2\sqrt{3} - 3\sqrt{5} = (8 + 2)\sqrt{3} + (1 - 3)\sqrt{5} = 10\sqrt{3} - 2\sqrt{5}$

3. $\sqrt{20} + \sqrt{45} = \sqrt{4}\sqrt{5} + \sqrt{9}\sqrt{5} = 2\sqrt{5} + 3\sqrt{5} = 5\sqrt{5}$

4. $\sqrt{50a^3} = \sqrt{25}\sqrt{a^2}\sqrt{2a} = 5a\sqrt{2a}$
$\sqrt{98a^3} = \sqrt{49}\sqrt{a^2}\sqrt{2a} = 7a\sqrt{2a}$
Thus, $\sqrt{50a^3} + \sqrt{98a^3} = 5a\sqrt{2a} + 7a\sqrt{2a} = 12a\sqrt{2a}$

5. $\dfrac{\sqrt{5}}{2} + \dfrac{\sqrt{5}}{2} = \dfrac{\sqrt{5} + \sqrt{5}}{2} = \dfrac{2\sqrt{5}}{2} = \sqrt{5}$

6. $\dfrac{\sqrt{2}}{2} + \dfrac{\sqrt{2}}{5} = \dfrac{5\sqrt{2} + 2\sqrt{2}}{10} = \dfrac{7\sqrt{2}}{10}$

7. $\dfrac{4\sqrt{7}}{3} - \dfrac{2\sqrt{7}}{5} + \dfrac{\sqrt{7}}{15} = \dfrac{20\sqrt{7} - 6\sqrt{7} + \sqrt{7}}{15} = \dfrac{15\sqrt{7}}{15} = \sqrt{7}$

HOME EXERCISES

Simplify each expression. Assume $a > 0,\ b > 0,\ x > 0,\ y > 0.$

1. $\sqrt{2} + \sqrt{2}$
2. $2\sqrt{3} + 4\sqrt{3}$
3. $5\sqrt{2} - \sqrt{2}$
4. $\sqrt{7} - 2\sqrt{7}$
5. $3\sqrt{6} + 2\sqrt{6}$
6. $7\sqrt{10} - 4\sqrt{10}$
7. $\sqrt{11} + 4\sqrt{11} + 3\sqrt{11}$
8. $6\sqrt{13} - \sqrt{13} + 4\sqrt{13}$
9. $2\sqrt{5} + 6\sqrt{5} - \sqrt{5}$
10. $\sqrt{7} - 2\sqrt{7} - 5\sqrt{7}$
11. $12\sqrt{2} - (7\sqrt{2} + 2\sqrt{2})$
12. $3\sqrt{7} - (2\sqrt{7} - 5\sqrt{7})$
13. $\sqrt{2} + \sqrt{3} + \sqrt{2} + 2\sqrt{3}$
14. $3\sqrt{5} - (\sqrt{2} + \sqrt{5})$
15. $\sqrt{7} - \sqrt{5} + 4\sqrt{7} + 3\sqrt{5}$
16. $\sqrt{11} - (\sqrt{2} + \sqrt{11} - \sqrt{2})$
17. $\sqrt{2} + \sqrt{8}$
18. $\sqrt{6} + \sqrt{24}$
19. $\sqrt{18} - 3\sqrt{2}$
20. $\sqrt{50} + \sqrt{32}$
21. $\sqrt{7} + \sqrt{28} - 2\sqrt{7}$
22. $\sqrt{18} - \sqrt{2} + \sqrt{50}$
23. $\sqrt{300} + \sqrt{12} - \sqrt{75}$
24. $\sqrt{44} + \sqrt{99} - 5\sqrt{11}$
25. $\sqrt{4a} + \sqrt{9a}$
26. $\sqrt{64x} - \sqrt{49x}$
27. $\sqrt{100y^3} + \sqrt{81y^3}$
28. $\sqrt{a^5} - \sqrt{16a^5}$
29. $\sqrt{144ab^2} + \sqrt{121ab^2}$
30. $\sqrt{36x} + \sqrt{64x} - \sqrt{81x}$
31. $\sqrt{12a} + \sqrt{3a}$
32. $\sqrt{20b} - \sqrt{5b}$
33. $\sqrt{98x^3} + \sqrt{50x^3}$
34. $\sqrt{288yz^4} - \sqrt{72yz^4}$
35. $\sqrt{7ab^3} + \sqrt{28ab^3} + \sqrt{63ab^3}$
36. $\sqrt{125x^5y^3} - \sqrt{20x^5y^3} + \sqrt{45x^5y^3}$
37. $\frac{2}{5} + \frac{\sqrt{5}}{5}$
38. $\frac{1}{8} - \frac{\sqrt{2}}{8}$
39. $\frac{\sqrt{2}}{3} - \frac{\sqrt{5}}{3} + \frac{\sqrt{7}}{3}$
40. $\frac{\sqrt{2}}{2} + \frac{\sqrt{5}}{4}$
41. $\frac{\sqrt{2}}{5} - \frac{\sqrt{3}}{7}$
42. $\frac{6\sqrt{3}}{5} - \frac{12\sqrt{3}}{15}$
43. $\frac{\sqrt{2}}{4} + \frac{\sqrt{3}}{8} - \frac{2\sqrt{2}}{12}$
44. $\frac{\sqrt{2}}{3} + \frac{\sqrt{3}}{6} + \frac{1}{12}$

8.4 MULTIPLICATION AND DIVISION OF ROOTS

$\sqrt{a}\sqrt{b}$ Recall that the square root of a product equals the product of the square roots.

$$\sqrt{ab} = \sqrt{a}\sqrt{b}, \text{ where } a > 0,\ b > 0$$

In Section 8.2, you used this to simplify the square root of a product such as $\sqrt{9x^2}$. Thus, if $x > 0$,

$$\sqrt{9x^2} = \sqrt{9}\sqrt{x^2} = 3x$$

Now, you will use the above law, in reverse, to simplify the product of square roots. Thus,

$$\sqrt{2}\sqrt{3} = \sqrt{2 \cdot 3} = \sqrt{6}$$

Example 1 Let $a > 0,\ x > 0$. Multiply and simplify.

a. $\sqrt{a^3}\sqrt{a}$ **b.** $\sqrt{2x}\sqrt{8x}$

Solution **a.**
$$\sqrt{a^3}\sqrt{a} = \sqrt{a^3 \cdot a} = \sqrt{a^4} = a^2$$

b.
$$\sqrt{2x}\sqrt{8x} = \sqrt{(2x)(8x)} = \sqrt{16x^2} = \sqrt{16}\sqrt{x^2} = 4x$$

CLASS EXERCISES *Multiply, as indicated. Then simplify, if possible. Assume $a > 0$, $b > 0$.*

1. $\sqrt{5}\sqrt{5}$ **2.** $\sqrt{3a}\sqrt{6a}$ **3.** $\sqrt{3a}\sqrt{27b}\sqrt{ab}$

DISTRIBUTIVE LAWS The Distributive Laws apply to products involving roots. Recall that

$$a(b + c) = ab + ac$$

Thus,

$$\begin{aligned}\sqrt{2}(\sqrt{2} + \sqrt{5}) &= \sqrt{2}\sqrt{2} + \sqrt{2}\sqrt{5}\\ &= \sqrt{2 \cdot 2} + \sqrt{2 \cdot 5}\\ &= 2 + \sqrt{10}\end{aligned}$$

Example 2 Multiply. **a.** $(1 + \sqrt{7})(2 + \sqrt{7})$ **b.** $(\sqrt{3} + \sqrt{2})(\sqrt{3} - \sqrt{2})$

Solution

a.

$$\begin{array}{l} 1 + \sqrt{7}\\ \underline{2 + \sqrt{7}}\\ 2 + 2\sqrt{7}\\ \underline{\quad + \sqrt{7} + \sqrt{7}\sqrt{7}}\\ 2 + 3\sqrt{7} + 7 \end{array}$$

($2 + 7 = 9$)

Thus, $(1 + \sqrt{7})(2 + \sqrt{7})$
$= 9 + 3\sqrt{7}$

b.

$$\begin{array}{l} \sqrt{3} + \sqrt{2}\\ \underline{\sqrt{3} - \sqrt{2}}\\ 3 + \sqrt{3}\sqrt{2}\\ \underline{\quad - \sqrt{3}\sqrt{2} - 2}\\ 3 \qquad\qquad - 2 \end{array}$$

($3 - 2 = 1$)

Thus, $(\sqrt{3} + \sqrt{2})(\sqrt{3} - \sqrt{2}) = 1$

Example 3 Multiply.

$$\begin{array}{l} 5\sqrt{7} + \sqrt{2}\\ \underline{2\sqrt{7} - 3\sqrt{2}}\\ 10 \cdot 7 + 2\sqrt{7}\sqrt{2}\\ \underline{\qquad - 15\sqrt{7}\sqrt{2} - 3 \cdot 2}\\ 70 \quad - 13\sqrt{14} - 6 \end{array}$$

($70 - 6 = 64$)

Thus,

$$(5\sqrt{7} + \sqrt{2})(2\sqrt{7} - 3\sqrt{2}) = 64 - 13\sqrt{14}$$

CLASS EXERCISES *Multiply and simplify.*

4. $\sqrt{3}(\sqrt{2} - \sqrt{3})$ **5.** $(5 + \sqrt{3})(5 - \sqrt{3})$ **6.** $(2\sqrt{3} + \sqrt{5})(\sqrt{3} + 3\sqrt{5})$

$\frac{\sqrt{a}}{\sqrt{b}}$ The square root of a quotient equals the quotient of the square roots.

$$\sqrt{\frac{a}{b}} = \frac{\sqrt{a}}{\sqrt{b}}, \text{ where } a > 0,\ b > 0$$

In Section 8.2, you simplified *the square root of a quotient*. For example,

$$\sqrt{\frac{4}{9}} = \frac{\sqrt{4}}{\sqrt{9}} = \frac{2}{3}$$

Now, you will use the above law, together with the product law

$$\sqrt{ab} = \sqrt{a}\sqrt{b},\ a > 0,\ b > 0$$

to simplify *the quotient of square roots*. For example,

$$\frac{\sqrt{48}}{\sqrt{3}} = \sqrt{\frac{48}{3}} = \sqrt{16} = 4$$

Example 4 Let $x > 0$. Simplify.

a. $\frac{\sqrt{28x}}{\sqrt{7x}}$ b. $\frac{\sqrt{3}\sqrt{14}}{\sqrt{6}}$

Solution a. $\frac{\sqrt{28x}}{\sqrt{7x}} = \sqrt{\frac{28x}{7x}} = \sqrt{4} = 2$ b. $\frac{\sqrt{3}\sqrt{14}}{\sqrt{6}} = \frac{\overset{1}{\cancel{\sqrt{3}}}\cancel{\sqrt{2}}\sqrt{7}}{\cancel{\sqrt{3}}\underset{1}{\cancel{\sqrt{2}}}} = \sqrt{7}$

CLASS EXERCISES *Assume* $x > 0,\ y > 0$. *Simplify.*

7. $\frac{\sqrt{40}}{\sqrt{10}}$ 8. $\frac{\sqrt{144x^3y^7}}{\sqrt{9xy}}$ 9. $\frac{\sqrt{3}\sqrt{20}}{\sqrt{15}}$

FACTORING AND DIVIDING You can often "isolate the common factor" of an expression containing radicals, just as you isolated the common factor of a polynomial. When this expression is the numerator of a fraction, you may be able to divide by factors common to numerator and denominator.

Example 5 a. Factor. $2 + \sqrt{12}$ b. Use part **a** to simplify the fraction $\frac{2 + \sqrt{12}}{4}$.

Solution a.
$$2 + \sqrt{12} = 2 + \sqrt{4 \cdot 3} = 2 + \sqrt{4}\sqrt{3} = 2 + 2\sqrt{3} = 2(1 + \sqrt{3})$$

b.
$$\frac{2 + \sqrt{12}}{4} = \frac{\overset{1}{\cancel{2}}(1 + \sqrt{3})}{\underset{2}{\cancel{4}}} = \frac{1 + \sqrt{3}}{2}$$

Example 6 Let $x > 0$, $y > 0$.

a. Factor. $\sqrt{xy} - \sqrt{x^3}$

b. Use part **a** to simplify the expression $\frac{\sqrt{xy} - \sqrt{x^3}}{\sqrt{x^5}}$.

Solution a. $\sqrt{xy} - \sqrt{x^3} = \sqrt{x}\sqrt{y} - \sqrt{x}\sqrt{x^2} = \sqrt{x}(\sqrt{y} - x)$

b.
$$\frac{\sqrt{xy} - \sqrt{x^3}}{\sqrt{x^5}} = \frac{\overset{1}{\cancel{\sqrt{x}}}(\sqrt{y} - x)}{\underset{1}{\cancel{\sqrt{x}}}\sqrt{x^4}} = \frac{\sqrt{y} - x}{x^2}$$

Example 7 Let $a > 0$, $x > 0$, $y > 0$, $x \neq y$.

a. Factor. $\sqrt{a^2x^2} - \sqrt{a^2y^2}$

b. Use part **a** to simplify the expression $\frac{\sqrt{a^2x^2} - \sqrt{a^2y^2}}{x^2 - y^2}$.

Solution a.
$$\sqrt{a^2x^2} - \sqrt{a^2y^2} = \sqrt{a^2}\sqrt{x^2} - \sqrt{a^2}\sqrt{y^2} = ax - ay = a(x - y)$$

b.
$$\frac{\sqrt{a^2x^2} - \sqrt{a^2y^2}}{x^2 - y^2} = \frac{a(x - y)}{(x + y)(x - y)} = \frac{a}{x + y}$$

CLASS EXERCISES *Assume* $x > 0$, $y > 0$.

10. a. Factor $2 + \sqrt{20}$. b. Use part **a** to simplify the fraction $\frac{2 + \sqrt{20}}{12}$.

11. a. Factor $\sqrt{x^3y^3} - \sqrt{xy}$.

    b. Use part **a** to simplify the expression $\frac{\sqrt{x^3y^3} - \sqrt{xy}}{\sqrt{x}}$.

12. a. Factor $\sqrt{x^2y^2} - \sqrt{x^2}$.

b. Use part a to simplify the expression $\dfrac{\sqrt{x^2y^2} - \sqrt{x^2}}{y^2 - 2y + 1}$, $y \neq 1$.

MULTIPLYING AND DIVIDING FRACTIONS

Recall that to multiply fractions, first divide by factors common to the numerators and denominators. Then, multiply the simplified numerators and denominators separately.

Example 8

$$\frac{\sqrt{12}}{5} \cdot \frac{\sqrt{6}}{4} = \frac{\overset{1}{\cancel{2}}\sqrt{3}}{5} \cdot \frac{\sqrt{2}\sqrt{3}}{\underset{2}{\cancel{4}}} = \frac{3\sqrt{2}}{10}$$

Recall that
$$\frac{a}{b} \div \frac{c}{d} = \frac{a}{b} \cdot \frac{d}{c}$$

Thus, *to divide fractions, invert the second fraction and multiply.*

Example 9

$$\frac{\sqrt{10}}{3} \div \frac{\sqrt{5}}{9} = \frac{\sqrt{10}}{\underset{1}{\cancel{3}}} \cdot \frac{\overset{3}{\cancel{9}}}{\sqrt{5}} = \sqrt{\overset{1}{\cancel{5}}}\sqrt{2} \cdot \frac{3}{\sqrt{\underset{1}{\cancel{5}}}} = 3\sqrt{2}$$

CLASS EXERCISES *Multiply or divide, and simplify.*

13. $\dfrac{\sqrt{7}}{4} \cdot \dfrac{2\sqrt{7}}{7}$

14. $\dfrac{\sqrt{8}}{2} \div \dfrac{\sqrt{18}}{10}$

SOLUTIONS TO CLASS EXERCISES

1. $\sqrt{5}\sqrt{5} = \sqrt{5 \cdot 5} = \sqrt{25} = 5$

2. $\sqrt{3a}\sqrt{6a} = \sqrt{18a^2} = \sqrt{9}\sqrt{a^2}\sqrt{2} = 3a\sqrt{2}$

3. $\sqrt{3a}\sqrt{27b}\sqrt{ab} = \sqrt{(3a)(27b)(ab)} = \sqrt{81a^2b^2} = 9ab$

4. $\sqrt{3}(\sqrt{2} - \sqrt{3}) = \sqrt{3}\sqrt{2} - \sqrt{3}\sqrt{3} = \sqrt{6} - 3$

5. Multiply.
$$\begin{array}{r} 5 + \sqrt{3} \\ 5 - \sqrt{3} \\ \hline 25 + 5\sqrt{3} \qquad \\ -5\sqrt{3} - 3 \\ \hline 25 \qquad\quad - 3 \end{array}$$
$(5 + \sqrt{3})(5 - \sqrt{3}) = 22$

6. Multiply.
$$\begin{array}{r} 2\sqrt{3} + \sqrt{5} \qquad\qquad \\ \sqrt{3} + 3\sqrt{5} \qquad\qquad \\ \hline 2 \cdot 3 + \sqrt{15} \qquad\qquad \\ + 6\sqrt{15} + 3 \cdot 5 \\ \hline 6 + 7\sqrt{15} + 15 \end{array}$$
$(2\sqrt{3} + \sqrt{5})(\sqrt{3} + 3\sqrt{5}) = 21 + 7\sqrt{15}$

7. $\dfrac{\sqrt{40}}{\sqrt{10}} = \sqrt{\dfrac{40}{10}} = \sqrt{4} = 2$

8. $\dfrac{\sqrt{144x^3y^7}}{\sqrt{9xy}} = \sqrt{\dfrac{144x^3y^7}{9xy}} = \sqrt{16x^2y^6} = 4xy^3$

9. $\dfrac{\sqrt{3}\sqrt{20}}{\sqrt{15}} = \dfrac{\sqrt{3}\sqrt{5}\sqrt{4}}{\sqrt{3}\sqrt{5}} = \sqrt{4} = 2$

10. **a.** $2 + \sqrt{20} = 2 + \sqrt{4 \cdot 5} = 2 + 2\sqrt{5} = 2(1 + \sqrt{5})$
b. $\dfrac{2 + \sqrt{20}}{12} = \dfrac{2(1 + \sqrt{5})}{12} = \dfrac{1 + \sqrt{5}}{6}$

11. **a.** $\sqrt{x^3y^3} - \sqrt{xy} = \sqrt{xy}\sqrt{x^2y^2} - \sqrt{xy} = \sqrt{xy}(\sqrt{x^2y^2} - 1) = \sqrt{xy}(xy - 1)$
b. $\dfrac{\sqrt{x^3y^3} - \sqrt{xy}}{\sqrt{x}} = \dfrac{\sqrt{xy}(xy - 1)}{\sqrt{x}} = \dfrac{\sqrt{x}\sqrt{y}(xy - 1)}{\sqrt{x}} = \sqrt{y}(xy - 1)$

12. **a.** $\sqrt{x^2y^2} - \sqrt{x^2} = \sqrt{x^2}\sqrt{y^2} - \sqrt{x^2} = \sqrt{x^2}(\sqrt{y^2} - 1) = x(y - 1)$
b. $\dfrac{\sqrt{x^2y^2} - \sqrt{x^2}}{y^2 - 2y + 1} = \dfrac{x(y - 1)}{(y - 1)^2} = \dfrac{x}{y - 1}$

13. $\dfrac{\sqrt{7}}{4} \cdot \dfrac{2\sqrt{7}}{7} = \dfrac{2 \cdot 7}{4 \cdot 7} = \dfrac{1}{2}$

14. $\dfrac{\sqrt{8}}{2} \div \dfrac{\sqrt{18}}{10} = \dfrac{\sqrt{8}}{2} \cdot \dfrac{10}{\sqrt{18}} = \dfrac{2\sqrt{2}}{2} \cdot \dfrac{10}{3\sqrt{2}} = \dfrac{10}{3}$

HOME EXERCISES

Assume $a > 0$, $b > 0$, $c > 0$, $x > 0$, $y > 0$, $z > 0$.

In Exercises 1–28, multiply and simplify.

1. $\sqrt{2}\sqrt{5}$
2. $\sqrt{3}\sqrt{5}$
3. $\sqrt{7}\sqrt{8}$
4. $\sqrt{2}\sqrt{2}$
5. $\sqrt{7}\sqrt{7}$
6. $\sqrt{12}\sqrt{3}$
7. $\sqrt{50}\sqrt{2}$
8. $\sqrt{2}\sqrt{6}\sqrt{5}$
9. $\sqrt{a}\sqrt{a}$
10. $\sqrt{ax}\sqrt{ax}$
11. $\sqrt{b}\sqrt{b^5}$
12. $\sqrt{a}\sqrt{ab}\sqrt{b}$
13. $\sqrt{5x}\sqrt{20x}$
14. $\sqrt{2b^3}\sqrt{12b}$
15. $\sqrt{xyz^3}\sqrt{x^2y}\sqrt{zy^2}$
16. $\sqrt{3a^2b}\sqrt{2bc}\sqrt{6c^3}$
17. $\sqrt{2}(\sqrt{2} + \sqrt{3})$
18. $\sqrt{a}(\sqrt{a} + \sqrt{b})$
19. $\sqrt{3}(\sqrt{27} - \sqrt{12})$
20. $2\sqrt{3}(\sqrt{3} + 12)$
21. $(7 + \sqrt{2})(7 - \sqrt{2})$
22. $(2\sqrt{5} + 3\sqrt{3})(2\sqrt{5} - 3\sqrt{3})$
23. $(2 + 3\sqrt{5})(4 + 2\sqrt{5})$
24. $(4\sqrt{7} - 2\sqrt{5})(2\sqrt{7} + 4\sqrt{5})$
25. $(\sqrt{5} - 3\sqrt{11})(2\sqrt{5} + \sqrt{11})$
26. $(6\sqrt{3} + 9\sqrt{5})(2\sqrt{3} + \sqrt{5})$
27. $(1 + \sqrt{2})^2$
28. $(2\sqrt{3} + \sqrt{5})^2$

In Exercises 29–40, simplify. If necessary, rationalize the denominator.

29. $\dfrac{\sqrt{27}}{\sqrt{3}}$
30. $\dfrac{\sqrt{2}}{\sqrt{8}}$
31. $\dfrac{-\sqrt{12}}{\sqrt{3}}$
32. $\dfrac{\sqrt{24}}{\sqrt{6}}$
33. $\dfrac{6}{\sqrt{3}}$
34. $\dfrac{\sqrt{98}}{\sqrt{2}}$
35. $\dfrac{\sqrt{18x}}{\sqrt{81x}}$
36. $\dfrac{\sqrt{9x}}{\sqrt{25z}}$
37. $\dfrac{\sqrt{48xyz}}{\sqrt{27xz^3}}$
38. $\dfrac{\sqrt{30}}{\sqrt{2}\sqrt{15}}$
39. $\dfrac{\sqrt{27}}{\sqrt{6}\sqrt{50}}$
40. $\dfrac{\sqrt{45}\sqrt{14}}{\sqrt{35}\sqrt{18}}$

In Exercises 41–50, **a** *factor and* **b** *use part* **a** *to simplify the fraction or rational expression.*

41. **a.** Factor. $3 + \sqrt{18}$
b. Simplify. $\dfrac{3 + \sqrt{18}}{9}$

42. **a.** Factor. $2 - \sqrt{32}$
b. Simplify. $\dfrac{2 - \sqrt{32}}{8}$

43. **a.** Factor. $5 + \sqrt{50}$
b. Simplify. $\dfrac{5 + \sqrt{50}}{-15}$

44. **a.** Factor. $7 + \sqrt{98}$
b. Simplify. $\dfrac{7 + \sqrt{98}}{1 + \sqrt{2}}$

45. **a.** Factor. $\sqrt{2} + \sqrt{6}$
b. Simplify. $\dfrac{\sqrt{2} + \sqrt{6}}{\sqrt{8}}$

46. **a.** Factor. $\sqrt{6} - \sqrt{18}$
b. Simplify. $\dfrac{\sqrt{6} - \sqrt{18}}{3 - \sqrt{3}}$

47. **a.** Factor. $\sqrt{a^3b} + \sqrt{abc}$
b. Simplify. $\dfrac{\sqrt{a^3b} + \sqrt{abc}}{\sqrt{a^3b^3}}$

48. **a.** Factor. $\sqrt{4a^2} - \sqrt{4b^2}$
b. Simplify. $\dfrac{\sqrt{4a^2} - \sqrt{4b^2}}{10a - 10b}$

49. **a.** Factor. $\sqrt{9a^2b^2} - \sqrt{16a^2b^2}$
b. Simplify. $\dfrac{15ab - 20ab}{\sqrt{9a^2b^2} - \sqrt{16a^2b^2}}$

50. **a.** Factor. $\sqrt{a^2c^4} + \sqrt{25a^2c^2}$
b. Simplify. $\dfrac{\sqrt{a^2c^4} + \sqrt{25a^2c^2}}{c^2 + 10c + 25}$

In Exercises 51–58, multiply or divide. Assume $s > 0$, $t > 0$.

51. $\dfrac{\sqrt{5}}{2} \cdot \dfrac{\sqrt{5}}{3}$

52. $\dfrac{\sqrt{8}}{5} \cdot \dfrac{\sqrt{3}}{\sqrt{2}}$

53. $\dfrac{\sqrt{12}}{\sqrt{5}} \cdot \dfrac{\sqrt{6}}{2} \cdot \dfrac{\sqrt{10}}{3}$

54. $\dfrac{\sqrt{3}}{\sqrt{2}} \cdot \dfrac{\sqrt{32}}{3} \cdot \dfrac{\sqrt{27}}{4}$

55. $\dfrac{\sqrt{2}}{3} \div \dfrac{\sqrt{8}}{6}$

56. $\dfrac{\sqrt{10}}{7} \div \dfrac{\sqrt{40}}{14}$

57. $\dfrac{\sqrt{st}}{\sqrt{48}} \div \dfrac{\sqrt{3}}{\sqrt{25s^2t}}$

58. $\dfrac{\sqrt{12}}{\sqrt{15}} \div \dfrac{\sqrt{45}}{\sqrt{144}}$

8.5 ZERO AND NEGATIVE EXPONENTS

DEFINITION Zero and negative exponents can be defined so as to simplify our notation. Consider the pattern in Table 8.1, and fill in the blanks.

Every time the exponent is decreased by 1, the corresponding number in the right-hand column is divided by 10. This suggests completing the table as follows:

TABLE 8.1

$10^4 =$	10,000
$10^3 =$	1000
$10^2 =$	100
$10^1 =$	10
$10^0 =$	
$10^{-1} =$	
$10^{-2} =$	
$10^{-3} =$	
$10^{-4} =$	

$$10^1 = 10$$

$$10^0 = 1$$

$$10^{-1} = \frac{1}{10} = \frac{1}{10^1}$$

$$10^{-2} = \frac{1}{100} = \frac{1}{10^2}$$

$$10^{-3} = \frac{1}{1000} = \frac{1}{10^3}$$

$$10^{-4} = \frac{1}{10,000} = \frac{1}{10^4}$$

Thus, the following definition seems natural.

DEFINITION

> Let $a \neq 0$ and let n be a positive integer (so that $-n$ is a negative integer). Define
>
> $$a^{-n} = \frac{1}{a^n} \quad \text{and} \quad a^0 = 1$$

Example 1 a. $2^{-1} = \dfrac{1}{2}$ b. $2^{-2} = \dfrac{1}{2^2} = \dfrac{1}{4}$ c. $2^{-3} = \dfrac{1}{2^3} = \dfrac{1}{8}$ d. $2^{-4} = \dfrac{1}{2^4} = \dfrac{1}{16}$

Example 2 a. $8^0 = 1$ b. $x^0 = 1$, if $x \neq 0$
c. $(10a)^0 = 1$, if $a \neq 0$, because any nonzero number raised to the 0th power equals 1.

Observe that 0^{-m} is *undefined* because $\frac{1}{0^m}$ would involve division by 0 (which is undefined). *By convention,* 0^0 *is undefined.*

CLASS EXERCISES *Assume $a \neq 0$. Find each value.*
1. 6^{-1} 2. $(-3)^{-2}$ 3. 2^{-5} 4. $(-12a)^0$

FRACTIONAL BASES The base a can be a fraction.

Example 3

a. $$\left(\frac{2}{3}\right)^{-1} = \frac{1}{\frac{2}{3}} = 1 \div \frac{2}{3} = \frac{3}{2}$$

b. $$\left(\frac{2}{3}\right)^{-2} = \frac{1}{\left(\frac{2}{3}\right)^2} = \frac{1}{\frac{4}{9}} = \frac{9}{4}$$

c. Let $x \neq 0$.
$$\left(\frac{5}{x}\right)^{-3} = \frac{1}{\left(\frac{5}{x}\right)^3} = \frac{1}{\frac{125}{x^3}} = \frac{x^3}{125}$$

Observe that in each case

$$\left(\frac{a}{b}\right)^{-n} = \frac{b^n}{a^n}, \text{ for } a \neq 0,\ b \neq 0, \text{ and for } n \text{ a positive integer}$$

CLASS EXERCISES *Assume $a \neq 0$. Find each value.*
5. $\left(\frac{3}{10}\right)^{-1}$ 6. $\left(\frac{4}{5}\right)^{-2}$ 7. $\left(\frac{10}{a}\right)^{-3}$

RULES FOR EXPONENTS Recall that for *positive integral exponents* m and n,

$$a^m \cdot a^n = a^{m+n} \quad \text{and} \quad \frac{a^m}{a^n} = a^{m-n}, \text{ where } m > n \text{ and } a \neq 0$$

These rules carry over to *zero and negative integral exponents*. Also, the second rule now applies to $m = n$ and to $m < n$. For example, if $a \neq 0$,

$$a^5 \cdot a^{-2} = a^5 \cdot \frac{1}{a^2} = a^3$$

Thus,

$$a^5 \cdot a^{-2} = a^{5 + (-2)}$$

And

$$\frac{a^5}{a^{-2}} = \frac{a^5}{\frac{1}{a^2}} = a^5 \cdot \frac{a^2}{1} = a^7$$

Thus,

$$\frac{a^5}{a^{-2}} = a^{5 - (-2)}$$

Finally,

$$a^{-5} \cdot a^{-2} = \frac{1}{a^5} \cdot \frac{1}{a^2} = \frac{1}{a^7} = a^{-7}$$

Thus,

$$a^{-5} \cdot a^{-2} = a^{(-5) + (-2)}$$

CLASS EXERCISES *Suppose $x \neq 0$. Simplify. Express your result without writing a fraction. Use negative exponents, if necessary.*

8. $x^7 \cdot x^{-4}$ **9.** $\frac{x^{-2}}{x^5}$ **10.** $\frac{x^{-2}}{x^{-1}}$

SOLUTIONS TO CLASS EXERCISES

1. $\frac{1}{6}$ **2.** $\frac{1}{9}$ **3.** $\frac{1}{32}$ **4.** 1 **5.** $\frac{10}{3}$ **6.** $\frac{25}{16}$ **7.** $\frac{a^3}{1000}$ **8.** x^3 **9.** x^{-7}

10. $\frac{x^{-2}}{x^{-1}} = x^{-2-(-1)} = x^{-2+1} = x^{-1}$

HOME EXERCISES

In Exercises 1–32, find each value.

1. 4^{-1} **2.** 1^{-1} **3.** $(-2)^{-1}$ **4.** 5^{-2} **5.** 8^{-2} **6.** $(-4)^{-2}$ **7.** -4^{-2} **8.** 3^{-3}
9. $(-3)^{-3}$ **10.** 4^{-3} **11.** 5^{-3} **12.** 3^{-4} **13.** 10^{-4} **14.** 10^{-5} **15.** 10^0 **16.** -10^0
17. $(-10)^0$ **18.** $\left(\frac{1}{2}\right)^{-1}$ **19.** $\left(\frac{3}{5}\right)^{-1}$ **20.** $\left(\frac{-7}{6}\right)^{-1}$ **21.** $\left(\frac{1}{2}\right)^{-2}$ **22.** $\left(\frac{3}{2}\right)^{-2}$
23. $\left(\frac{4}{5}\right)^{-2}$ **24.** $\left(\frac{-3}{4}\right)^{-2}$ **25.** $\left(\frac{3}{4}\right)^{-3}$ **26.** $\left(\frac{-3}{4}\right)^{-3}$ **27.** $\left(\frac{1}{10}\right)^{-5}$ **28.** $\left(\frac{1}{10}\right)^{-6}$
29. $\frac{2^{-2}}{4}$ **30.** $8 \cdot 2^{-4}$ **31.** $2^{-1} + 2^{-2}$ **32.** $2^{-1} + 3^{-1}$

In Exercises 33–41, assume $a \neq 0$, $b \neq 0$, $x \neq 0$, $y \neq 0$. Simplify and express your answer without writing a fraction. Use negative exponents, if necessary.

33. $a^4 \cdot a^{-3}$ **34.** $a^3 \cdot a^{-4}$ **35.** $y^{-7} \cdot y^{-3}$ **36.** $b^{-1} \cdot b^{-6} \cdot b^{-8}$ **37.** $b^3 \cdot b^{-2} \cdot b^{-4}$
38. $\frac{a^6}{a^{-4}}$ **39.** $\frac{b^{-5}}{b^5}$ **40.** $\frac{x^{-3}}{x^{-6}}$ **41.** $\frac{y^{-4}}{y^{-4}}$

In Exercises 42–52, determine the exponent m.

Sample $\frac{1}{9} = 3^m$	**Solution** $\frac{1}{9} = \frac{1}{3^2} = 3^{-2}$ Thus, $m = -2$

42. $\frac{1}{5} = 5^m$ **43.** $\frac{1}{4} = 2^m$ **44.** $\frac{1}{8} = 2^m$ **45.** $\frac{1}{1000} = 10^m$ **46.** $100 = \left(\frac{1}{10}\right)^m$ **47.** $\frac{2}{3} = \left(\frac{3}{2}\right)^m$

48. $\frac{9}{4} = \left(\frac{2}{3}\right)^m$ **49.** $1 = 5^m$ **50.** $-8 = \left(\frac{-1}{2}\right)^m$ **51.** $\frac{7}{10^2} = 7 \cdot 10^m$ **52.** $\frac{3}{10^5} = 3 \cdot 10^m$

8.6 SCIENTIFIC NOTATION

Scientific notation is a convenient way of expressing large numbers, such as 420,000,000,000, and small positive numbers, such as .000 003 . These numbers frequently occur in fields such as biology, chemistry, engineering, and astronomy.

Consider the **powers of 10**—positive, zero, and negative:

$10^1 = 10$ $\quad 10^0 = 1 \quad$ $10^{-1} = \frac{1}{10} = .1$

$10^2 = 100$ $\quad 10^{-2} = \frac{1}{100} = .01$

$10^3 = 1000$ $\quad 10^{-3} = \frac{1}{1000} = .001$

$10^4 = 10{,}000$ $\quad 10^{-4} = \frac{1}{10{,}000} = .0001$

$10^5 = 100{,}000$ $\quad 10^{-5} = \frac{1}{100{,}000} = .000\ 01$

$10^6 = 1{,}000{,}000$ $\quad 10^{-6} = \frac{1}{1{,}000{,}000} = .000\ 001$

EXPRESSING NUMBERS IN SCIENTIFIC NOTATION

Every positive number N can be expressed as the product of

a. a number between 1 and 10 and
b. a power of 10

This form of writing a number is called **scientific notation.**

Example 1 To express 240 in scientific notation, observe that

$$240 = \frac{240}{100} \times 100 = 2.40 \times 10^2$$

To divide by 100 (or 10^2), move the decimal point 2 digits to the left (to obtain 2.40). Balance this by multiplying by 10^2.

Example 2 To express .0051 in scientific notation, observe that

$$.0051 = (.0051 \times 1000) \times \frac{1}{1000} = 5.1 \times 10^{-3}$$

To multiply by 1000 (or 10^3), move the decimal point 3 digits to the right (to obtain 5.1). Balance this by multiplying by 10^{-3}.

Refer to Examples 1 and 2 as you read the rule for writing a number in scientific notation.

> Let N be a positive number, expressed as a decimal. To obtain the scientific notation,
>
> $$N = M \times 10^m \quad \text{or} \quad N = M \times 10^{-m}$$
>
> where m is a positive integer or 0:
>
> **1.** Place the decimal point after the first nonzero digit of N to obtain M.
>
> **2. a.** If you moved the decimal point m digits to the *left*, you *divided* by 10^m. To balance this, multiply by 10^m. Thus,
>
> $$N = M \times 10^m$$
>
> **b.** If you moved the decimal point m digits to the *right*, you *multiplied* by 10^m. To balance this, multiply by 10^{-m}. Thus,
>
> $$N = M \times 10^{-m}$$
>
> **c.** If N is between 1 and 10, the scientific notation for N is N itself. Note that
>
> $$N = N \times 10^0$$

Example 3 Express in scientific notation.
a 491.2 **b.** .007 32

Solution **a.** $N = 491.2$ Place the decimal point after the 4. Thus,

$$M = 4.912$$

You moved the decimal point 2 digits to the *left*:

$$491.2 \text{ to } 4.912$$

Therefore, $m = 2$ and

$$491.2 = 4.91 \times 10^2$$

b. $N = .007\ 32$ The first *nonzero* digit is 7. Thus,

$$M = 7.32$$

(The zeros to the left of 7 can be omitted.) You moved the decimal point 3 digits to the *right*:

$$.007\ 32 \text{ to } 007.\ 32$$

Therefore, $m = -3$ and

$$.007\ 32 = 7.32 \times 10^{-3}$$

CLASS EXERCISES *Express in scientific notation.*

1. 643 **2.** .067 **3.** 4.8 **4.** $\frac{1}{2}$ (*Hint*: Express $\frac{1}{2}$ as a decimal.)

SCIENTIFIC NOTATION TO DECIMAL NOTATION

To convert scientific notation,

$$M \times 10^m \quad \text{or} \quad M \times 10^{-m}\ ,\ m \neq 0$$

back into the usual decimal notation, reverse the preceding process:

1. Move the decimal point m digits to the right if $m > 0$.
2. Move the decimal point m digits to the left if $m < 0$.

Example 4 Convert scientific notation back into the usual decimal notation.

a. 3.52×10^5 **b.** 1.74×10^{-2}

Solution **a.** $3.52 \times 10^5 = 352{,}000$

In order to move the decimal point 5 digits to the right, add 3 zeros after the 2. Because 352,000 is an integer, the decimal point can be omitted.

b. $1.74 \times 10^{-2} = .0174$

In order to move the decimal point 2 digits to the left, add a zero before the 1.

CLASS EXERCISES *Convert scientific notation back into the usual decimal notation.*

5. 4.39×10^1 6. 2.72×10^6 7. 8.82×10^{-1} 8. 5.38×10^{-4}

COMPUTATIONS Some arithmetic computations can be simplified by first expressing numbers in terms of powers of 10. Instead of using scientific notation, express each number as the product of an *integer* and a power of 10. Then, write the final result in scientific notation.

Example 5 Find the following value.

$$\frac{75{,}000 \times 2400}{2 \times 120{,}000 \times .0125}$$

Solution

$$\frac{75{,}000 \times 2400}{2 \times 120{,}000 \times .0125} = \frac{\overset{3}{\cancel{75}} \times 10^3 \times \overset{\overset{1}{\cancel{2}}}{\cancel{24}} \times 10^2}{\underset{1}{\cancel{2}} \times \underset{1}{\cancel{12}} \times 10^4 \times \underset{5}{\cancel{125}} \times 10^{-4}}$$

$$= \frac{3}{5} \times 10^{3+2-(4-4)}$$

$$= .6 \times 10^5$$

$$= 6 \times 10^4 \qquad \textit{in scientific notation}$$

CLASS EXERCISES *Find each value. Express your result in scientific notation.*

9. $2000 \times 50{,}000 \times .08$ 10. $\dfrac{12{,}000 \times 5{,}000{,}000}{.003 \times .0025}$

USES OF SCIENTIFIC NOTATION

The *speed of light* is 3×10^{10} centimeters per second.
The *earth's mass* is approximately 6×10^{27} grams.
An electron has a charge of 4.8×10^{-10} electrostatic units.

All gases occupying equal volumes at the same temperature and pressure contain the same number of molecules (**Avogadro's Number**) which has been determined to be 6.023×10^{23} molecules per gram molecular weight.

SOLUTIONS TO CLASS EXERCISES

1. 6.43×10^2 **2.** $\not{0}6.7 \times 10^{-2}$ **3.** 4.8 **4.** $\frac{1}{2} = .5 = 5 \times 10^{-1}$

5. 43.9 **6.** $2,720,000$ **7.** $.882$ **8.** $.000\ 538$

9.
$$\begin{aligned} 2000 \times 50{,}000 \times .08 &= 2 \times 10^3 \times 5 \times 10^4 \times 8 \times 10^{-2} \\ &= (2 \times 5 \times 8) \times 10^{3+4-2} \\ &= 80 \times 10^5 \\ &= 8 \times 10^6 \end{aligned}$$

10.
$$\begin{aligned} \frac{12{,}000 \times 5{,}000{,}000}{.003 \times .0025} &= \frac{\overset{4}{\not{12}} \times 10^3 \times \overset{1}{\not{5}} \times 10^6}{\underset{1}{\not{3}} \times 10^{-3} \times \underset{5}{\not{25}} \times 10^{-4}} \\ &= \frac{4}{5} \times 10^{3+6-(-3+-4)} \\ &= .8 \times 10^{9+7} \\ &= 8 \times 10^{15} \end{aligned}$$

HOME EXERCISES

In Exercise 1–16, express each number in scientific notation.

1. 12.5 **2.** 465 **3.** 7028 **4.** 5.93 **5.** .497 **6.** .005 31 **7.** .8 **8.** 6.66
9. 100 **10.** 7048.2 **11.** 361,000 **12.** 1,011,000 **13.** .000 005 **14.** .030 003 **15.** 6 **16.** .01

In Exercises 17–30, convert scientific notation back into the usual decimal notation.

17. 6.52×10^1 **18.** 3.41×10^3 **19.** 3.23×10^6 **20.** 3×10^4 **21.** 5×10^{-1}
22. 3.7×10^{-2} **23.** 9.01×10^{-4} **24.** 1.53×10^{-6} **25.** 8.78×10^8 **26.** 8.78×10^{-8}
27. 2.001×10^2 **28.** 2.001×10^{-2} **29.** 5.4314×10^4 **30.** $3.612\ 54 \times 10^{-3}$

In Exercises 31–38, find each value. Express your result in scientific notation.

31. $10^4 \times 10^2 \times 10^{-3}$ **32.** $\frac{10^5 \times 10^8}{10^2 \times 10^6}$ **33.** $5000 \times 20{,}000 \times 4000$ **34.** $8000 \times .0009$

35. $.0062 \times .000\ 02$ **36.** $\frac{2000 \times 40{,}000}{.05 \times 800}$ **37.** $\frac{2500 \times 70{,}000}{.05 \times .014}$ **38.** $\frac{720{,}000 \times .09}{1{,}200{,}000 \times .0003}$

39. A light year is the distance over which light can travel in a year. One light year is approximately 6×10^{12} miles. A star is 180 light years away from earth. How many miles away is it?

40.
$$1 \text{ centimeter} \simeq .3937 \text{ inch}$$
Express this number of inches in scientific notation.

41. One electron volt is the kinetic energy an electron acquires when it is accelerated in an electric field produced by a difference of potential of one volt.
$$1 \text{ electron volt} = 1.60 \times 10^{-19} \text{ joule}$$
How many joules equal one million electron volts?

42. The human eye is normally capable of responding to light waves whose lengths are in the range of 3.8×10^{-5} to 7.6×10^{-5} centimeter. Which of the following centimeter readings lie within this range?
a. .000 54 centimeter **b.** .000 054 centimeter
c. 7×10^{-6} centimeter **d.** $.39 \times 10^{-4}$ centimeter

43. The coefficient of linear expansion indicates the fractional change in the length of rod per degree change of temperature. For tungsten, the coefficient is 4.4×10^{-6} per degree Celsius. What is the fractional change in the length of a tungsten rod when the temperature rises from 5° to 25° Celsius?

Let's Review Chapter 8

In Exercises 1–3, find each square root.

1. $\sqrt{64}$ **2.** $-\sqrt{900}$ **3.** $\sqrt{x^4}$

4. Assume $x > 0$ and $y < 0$. Find $\sqrt{(xy)^2}$.

5. Find consecutive integers N and $N + 1$ such that $N < \sqrt{55} < N + 1$.

6. Use $\sqrt{2} \approx 1.41$ to approximate $7\sqrt{2}$ to the nearest tenth.

In the remaining Exercises, assume $a > 0$, $b > 0$, $x > 0$.
In Exercises 7–9, find each square root.

7. $\sqrt{25a^2}$ **8.** $\sqrt{a^4 b^6}$ **9.** $\sqrt{\dfrac{81}{100}}$

In Exercises 10–20, simplify each expression.

10. $2\sqrt{3} + 5\sqrt{2} + 3\sqrt{3} - 3\sqrt{2}$ **11.** $5\sqrt{5} - (3\sqrt{5} - \sqrt{5})$

12. $\sqrt{9a} - \sqrt{4a}$ **13.** $\sqrt{64a^2b^4} + \sqrt{81a^2b^4}$

14. $\dfrac{4\sqrt{3}}{15} + \dfrac{2\sqrt{3}}{5}$ **15.** $\sqrt{10}\sqrt{5}\sqrt{2}$

16. $\sqrt{2}(\sqrt{8} - \sqrt{2})$ **17.** $\dfrac{\sqrt{8}}{\sqrt{3}} \cdot \dfrac{\sqrt{3}}{\sqrt{2}}$

18. $\dfrac{\sqrt{32}}{\sqrt{8}}$ **19.** $\dfrac{\sqrt{12x}}{\sqrt{75x}}$

20. $\dfrac{\sqrt{15}}{\sqrt{18}} \div \dfrac{\sqrt{5}}{\sqrt{32}}$

21. a. Factor $7 + \sqrt{98}$. **b.** Use part **a** to simplify $\dfrac{7 + \sqrt{98}}{21}$.

In Exercises 22–24, find each value.

22. 7^{-2} **23.** 8^0 **24.** $\left(\dfrac{2}{3}\right)^{-2}$

25. Express $\dfrac{x^{-4}}{x^3}$ without writing a fraction. Use a negative exponent.

In Exercises 26 and 27, express each number in scientific notation.

26. 265,000 **27.** .0257

28. Convert 3.45×10^{-4} back into the usual decimal notation.

29. Find the value of $\dfrac{16{,}000 \times .06}{.004 \times 600}$ Express your result in scientific notation.

9 QUADRATIC EQUATIONS

DIAGNOSTIC TEST

Perhaps you are already familiar with some of the material in Chapter 9. This test will indicate which sections in Chapter 9 you need to study. The question number refers to the corresponding section in Chapter 9. If you answer *all* parts of the question correctly, you may omit the section. But *if any part of your answer is wrong, you should study the section.* When you have completed the test, turn to page A25 for the answers.

9.1 Solutions by Factoring

Solve by factoring. a. $x^2 - 8x + 7 = 0$ b. $4x^2 + 7x - 2 = 0$

9.2 Equations of the Form $x^2 = a$

Solve each equation by using the method:

$$\text{If } x^2 = a, \text{ where } a > 0, \text{ then } x = \pm\sqrt{a}$$

a. $x^2 - 200 = 0$ b. $(3t + 4)^2 = 1$

9.3 The Quadratic Formula

Apply the quadratic formula to solve each equation.

a. $x^2 + 4x + 2 = 0$ b. $9t^2 - 6t + 1 = 0$ c. $y^2 + 4y + 2 = 0$

9.4 Word Problems

a. Find two integers whose sum is -2 and whose product is -120.
b. A stone rolls off a cliff 1936 feet high. As the stone falls, its height is given by the formula

$$h = 1936 - 16t^2 \text{ (feet)}$$

where t is the time falling (in seconds). How long does it take to hit the ground?

9.1 SOLUTIONS BY FACTORING

FORM OF A QUADRATIC EQUATION

A **quadratic**, or **second-degree**, **equation** (in a single variable, x) is an equation that can be expressed in the form

$$Ax^2 + Bx + C = 0 \qquad (A \neq 0)$$

The quadratic equation

$$x^2 - 3x + 2 = 0$$

is already in this form with $A = 1$, $B = -3$, $C = 2$. The quadratic equation

$$3x^2 + 5 = 2x - 3$$

can be transformed into the above form by adding $-2x + 3$ to both sides:

$$3x^2 - 2x + 8 = 0$$

Here, $A = 3$, $B = -2$, $C = 8$.

An equation may contain *second-degree terms*, such as x^2, and yet it need *not* be a *quadratic* (*second-degree*) *equation*. For example, by subtracting x^2 from both sides, the equation

$$x^2 + 3x = x^2 + 9$$

reduces to the linear (first-degree) equation

$$3x = 9$$

or

$$0x^2 + 3x - 9 = 0$$

The given equation cannot be expressed in the form

$$Ax^2 + Bx + C = 0 \qquad (A \neq 0)$$

Similarly, the second-degree terms can be eliminated from the equation of Example 1.

Example 1 Solve. $(x + 1)(x + 2) = (1 - x)^2 + 11$

Solution

$$(x + 1)(x + 2) = (1 - x)^2 + 11 \qquad \textit{Multiply out.}$$

$$x^2 + 3x + 2 = 1 - 2x + x^2 + 11$$

$$x^2 + 3x + 2 - x^2 + 2x - 2 = 1 - 2x + x^2 + 11 - x^2 + 2x - 2$$

$$5x = 10$$

$$x = 2$$

CLASS EXERCISES

1. Which of the following are quadratic equations?
   a. $3x + 2 = 0$ **b.** $3x^2 + x + 2 = 0$ **c.** $x^3 + 3x^2 + 2 = 0$
   d. $2x^2 + x - x^2 = x^2 + 1$ **e.** $(x - 1)^2 = (x + 1)^2$

d. Check each of these roots by replacing x first by 4 and then by -2 in the equation

$$x^2 - 2x - 8 = 0$$

check

for 4:	for −2:
$4^2 - 2 \cdot 4 - 8 \stackrel{?}{=} 0$	$(-2)^2 - 2(-2) - 8 \stackrel{?}{=} 0$
$16 - 8 - 8 \stackrel{?}{=} 0$	$4 + 4 - 8 \stackrel{?}{=} 0$
$0 \stackrel{\checkmark}{=} 0$	$0 \stackrel{\checkmark}{=} 0$

CLASS EXERCISES

4. Consider the equation $(x + 5)(x - 1) = 0$
   a. Show that this is a quadratic equation.
   b. Solve this quadratic equation.
   c. Check the roots by using the given equation.
   d. Check the roots by using the form of the equation obtained in part a.

FACTORING THE LEFT SIDE

If you can factor the left side of a quadratic equation of the form

$$Ax^2 + Bx + C = 0 \qquad A \neq 0$$

you can apply the above method.

Example 3 Solve. $x^2 - 3x + 2 = 0$

Solution $x^2 - 3x + 2 = (x - 1)(x - 2)$

Thus, $x^2 - 3x + 2 = 0$

can be expressed as

$$(x - 1)(x - 2) = 0$$

$x - 1 = 0$	$x - 2 = 0$
$x = 1$	$x = 2$

The roots are 1 and 2.

Example 4 Solve. $2x^2 + 5x + 2 = 0$

Solution $2x^2 + 5x + 2 = (2x + 1)(x + 2)$

Thus, $2x^2 + 5x + 2 = 0$

can be expressed as

$$(2x + 1)(x + 2) = 0$$

$2x + 1 = 0$	$x + 2 = 0$
$2x = -1$	$x = -2$
$x = \frac{-1}{2}$	

The roots are $\frac{-1}{2}$ and -2.

Observe that

$a^2 = 0$ means that $a = 0$

Example 5 Solve. $x^2 - 6x + 9 = 0$

Solution

$$x^2 - 6x + 9 = (x - 3)(x - 3) = (x - 3)^2$$

Thus, $x^2 - 6x + 9 = 0$

can be expressed as

$$(x - 3)^2 = 0$$

Thus, $x - 3 = 0$

$$x = 3$$

The *only* root is 3. This root is sometimes called a *double root*.

CLASS EXERCISES *Solve each equation.*

5. $t^2 + 4t + 3 = 0$ **6.** $x^2 + x - 6 = 0$

7. $2y^2 + 7y + 3 = 0$ **8.** $x^2 + 4x + 4 = 0$

SOLUTIONS TO CLASS EXERCISES

1. Only b. Note that d reduces to $x = 1$.
And c reduces to a linear equation as follows:

$$(x - 1)^2 = (x + 1)^2$$
$$x^2 - 2x + 1 = x^2 + 2x + 1$$
$$-2x = 2x$$
$$0 = 4x, \quad \text{or}$$
$$0 = x$$

2. $3x^2 + 2x - 5 = 0$

3. $(x - 2)(x + 1) = x^2$
$$x^2 - x - 2 = x^2$$
$$-2 = x$$

4. a. $x^2 + 4x - 5 = 0$

b. From the original equation,

$x + 5 = 0$, $x = -5$ | $x - 1 = 0$, $x = 1$

The roots are -5 and 1.

c. **check**

for -5:

$$(-5 + 5)(-5 - 1) \stackrel{?}{=} 0$$
$$0(-6) \stackrel{?}{=} 0$$
$$0 \stackrel{\checkmark}{=} 0$$

for 1:

$$(1 + 5)(1 - 1) \stackrel{?}{=} 0$$
$$6 \cdot 0 \stackrel{?}{=} 0$$
$$0 \stackrel{\checkmark}{=} 0$$

d. Use $x^2 + 4x - 5 = 0$.

check

for -5:

$$(-5)^2 + 4(-5) - 5 \stackrel{?}{=} 0$$
$$25 - 20 - 5 \stackrel{?}{=} 0$$
$$0 \stackrel{\checkmark}{=} 0$$

for 1:

$$1^2 + 4(1) - 5 \stackrel{?}{=} 0$$
$$1 + 4 - 5 \stackrel{?}{=} 0$$
$$0 \stackrel{\checkmark}{=} 0$$

5.
$$t^2 + 4t + 3 = 0$$
$$(t + 1)(t + 3) = 0.$$

$$t + 1 = 0 \qquad t = -1 \quad \Big| \quad t + 3 = 0 \qquad t = -3$$

The roots are -1 and -3.

6.
$$x^2 + x - 6 = 0$$
$$(x + 3)(x - 2) = 0$$

$$x + 3 = 0 \qquad x = -3 \quad \Big| \quad x - 2 = 0 \qquad x = 2$$

The roots are -3 and 2.

7.
$$2y^2 + 7y + 3 = 0$$
$$(2y + 1)(y + 3) = 0$$

$$2y + 1 = 0 \qquad 2y = -1 \qquad y = \frac{-1}{2} \quad \Big| \quad y + 3 = 0 \qquad y = -3$$

The roots are $\frac{-1}{2}$ and -3.

8.
$$x^2 + 4x + 4 = 0$$
$$(x + 2)^2 = 0$$
$$x + 2 = 0$$
$$x = -2$$

The only root is -2.

HOME EXERCISES

1. Which of the following are quadratic equations?

a. $4x - 1 = 0$ **b.** $x(4x - 1) = 0$ **c.** $3x^3 - 2x = x^2 - 1$
d. $x^2 - 2x + 5 = 3x - x^2$ **e.** $x^2 - 2x + 5 = 3x + x^2$ **f.** $(x + 3)^2 = (x - 2)^2$

2. Express each of the following quadratic equations in the form $Ax^2 + Bx + C = 0$.

a. $1 - 12x + x^2 = 0$ **b.** $5x^2 = 9 - 3x$ **c.** $10x^2 + 2x - 5 = 3x^2 + 2x + 3$ **d.** $(x + 4)^2 = (2x - 1)^2$

In Exercises 3–46, solve each equation. Check the roots, when indicated, by using the given equation.

3. $x^2 - (x + 1)^2 = 3$
4. $(x + 2)^2 - (x + 1)^2 = 9$
5. $(y + 1)^2 - 3 = (y - 2)^2$
6. $(y + 2)(y + 3) = (y + 1)(y - 2)$
7. $(x + 4)^2 = x^2$ (*Check.*)
8. $(2 - x)^2 + 5 = x^2 + 1$
9. $[4 - (6 - x)]^2 = x^2 - 8$
10. $x(x - 4) = 0$
11. $(x - 2)(x - 3) = 0$
12. $(x - 1)(x + 4) = 0$
13. $2(y + 3)(y - 8) = 0$ (*Check.*)
14. $\left(x + \frac{1}{2}\right)\left(x - \frac{1}{4}\right) = 0$
15. $(x + 2)^2 = 0$
16. $x^2 - 5x + 4 = 0$
17. $x^2 - 5x + 6 = 0$
18. $t^2 + 6t + 8 = 0$ (*Check.*)
19. $y^2 + y - 12 = 0$
20. $x^2 - 2x = 8$
21. $y^2 + 2y - 8 = 0$
22. $x^2 + 8x + 18 = 3$
23. $z^2 + 9z + 18 = 0$
24. $t^2 - t - 30 = 0$
25. $x^2 - 10x + 25 = 0$
26. $x^2 + 8x + 16 = 0$
27. $x^2 + 4x - 21 = 0$
28. $x^2 + 7x + 10 = 0$ (*Check.*)
29. $3x^2 + 24x - 27 = 0$
30. $x^2 + 4x - 12 = 0$
31. $2x^2 + 2x - 4 = 0$ (*Check.*)
32. $3x^2 + 6x - 9 = 0$
33. $2x^2 + 3x + 1 = 0$
34. $2x^2 + 7x + 3 = 0$
35. $2t^2 + 7t - 4 = 0$
36. $3y^2 + 7y + 2 = 0$ (*Check.*)
37. $4x^2 + 8x + 3 = 0$
38. $6y^2 + y - 1 = 0$
39. $4x^2 + 20x + 25 = 0$
40. $3x^2 + 3 = 10x$
41. $2y^2 + 6 = 7y$
42. $2z^2 - 13z = 7$
43. $2x^2 - 4x = 0$
44. $x^2 + 5x = 0$
45. $2x^2 = x$
46. $3x^2 + 2x = 0$

9.2 EQUATIONS OF THE FORM $x^2 = a$

Let n be a positive number, and let

$$\pm n \quad \text{stand for} \quad \text{both } n \text{ and } -n$$

For example, ± 2 stands for both 2 and -2

If a is positive, there are two numbers whose square is a. These numbers are

$$\pm\sqrt{a}$$

because $\sqrt{a}\sqrt{a} = a$ and $(-\sqrt{a})(-\sqrt{a}) = a$

Thus, *if a quadratic equation can be written in the form*

$$x^2 = a$$

its roots are $\pm\sqrt{a}$.

$a > 0$

Example 1 Solve. $x^2 - 25 = 0$

Solution 1

$$x^2 - 25 = 0$$
$$x^2 = 25$$
$$x = \pm 5$$

Solution 2 Factor the left side of the equation

$$x^2 - 25 = 0$$
$$(x + 5)(x - 5) = 0$$

$$x + 5 = 0 \quad\Big|\quad x - 5 = 0$$
$$x = -5 \quad\Big|\quad x = 5$$

The roots are ± 5.

Example 2 Solve. $x^2 - 24 = 0$

Solution

$$x^2 - 24 = 0$$
$$x^2 = 24$$
$$x = \pm\sqrt{24}$$
$$x = \pm\sqrt{4 \cdot 6}$$
$$x = \pm\sqrt{4}\sqrt{6}$$
$$x = \pm 2\sqrt{6}$$

CLASS EXERCISES

Solve for the indicated variable. Check the roots, when indicated, by using the given equation.

1. $y^2 - 144 = 0$ (Check.) **2.** $t^2 - 7 = 0$ **3.** $x^2 - 20 = 0$

FURTHER EXAMPLES

Example 3 Solve for u. $(2u + 3)^2 = 25$ *Check the roots.*

Solution This second-degree equation is of the form

$$x^2 = 25$$

Here, $$x = 2u + 3$$

Thus, $$2u + 3 = \pm 5$$

You obtain two first-degree equations:

$$\begin{array}{r|r} 2u + 3 = 5 & 2u + 3 = -5 \\ 2u = 2 & 2u = -8 \\ u = 1 & u = -4 \end{array}$$

The roots are 1 and -4.

check

$$\begin{array}{r|r} \text{for } 1: & \text{for } -4: \\ (2 \cdot 1 + 3)^2 \stackrel{?}{=} 25 & [2(-4) + 3]^2 \stackrel{?}{=} 25 \\ 5^2 \stackrel{?}{=} 25 & (-5)^2 \stackrel{?}{=} 25 \\ 25 \stackrel{\checkmark}{=} 25 & 25 \stackrel{\checkmark}{=} 25 \end{array}$$

Example 4 Solve for x. $(3x - 1)^2 = 5$ *Check the roots.*

Solution

$$(3x - 1)^2 = 5$$

$$3x - 1 = \pm\sqrt{5}$$

$$\begin{array}{r|r} 3x - 1 = \sqrt{5} & 3x - 1 = -\sqrt{5} \\ 3x = 1 + \sqrt{5} & 3x = 1 - \sqrt{5} \\ x = \dfrac{1 + \sqrt{5}}{3} & x = \dfrac{1 - \sqrt{5}}{3} \end{array}$$

The roots differ only in the sign of $\sqrt{5}$. Thus, the roots are $\dfrac{1 \pm \sqrt{5}}{3}$.

check

for $\frac{1+\sqrt{5}}{3}$:

$$\left[3\left(\frac{1+\sqrt{5}}{3}\right) - 1\right]^2 \stackrel{?}{=} 5$$

$$[(1+\sqrt{5}) - 1]^2 \stackrel{?}{=} 5$$

$$(\sqrt{5})^2 \stackrel{?}{=} 5$$

$$5 \stackrel{\checkmark}{=} 5$$

for $\frac{1-\sqrt{5}}{3}$:

$$\left[3\left(\frac{1-\sqrt{5}}{3}\right) - 1\right]^2 \stackrel{?}{=} 5$$

$$[(1-\sqrt{5}) - 1]^2 \stackrel{?}{=} 5$$

$$(-\sqrt{5})^2 \stackrel{?}{=} 5$$

$$5 \stackrel{\checkmark}{=} 5$$

CLASS EXERCISES *Solve for the indicated variable. Check the roots by using the given equation.*

4. $(2x + 1)^2 = 9$ **5.** $(t - 4)^2 = 2$

$a = 0$ *The equation*

$$x^2 = 0$$

has one root, 0.

Example 5 Solve for z. $(5z - 2)^2 = 0$

Solution Let $x = 5z - 2$. From

$$(5z - 2)^2 = 0$$

you obtain

$$5z - 2 = 0,$$

$$5z = 2,$$

and finally,

$$z = \frac{2}{5}.$$

CLASS EXERCISES *Solve for the indicated variable.*

6. $(x - 4)^2 = 0$ **7.** $(2t + 5)^2 = 0$

$a < 0$ When a is *negative*, the equation

$$x^2 = a$$

does not have any *real number* as a root. In fact, the square of a real number is always *at least* 0. In more advanced courses, you will learn about other numbers, known as "**complex numbers**." *If a is negative, the roots of the above equation are complex numbers.* For the present, you need only consider real numbers.

To sum up:

The equation	$x^2 = a$
has	$\begin{cases} 2 \text{ (real) roots if } a > 0 \\ 1 \text{ (real) root if } a = 0 \\ 0 \text{ (real) roots if } a < 0 \end{cases}$

SOLUTIONS TO CLASS EXERCISES

1. ± 12

check

for 12:	for −12:
$12^2 - 144 \overset{?}{=} 0$	$(-12)^2 - 144 \overset{?}{=} 0$
$0 \overset{\checkmark}{=} 0$	$0 \overset{\checkmark}{=} 0$

2. $\pm\sqrt{7}$

3. $\pm 2\sqrt{5}$

4. $(2x + 1)^2 = 9$

$2x + 1 = \pm 3$

$2x + 1 = 3$	$2x + 1 = -3$
$2x = 2$	$2x = -4$
$x = 1$	$x = -2$

The roots are 1 and -2.

check

for 1:	for −2:
$(2 \cdot 1 + 1)^2 \overset{?}{=} 9$	$(2(-2) + 1)^2 \overset{?}{=} 9$
$3^2 \overset{?}{=} 9$	$(-3)^2 \overset{?}{=} 9$
$9 \overset{\checkmark}{=} 9$	$9 \overset{\checkmark}{=} 9$

5. $(t - 4)^2 = 2$

$t - 4 = \pm\sqrt{2}$

$t - 4 = \sqrt{2}$	$t - 4 = -\sqrt{2}$
$t = 4 + \sqrt{2}$	$t = 4 - \sqrt{2}$

The roots are $4 \pm \sqrt{2}$.

check

for $4 + \sqrt{2}$:	for $4 - \sqrt{2}$:
$(4 + \sqrt{2} - 4)^2 \overset{?}{=} 2$	$(4 - \sqrt{2} - 4)^2 \overset{?}{=} 2$
$(\sqrt{2})^2 \overset{?}{=} 2$	$(-\sqrt{2})^2 \overset{?}{=} 2$
$2 \overset{\checkmark}{=} 2$	$2 \overset{\checkmark}{=} 2$

6. $(x - 4)^2 = 0$

$x - 4 = 0$

$x = 4$

The only root is 4.

7. $(2t + 5)^2 = 0$

$2t + 5 = 0$

$2t = -5$

$t = \dfrac{-5}{2}$

The only root is $\dfrac{-5}{2}$.

HOME EXERCISES

Solve for the indicated variable. Check the roots, when indicated, by using the given equation.

1. $x^2 = 16$ **2.** $x^2 - 36 = 0$ **3.** $y^2 - 49 = 0$ **4.** $t^2 - 100 = 0$ **5.** $x^2 = 6$

6. $x^2 - 11 = 0$ **7.** $y^2 - 13 = 0$ (*Check.*) **8.** $z^2 - 15 = 0$ **9.** $x^2 - 8 = 0$

10. $t^2 - 12 = 0$ **11.** $u^2 - 18 = 0$ **12.** $y^2 - 20 = 0$ **13.** $u^2 - 45 = 0$ **14.** $x^2 - 40 = 0$

15. $x^2 - \dfrac{4}{9} = 0$ **16.** $u^2 - \dfrac{25}{16} = 0$ (*Check.*) **17.** $x^2 - \dfrac{2}{3} = 0$ **18.** $t^2 - \dfrac{12}{25} = 0$

19. $(2u + 1)^2 = 49$ (*Check.*) **20.** $(2x - 3)^2 = 1$ (*Check.*) **21.** $(5x - 2)^2 = 16$

22. $\left(\frac{y}{2} + 1\right)^2 = 9$ **23.** $(3z + 4)^2 = 64$ **24.** $(5x - 2)^2 = 81$ **25.** $(3x - 2)^2 = 3$ (*Check.*)

26. $(2x + 1)^2 = 5$ **27.** $(5t + 4)^2 = 7$ **28.** $(u - 4)^2 = 11$ **29.** $(3y + 5)^2 = 6$ **30.** $(4z - 1)^2 = 13$

31. $(2x - 1)^2 = 8$ **32.** $(2t - 5)^2 = 12$ **33.** $(z + 3)^2 = 32$ **34.** $\left(\frac{t}{2} + 1\right)^2 = 20$ (*Check.*)

35. $\left(\frac{x - 1}{2}\right)^2 = 98$ **36.** $\left(\frac{2x + 7}{3}\right)^2 = 80$ **37.** $(x - 2)^2 = 0$ **38.** $(2t + 1)^2 = 0$ **39.** $(5z - 3)^2 = 0$

40. $\left(\frac{4u + 5}{3}\right)^2 = 0$

9.3 THE QUADRATIC FORMULA

APPLYING THE FORMULA There is a mechanical procedure for finding the roots of *any quadratic equation*

$$Ax^2 + Bx + C = 0 \qquad (A \neq 0)$$

This method is particularly useful when the previous methods do not easily apply. First, write the equation in the above form. *The roots are then given by the* **quadratic formula:**

$$\boxed{x = \frac{-B \pm \sqrt{B^2 - 4AC}}{2A}}$$

Example 1 Use the quadratic formula to find the roots of the equation

$$x^2 + 5x + 3 = 0$$

Check the roots.

Solution

$$x^2 + 5x + 3 = 0$$

is of the form

$$Ax^2 + Bx + C = 0$$

where

$$A = 1, B = 5, C = 3$$

Apply the quadratic formula.

$$x = \frac{-B \pm \sqrt{B^2 - 4AC}}{2A}$$

$$x = \frac{-5 \pm \sqrt{5^2 - 4 \cdot 1 \cdot 3}}{2 \cdot 1}$$

$$= \frac{-5 \pm \sqrt{25 - 12}}{2}$$

$$= \frac{-5 \pm \sqrt{13}}{2}$$

check for $\dfrac{-5 + \sqrt{13}}{2}$:

$$\left(\frac{-5 + \sqrt{13}}{2}\right)^2 + 5\left(\frac{-5 + \sqrt{13}}{2}\right) + 3 \stackrel{?}{=} 0$$

$$\frac{25 - 10\sqrt{13} + 13}{4} + \frac{-25 + 5\sqrt{13}}{2} + 3 \stackrel{?}{=} 0$$

$$\frac{25 - 10\sqrt{13} + 13 - 50 + 10\sqrt{13} + 12}{4} \stackrel{?}{=} 0$$

$$\frac{\overbrace{(25 + 13 + 12 - 50)}^{0} + \overbrace{(-10\sqrt{13} + 10\sqrt{13})}^{0}}{4} \stackrel{?}{=} 0$$

$$0 \stackrel{\checkmark}{=} 0$$

check for $\dfrac{-5 - \sqrt{13}}{2}$:

$$\left(\frac{-5 - \sqrt{13}}{2}\right)^2 + 5\left(\frac{-5 - \sqrt{13}}{2}\right) + 3 \stackrel{?}{=} 0$$

$$\frac{25 + 10\sqrt{13} + 13}{4} + \frac{-25 - 5\sqrt{13}}{2} + 3 \stackrel{?}{=} 0$$

$$\frac{25 + 10\sqrt{13} + 13 - 50 - 10\sqrt{13} + 12}{4} \stackrel{?}{=} 0$$

$$\frac{\overbrace{(25 + 13 + 12 - 50)}^{0} + \overbrace{(10\sqrt{13} - 10\sqrt{13})}^{0}}{4} \stackrel{?}{=} 0$$

$$0 \stackrel{\checkmark}{=} 0$$

Example 2 Solve the quadratic equation

$$x^2 - 7x + 10 = 0$$

a. by applying the quadratic formula and
b. by factoring the left side.

Solution a. $x^2 - 7x + 10 = 0$

Here, $A = 1$, $B = -7$, $C = 10$. Apply the quadratic formula:

$$x = \frac{-B \pm \sqrt{B^2 - 4AC}}{2A}$$

$$x = \frac{-(-7) \pm \sqrt{(-7)^2 - 4 \cdot 1 \cdot 10}}{2 \cdot 1}$$

$$= \frac{7 \pm \sqrt{49 - 40}}{2}$$

$$= \frac{7 \pm \sqrt{9}}{2}$$

$$= \frac{7 \pm 3}{2}$$

$$x = \frac{7 + 3}{2} \quad \Big| \quad x = \frac{7 - 3}{2}$$

$$x = 5 \quad \Big| \quad x = 2$$

The roots are 2 and 5.

b. $x^2 - 7x + 10 = 0$

$$(x - 2)(x - 5) = 0$$

$$x - 2 = 0 \quad \Big| \quad x - 5 = 0$$

$$x = 2 \quad \Big| \quad x = 5$$

As you see in Example 2, if you can easily factor the left side of an equation

$$Ax^2 + Bx + C = 0$$

you obtain the roots more readily than by applying the quadratic formula.

CLASS EXERCISES

1. Use the quadratic formula to find the roots of the equation $x^2 + 4x + 1 = 0$. *Check the roots.*
2. **a.** Apply the quadratic formula to solve the equation $(y + 3)^2 = 3$.
   b. Solve the equation by some other method.

THE DISCRIMINANT In the quadratic formula,

$$x = \frac{-B \pm \sqrt{B^2 - 4AC}}{2A}$$

the expression within the radical sign

$$B^2 - 4AC$$

is known as the **discriminant**. For the equation

$$2x^2 - 6x + 1 = 0$$

the discriminant is given by

$$B^2 - 4AC = (-6)^2 - 4 \cdot 2 \cdot 1 = 36 - 8 = 28$$

When you evaluate the discriminant, you can tell how many *real* numbers are roots of the given equation. (See Table 9.1 .)

Table 9.1

$B^2 - 4AC$ (discriminant)	Number of Roots
positive	2 (real) roots
0	1 (real) root
negative	0 (real) roots

For, consider the quadratic formula,

$$x = \frac{-B \pm \sqrt{B^2 - 4AC}}{2A}$$

Suppose the discriminant, $B^2 - 4AC$, is positive. In the numerator, you can add its principal square root, $\sqrt{B^2 - 4AC}$, or its negative square root, $-\sqrt{B^2 - 4AC}$. This amounts to adding or subtracting the principal square root in the numerator. Thus, you obtain two different roots of the quadratic equation.

If the discriminant is 0, then because $\sqrt{0} = 0$, the only root of the quadratic equation is $\frac{-B}{2A}$.

If the discriminant is negative, the quadratic equation has no *real* square root. (The roots obtained by applying the formula are *complex numbers*.)

Example 3 Consider the equation

$$x^2 + 7x + 1 = 0$$

a. Evaluate the discriminant.
b. How many (real) roots are there?
c. Apply the quadratic formula to solve the equation.

Solution **a.** $A = 1$, $B = 7$, $C = 1$. The discriminant is given by

$$B^2 - 4AC = 49 - 4 = 45$$

b. There are two (real) roots because the discriminant is positive.

c. $x = \dfrac{-B \pm \sqrt{B^2 - 4AC}}{2A}$

$= \dfrac{-7 \pm \sqrt{45}}{2}$

$= \dfrac{-7 \pm 3\sqrt{5}}{2}$

Example 4 Consider the equation

$$9y^2 + 12y + 4 = 0$$

a. Evaluate the discriminant.
b. How many (real) roots are there?
c. Apply the quadratic formula to solve the equation.

Solution **a.** $A = 9$, $B = 12$, $C = 4$. The discriminant is given by

$$B^2 - 4AC = 12^2 - 4 \cdot 9 \cdot 4$$
$$= 144 - 144$$
$$= 0$$

b. There is one root because the discriminant is 0.

c. $y = \dfrac{-B \pm \sqrt{B^2 - 4AC}}{2A}$

$= \dfrac{-12 \pm \sqrt{0}}{2 \cdot 9}$

$= \dfrac{-2}{3}$

Example 5 Consider the equation

$$2x^2 - 3x + 2 = 0$$

a. Evaluate the discriminant.
b. How many (real) roots are there?

Solution **a.** $A = 2$, $B = -3$, $C = 2$

$$B^2 - 4AC = (-3)^2 - 4 \cdot 2 \cdot 2$$
$$= 9 - 16$$
$$= -7$$

b. There are no (real) roots because the discriminant is negative.

In later courses you will learn to apply the quadratic formula in the case where the discriminant is negative. You will also verify that the formula is correct in all cases.

CLASS EXERCISES **a.** *Evaluate the discriminant.* **b.** *How many (real) roots are there?*
c. *If there are (real) roots, use the quadratic formula to solve the equation.*

3. $t^2 + 10t + 3 = 0$ **4.** $y^2 + 3y + 10 = 0$ **5.** $z^2 + 4 = 4z$

SOLUTIONS TO CLASS EXERCISES

1. $A = 1, B = 4, C = 1$

$$x = \frac{-4 \pm \sqrt{16-4}}{2} = \frac{-4 \pm \sqrt{12}}{2} = \frac{-4 \pm 2\sqrt{3}}{2} = -2 \pm \sqrt{3}$$

check

for $-2 + \sqrt{3}$:

$$(-2+\sqrt{3})^2 + 4(-2+\sqrt{3}) + 1 \stackrel{?}{=} 0$$
$$(4 - 4\sqrt{3} + 3) + (-8 + 4\sqrt{3}) + 1 \stackrel{?}{=} 0$$
$$\overbrace{(4 + 3 - 8 + 1)}^{0} + \overbrace{(-4\sqrt{3} + 4\sqrt{3})}^{0} \stackrel{?}{=} 0$$
$$0 \stackrel{\checkmark}{=} 0$$

for $-2 - \sqrt{3}$:

$$(-2-\sqrt{3})^2 + 4(-2-\sqrt{3}) + 1 \stackrel{?}{=} 0$$
$$(4 + 4\sqrt{3} + 3) + (-8 - 4\sqrt{3}) + 1 \stackrel{?}{=} 0$$
$$\overbrace{(4 + 3 - 8 + 1)}^{0} + \overbrace{(4\sqrt{3} - 4\sqrt{3})}^{0} \stackrel{?}{=} 0$$
$$0 \stackrel{\checkmark}{=} 0$$

2. a.
$$(y+3)^2 = 3$$
$$y^2 + 6y + 9 = 3$$
$$y^2 + 6y + 6 = 0$$
$$A = 1, B = C = 6$$

$$y = \frac{-6 \pm \sqrt{36-24}}{2} = \frac{-6 \pm \sqrt{12}}{2} = \frac{-6 \pm 2\sqrt{3}}{2} = -3 \pm \sqrt{3}$$

b.
$$(y+3)^2 = 3$$
$$y + 3 = \pm\sqrt{3}$$
$$y = -3 \pm \sqrt{3}$$

3. a. $B^2 - 4AC = 10^2 - 4 \cdot 1 \cdot 3 = 88$ **b.** 2

c. $$x = \frac{-10 \pm \sqrt{88}}{2} = \frac{-10 \pm 2\sqrt{22}}{2} = -5 \pm \sqrt{22}$$

4. a. $B^2 - 4AC = 3^2 - 4 \cdot 1 \cdot 10 = -31$ **b.** 0

5. a. $z^2 - 4z + 4 = 0$

$B^2 - 4AC = (-4)^2 - 4 \cdot 1 \cdot 4 = 0$

b. 1 **c.** $x = \dfrac{-(-4) \pm \sqrt{0}}{2} = 2$

HOME EXERCISES

In Exercises 1–16: **a.** *Evaluate the discriminant.* **b.** *How many (real) roots are there?* **c.** *If there are (real) roots, use the quadratic formula to solve the equation.*

1. $x^2 + 4x + 2 = 0$ **2.** $x^2 + 6x + 9 = 0$ **3.** $2x^2 + 3 = 5x$ **4.** $5x^2 + 10x + 8 = 0$
5. $t^2 - 3t = 1$ **6.** $t^2 - 8t + 15 = 0$ **7.** $2y^2 - 6y + 3 = 0$ **8.** $z^2 + 20 = 9z$
9. $4x^2 - 4x + 1 = 0$ **10.** $2u^2 + 7u + 9 = 0$ **11.** $4z^2 - 20z + 5 = 0$ **12.** $u^2 + u + 1 = 0$
13. $y^2 + y - 1 = 0$ **14.** $x^2 - 2x + 1 = 0$ **15.** $3z^2 + z + 3 = 0$ **16.** $y^2 - 5y + 1 = 0$

In Exercises 17–28, apply the quadratic formula to solve the equation. In Exercises 17–20, check the roots.

17. $x^2 + 5x - 2 = 0$ **18.** $y^2 + 9y + 1 = 0$ **19.** $3t^2 - 5t + 1 = 0$ **20.** $u^2 + 12u + 36 = 0$
21. $x^2 + 10x + 5 = 0$ **22.** $2z^2 - 9z + 6 = 0$ **23.** $u^2 + 3u - 1 = 0$ **24.** $u^2 + 3u + 1 = 0$
25. $2t^2 + 5t + 3 = 0$ **26.** $3x^2 - 7x + 4 = 0$ **27.** $3y^2 = 10y - 3$ **28.** $(y - 2)^2 = y + 3$

In Exercises 29–36: **a.** *Apply the quadratic formula to solve the equation.* **b.** *Solve the equation by some other method.*

29. $x^2 - 3x + 2 = 0$ **30.** $x^2 + 8x + 16 = 0$ **31.** $y^2 + y - 20 = 0$ **32.** $z^2 - 81 = 0$
33. $(t + 1)^2 = 5$ **34.** $2x^2 + 3x + 1 = 0$ **35.** $6y^2 - y = 1$ **36.** $10u^2 + 11u + 2 = 0$

9.4 WORD PROBLEMS

Stated problems can often be translated into quadratic equations. Your first task is to translate the problem into algebraic symbols. You then apply the techniques of this chapter to solve the corresponding quadratic equation.

NUMBER PROBLEMS

Example 1 The sum of two numbers is 16 and the product is 48. Find these numbers.

Solution Let x be one of these numbers. Then, the other is $16 - x$ because

$$\underbrace{\text{the sum of (the) two numbers}}_{x + (16 - x)} \underset{=}{\text{is}} \underset{16}{16}$$

The quadratic equation is obtained from the second piece of information:

$$\underbrace{\text{The product (of the two numbers)}}_{x(16 - x)} \underset{=}{\text{is}} \underset{48}{48}$$

Multiply, as indicated, to obtain the quadratic equation

$$16x - x^2 = 48 \quad \text{or}$$

$$x^2 - 16x + 48 = 0 \qquad \textit{Factor the left side.}$$

$$(x - 4)(x - 12) = 0$$

$x - 4 = 0$	$x - 12 = 0$
$x = 4$	$x = 12$
$16 - x = 12$	$16 - x = 4$

In either case, the two numbers are 4 and 12.

CLASS EXERCISES

1. The sum of two numbers is -7 and the product is -60. Find these numbers.
2. The square of a certain *positive* number is six more than four times the number. Find this number.

MOTION PROBLEMS There are problems that involve an object traveling at a *variable rate* and that can be expressed in terms of quadratic equations. Because the rate is *not constant*, the distance formula $d = r \cdot t$ no longer applies.

Example 2 An object falls from a tower 320 feet high. As it is falling, its height is given by the formula

$$h = 320 - 16t^2 \text{ (feet)}$$

where t, the time falling, is measured in seconds. How long does it take for the object to hit the ground?

Solution

$$h = 320 - 16t^2$$

Because h represents height, the object hits the ground when $h = 0$ (see Figure 9.1). Thus, replace h by 0 and solve for t.

$$\begin{aligned} 0 &= 320 - 16t^2 \\ 16t^2 &= 320 \\ t^2 &= 20 \\ t &= \pm\sqrt{20} \\ &= \pm\sqrt{4 \cdot 5} \\ &= \pm\sqrt{4}\sqrt{5} \\ &= \pm 2\sqrt{5} \end{aligned}$$

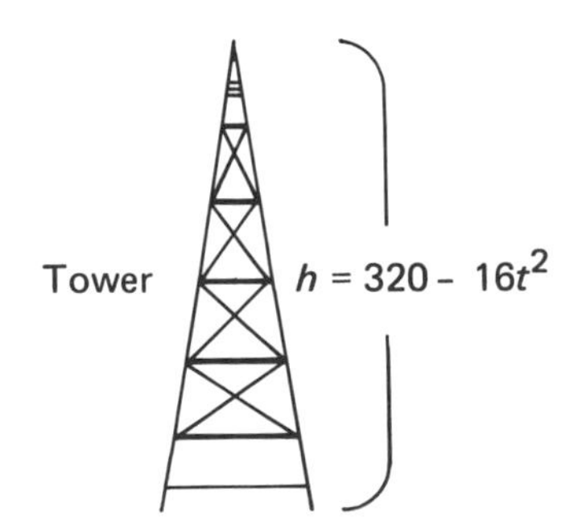

Fig. 9.1

Because of the physical nature of the problem, t must be positive. Therefore, $t = 2\sqrt{5}$. Note that

$$\sqrt{5} \approx 2.23$$

$$2\sqrt{5} \approx 4.46$$

Thus, it takes slightly less than $4\frac{1}{2}$ seconds for the object to hit the ground.

CLASS EXERCISES

3. An object falls from a tower 400 feet high. As it falls, its height is given by the formula

$$h = 400 - 16t^2 \text{ (feet)}$$

where t is the time falling (in seconds). How long does it take to hit the ground?

PROFIT PROBLEMS

The profit derived from manufacturing certain items can be described by means of a quadratic equation. Because of limited plant facilities, the profit will rise to a **maximum** and then taper off. This profit is often described by a curve called a **parabola** (see Figure 9.2).

Example 3 The annual profit P (in thousands of dollars) derived from manufacturing industrial machines is given by the formula

$$P = 100 - x^2 + 20x$$

How many machines must be made in order to derive a profit of \$200,000?

Solution Let $P = 200$ in the equation

$$P = 100 - x^2 + 20x$$

(Note that P is measured in thousands of dollars.) Therefore,

$$200 = 100 - x^2 + 20x$$

$$x^2 - 20x + 100 = 0$$

$$(x - 10)^2 = 0$$

$$x - 10 = 0$$

$$x = 10$$

Thus, 10 machines must be produced.

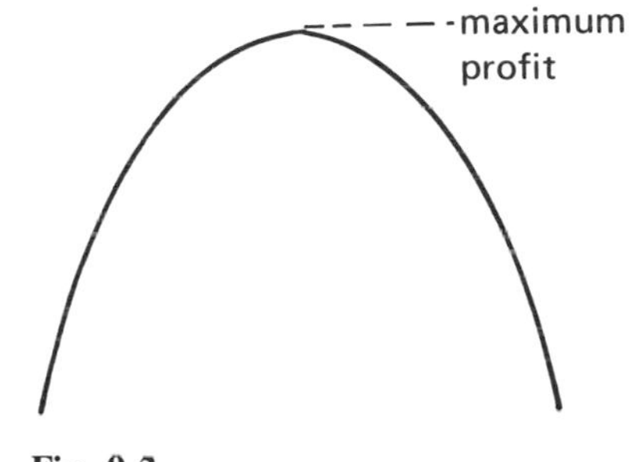

Fig. 9.2

CLASS EXERCISES

4. In Example 3 of the text, how many machines must be made in order to derive a profit of \$175,000?

SOLUTIONS TO CLASS EXERCISES

1. Let x be one of these numbers. Then the other is

$-7 - x$ because $x + (-7 - x) = -7$

The product of the two numbers is -60

$$x(-7 - x) = -60$$

$$-7x - x^2 = -60$$

Multiply both sides by -1.

$$\begin{aligned} 7x + x^2 &= 60 \\ x^2 + 7x - 60 &= 0 \\ (x + 12)(x - 5) &= 0 \end{aligned}$$

$x + 12 = 0$	$x - 5 = 0$
$x = -12$	$x = 5$
$-7 - x = 5$	$-7 - x = -12$

In either case, the two numbers are -12 and 5.

2. Let x be this *positive* number.

The square of a certain positive number is six more than four times the number.

$$\begin{aligned} x^2 &= 6 + 4x \\ x^2 - 4x - 6 &= 0 \\ x &= \frac{4 \pm \sqrt{16 - 4 \cdot 1(-6)}}{2} \\ &= \frac{4 \pm \sqrt{40}}{2} \\ &= 2 \pm \sqrt{10} \end{aligned}$$

The number is $2 + \sqrt{10}$. (Note that $2 - \sqrt{10} < 0$ because $\sqrt{10} > 3$ and thus, $2 - \sqrt{10} < 2 - 3 < 0$.)

3. Let $h = 0$ in the given equation because h measures the distance above the ground.

$$\begin{aligned} 0 &= 400 - 16t^2 \\ 16t^2 &= 400 \\ t^2 &= 25 \\ t &= \pm 5 \end{aligned}$$

(Note that t, the number of seconds, must be positive.) Thus, it takes 5 seconds for the object to hit the ground.

4. $P = 100 - x^2 + 20x$

Let $P = 175$.

$$\begin{aligned} 175 &= 100 - x^2 + 20x \\ x^2 - 20x + 75 &= 0 \\ (x - 15)(x - 5) &= 0 \end{aligned}$$

$x - 15 = 0$	$x - 5 = 0$
$x = 15$	$x = 5$

If either 15 machines or 5 machines are made, the profit is \$175,000.

HOME EXERCISES

1. The sum of two numbers is nine and the product is 20. Find these numbers.
2. The sum of two numbers is 20 and the product is 96. Find these numbers.
3. The sum of two numbers is three and the product is -70. Find these numbers.
4. The sum of two numbers is -14 and the product is 48. Find these numbers.
5. The sum of two numbers is 0 and the product is -36. Find these numbers.
6. The product of two consecutive negative integers is 72. Find these integers.

7. Four times a certain positive number is five less than its square. Find this number.
8. The square of a certain negative number is three more than twice the number. Find this number.
9. An object falls from a window 576 feet high. As it is falling, its height is given by the formula $h = 576 - 16t^2$ (feet), where t is the time falling (in seconds). How long does it take to hit the ground?
10. A flower pot falls from a window 192 feet high. As it falls, its height is given by the formula $h = 192 - 16t^2$ (feet), where t is the time falling (in seconds). How long does it take to hit the ground?
11. A baseball is thrown straight up from ground level. Its height, h feet, after t seconds is given by the formula $h = 128t - 16t^2$.
    a. How long does it take to rise 112 feet? **b.** At what time, t, is it 112 feet above the ground on its way down?
12. The monthly profit P (in *thousands* of dollars) derived from manufacturing prefabricated houses is given by the formula $P = -x^2 + 16x - 14$, where x is the number of houses constructed. How many houses must be constructed in order to make a monthly profit of \$50,000?
13. In Exercise 12, how many houses must be constructed in order to make a monthly profit of \$46,000?
14. In Exercise 12, because of limited plant facilities, it is unprofitable to build too many houses per month. *To the nearest integer*, what is the largest number of houses the firm can construct per month before it loses money? (*Hint*: Replace P by 0 in the equation of Exercise 12 and apply the quadratic formula.)
15. The cost C (in dollars) of building a bookcase is given by the formula $C = \frac{l^2}{10} - 3l$, where l is the *total* length of the shelves (in yards). Suppose there are to be five equally long shelves. How long will *each* shelf be if the cost is to be \$100?

Let's Review Chapter 9

Solve the equations in Exercises 1–3 by factoring the left side. Check the roots in Exercise 1.

1. $x^2 + 7x + 12 = 0$ **2.** $x^2 - 6x + 9 = 0$ **3.** $3x^2 + 5x - 2 = 0$

Solve the equations in Exercises 4–7. Use the method:

$$\text{If } x^2 = a, \text{ where } a > 0, \text{ then } x = \pm\sqrt{a}$$

Check Exercise 6.

4. $x^2 - 17 = 0$ **5.** $x^2 - 80 = 0$ **6.** $(2u + 1)^2 = 9$

7. $(3y - 5)^2 = 2$

8. **a.** Evaluate the discriminant of the equation $x^2 + 3x + 5 = 0$.
b. Without solving, indicate how many (real) roots there are.

In Exercises 9–11, apply the quadratic formula to solve the equation. Check the root(s) in Exercise 9.

9. $x^2 + 5x + 5 = 0$ **10.** $4x^2 + 4x + 1 = 0$ **11.** $y^2 + 6y - 1 = 0$

12. Find two integers whose sum is four and whose product is -45.

13. A stone falls from the top of a monument 288 feet high. As the stone falls, its height is given by the formula $h = 288 - 16t^2$ (feet), where t is the time falling (in seconds). How long does it take for the stone to hit the ground?

10 LENGTH, AREA, AND VOLUME

DIAGNOSTIC TEST

Perhaps you are already familiar with some of the material in Chapter 10. This test will indicate which sections in Chapter 10 you need to study. The question number refers to the corresponding section in Chapter 10. If you answer *all* parts of the question correctly, you may omit the section. *But if any part of your answer is wrong, you should study the section*. When you have completed the test, turn to page A29 for the answers.

10.1 Length: American and Metric Systems

Convert from one unit to another, as indicated.

a. 7 feet into inches b. 6 centimeters into meters

c. 40 kilometers into miles (Express your result to the nearest mile.)

10.2 Rectangles

a. Find the perimeter of a rectangle of length 5 centimeters and width 2 centimeters.
b. The perimeter of a square is 100 inches. Find the length of a side.
c. Find the area of a rectangle with base 7 inches and height 5 inches.
d. The perimeter of a rectangle is 24 meters and the area is 35 square meters. Find the dimensions of the rectangle.

10.3 Triangles, Parallelograms, Trapezoids

a. Find the area of a triangle with base 6 centimeters and height 5 centimeters.
b. Find the base of a parallelogram with area 4 square feet and height 1 foot.
c. Find the area of a trapezoid with bases 10 centimeters and 30 centimeters and height 15 centimeters.

10.4 Circles

Express the following in terms of π.

a. the area of a circle with diameter 6 centimeters
b. the area of a circle with circumference 12π inches
c. the circumference of a circle with area 49π square meters

10.5 Volume and Surface Area

a. Suppose a rectangular box has length 2 feet, width 1 foot, and height 3 feet. Find its volume and surface area.

In parts b and c, express your results in terms of π.

b. Suppose the radius of a sphere is 5 inches. Find its volume and surface area.
c. Suppose a cylinder has base radius 3 centimeters and height 5 centimeters. Find its volume and lateral surface area.
d. Find the base radius of a cone that has height 9 feet and volume 300π cubic feet.

10.1 LENGTH: AMERICAN AND METRIC SYSTEMS

There are two common systems of measurements: the *American system* and the *metric system*.

AMERICAN SYSTEM

To measure *length* in the American system, we begin with the basic unit of 1 *inch* (*in.*). Thus, a 5-inch wire is 5 times as long as a 1-inch wire. (See Figure 10.1.)

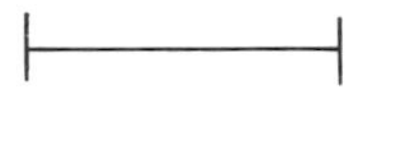

Fig. 10.1(a) 1 inch.

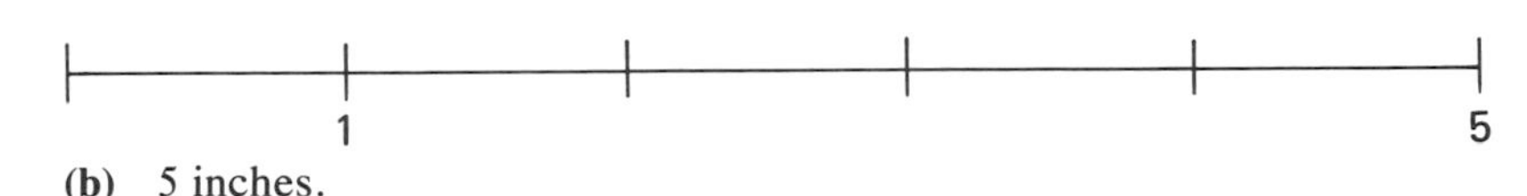

(b) 5 inches.

To describe larger lengths, use *feet* (*ft.*), *yards* (*yd.*), and *miles* (*mi.*). To change units, use the following:

1 foot = 12 inches
1 yard = 3 feet = 36 inches
1 mile = 1760 yards = 5280 feet

To convert *from* a *larger* unit *to* a *smaller* unit, *multiply*:

feet into inches, multiply by 12
yards into inches, multiply by 36
miles into feet, multiply by 5280

Then, change units, as indicated.

Example 1 **a.** Convert 5 feet into inches. **b.** Convert 3.2 yards into feet.

Solution **a.** 5 ft. = 5 × 12 in. = 60 in. **b.** 3.2 yd. = (3.2) (3) ft. = 9.6 ft.

To convert *from* a *smaller* unit *to* a *larger* unit, *divide*:

inches into feet, divide by 12
feet into yards, divide by 3
yards into miles, divide by 1760

Then, change units, as indicated.

Example 2 **a.** Convert 54 inches into feet. **b.** Convert 144 inches into yards.

Solution **a.** $54 \text{ in.} = \frac{\cancel{54}^{9}}{\cancel{12}_{2}} \text{ ft.} = 4\frac{1}{2} \text{ ft.}$ **b.** $144 \text{ in.} = \frac{\cancel{144}^{\cancel{12}^{4}}}{\cancel{36}_{\cancel{3}_{1}}} \text{ yd.} = 4 \text{ yd.}$

CLASS EXERCISES *Convert from one unit to another.*

1. 7 feet into inches
2. 1.5 miles into feet
3. 72 inches into feet
4. 72 inches into yards

METRIC SYSTEM

Powers of 10 play a central role in the metric system. Because 10 is the base of our number system, it is relatively easy to convert units within this system. Most countries use the metric system, and the United States is moving in this direction. First, you will consider various units of length in the metric system. Then, you will compare the American and metric systems.

The basic unit of length in the metric system is the *meter* (*m.*). *1 meter is approximately 39.37 inches*, or a little more than a yard (36 inches). (See Figure 10.2 .)

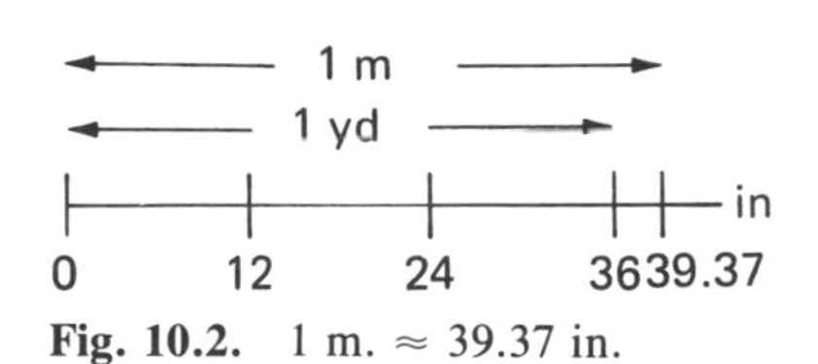

Fig. 10.2. 1 m. ≈ 39.37 in.

1 *kilo*meter (*km.*) = 1000 meters
1 *hecto*meter (*hm.*) = 100 meters
1 *deka*meter (*dkm.*) = 10 meters

There are also units for fractional parts of a meter.

1 *deci*meter (*dm.*) = $\frac{1}{10}$ meter (or .1 meter)
1 *centi*meter (*cm.*) = $\frac{1}{100}$ meter (or .01 meter)
1 *milli*meter (*mm.*) = $\frac{1}{1000}$ meter (or .001 meter)

To convert from one metric unit to another, multiply or divide by the appropriate power of 10. If you convert

from a *larger to* a *smaller* unit, *multiply*
from a *smaller to* a *larger* unit, *divide*

Then, change units, as called for. For example, to change

kilometers into meters, multiply by 1000
meters into centimeters, multiply by 100
millimeters into meters, divide by 1000
meters into hectometers, divide by 100

When working with decimals, you can remember how to move the decimal point by thinking of the following chart.

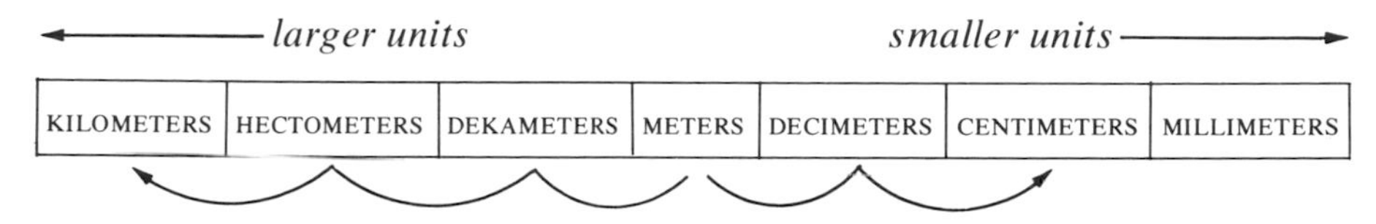

Note that KILOMETERS lies three spaces to the *left* of METERS. To convert from meters to kilometers, move the decimal point three digits to the *left*.

CENTIMETERS lies two spaces to the *right* of METERS. To convert from meters to centimeters, move the decimal point two digits to the *right*.

Example 3 **a.** Convert 5.3 kilometers into meters.
b. Convert 425 decimeters to meters.
c. Convert 7.5 meters into centimeters.

Solution **a.** To change from kilometers to meters, multiply by 1000. Here, you must insert two 0's before moving the decimal point three digits to the *right*. (Think of the chart.)

5.300 to 53**00**.

Therefore, 5.3 km. = 53**00** m.

b. To convert decimeters to meters, divide by 10. Thus, move the decimal point one digit to the *left*.

425. to 42.5

Therefore, 425 dm. = 42.5 m.

c. To convert from meters to centimeters, multiply by 100. Here, you must add a zero before moving the decimal point two digits to the *right*.

7.50 to 75**0**.

Therefore, 7.5 m. = 75**0** cm.

CLASS EXERCISES *Convert from one unit to another.*

5. 3 kilometers into meters **6.** 958 meters into decimeters

7. 5.2 decimeters into meters

AMERICAN AND METRIC SYSTEMS

Recall that 1 meter is approximately 39.37 inches.

1 m. ≈ 39.37 in.

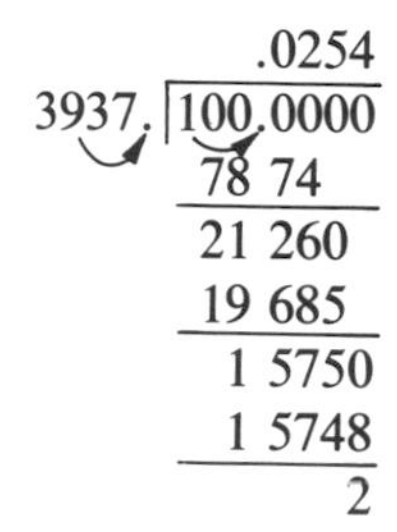

To convert inches to meters, divide by 39.37 .

1 inch ≈ .0254 meter (or 2.54 centimeters)

The comparison is usually with centimeters. Thus,

1 in. ≈ 2.54 cm.

Example 4 **a.** Convert 12 meters into inches. Express your result to the nearest inch.
b. Convert 60 inches into centimeters. Express your result to the nearest centimeter.

Solution **a.** $39.37 \times 12 = 472.44$ To the nearest inch, 12 m. ≈ 472 in.

b. $2.54 \times 60 = 152.40$ To the nearest centimeter, 60 in. ≈ 152 cm.

There is also a useful conversion between kilometers and miles.

1 km. ≈ .62 mi.
1 mi. ≈ 1.61 km.

Example 5 If you are 58 kilometers from Mexico City, what is the distance to Mexico City in miles?

Solution

$$\begin{array}{r} 58 \\ \times\ .62 \\ \hline 1\ 16 \\ 34\ 8 \\ \hline 35.96 \end{array}$$

The distance to Mexico City is about 36 miles.

CLASS EXERCISES *Convert from one unit to another. Express your result to the nearest unit.*

8. 9.2 inches into centimeters **9.** 38 kilometers into miles
10. 20 miles into kilometers

SOLUTIONS TO CLASS EXERCISES

1. 7 ft. = 7 × 12 in. = 84 in. **2.** 1.5 mi. = (1.5) 5280 ft. = 7920 ft. **3.** 72 in. = $\frac{72}{12}$ ft. = 6 ft.

4. 72 in. = $\frac{72}{36}$ yd. = 2 yd. **5.** 3 km. = 3 × 1000 m. = 3000 m. **6.** 958 m. = 958 × 10 dm. = 9580 dm.

7. 5.2 dm. = $\frac{5.2}{10}$ m. = .52 m. **8.** 9.2 in. ≈ 9.2 × 2.54 cm. = 23.368 cm.
To the nearest centimeter, 9.2 in. ≈ 23 cm.

9. 38 km. ≈ 38 × .62 mi. = 23.56 mi.
To the nearest mile, 38 km. ≈ 24 mi.

10. 20 mi. ≈ 20 × 1.6 km. = 32.2 km.
To the nearest kilometer, 20 mi. ≈ 32 km.

HOME EXERCISES

In Exercises 1–24, convert from one unit to another.

1. 96 inches into feet **2.** 108 inches into yards **3.** 18 feet into yards
4. 10,560 feet into miles **5.** 12,320 yards into miles **6.** 100 inches into feet
7. 32 inches into yards **8.** 25 feet into yards **9.** 13.5 feet into yards
10. 6 feet into inches **11.** 3 miles into feet **12.** 3 miles into yards
13. $9\frac{1}{3}$ yards into feet **14.** $2\frac{1}{2}$ feet into inches **15.** $\frac{1}{10}$ mile into feet
16. 1 mile into inches **17.** 5 kilometers into meters **18.** 7 decimeters into meters
19. 405 centimeters into meters **20.** 3826 millimeters into meters **21.** 2440 meters into kilometers
22. 8.5 meters into decimeters **23.** 3.58 meters into centimeters **24.** 1.2 meters into millimeters

In Exercises 25–38, convert from one unit to another, where indicated. Express your result to the nearest unit.

25. 50 meters into inches **26.** 23 meters into inches **27.** 10 inches into centimeters
28. 1 foot into centimeters **29.** 50 kilometers into miles **30.** 12 kilometers into miles
31. 30 miles into kilometers **32.** 54 miles into kilometers
33. The distance between New York City and Boston is 210 miles. Express this distance in kilometers. *Use 1 mi. ≈ 1.609 km.*
34. The distance between Paris and Brussels is 295 kilometers. Express this distance in miles. *Use 1 km. ≈ .621 mi.*
35. A football field is 100 yards long. **a.** How many inches long is it? **b.** How many centimeters long is it?
36. Which is the longest? **a.** 50 yards **b.** 140 feet **c.** 45 meters
37. A car gets 31 miles to a gallon. How many gallons does it need to travel 100 kilometers?
38. The center for the Yugoslavian Olympic basketball team is 203.2 centimeters tall. What is his height in feet and inches?

10.2 RECTANGLES

PERIMETER

DEFINITION

> A **polygon** is a closed figure on a plane, composed of three or more line segments. A three-sided polygon is called a **triangle** and a four-sided polygon is called a **quadrilateral**.

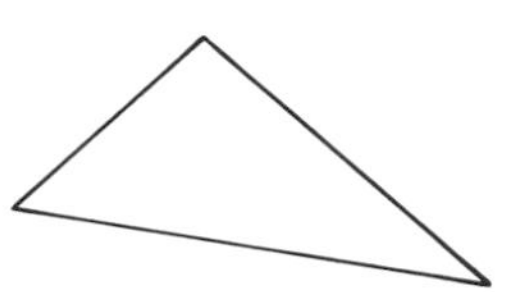

Fig. 10.3. A triangle.

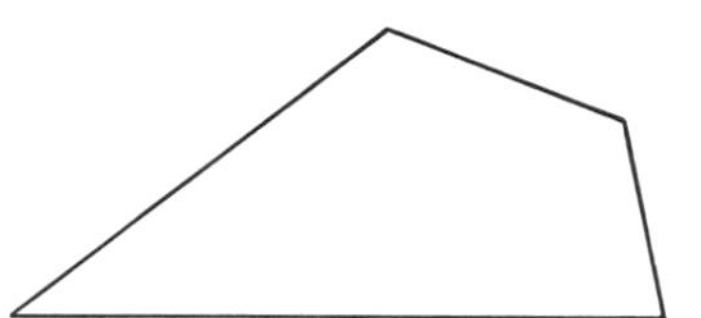

Fig. 10.4. A quadrilateral.

DEFINITION

> The **perimeter** of a polygon is the sum of the lengths of its sides.

The perimeter of a polygon can be thought of as the distance around it.

Example 1 Find the perimeter of the polygon in Figure 10.5 .

Solution In each case, the expressions to be added involve the same units. Add in columns, as indicated.

2 cm.
3 cm.
3 cm.
2 cm.
3 cm.
3 cm.
______
16 cm.

2 cm, 3 cm, 3 cm, 3 cm, 3 cm, 2 cm

Fig. 10.5

CLASS EXERCISES *Find the perimeter of each polygon.*

1. Fig. 10.6 **2.** Fig. 10.7

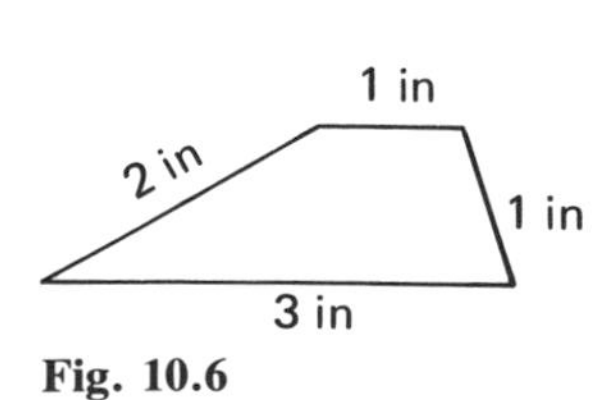

Fig. 10.6

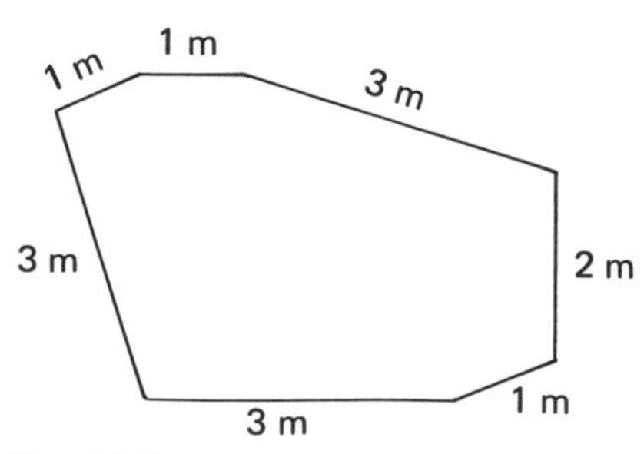

Fig. 10.7

RECTANGLES **Angles** are formed when lines intersect. The **vertex** (plural: *vertices*) of an angle is the point of intersection of the lines. The vertices of a polygon are the intersections of the line segments of the polygon.

An angle is measured in **degrees**; normally, an angle measures between 0° (0 degrees) and 360°. In Figure 10.8, several angles and their measurements are shown. The 90° angle in Figure 10.8(b) is called a **right angle** and the intersecting lines are said to be **perpendicular.** Write $\ell_1 \perp \ell_2$ when lines ℓ_1 and ℓ_2 are perpendicular.

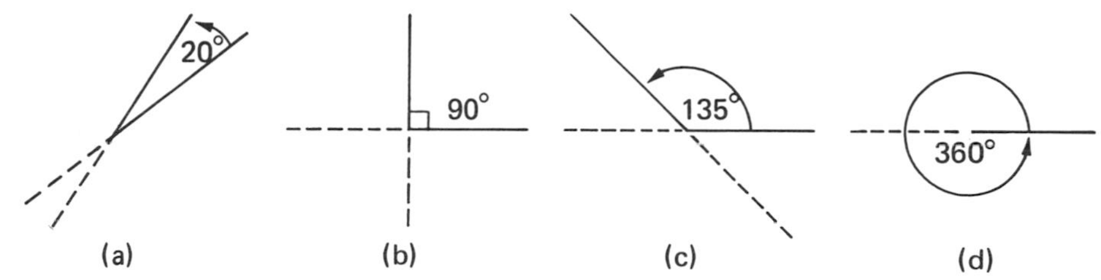

Fig. 10.8 **(b)** Right angle. **(d)** A 360° angle. The two "sides" of the angle coincide.

DEFINITION

> A **rectangle** is a quadrilateral, or four-sided polygon, in which each (interior) angle is a right angle.

In Figure 10.9, different quadrilaterals are shown. Only part **(a)** depicts a rectangle. Note that *the opposite sides of a rectangle are of equal length.*

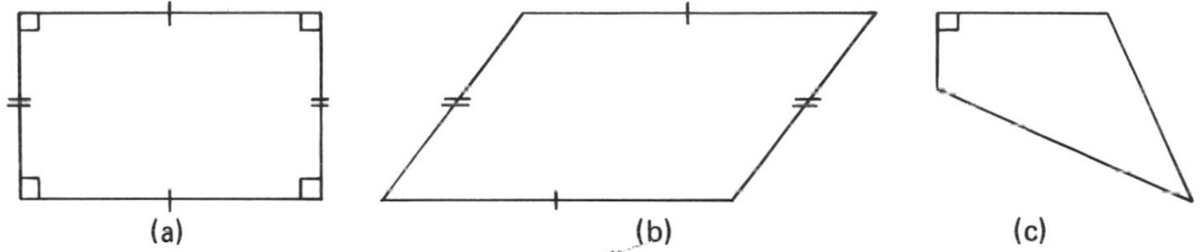

Fig. 10.9(a) This quadrilateral is a rectangle. Opposite sides are of equal length and are marked with the same number of bars. Note that each angle is a right angle. **(b)** This quadrilateral (which is called a *parallelogram*) is not a rectangle, even though opposite sides are of equal length. The angles are not right angles. **(c)** This quadrilateral is not a rectangle. Only one angle is a right angle.

The perimeter of a rectangle is twice the sum of its length and width.

> *Perimeter of rectangle* = 2 (*length* + *width*)

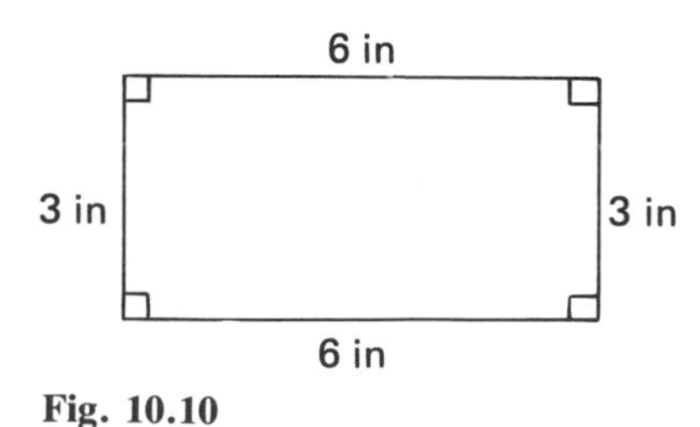

Fig. 10.10

Thus, in Figure 10.10,

```
   length 6 in.
 + width  3 in.
 -------------
          9 in.
        × 2
 -------------
         18 in.
```

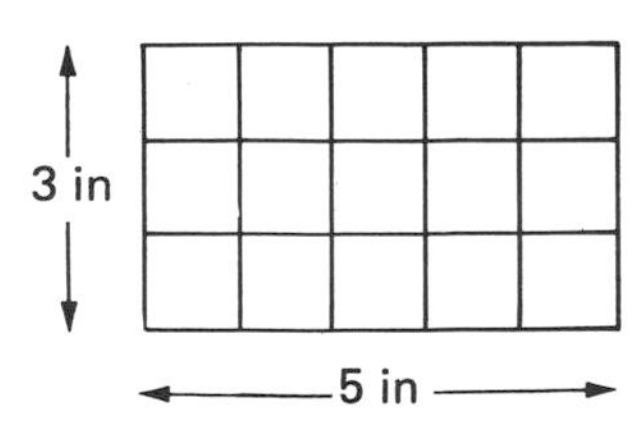

Fig. 10.12. The area of the rectangle is 15 square inches.

When you multiply 9 *inches* by (the number) 2, the product is (9 × 2) *inches*. *The area of the rectangle*† *is the product of length and width.*

$$\text{Area of Rectangle} = \text{length} \times \text{width}$$

In Figure 10.12, the length of the rectangle is 5 inches and the width is 3 inches. The area is

$$\underbrace{5 \text{ inches}}_{\text{length}} \times \underbrace{3 \text{ inches}}_{\text{width}} = 15 \text{ inches} \times \text{inches}, \quad \text{or } 15 \textit{ square inches}$$

Note that the unit of area is *square inches*.

$$\text{inches} \times \text{inches} = \text{square inches}$$

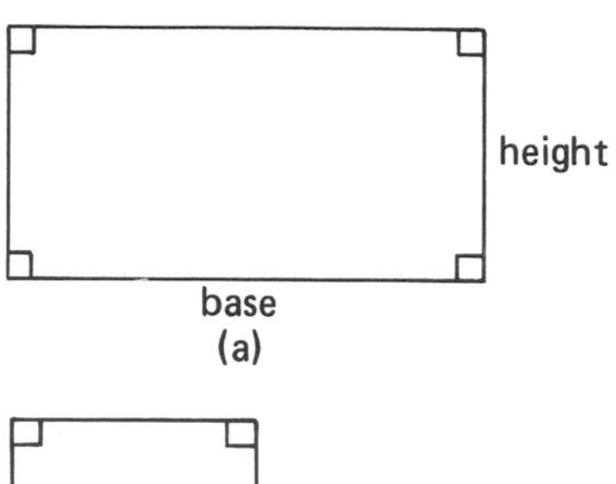

Write

in.^{2} for *square inches*

Similarly, write *ft.*^{2} for *square feet*
yd.^{2} for *square yards*
mi.^{2} for *square miles*

and in the metric system, *cm.*^{2} for *square centimeters*
m.^{2} for *square meters*
km.^{2} for *square kilometers*, and so on.

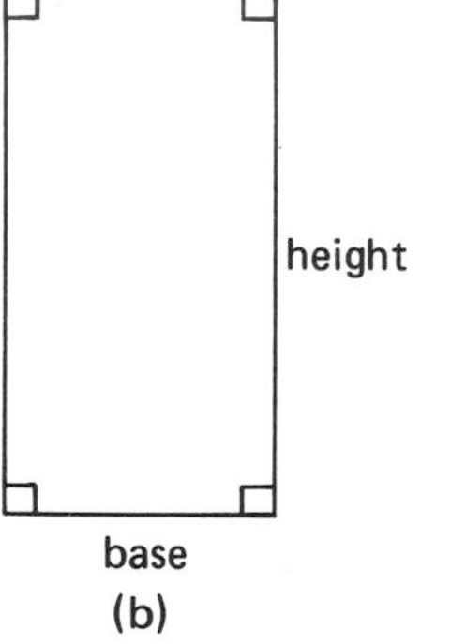

Fig. 10.13(a) base = length; height = width. (b) base = width; height = length.

When rectangles are depicted as in Figure 10.13, it is customary to call the horizontal side of each the **base** and the vertical side the **height**, regardless of which side is longer.‡ Thus,

$$\text{Area of rectangle} = \text{base} \times \text{height}$$

Let $A = area$, $b = base$, $h = height$

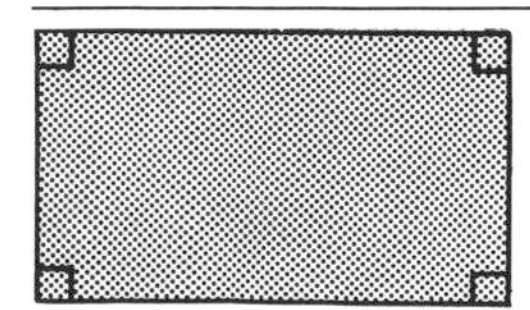

Fig. 10.11. The rectangle is outlined. The rectangular region is shaded.

†In more precise language, we would speak of the area of a *rectangular region*, or the area enclosed by a rectangle. (See Figure 10.11 .) For brevity, we will speak of the area of a rectangle. Similarly, we will refer to the area of a triangle (rather than of a triangular region), the area of a circle (rather than of a circular region), and so on.

‡As is the usual custom, we will use the terms "base" and "height" both for the sides of a rectangle as well as for the lengths of these sides. Similarly, we will use terms such as "radius," "diameter," and so on, both for line segments and for the lengths of these line segments.

In symbols, *the formula for area is*

$$\boxed{A = b \cdot h}$$

Example 2 Find the area of the rectangle with base 20 centimeters and height 10 centimeters.

(See Figure 10.14 .)

Solution

$$\begin{aligned} A &= b \cdot h \\ &= 20 \text{ cm.} \times 10 \text{ cm.} \\ &= 200 \text{ cm.}^2 \end{aligned}$$

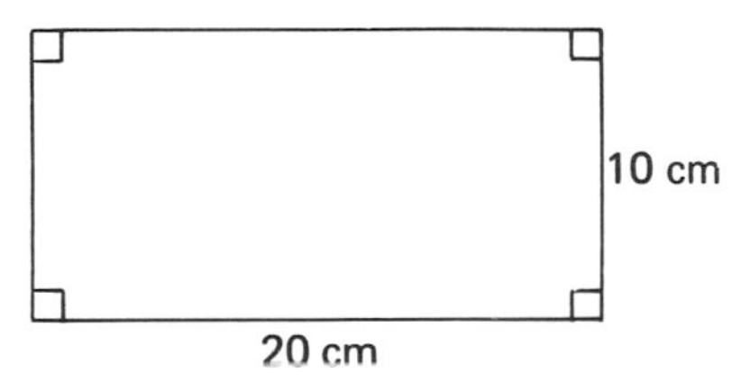

Fig. 10.14

Example 3 The perimeter of a rectangle is 36 inches. The area is 80 square inches. Find the dimensions.

Solution Let ℓ and w be the dimensions of the rectangle (in inches). Then,

$$2\ell + 2w = 36$$

and

$$\ell w = 80$$

From the first equation,

$$\begin{aligned} \ell + w &= 18 \\ w &= 18 - \ell \end{aligned}$$

The second equation, $\ell w = 80$, becomes

$$\begin{aligned} \ell\,(18 - \ell) &= 80 \\ 18\ell - \ell^2 &= 80 \\ \ell^2 - 18\ell + 80 &= 0 \\ (\ell - 10)\,(\ell - 8) &= 0 \end{aligned}$$

For the product to be 0, one of the factors must be 0.

$$\begin{array}{rcl|rcl} \ell - 10 &=& 0 & \ell - 8 &=& 0 \\ \ell &=& 10 & \ell &=& 8 \\ w &=& 18 - \ell & w &=& 18 - \ell \\ w &=& 8 & w &=& 10 \end{array}$$

The rectangle is 8 inches by 10 inches.

CLASS EXERCISES

3. Find the area of a rectangle with base 8 inches and height 3 inches.
4. Find the height of a rectangle with base 4 inches and area 20 square inches.
5. The perimeter of a rectangle is 64 feet and the area is 240 square feet. Find the dimensions of the rectangle.

SQUARES

DEFINITION

> A **square** is a rectangle in which each side has the same length.

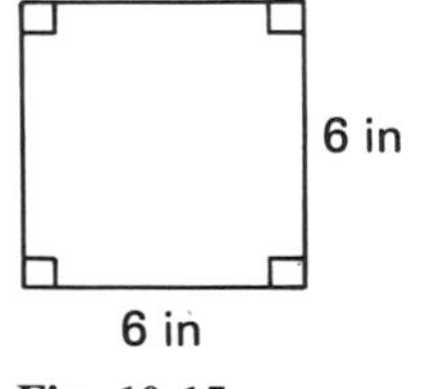

Fig. 10.15

In Figure 10.15, the length of each side of the square is 6 inches. The perimeter is 24 inches.

$$\begin{array}{r} 6 \text{ in.} \\ 6 \text{ in.} \\ 6 \text{ in.} \\ \underline{6 \text{ in.}} \\ 24 \text{ in.} \end{array}$$

The length and width of a square are equal. Therefore,

$$\text{Perimeter} = 2\,(\text{length} + \overset{\text{length}}{\cancel{\text{width}}}) = 2\,(2 \times \text{length}) = 4 \times \text{length}$$

Also,

$$\text{Area} = \text{length} \times \overset{\text{length}}{\cancel{\text{width}}} = \text{length}^2$$

16 cm

16 cm

Fig. 10.16

Thus, *let s be the length of a side of a square, let P be the perimeter, and let A be the area. Then,*

$$\boxed{P = 4s \qquad \text{and} \qquad A = s^2}$$

Example 4 Find **a.** the perimeter and **b.** the area of a square of side length 16 centimeters.

Solution **a.** Perimeter $= 4 \times 16$ cm.
$= 64$ cm.

b. $A = (16 \text{ cm.})^2 = 256 \text{ cm.}^2$
(See Figure 10.16.)

CLASS EXERCISES **6.** Find **a.** the perimeter, and **b.** the area, of a square of side length 5 inches.

SOLUTIONS TO CLASS EXERCISES

1. 7 in. **2.** 14 m. **3.** A = 8 in. × 3 in. = 24 in.^{2}

4. Area = base × height
20 in.^{2} = 4 in. × h

$$\frac{20 \text{ in.}^2}{4 \text{ in.}} = h$$

To divide expressions with units:
1. Divide the numbers.
2. Divide the units.

$$\frac{\overset{5 \text{ in.}}{\cancel{20} \cancel{\text{in.}^2}}}{\underset{1}{\cancel{4} \cancel{\text{in.}}}} = h$$

5 in. = h

5. Let ℓ and w be the dimensions of the rectangle (in feet). Then,

$2\ell + 2w = 64$ *The perimeter is 64 feet.*
$\ell + w = 32$
and $w = 32 - \ell$

Also, $\ell w = 240$ *The area is 240 square feet.*
$\ell(32 - \ell) = 240$
$32\ell - \ell^2 = 240$
$\ell^2 - 32\ell - 240 = 0$
$(\ell - 20)(\ell - 12) = 0$

$\ell - 20 = 0$	$\ell - 12 = 0$
$\ell = 20$	$\ell = 12$
$w = 32 - \ell$	$w = 32 - \ell$
$w = 12$	$w = 20$

The rectangle is 20 feet by 12 feet.

6. a. Perimeter = 4 × 5 in. = 20 in.
b. Area = (5 in.)^2 = 25 in.^{2}

HOME EXERCISES

In Exercises 1–4, find the perimeter of each polygon.

1. Fig. 10.17 **2.** Fig. 10.18 **3.** Fig. 10.19 **4.** Fig. 10.20

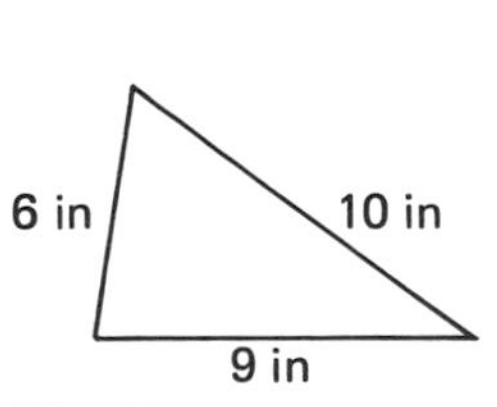

Fig. 10.17

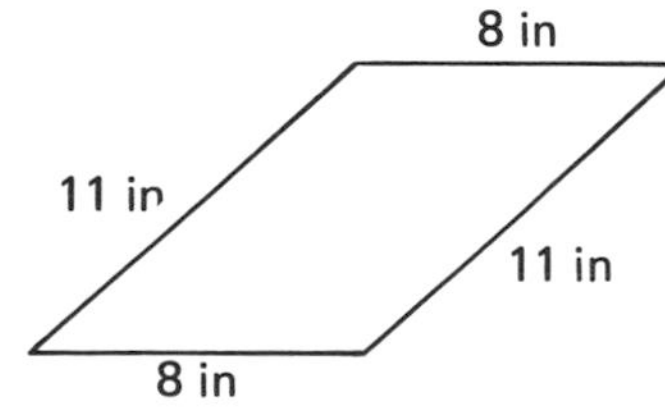

Fig. 10.18

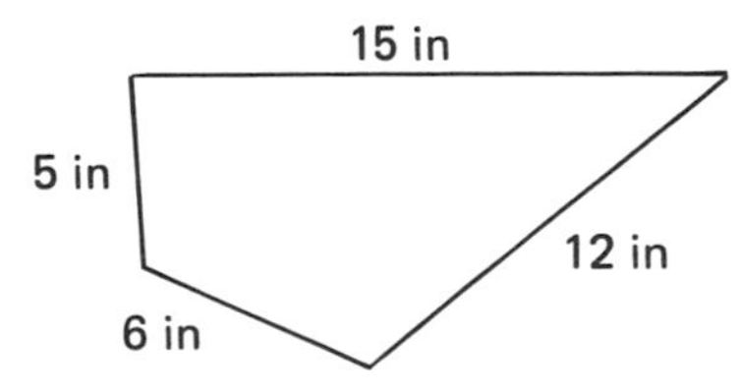

Fig. 10.19

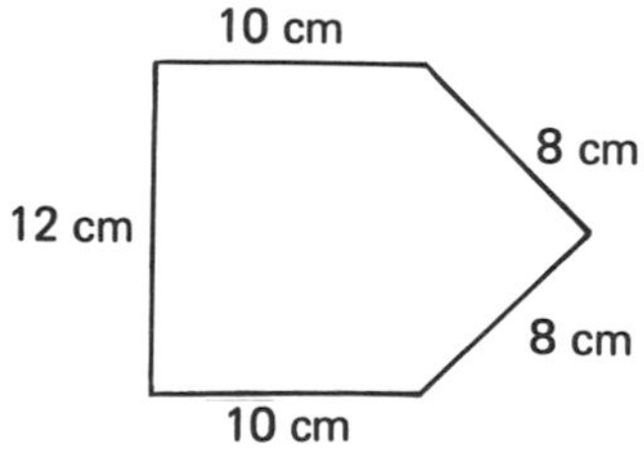

Fig. 10.20

In Exercises 5–7, find the perimeter of each rectangle.

5. length = 9 inches, width = 2 inches **6.** length = 6 centimeters, width = 5 centimeters

7. length = 4 miles, width = $\frac{1}{2}$ mile

In Exercises 8–10, find the perimeter of the square whose side length is as indicated.

8. 7 centimeters **9.** 1.5 miles **10.** $\frac{2}{3}$ mile

11. The perimeter of a square is 40 inches. Find the length of a side.

12. The perimeter of a square is 2 feet. Find the length of a side.

13. How much rope is needed to enclose a rectangular field 90 feet by 50 feet?

14. How much rope is needed to enclose a rectangular field 90 feet by 50 feet and to divide it in two, **a.** lengthwise? **b.** widthwise?

15. A baseball diamond is a square with side length 90 feet. What is the minimum distance a player must run when he hits a home run?

16. In Figure 10.21, a square is cut from each corner of a 14 inch by 10 inch rectangle. If a side of each square is 2 inches long, find the perimeter of the figure.

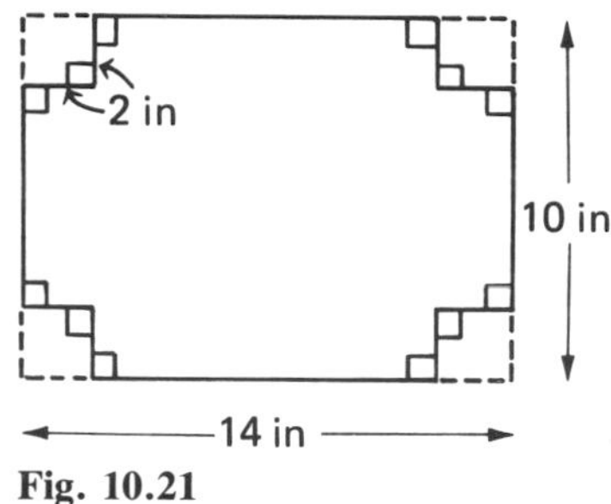

Fig. 10.21

In Exercises 17–20, find the area of each rectangle.

17. base 4 inches, height 3 inches

18. base 25 centimeters, height 12 centimeters

19. base $\frac{3}{4}$ inch, height 2 inches

20. base $\frac{2}{3}$ inch, height $\frac{9}{10}$ inch

In Exercises 21 and 22, find the base of each rectangle.

21. height 5 inches, area 20 square inches

22. height 25 centimeters, area 75 square centimeters

In Exercises 23 and 24, find the height of each rectangle.

23. base 4 yards, area 24 square yards

24. base 9 centimeters, area 108 square centimeters

25. Find the area of a square of side length 9 inches.

26. Find the length of a side of a square whose area is 49 square yards.

27. How many square feet of material are needed to make a tablecloth that is 6 feet by 4 feet?

28. A piece of typing paper measures $8\frac{1}{2}$ inches by 11 inches. What is its area?

29. A football field measures 100 yards by 53.333 yards. To the nearest square yard, what is its area?

30. One square has side length twice that of another square. If the smaller square has area A, express the area of the larger square in terms of A.

31. The floor of a room measures 20 feet by 14 feet. How much does it cost to varnish the floor at 10 cents per square foot?

32. Suppose the ceiling of the room in Exercise **31** is 9 feet high. How much does it cost to paint the four sides of the room and the ceiling at eight cents per square foot?

33. The perimeter of a rectangle is 70 inches and the area is 300 square inches. Find the dimensions of the rectangle.

34. The area of a garden is 1200 square feet. If the length of the garden were increased by 5 feet and the width decreased by 5 feet, the area would be decreased by 75 square feet. Find the dimensions of the garden.

35. The area of a Chinese rug is 360 square feet. A Persian rug, which is 4 feet longer but 3 feet narrower, has the same area. Find the dimensions of the Chinese rug.

In Exercises 36–38, find the areas of the indicated figures, which are composed of rectangles.

36. Fig. 10.22 **37.** Fig. 10.23 **38.** Fig. 10.24

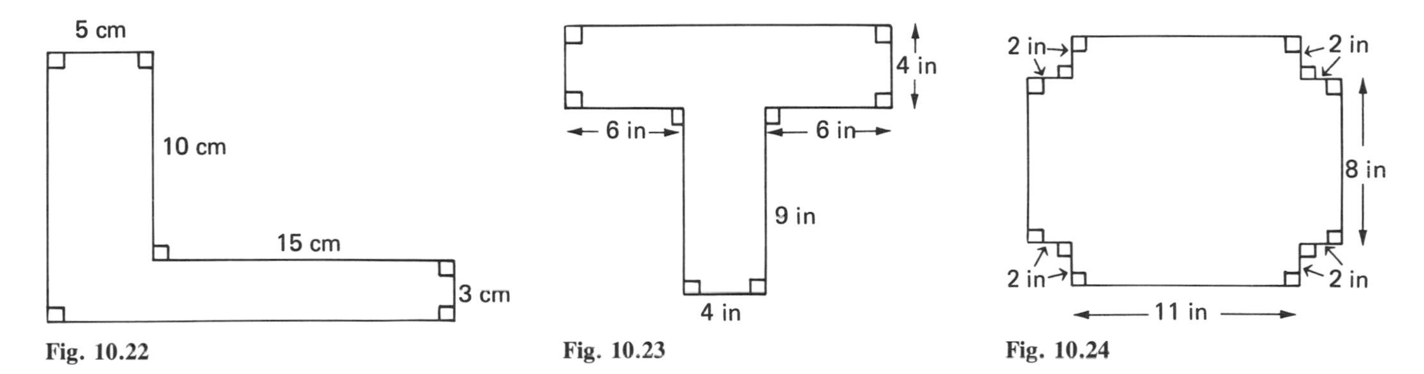

Fig. 10.22 **Fig. 10.23** **Fig. 10.24**

10.3 TRIANGLES, PARALLELOGRAMS, TRAPEZOIDS

TRIANGLES Recall that a triangle is a three-sided polygon. The line segment from any vertex drawn perpendicular to the opposite side is called a **height** of the triangle. The corresponding perpendicular side of the triangle is called a **base.**

Several triangles are shown in Figure 10.25. A base, b, and corresponding height, h, are indicated in each case. The triangle in Figure 10.25(a) is known as a **right triangle.** One of its angles is a right angle.

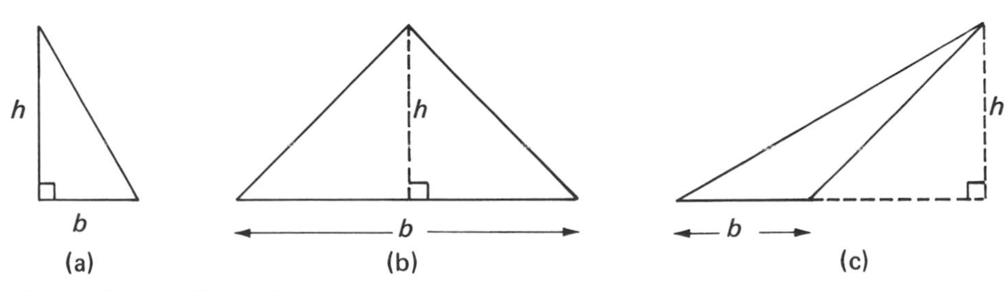

Fig. 10.25(a) Right triangle.

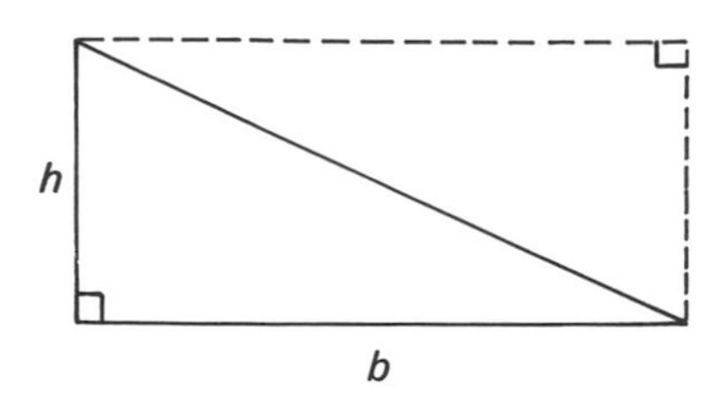

Fig. 10.26

The formula for the area of a triangle can be obtained from the formula for the area of a rectangle. A right triangle can be thought of as "half of a rectangle." (See Figure 10.26.) Therefore, its area is half that of the corresponding rectangle.

$$\text{Area of (right) triangle} = \frac{1}{2}\text{ Area of rectangle}$$
$$= \frac{1}{2} b \cdot h$$
$$= \frac{b \cdot h}{2}$$

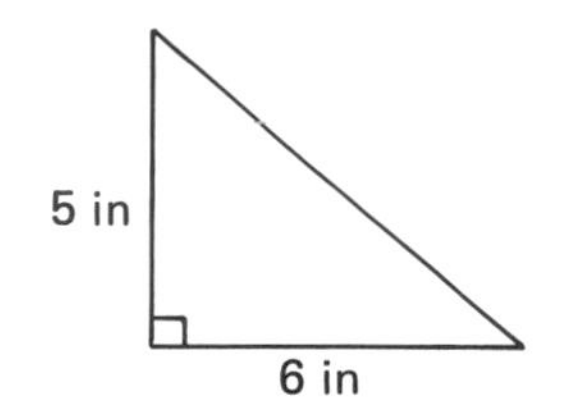

Fig. 10.27

Example 1 Find the area of a right triangle with base 6 inches and height 5 inches. (See Figure 10.27.)

Solution

$$\text{Area} = \frac{b \cdot h}{2}$$
$$= \frac{\overset{3\text{ in.}}{\cancel{6\text{ in.}}} \times 5\text{ in.}}{\underset{1}{\cancel{2}}}$$
$$= 15\text{ in.}^2$$

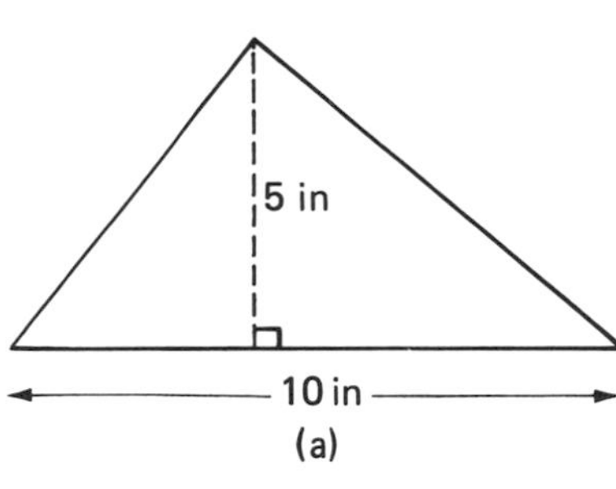

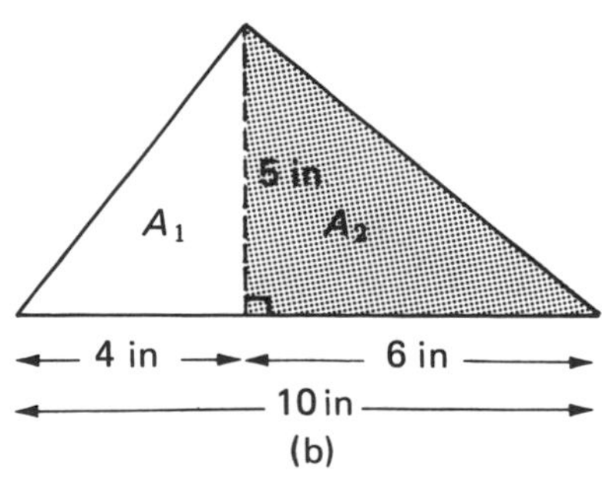

Fig. 10.28

For other types of triangles, the formula for area is the same. For example, consider the triangle in Figure 10.28(a), with base 10 inches and height 5 inches. Its area is the sum of the areas of the two *right* triangles in Figure 10.28(b), with bases 4 inches and 6 inches and with common height 5 inches. Note that

$$A_1 = \frac{\overset{2\text{ in.}}{\cancel{4\text{ in.}}} \times 5\text{ in.}}{\underset{1}{\cancel{2}}} = 10\text{ in.}^2 \quad \text{and} \quad A_2 = 15\text{ in.}^2$$

(as in Example 1). Let A be the area of the triangle in Figure 10.28(a).

$$A = A_1 + A_2$$
$$= 10\text{ in.}^2 + 15\text{ in.}^2 \qquad \text{[See Figure 10.28(b).]}$$
$$= 25\text{ in.}^2$$

Observe that $$25\text{ in.}^2 = \frac{50\text{ in.}^2}{2} = \frac{10\text{ in.} \times 5\text{ in.}}{2}$$

that is, $\dfrac{\text{base} \times \text{height}}{2}$ [*See Figure 10.28(a)*.]

To sum up, *for any triangle*,

$$\boxed{\textit{Area} = \frac{1}{2}\ \textit{base} \cdot \textit{height}}$$

or in symbols,

$$\boxed{A = \frac{1}{2} b \cdot h = \frac{b \cdot h}{2}}$$

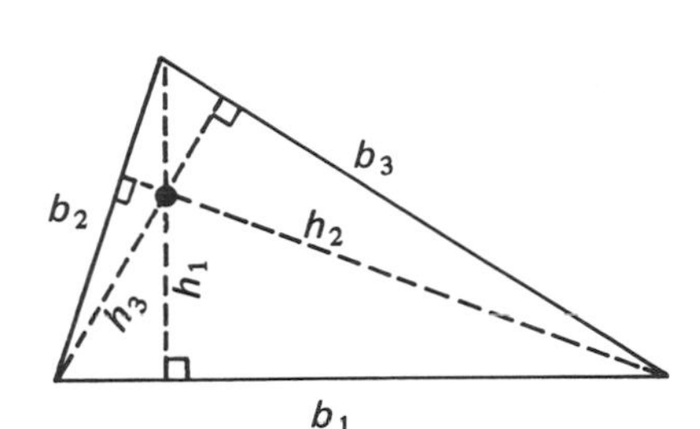

Fig. 10.29

Actually, the area of a triangle can be determined by using *any* side as base, b, together with the corresponding height, h. Thus, in Figure 10.29,

$$A = \frac{b_1 \cdot h_1}{2} = \frac{b_2 \cdot h_2}{2} = \frac{b_3 \cdot h_3}{2}$$

Note, further, that *the three heights intersect at a common point.*

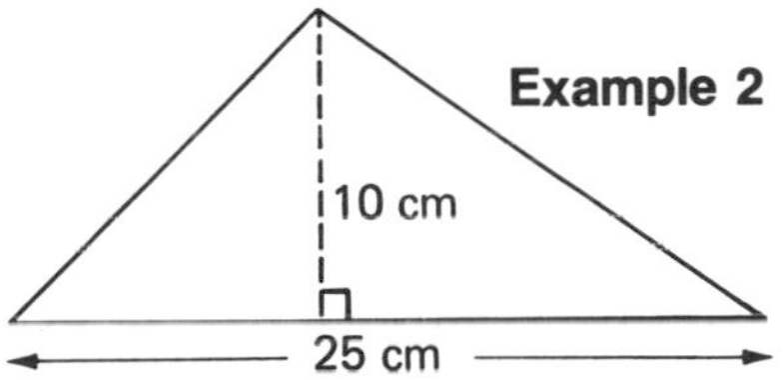

Fig. 10.30

Example 2 Find the area of the triangle in Figure 10.30.

Solution

$$A = \frac{b \cdot h}{2}$$

$$= \frac{25 \text{ cm.} \times \overset{5 \text{ cm.}}{\cancel{10 \text{ cm.}}}}{\underset{1}{\cancel{2}}}$$

$$= 125 \text{ cm.}^2$$

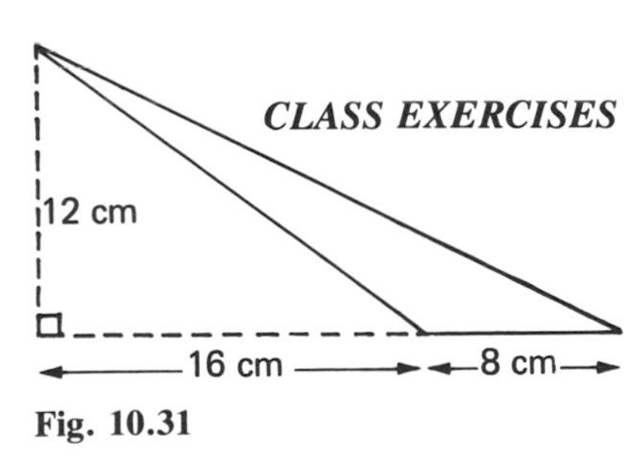

Fig. 10.31

CLASS EXERCISES

1. Find the area of a triangle with base 8 inches and height 4 inches.
2. In Figure 10.31, note that there are two right triangles: the smaller one with base 16 centimeters and the larger one with base 24 centimeters (24 cm. = 16 cm. + 8 cm.)
   a. Find the area of the smaller right triangle.
   b. Find the area of the larger right triangle.
   c. Find the area of the third triangle (with base 8 centimeters) and show that this is the area of the larger right triangle minus the area of the smaller right triangle.

PYTHAGOREAN THEOREM

In a right triangle, the side opposite the right angle is known as the **hypotenuse.** Let

$$b = \textit{base}, \qquad h = \textit{height}, \qquad c = \textit{hypotenuse}$$

as in Figure 10.32 on page 330.

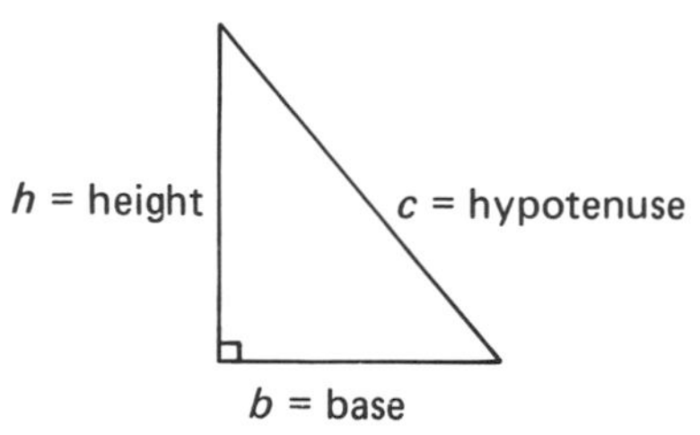

Fig. 10.32

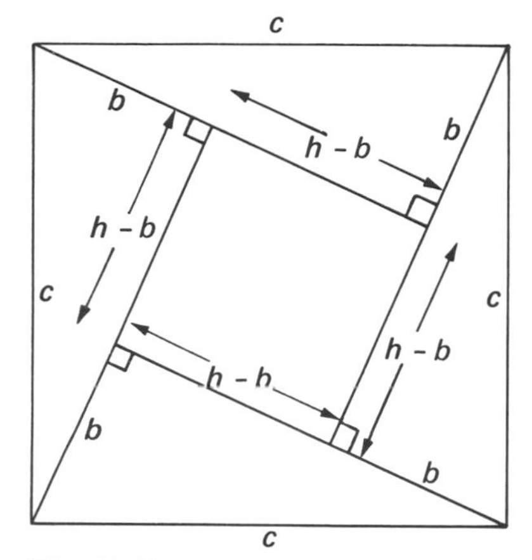

Fig. 10.33

The **Pythagorean Theorem** states that

$$\boxed{b^2 + h^2 = c^2}$$

To see this, consider Figure 10.33 . A square of side length c is divided up to form four right triangles and a smaller square. For each right triangle, the base is b, the height is $b + (h - b)$, or h, and the hypotenuse is c.†

Area of four right triangles + *area of small square* = *area of large square*

$$4 \cdot \frac{b \cdot h}{2} + (h - b)^2 = c^2$$

$$\frac{\overset{2}{\cancel{4}}\, bh}{\underset{1}{\cancel{2}}} + h^2 - 2bh + b^2 = c^2$$

$$h^2 + b^2 = c^2 \quad \text{or}$$

$$b^2 + h^2 = c^2$$

Example 3 Find the hypotenuse of a right triangle with base 3 inches and height 4 inches.

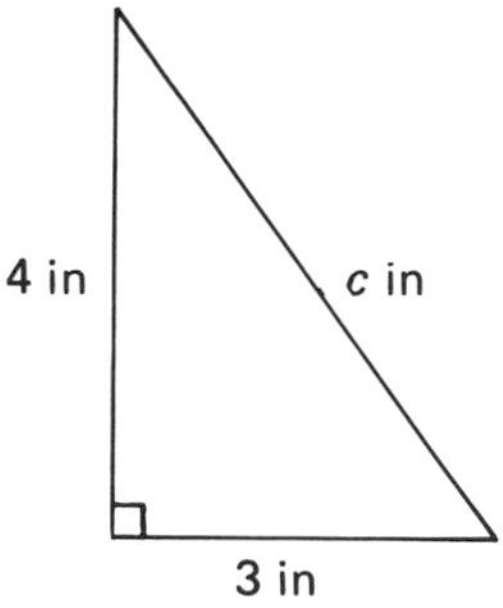

Fig. 10.34

Solution Let b, h, and c be as above. Then c, the (*length* of the) hypotenuse is positive.

$$b^2 + h^2 = c^2$$

$$3^2 + 4^2 = c^2$$

$$9 + 16 = c^2$$

$$25 = c^2$$

$$\sqrt{25} = \sqrt{c^2}$$

$$5 = c$$

The hypotenuse is 5 inches.

In the following example, the hypotenuse is an *irrational number*.

†This construction uses the fact that the sum of the measures of the angles of a triangle is 180°. See Section 11.2 .

Example 4 Find the hypotenuse of the triangle in Figure 10.35.

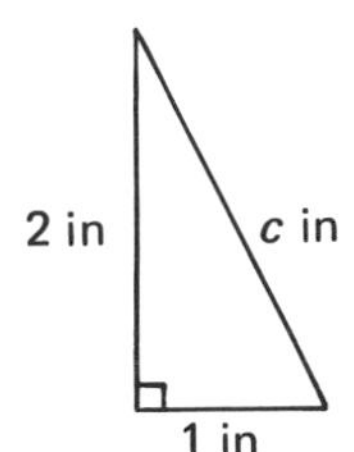

Fig. 10.35

Solution

$$b^2 + h^2 = c^2$$
$$1^2 + 2^2 = c^2$$
$$1 + 4 = c^2$$
$$5 = c^2 \qquad \textit{Find the positive root of this equation.}$$
$$\sqrt{5} = c$$

The *exact* measurement of the hypotenuse is $\sqrt{5}$ inches. (Usually, you will leave your result in *radical form*, as $\sqrt{5}$ inches.) Because $\sqrt{5} \approx 2.2$, the hypotenuse is *approximately* 2.2 inches.

Example 5 Find the height of the triangle in Figure 10.36.

h cm
10 cm
6 cm

Fig. 10.36

Solution

$$b^2 + h^2 = c^2$$
$$6^2 + h^2 = 10^2$$
$$36 + h^2 - 36 = 100 - 36$$
$$h^2 = 64$$
$$h = 8$$

The height is 8 centimeters.

CLASS EXERCISES

3. Find the hypotenuse of a right triangle with base 9 centimeters and height 12 centimeters.
4. Find the height of a right triangle with base 12 centimeters and hypotenuse 13 centimeters.
5. Find the base of a right triangle with height 5 inches and hypotenuse 6 inches. (Leave your answer in radical form.)

PARALLELOGRAMS

Lines extend indefinitely.

DEFINITION

> **Parallel lines** are lines (in a plane) that do not intersect.

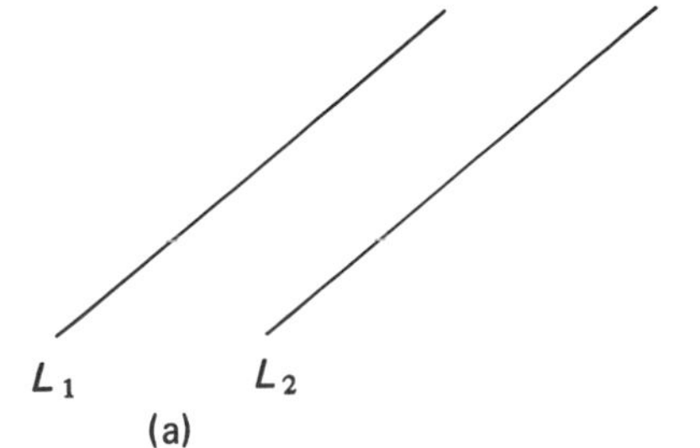

Fig. 10.37(a) Lines L_1 and L_2 are parallel. **(b)** Lines L_3 and L_4 intersect and, therefore, are not parallel.

DEFINITION

> A **parallelogram** is a quadrilateral, or four-sided polygon, in which *both* pairs of opposite sides are parallel. The line segment from any vertex drawn perpendicular to an "opposite side" is called a **height** of the parallelogram. The corresponding perpendicular side of the parallelogram is called a **base**.

[See Figures 10.38(a) and 10.38(b).]

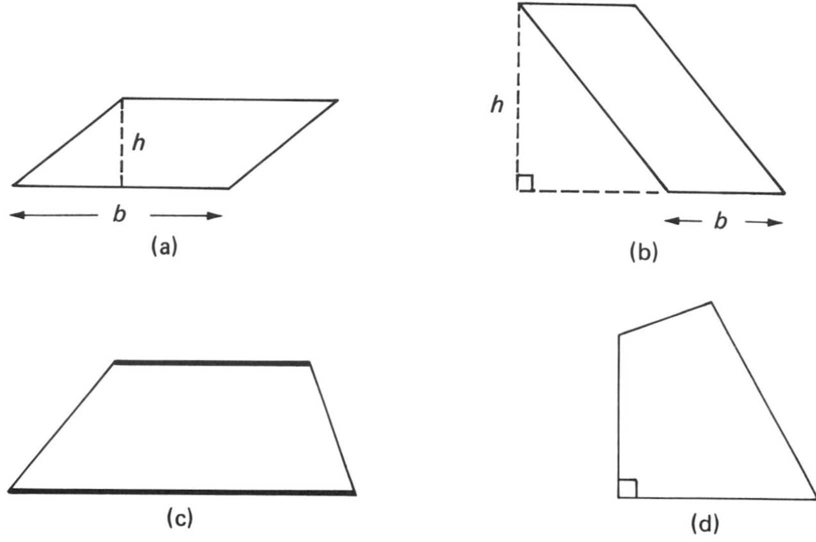

Fig. 10.38(a) A parallelogram with base b and height h. **(b)** A parallelogram with base b and height h. **(c)** This quadrilateral (which is called a *trapezoid*) is not a parallelogram. The sides drawn with thickened lines are parallel, but the other sides are not. **(d)** This quadrilateral is not a parallelogram.

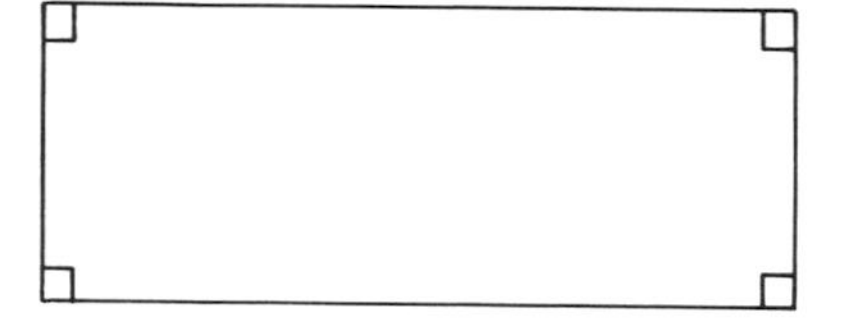

Fig. 10.39. A rectangle is a parallelogram.

It can be shown that *in a parallelogram opposite sides are of equal length. A rectangle is a parallelogam in which adjacent sides form right angles.* (See Figure 10.39 .)

To find a formula for the area of a parallelogram, consider Figure 10.40 . Break up the parallelogram into a rectangle and two right triangles. (We neglect units.)

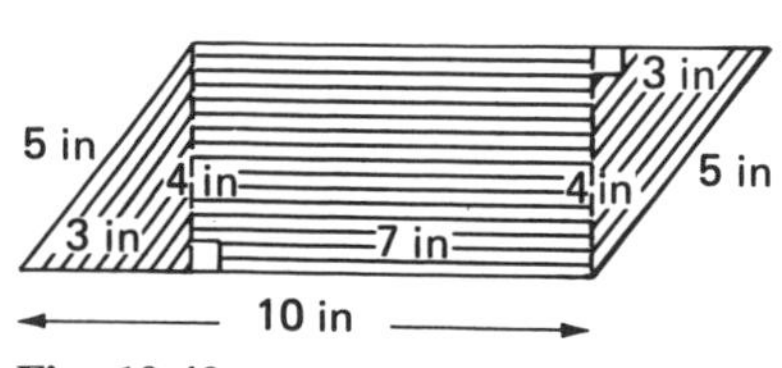

Fig. 10.40

Area of parallelogram = *Area of rectangle* + 2 × *Area of each one of the triangles*

$$= 7 \cdot 4 + \frac{\overset{1}{\cancel{2}}(3 \cdot 4)}{\underset{1}{\cancel{2}}}$$

$= (7 + 3)\,4$ *by the Distributive Laws*

$= 10 \cdot 4$

$= 40$

Observe that in the next to last line, 10 represents the base and 4 the height of the parallelogram. In general,

$$\textit{Area of parallelogram} = \textit{base} \times \textit{height}$$

In symbols,

$$A = b \cdot h$$

Note that this is the same as the formula for the area of a rectangle with base b and height h. (See Figure 10.41 .)

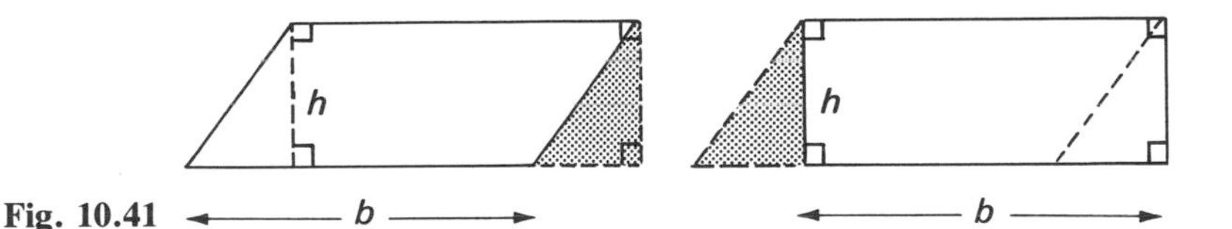

Fig. 10.41

Example 6 Find the area of the parallelogram in Figure 10.42 .

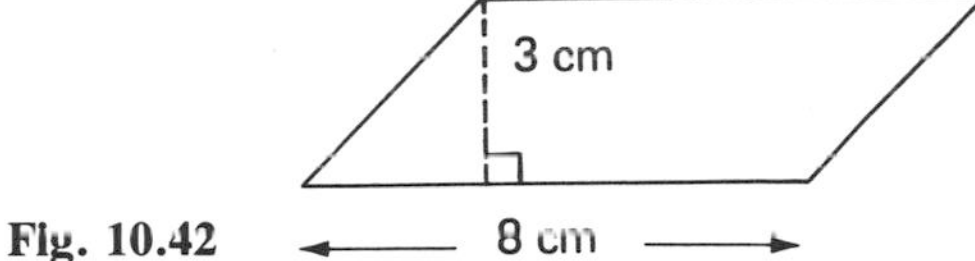

Fig. 10.42

Solution

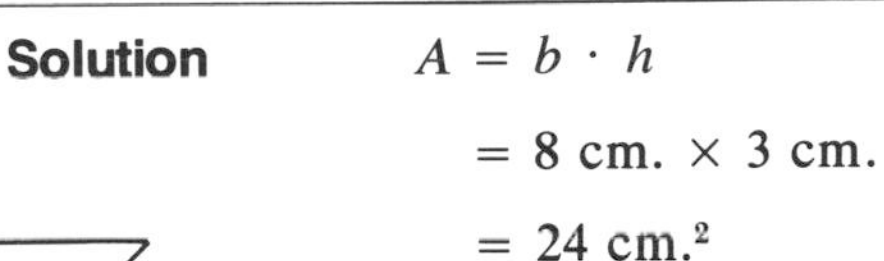

$A = b \cdot h$

$= 8 \text{ cm.} \times 3 \text{ cm.}$

$= 24 \text{ cm.}^2$

CLASS EXERCISES **6.** Find the area of the parallelogram with base 9 inches and height 7 inches.

TRAPEZOIDS

DEFINITION A **trapezoid** is a quadrilateral in which exactly one pair of opposite sides is parallel.†

†Note that, according to this definition, *a parallelogram is not a trapezoid.* In contrast, some authors define a trapezoid to be a quadrilateral in which *at least one* pair of opposite sides is parallel, so that a parallelogram is then a special type of trapezoid.

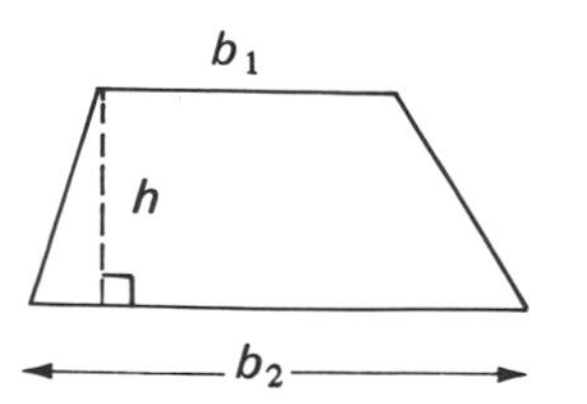

Fig. 10. 43. A trapezoid with bases b_1 and b_2 and height h.

Figure 10.43 depicts a trapezoid. The parallel sides are called its **bases**. *The bases are unequal in length.* Consider the longer base. A line segment from an "opposite vertex" drawn perpendicular to this base is called a **height** of the trapezoid.

To find the formula for the area of a trapezoid, label the *vertices* of the trapezoid. Then, rotate the trapezoid, as indicated in Figure 10.44(a). Join the two trapezoids to form a parallelogram with base $b_1 + b_2$ and with height h, as shown in Figure 10.44(b). The area of the parallelogram is

$$(b_1 + b_2)\,h$$

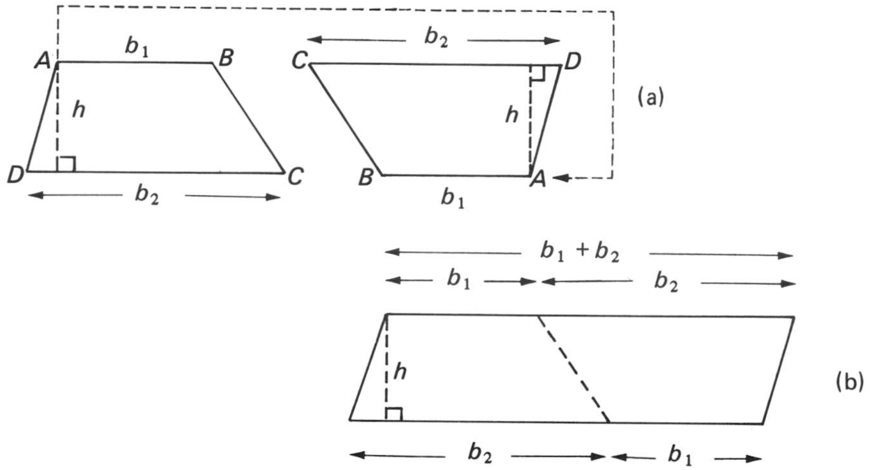

Fig. 10.44(a) The second trapezoid is the first one rotated. Thus, both have the same area. **(b)** The two trapezoids are joined to form a parallelogram with base $b_1 + b_2$ and height h.

The *area of the* original *trapezoid* is *half* of that of the parallelogram. Consequently,

$$\textit{Area of trapezoid} = \frac{(b_1 + b_2)\,h}{2}$$

In other words,

the area of a trapezoid is half the sum of the bases times the height.

Note that $\frac{1}{2}(b_1 + b_2)$ is the average of the two bases. Thus,

the area of a trapezoid is the average of the bases times the height.

Example 7 Find the area of the trapezoid in Figure 10.45.

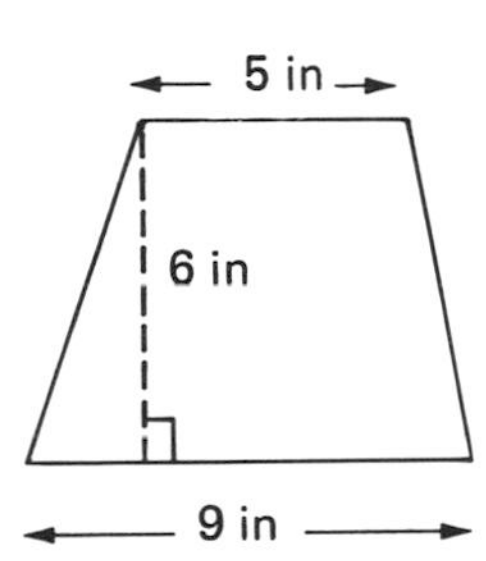

Fig. 10.45

Solution

$$A = \frac{(b_1 + b_2)h}{2}$$

$$= \frac{(5 \text{ in.} + 9 \text{ in.})\ 6 \text{ in.}}{2}$$

$$= \frac{\overset{7 \text{ in.}}{\cancel{14 \text{ in.}}} \times 6 \text{ in.}}{\underset{1}{\cancel{2}}}$$

$$= 42 \text{ in.}^2$$

CLASS EXERCISES *Find the area of each trapezoid.*

7. bases 12 centimeters and 10 centimeters, height 9 centimeters

8. average of bases 8 inches, height 10 inches

SOLUTIONS TO CLASS EXERCISES

1. 16 in.^{2} **2. a.** 96 cm.^{2} **b.** 144 cm.^{2} **c.** From parts **a.** and **b.**, note that 144 cm.^{2} − 96 cm.^{2} = 48 cm.^{2}

3. $9^2 + 12^2 = c^2$
$225 = c^2$
$15 = c$
The hypotenuse is 15 centimeters.

4. $12^2 + h^2 = 13^2$
$144 + h^2 = 169$
$h^2 = 25$
$h = 5$
The height is 5 centimeters.

5. $b^2 + 5^2 = 6^2$
$b^2 + 25 = 36$
$b^2 = 11$
$b = \sqrt{11}$
The base is $\sqrt{11}$ inches.

6. 63 in.^{2} **7.** $A = \frac{(12 \text{ cm.} + 10 \text{ cm.})\ 9 \text{ cm.}}{2} = 99 \text{ cm.}^2$ **8.** 80 in.^{2}

HOME EXERCISES

In Exercises 1–3, find the area of each triangle.

1. base 4 inches, height 4 inches **2.** base 9 inches, height 6 inches **3.** base $\frac{3}{4}$ inch, height 6 inches

In Exercises 4 and 5, find the height of each triangle.

4. base 10 inches, area 100 square inches

5. base 5 centimeters, area 30 square centimeters

In Exercises 6 and 7, find the base of each triangle.

6. height 6 inches, area 9 square inches

7. height 10 feet, area 40 square feet

In Exercises 8 and 9, find the hypotenuse of each right triangle. If necessary, leave your answer in radical form.

8. base 40 yards, height 30 yards

9. base 8 meters, height 8 meters

In Exercises 10 and 11, find the height of each right triangle. If necessary, leave your answer in radical form.

10. base 24 inches, hypotenuse 25 inches

11. base 1 centimeter, hypotenuse 4 centimeters

In Exercises 12 and 13, find the base of each right triangle. If necessary, leave your answer in radical form.

12. height 16 inches, hypotenuse 20 inches

13. height 3 feet, hypotenuse 6 feet

In Exercises 14–16, find the area of each parallelogram.

14. base 10 inches, height 8 inches

15. base 6 inches, height 6 inches

16. base 24 centimeters, height 5 centimeters

In Exercises 17 and 18, find the height of each parallelogram.

17. base 5 inches, area 45 square inches

18. base 8 decimeters, area 32 square decimeters

In Exercises 19 and 20, find the base of each parallelogram.

19. height 9 inches, area 72 square inches

20. height 15 centimeters, area 75 square centimeters

In Exercises 21–23, find the area each of trapezoid.

21. bases 4 inches and 6 inches, height 5 inches

22. average of bases 4 meters, height 1 meter

23. bases 10 inches and 1 foot, height 10 inches

In Exercises 24 and 25, find the height of each trapezoid.

24. bases 10 centimeters and 20 centimeters, area 30 square centimeters

25. average of bases 9 inches, area 70 square inches

26. The height of a trapezoid is 5 inches and its area is 40 square inches. Find the sum of the bases.

27. The height of a trapezoid is 7 inches and its area is 28 square inches. Find the average of the bases.

28. The height of a trapezoid is 12 centimeters, the area is 60 square centimeters, and one base is 6 centimeters. Find the other base.

In Exercises 29–39, find the area of each shaded figure. (These figures consist of rectangles, triangles, parallelograms, and trapezoids.)

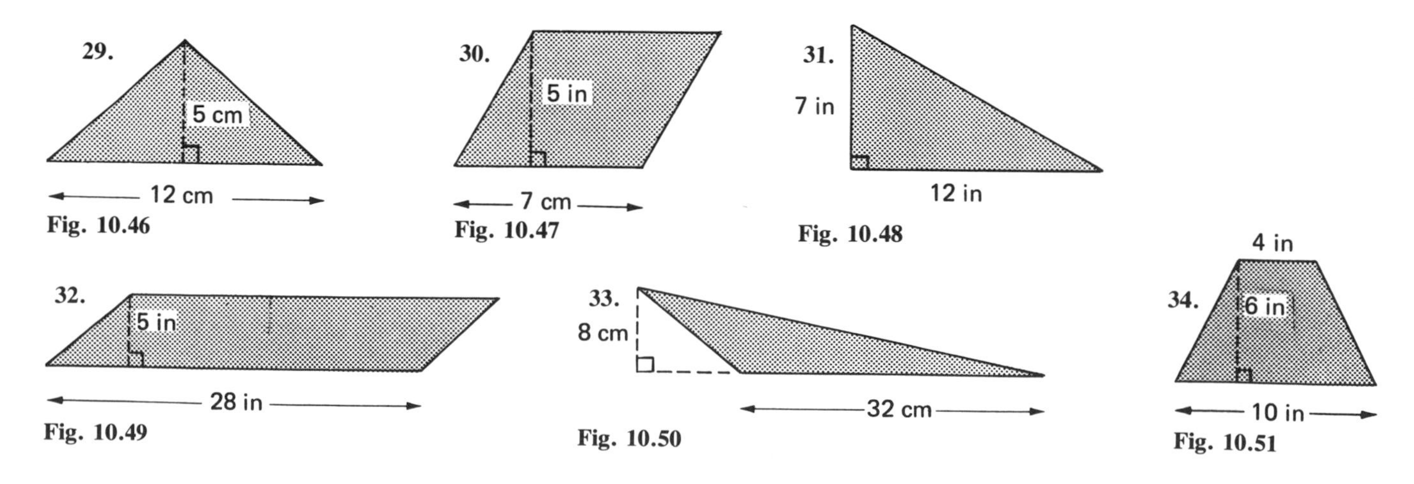

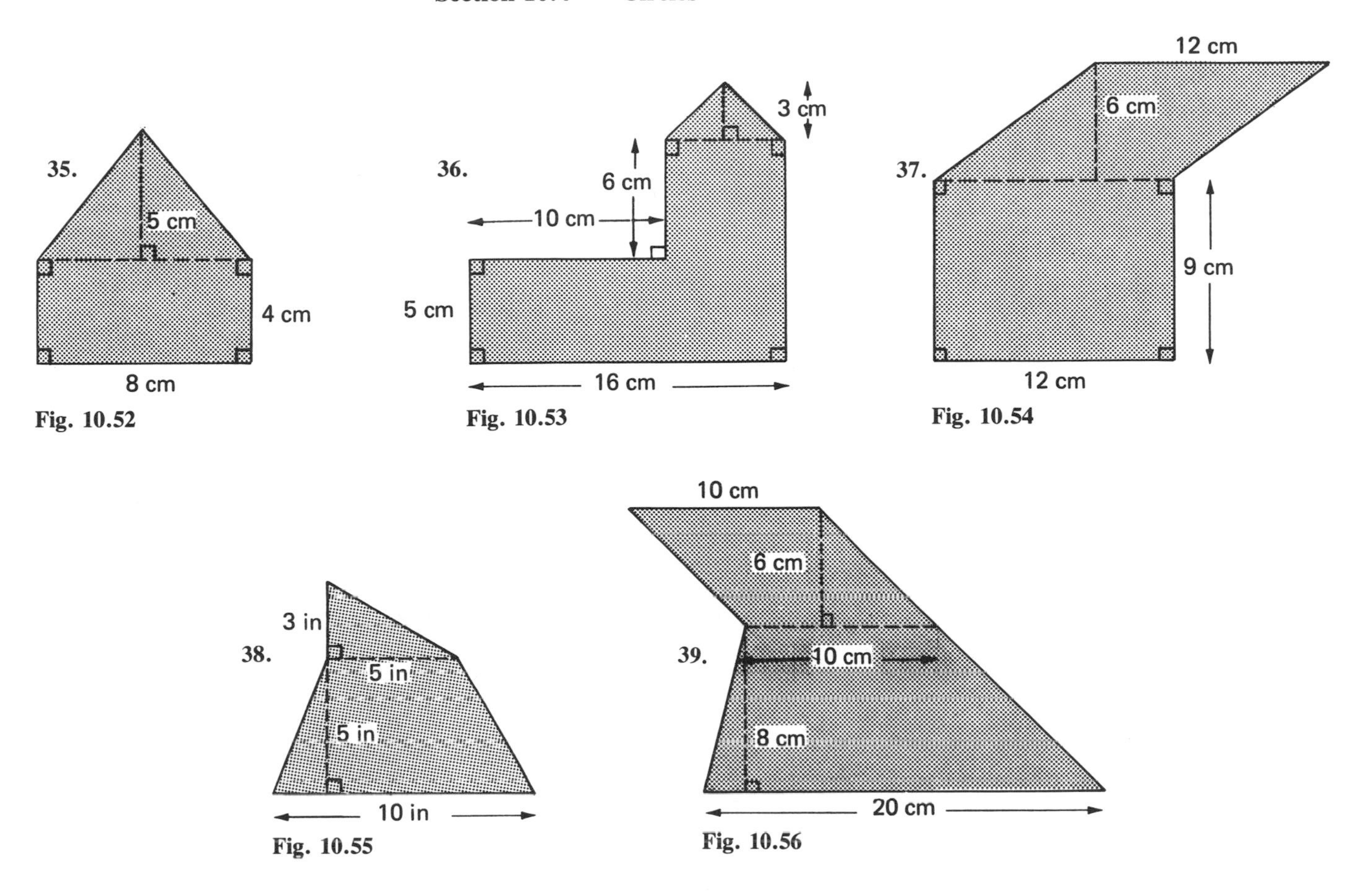

Fig. 10.52

Fig. 10.53

Fig. 10.54

Fig. 10.55

Fig. 10.56

10.4 CIRCLES

RADIUS AND DIAMETER

DEFINITION

> A **circle** is a closed figure on a plane every point of which is at a *fixed* distance from a given point, known as the **center** of the circle. The fixed distance is called the **radius** of the circle. (The plural of "radius" is "*radii.*")

Consider any line through the center of a circle. It cuts the circle at two points, P and Q. (See Figure 10.58 .) The length of the line segment from P to Q is called the **diameter** of the circle. Clearly, *the diameter is twice the radius.* In symbols, let

$$d = \textit{diameter} \quad \text{and} \quad r = \textit{radius}$$

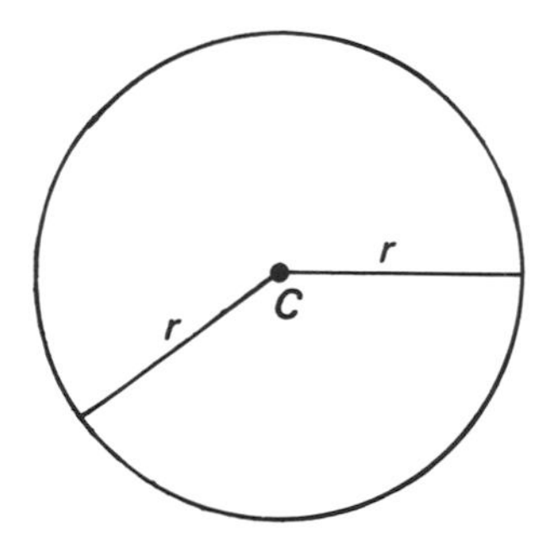

Fig. 10.57. A circle with center C and radius r.

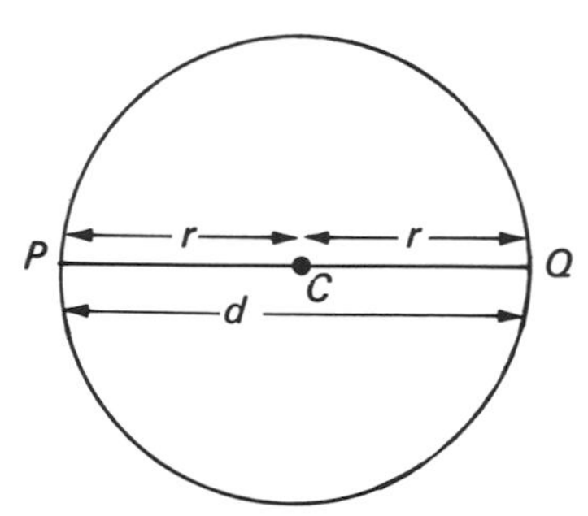

Fig. 10.58. A circle with diameter d.

$d = 2r$

Then,

$$\boxed{d = 2r}$$

Thus, if the radius of a circle is 7 inches, its diameter is 14 inches.

Example 1 Find the radius of the circle in Figure 10.59.

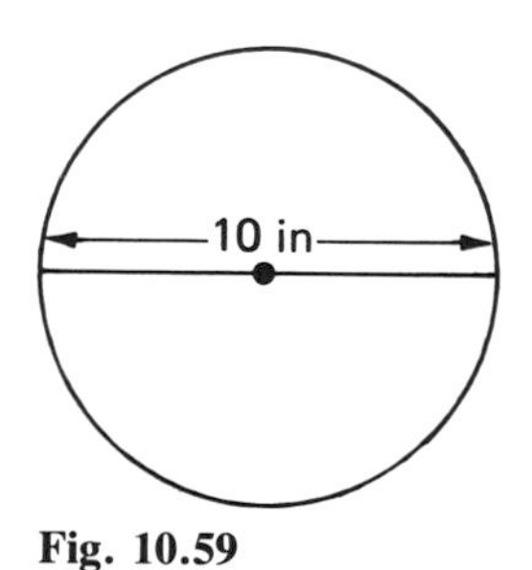

Fig. 10.59

Solution

$$d = 2r$$

$$10 \text{ in.} = 2r$$

$$\frac{10 \text{ in.}}{2} = \frac{2r}{2}$$

$$5 \text{ in.} = r$$

CLASS EXERCISES

1. Find the diameter of the circle with radius 6 inches.
2. Find the radius of the circle with diameter 6 inches.

CIRCUMFERENCE

Recall that the perimeter of a polygon can be thought of as the distance around it. The **circumference** of a circle is defined to be the distance around the circle. The ancient Greeks discovered that *the circumference of any circle is always equal to the diameter times the irrational number* π. To the nearest hundredth,

$$\pi \approx 3.14$$

Let C = circumference and d = diameter

Then, $\boxed{C = \pi d}$

For a circle of diameter 5 inches, the circumference, C, is 5π inches.

$$C \approx 5 \times 3.14 \text{ inches} = 15.70 \text{ inches}$$

To the nearest tenth of an inch, $C \approx 15.7$ inches (See Figure 10.60 .)

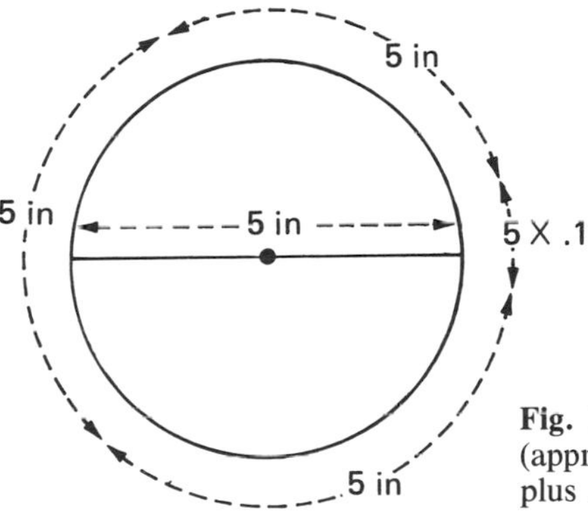

Fig. 10.60. The circumference is (approximately) 3 times the diameter plus .14 times the diameter.

Because $C = \pi d$ and $d = 2r$ replace d by $2r$ in the first formula to obtain

$$C = \pi \cdot 2r$$

or $\boxed{C = 2\pi r}$

Thus,

> *the circumference of a circle is 2π times the radius.*

When approximations involving π are called for, if nothing to the contrary is stated, use

$$\pi \approx 3.14$$

Example 2 Find the circumference of a circle of radius 10 centimeters.

a. Express your answer in terms of π.
b. Approximate to the nearest centimeter.

Solution a. $C = 2\pi r = 2\pi \times 10$ cm. $= 20\pi$ cm.

b. $C \approx 20 \times 3.14$ cm.
$= 62.80$ cm.
To the nearest centimeter,
$C \approx 63$ cm.

CLASS EXERCISES *Find the circumference of the circle.* **a.** *Express your answer in terms of* π.
b. *Approximate to the nearest tenth of a unit.*

3. diameter 4 inches **4.** radius 5 centimeters

AREA Consider a circle of radius r. Draw the square that has side length r, as shown in Figure 10.61(a). About $\frac{3}{4}$ of the square region lies within the circle. Because the area of the square is r^2, the area of this pie-shaped circular region in Figure 10.61(a) is *approximately* $\frac{3}{4} r^2$. Now, draw four of these squares, as in Figure 10.61(b). The total area of the four-square region is $4r^2$. The area of the circular region is therefore *approximately*

$$\frac{3}{4} \cdot 4r^2, \quad \text{or} \quad 3r^2$$

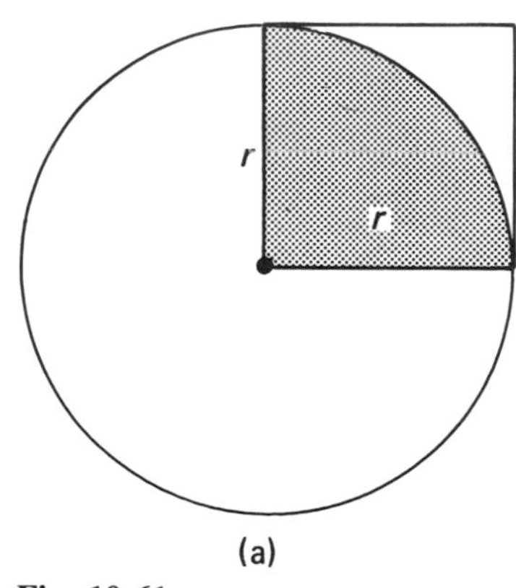

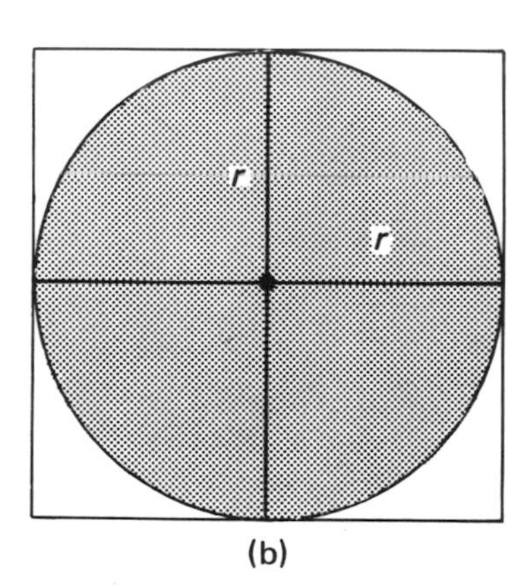

Fig. 10.61

The Greeks discovered that *the exact area, A, of a circle is given by the formula*

$$\boxed{A = \pi r^2}$$

Thus,

$$\boxed{A \approx 3.14r^2}$$

Example 3 Find the area of a circle with radius 10 centimeters.

a. Express your answer in terms of π.

b. Approximate to the nearest centimeter, using $\pi \approx 3.14$.

Solution **a.** $A = \pi r^2$

$= \pi(10 \text{ cm.})^2$

$= \pi \times 100 \text{ cm.}^2$

$= 100\,\pi \text{ cm.}^2$

b. $A \approx 100 \times 3.14 \text{ cm.}^2$

$= 314 \text{ cm.}^2$

CLASS EXERCISES *Find the area of each circle.*

a. Express your answer in terms of π. **b.** Approximate to the nearest unit.

5. radius = 7 in. **6.** diameter = 18 cm.

SOLUTIONS TO CLASS EXERCISES

1. 12 in. **2.** 3 in. **3. a.** 4π in. **b.** 12.6 in. **4. a.** 10π cm. **b.** 31.4 cm.

5. a. 49π in.^{2} **b.** 154 in.^{2} **6. a.** $r = 9$ cm., $A = 81\pi$ cm.^{2} **b.** 254 cm.^{2}

HOME EXERCISES

In approximations, unless otherwise specified, use $\pi \approx 3.14$.

In Exercises 1 and 2, find the diameter of the given circle.

1. radius 10 inches **2.** radius 5 feet

In Exercises 3 and 4 find the radius of the given circle.

3. diameter 4 inches **4.** diameter 2 miles

In Exercises 5–11 find the circumference of the given circle.

a. Express your answer in terms of π. **b.** Approximate to the nearest unit.

5. diameter 1 inch **6.** diameter 12 centimeters **7.** diameter 9 yards **8.** radius 50 inches

9. radius 1 foot **10.** radius 20 meters **11.** radius $\frac{1}{2}$ mile

In Exercises 12–18, find the area of each circle.

a. Express your answer in terms of π. **b.** Approximate to the nearest unit.

12. radius 3 yards **13.** radius 5 feet **14.** radius 20 inches [In part **b.**, *use* $\pi \approx 3.142$.] **15.** radius 50 centimeters [In part **b.**, *use* $\pi \approx 3.1416$.]

16. diameter 12 dekameters **17.** diameter 40 feet [In part **b.**, *use* $\pi \approx 3.142$.] **18.** diameter 200 miles [In part (**b.**), *use* $\pi \approx 3.1416$.]

19. Find the area of a circle whose circumference is 6π inches.

20. Find the circumference of a circle whose area is 25π square feet.

21. Suppose the circumference of a circle is 10π centimeters. Find **a.** the diameter, **b.** the radius, **c.** the area.

22. Suppose the area of a circle is 9π square feet. Find **a.** the radius, **b.** the diameter, **c.** the circumference.

23. A circular swimming pool is 60 feet in diameter. How far is it around the pool? Approximate to the nearest foot.

24. A small (circular) pizza pie has a radius of 6 inches and costs one dollar. A large pizza pie has a radius of 9 inches and costs two dollars. Which is the better buy?

In Exercises 25–28, find the area of each shaded region. Leave your answers in terms of π.

25. Fig. 10.62 **26.** Fig. 10.63 **27.** Fig. 10.64 **28.** Fig. 10.65

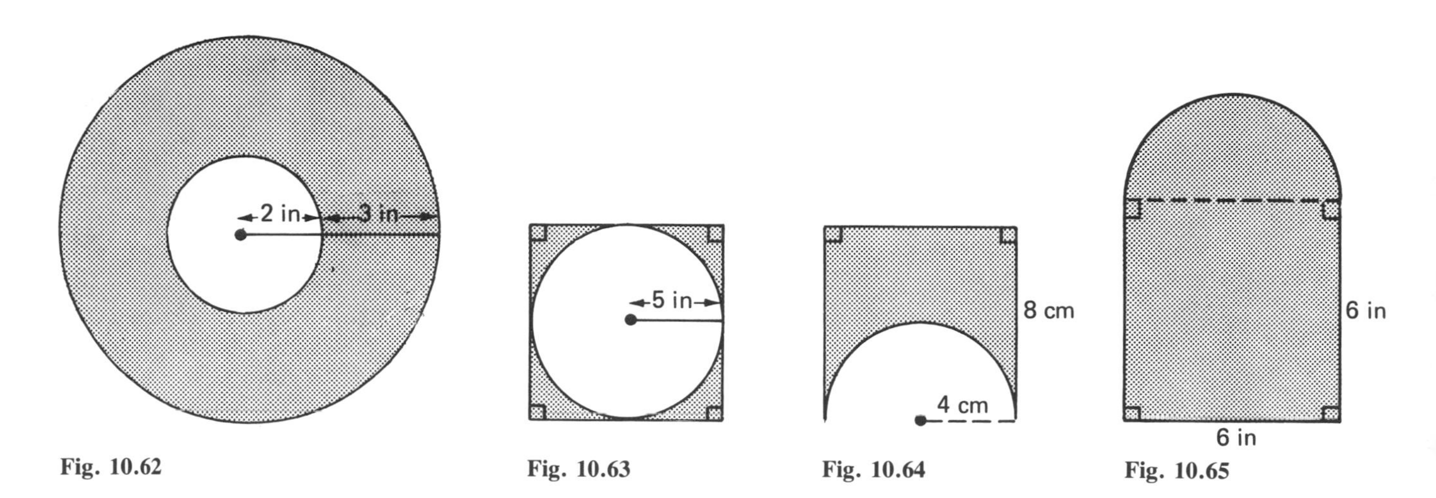

Fig. 10.62 Fig. 10.63 Fig. 10.64 Fig. 10.65

29. The diameter of a sink faucet is $\frac{1}{2}$ inch, whereas the diameter of a bathtub faucet is 1 inch. The water pressure is the same in both cases. How much faster does the water flow from the bathtub faucet than from the sink faucet?

10.5 VOLUME AND SURFACE AREA

RECTANGULAR BOXES

The **volume** of a solid figure, such as a rectangular box, is a measure of how many (cubic) units it can hold.

As you know, the area of a rectangle is the product of its length and width.

$$\textit{Area of rectangle} = \textit{length} \times \textit{width}$$

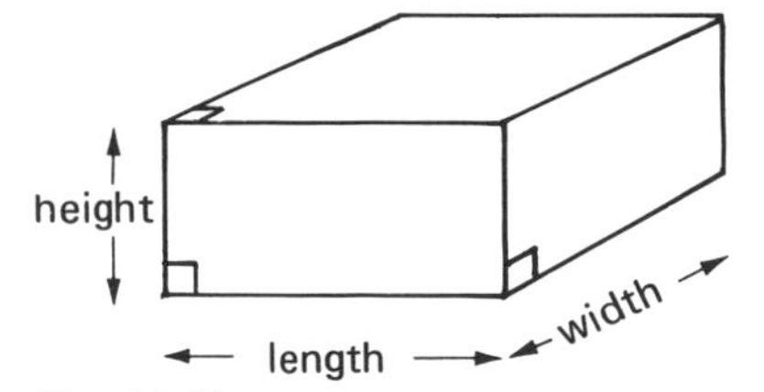

Fig. 10.66

Now, consider a box that has all rectangular faces, as shown in Figure 10.66. The volume of such a **rectangular box**† is the product of its length, width, and height.

†Specifically, the volume of the region enclosed by the box.

$$\textit{Volume of rectangular box} = \textit{length} \times \textit{width} \times \textit{height}$$

In symbols, let

V = *volume of a rectangular box*, ℓ = *length*, w = *width*, h = *height*

Then, $$\boxed{V = \ell \cdot w \cdot h}$$

The **surface area** of a rectangular box, that is, the sum of the areas of its six rectangular faces, is obtained by considering Figure 10.66 .

$$\begin{aligned} &2\ell w && \textit{(top and bottom)} \\ +\ &2\ell h && \textit{(front and back)} \\ +\ &2wh && \textit{(left and right sides)} \end{aligned}$$

Let S = *surface area of a rectangular box. Then,*

$$\boxed{S = 2\ell w + 2\ell h + 2wh \quad \text{or} \quad S = 2(\ell w + \ell h + wh)}$$

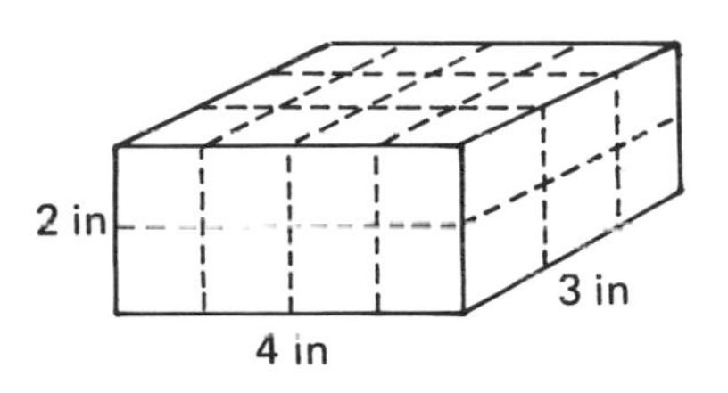

Fig. 10.67

In Figure 10.67, the length of the box is 4 inches, the width is 3 inches, and the height is 2 inches. The volume is given by

$$\begin{aligned} V &= \ell \cdot w \cdot h \\ &= 4 \text{ inches} \times 3 \text{ inches} \times 2 \text{ inches} \\ &= 24 \text{ inches} \times \text{inches} \times \text{inches}, \quad \text{or 24 cubic inches} \end{aligned}$$

The surface area is given by

$$\begin{aligned} S &= 2(\ell w + \ell h + wh) \\ &= 2(4 \text{ inches} \times 3 \text{ inches} + 4 \text{ inches} \times 2 \text{ inches} + 3 \text{ inches} \times 2 \text{ inches}) \\ &= 2(12 \text{ square inches} + 8 \text{ square inches} + 6 \text{ square inches}) \\ &= 2(26 \text{ square inches}) \\ &= 52 \text{ square inches} \end{aligned}$$

Note that the unit of volume is cubic inches, whereas the unit of surface area is square inches.

$$\boxed{\textit{inches} \times \textit{inches} \times \textit{inches} = \textit{cubic inches}}$$

Write

in.^{3} for *cubic inches*
ft.^{3} for *cubic feet*
cm.^{3} for *cubic centimeters*, and so on

Example 1 Find **a.** the volume and **b.** the surface area of the rectangular box shown in Figure 10.68 .

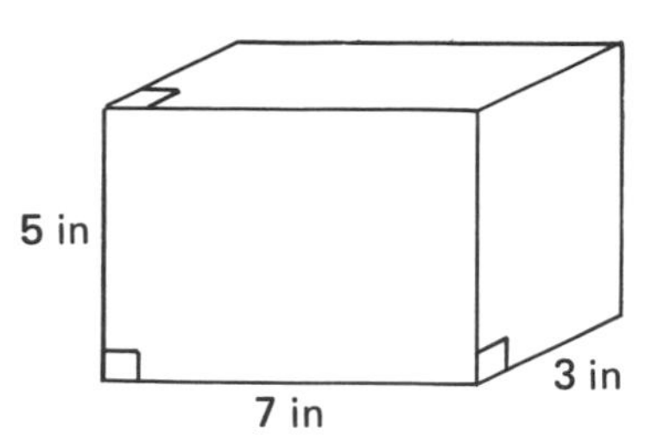

Fig. 10.68

Solution **a.** $V = \ell \cdot w \cdot h$
$= 7 \text{ in.} \times 3 \text{ in.} \times 5 \text{ in.}$
$= 105 \text{ in.}^3$

b. $S = 2(\ell w + \ell h + wh)$
$= 2(7 \text{ in.} \times 3 \text{ in.} + 7 \text{ in.} \times 5 \text{ in.} + 3 \text{ in.} \times 5 \text{ in.})$
$= 2(21 \text{ in.}^2 + 35 \text{ in.}^2 + 15 \text{ in.}^2)$
$= 142 \text{ in.}^2$

A **cube** is a rectangular box in which length, width, and height are equal.

> Volume of cube $= s^3$
> Surface Area of cube $= 2(s^2 + s^2 + s^2) = 6s^2$

where s is the length of a side.

Example 2 Find **a.** the volume and **b.** the surface area of a cube if the length of a side is 3 centimeters.

Solution (See Figure 10.69 .)

a. $V = s^3$
$= (3 \text{ cm.})^3$
$= 27 \text{ cm.}^3$

b. $S = 6s^2$
$= 6(3 \text{ cm.})^2$
$= 6 \times 9 \text{ cm.}^2$
$= 54 \text{ cm.}^2$

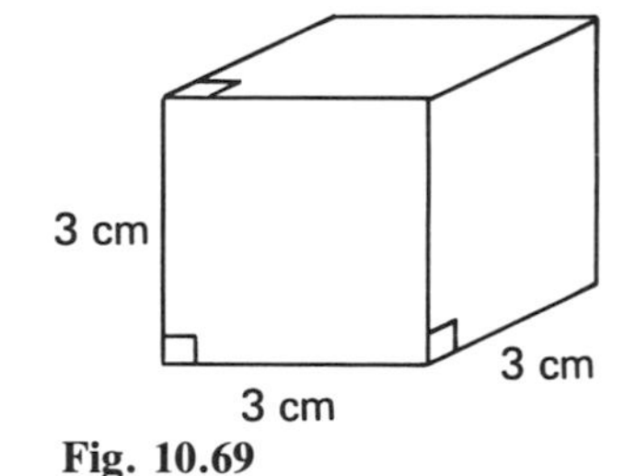

Fig. 10.69

CLASS EXERCISES

1. Find **a.** the volume and **b.** the surface area of a rectangular box of length 3 inches, width 1 inch, and height 4 inches.
2. Find **a.** the volume and **b.** the surface area of a cube if the length of each side is 4 inches.

SPHERES Recall that a circle is a closed figure *in a plane* every point of which is at a fixed distance from a given point. We will now consider the "three-dimensional" counterpart of a circle.

DEFINITION

> A **sphere** (or **ball**) is a surface *in space* every point of which is at a fixed distance from a given point, known as the **center** of the sphere. The fixed distance is called the **radius** of the sphere.

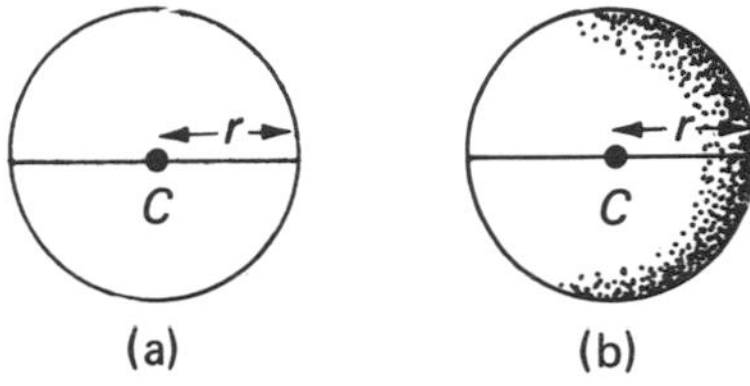

Fig. 10.70(a) A circle with center C and radius r. **(b)** A sphere with center C and radius r.

The volume, V, of a sphere† *is given by the formula*

$$V = \frac{4}{3}\pi r^3, \qquad \text{or} \qquad V = \frac{4\pi r^3}{3}$$

where r is the radius. The surface area, S, is given by

$$S = 4\pi r^2$$

Example 3 A sphere has radius 3 inches. (See Figure 10.71 .)

a. Express the volume in terms of π.

b. Approximate the volume to the nearest cubic inch.

c. Express the surface area in terms of π.

Fig. 10.71

Solution

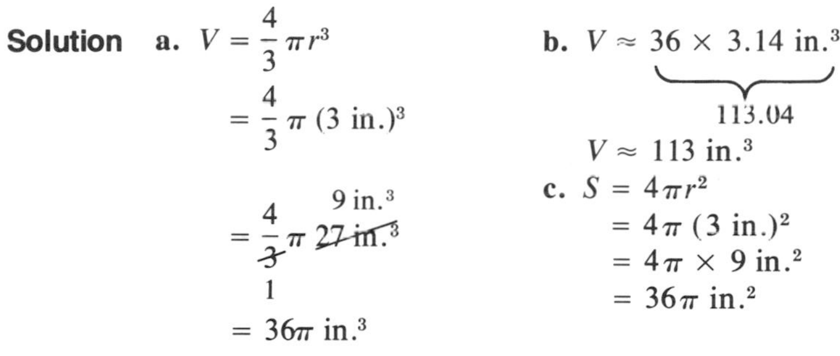

a. $V = \frac{4}{3}\pi r^3$

$= \frac{4}{3}\pi\,(3 \text{ in.})^3$

$= \frac{4}{\cancel{3}_{1}}\pi\,\cancel{27 \text{ in.}^3}^{\,9 \text{ in.}^3}$

$= 36\pi \text{ in.}^3$

b. $V \approx \underbrace{36 \times 3.14}_{113.04} \text{ in.}^3$

$V \approx 113 \text{ in.}^3$

c. $S = 4\pi r^2$

$= 4\pi\,(3 \text{ in.})^2$

$= 4\pi \times 9 \text{ in.}^2$

$= 36\pi \text{ in.}^2$

CLASS EXERCISES

3. Consider a sphere with radius 10 feet. **a.** Express the volume in terms of π. **b.** Approximate the volume to the nearest unit. *Use* $\pi \approx 3.1416$. **c.** Express the surface area in terms of π.

CYLINDERS Figure 10.72(a) shows a (**right circular) cylinder.** Its **bases** (both bottom and top) are circular regions, each of which has **radius** r. Its **height** is h. Note that *the radius and the corresponding height form a right angle.*

†We will speak of the volume of a sphere rather than of a spherical region. Similarly, we will speak of the volume of a cylinder rather than of a cylinderical region, and so on.

The ***volume,*** *V, of a cylinder equals the area of a base times the height.* Because the area of a base is πr^2, it follows that

$$V = \pi r^2 h$$

A rectangular sheet of paper can be rolled to form the **lateral surface** of a cylinder of base radius r and height h. [See Figure 10.72(b).] The lateral surface area is then equal to the area of the sheet. The length of the sheet equals $2\pi r$, the circumference of the circular base, and the width of the sheet equals h, the height of the cylinder. Therefore,

$$\textit{lateral surface area of cylinder} = \textit{area of sheet} = 2\pi r \cdot h$$

Let S be the lateral surface area. Then,

$$S = 2\pi rh$$

(If the top and bottom are also included, the *total surface area* would be

$$\underbrace{2\pi rh}_{\textit{lateral surface area}} + \underbrace{2\pi r^2}_{\substack{2 \times \textit{area of circular base} \\ (\textit{top and bottom})}}$$

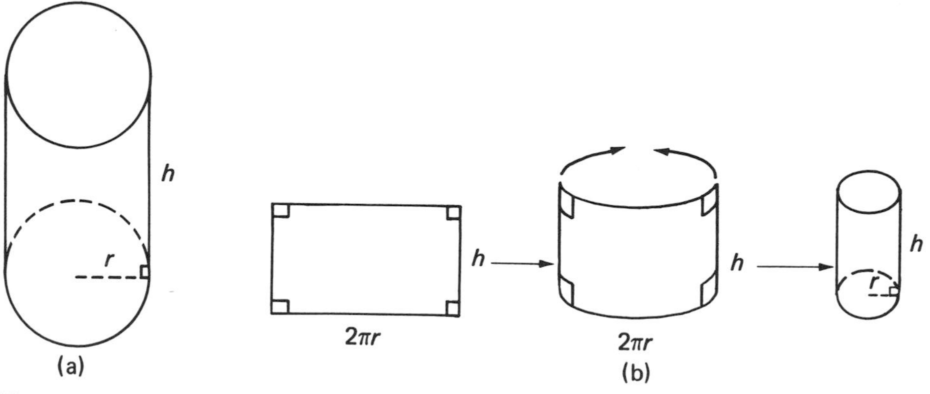

Fig. 10.72 A cylinder with base radius r and height h.

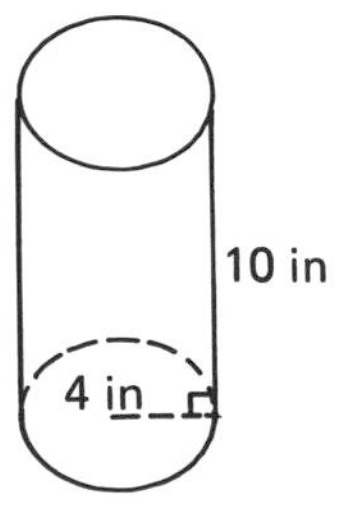

Fig. 10.73

Example 4 Consider the cylinder of Figure 10.73.

a. Express the volume in terms of π.
b. Approximate the volume to the nearest cubic inch. *Use* $\pi \approx 3.142$.
c. Express the lateral surface area in terms of π.

Solution

a. $V = \pi r^2 h$
$= \pi\,(4\text{ in.})^2 \times 10\text{ in.}$
$= \pi \times 16\text{ in.}^2 \times 10\text{ in.}$
$= 160\pi\text{ in.}^3$

b. $V \approx 160 \times 3.142\text{ in.}^3$
$= 502.72\text{ in.}^3$
To the nearest cubic inch,
$V \approx 503\text{ in.}^3$

c. $S = 2\pi rh$
$= 2\pi \times 4\text{ in.} \times 10\text{ in.}$
$= 80\pi\text{ in.}^2$

Example 5 Find the base radius of a cylinder with height 8 centimeters and volume 72π cubic centimeters.

Solution

$$V = \pi r^2 h$$

$$\frac{\overset{9\text{ cm.}^2}{\cancel{72\pi\text{ cm.}}^3}}{\underset{1}{\cancel{8\pi\text{ cm.}}}} = \frac{\overset{1}{\cancel{\pi}}r^2 \times \overset{1}{\cancel{8\text{ cm.}}}}{\underset{1}{\cancel{8\pi\text{ cm.}}}}$$

$$9\text{ cm.}^2 = r^2$$

$$3\text{ cm.} = r$$

CLASS EXERCISES

4. Consider a cylinder with base radius 5 centimeters and height 8 centimeters.
   a. Express the volume in terms of π.
   b. Approximate the volume to the nearest unit. *Use* $\pi \approx 3.142$.
   c. Express the lateral surface area in terms of π.
5. Find the base radius of a cylinder with height 5 feet and volume 20π cubic feet.

CONES Figure 10.74 shows a (**right circular**) **cone**. Its **base** is a circle with radius r. Its **height** is h and its **slant height** is s. *The height together with a radius and the corresponding slant height form a right triangle*. By the Pythagorean Theorem,

$$\boxed{s^2 - r^2 + h^2}$$

Fig. 10.74 . A cone with base radius r, height h, and slant height s.
$s^2 = r^2 + h^2$

The ***volume****, V, of a cone equals* $\frac{1}{3}$ *the area of the base,* πr^2*, times the height, h.*

$$\boxed{V = \frac{1}{3}\pi r^2 h = \frac{\pi r^2 h}{3}}$$

The ***lateral surface area****, S, is given by*

$$\boxed{S = \pi rs}$$

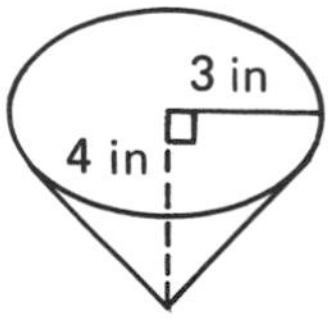

Fig. 10.75

Example 6 Consider a cone with base radius 3 inches and height 4 inches, as in Figure 10.75.

a. Express the volume in terms of π.
b. Approximate the volume to the nearest cubic inch.
c. Find the slant height.
d. Express the lateral surface area in terms of π.

Solution

a.
$$V = \frac{1}{3}\pi r^2 h$$
$$= \frac{1}{3}\pi\,(3\ in.)^2 \times 4\ in.$$
$$= \frac{1}{\cancel{3}_1}\pi\,\overset{3\ in.^2}{\cancel{9\ in.^2}} \times in.$$
$$= 12\,\pi\ in.^3$$

b.
$$V \approx 12 \times 3.14\ \text{in.}^3$$
$$= 37.68\ \text{in.}^3$$
To the nearest cubic inch,
$$V \approx 38\ \text{in.}^3$$

c. By the Pythagorean Theorem,
$$s^2 = 3^2\ \text{in.}^2 + 4^2\ \text{in.}^2$$
$$= 9\ \text{in.}^2 + 16\ \text{in.}^2$$
$$= 25\ \text{in.}^2$$
$$s = 5\ \text{in.}$$

d.
$$S = \pi r s$$
$$= \pi(3\ \text{in.})(5\ \text{in.})$$
$$= 15\pi\ \text{in.}^2$$

CLASS EXERCISES

6. Consider a cone with base radius 6 inches and height 8 inches.
   a. Express the volume in terms of π.
   b. Approximate the volume to the nearest unit. *Use* $\pi \approx 3.142$.
   c. Find the slant height. **d.** Express the lateral surface area in terms of π.
7. Find the height of a cone with base radius 2 inches and volume 28π cubic inches.

SOLUTIONS TO CLASS EXERCISES

1. a. 12 in.^{3} **b.** 38 in.^{2} **2. a.** 4 ft.^{3} **b.** 16 ft.^{2}

3. a. $\frac{4000\pi}{3}$ ft.^{3} **b.** 4189 ft.^{3} **c.** 400π ft.^{2} **4. a.** 200π cc. **b.** 628 cc. **c.** 80π cm.^{2}

5.
$$V = \pi r^2 h$$
$$20\pi\ ft.^3 = \pi r^2 \times 5\ ft.$$
$$4\ ft.^2 = r^2$$
$$2\ ft. = r$$

6. a. 96π in.^{3} **b.** 302 in.^{3} **c.** 10 in. **d.** 60π in.^{2}

7.
$$V = \frac{1}{3}\pi r^2 h$$
$$28\pi\ \text{in.}^3 = \frac{1}{3}\pi\,(2\ \text{in.}^2) \times h$$
$$84\pi\ \text{in.}^3 = 4\pi\ \text{in.}^2 \times h$$
$$21\ \text{in.} = h$$

HOME EXERCISES

*In Exercises 1–6, find **a.** the volume and **b.** the surface area of each rectangular box.*

1. length 5 inches, width 2 inches, height 2 inches **2.** length 4 feet, width 2 feet, height 8 feet

3. length 12 centimeters, width 3 centimeters, height 5 centimeters

4. Fig. 10.76 **5.** Fig. 10.77 **6.** Fig. 10.78

*In Exercises 7–9, find **a.** the volume and **b.** the surface area of each cube.*

7. side length 2 centimeters **8.** side length 10 inches **9.** Fig. 10.79

10. Find the volume of the solid in Figure 10.80.

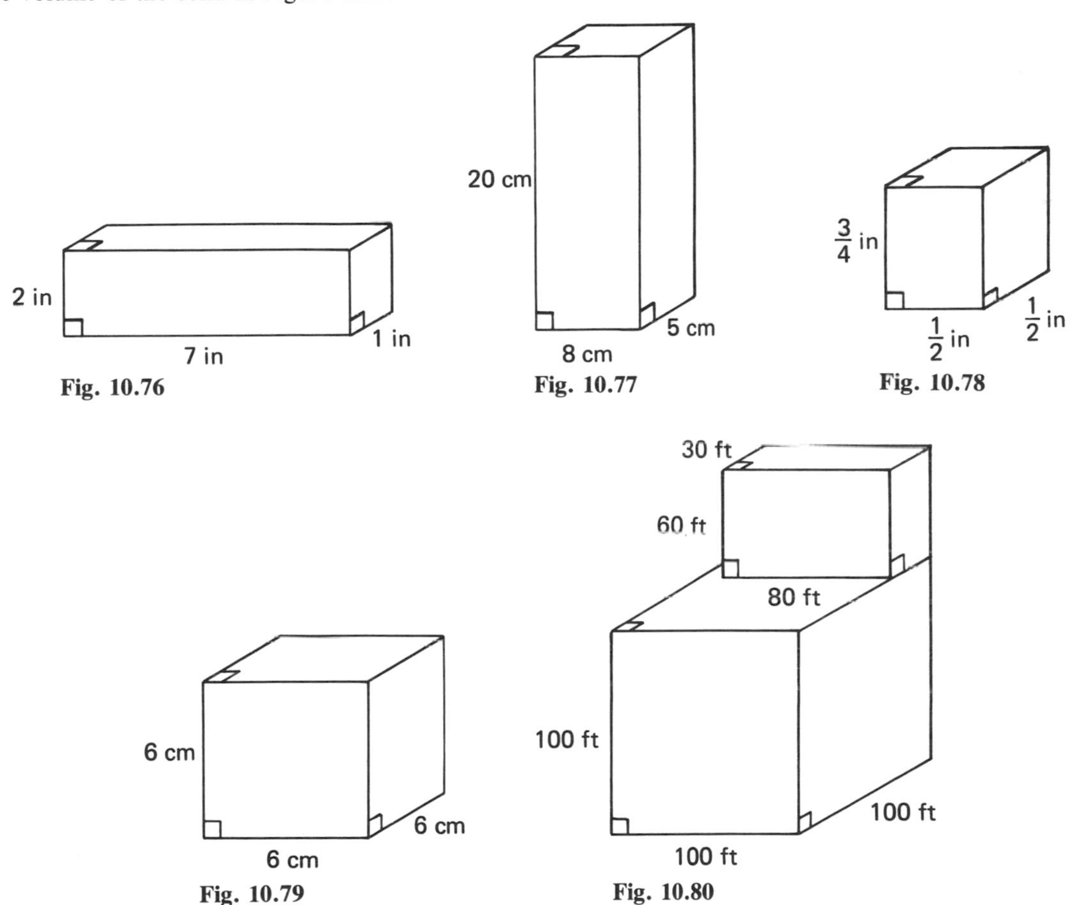

Fig. 10.76 **Fig. 10.77** **Fig. 10.78** **Fig. 10.79** **Fig. 10.80**

11. Find the height of a rectangular box that has length 4 inches, width 2 inches, and volume 40 cubic inches.

12. Find the width of a rectangular box that has length 7 feet, height 10 feet, and volume 140 cubic feet.

13. Find the length of a rectangular box that has width 3 inches, height 8 inches, and volume 96 cubic inches.

14. Find the side length of a cube that has volume 27 cubic inches.

15. Find the volume of air in a room that measures 14 feet by 10 feet by 9 feet.

16. A rectangular swimming pool measures 25 feet by 20 feet by 6 feet. How much water can it hold?

17. a. What is the volume of a cigar box of length 11 inchs, width 6 inches, and height 3 inches?
b. How much paper is needed to line the inside of the box?

18. A trunk measures 3 feet by 2 feet by $1\frac{1}{2}$ feet. What is its volume?

In Exercises 19–21, consider a sphere with the given radius.
a. Express the volume in terms of π. **b.** Approximate the volume to the nearest unit, *using* $\pi \approx 3.142$.
c. Express the surface area in terms of π.

19. 4 kilometers **20.** 6 centimeters **21.** 9 meters

22. Find the radius of a sphere whose volume is $\frac{4\pi}{3}$ cubic centimeters.

23. Find the radius of a sphere whose surface area is 16π square yards.

In Exercises 24–27, consider the given cylinder.
a. Express the volume in terms of π. **b.** Approximate the volume to the nearest unit, *using* $\pi \approx 3.142$.
c. Express the lateral surface area in terms of π.

24. base radius 4 inches, height 10 inches

25. base radius 5 centimeters, height 5 centimeters

26.

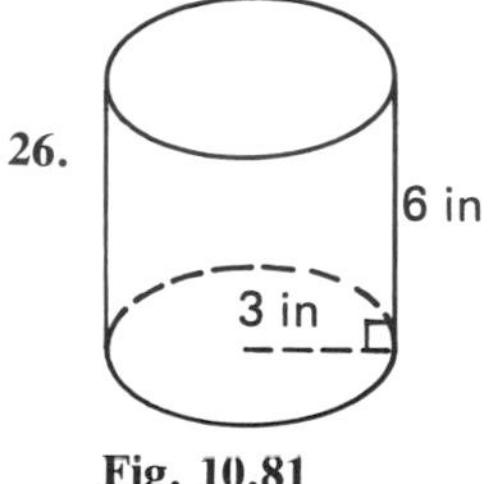

Fig. 10.81

27.

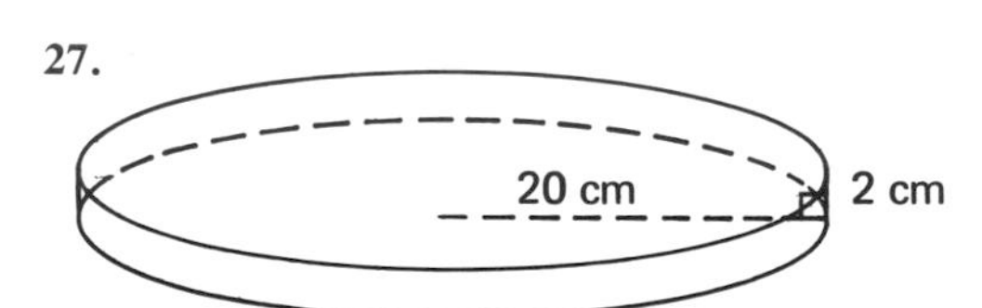

Fig. 10.82

In Exercises 28 and 29, find the height of the given cylinder.

28. base radius 5 inches, volume 150π cubic inches

29. base radius 6 centimeters, volume 72π cubic centimeters

In Exercises 30 and 31, find the base radius of the given cylinder.

30. height 4 inches, volume 400π cubic inches

31. height 5 centimeters, volume 20π cubic centimeters

In Exercises 32–35, consider the given cone.
a. Express the volume in terms of π. **b.** Approximate the volume to the nearest unit. *Use* $\pi \approx 3.14$.
c. Express the lateral surface area in terms of π.

32. base radius 3 inches, height 10 inches

33. base radius 7 centimeters, height 6 centimeters

34.

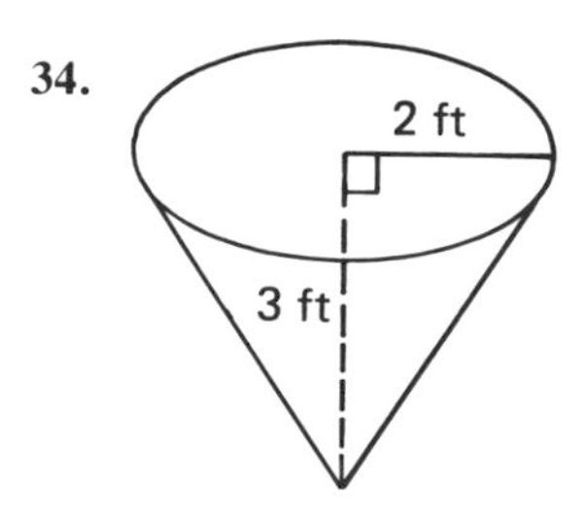

Fig. 10.83

35.

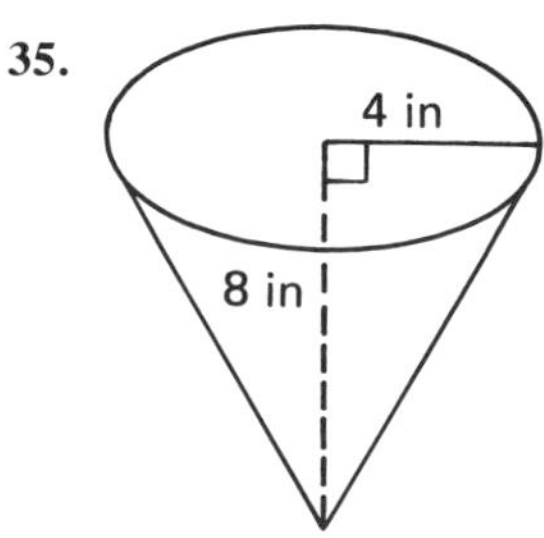

Fig. 10.84

In Exercises 36 and 37, find the height of the given cone.

36. base radius 3 inches, volume 18π cubic inches

37. base radius 2 centimeters, volume 4π cubic centimeters

In Exercises 38 and 39, find the base radius of the given cone.

38. height 5 inches, volume 15π cubic inches

39. height 2 centimeters, volume 18π cubic centimeters

40. Suppose the volume of the cylinder in Figure 10.85 is 300π cubic inches. Find the volume of the cone.

41. Suppose the volume of the cone in Figure 10.86 is 100π cubic centimeters. Find the volume of the cylinder.

42. A volleyball has a 5-inch radius. Find its volume.

43. A cylindrical tin can has base radius 2 inches and height 5 inches. Find its volume.

44. A conical sand pile has base radius 4 inches and height 7 inches. (See Figure 10.87.) Find its volume.

45. A cone is filled with ice cream and is topped with a *hemisphere* (half of a sphere) of ice cream. (See Figure 10.88.) The radius of the hemisphere and of the base of the cone is 2 inches. The height of the cone is 4 inches. Find the volume of ice cream.

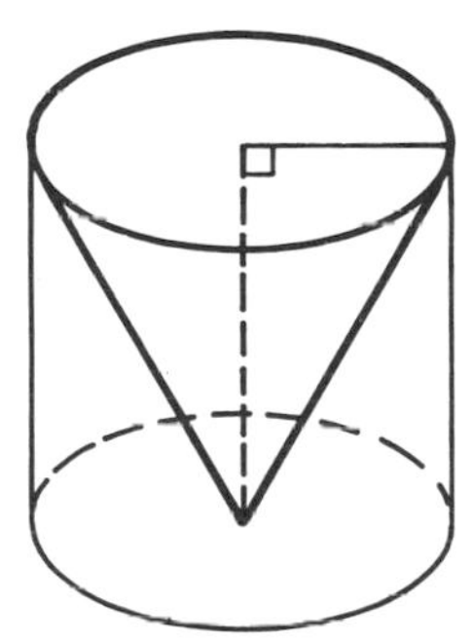

Fig. 10.85

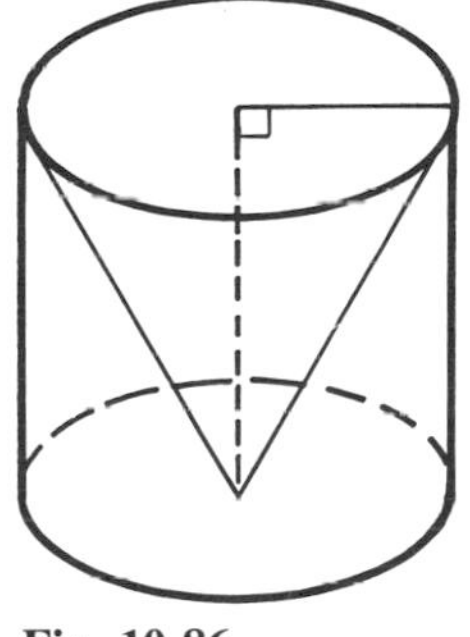

Fig. 10.86

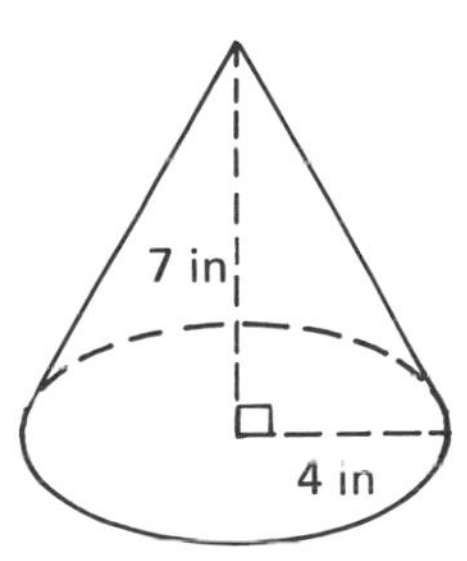

Fig. 10.87. Conical sand pile.

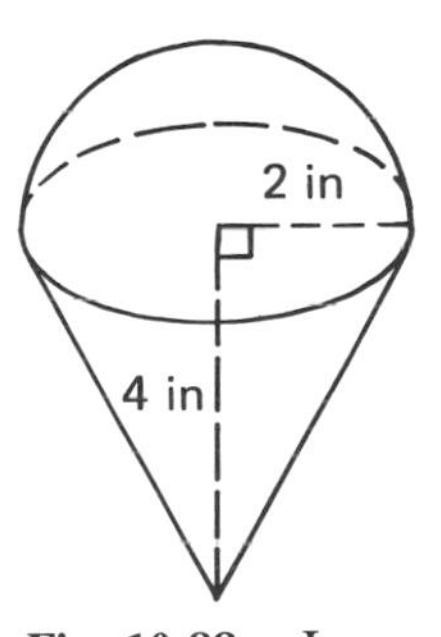

Fig. 10.88. Ice cream cone.

Let's Review Chapter 10

In Exercises 1–4, convert from one unit to another.

1. $3\frac{1}{2}$ feet into inches

2. 5.67 meters into centimeters

3. 10 meters into inches (Express your result to the nearest tenth of an inch.)

4. 32 miles into kilometers (Express your result to the nearest kilometer.)

5. Find **a.** the perimeter and **b.** the area of a rectangle that has length 5 centimeters and width 4 centimeters.

6. Find **a.** the perimeter and **b.** the area of a square whose side length is 9 inches.

7. Find the height of a rectangle with base 6 inches and area 42 square inches.

8. Find the area of the indicated figure, which consists of a rectangle and four squares. (See Fig. 10.89.)

9. Find the area of a triangle with base 9 centimeters and height 10 centimeters.

10. Find the area of a parallelogram with base 7 inches and height 6 inches.

11. Find the base of a parallelogram with height 9 meters and area 108 square meters.

12. Find the area of a trapezoid with bases 8 inches and 10 inches and height 6 inches.

13. Find the radius of a circle with diameter 9 centimeters.

14. Find the area of a circle with radius 2 feet.

a. Express your answer in terms of π.

b. Approximate to the nearest tenth of a square foot.

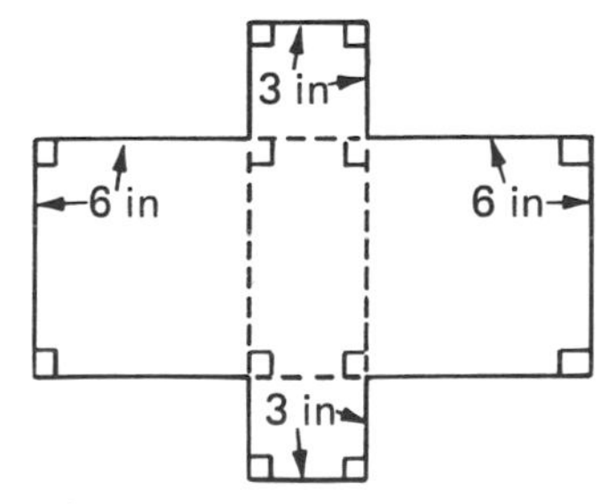

Fig. 10.89

15. Suppose the circumference of a circle is 8π centimeters. Find **a.** the diameter, **b.** the radius, **c.** the area.
16. Find **a.** the volume and **b.** the surface area of a rectangular box with length 4 inches, width 3 inches, and height 5 inches.
17. Find **a.** the volume and **b.** the surface area of a cube of side length 4 centimeters.

In Exercises 18–20, for each figure:

a. Express the volume in terms of π.
b. Approximate the volume to the nearest cubic inch.
c. Express the surface area in terms of π.

18. a sphere of radius 4 inches
19. a cylinder with base radius 3 inches and height 4 inches
20. a cone with base radius 4 inches and height 6 inches
21. Find the base radius of a cylinder with height 8 centimeters and volume 32π cubic centimeters.

11 ANGLES

DIAGNOSTIC TEST Perhaps you are already familiar with some of the material in Chapter 11. This test will indicate which sections in Chapter 11 you need to study. The question number refers to the corresponding section in Chapter 11. If you answer *all* parts of the question correctly, you may omit the section. *But if any part of your answer is wrong, you should study the section*. When you have completed the test, turn to page A31 for the answers.

11.1 Intersecting Lines

a. Convert 25.4° into degrees and minutes.
b. Find the measure of the complement of a 36°22′ angle.
c. Find the measure of the supplement of a 36°22′ angle.
d. In Figure 11A, find the measure of $\angle ABC$.

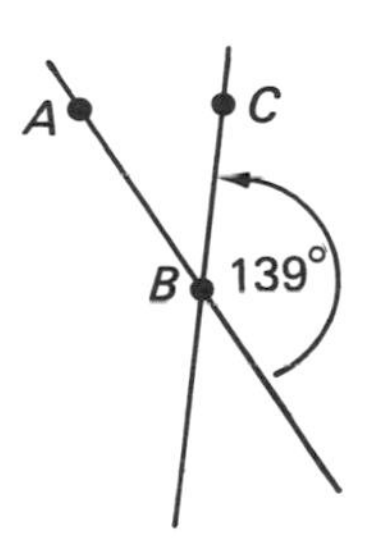

Fig. 11A

11.2 More on Triangles

a. Each of the base angles of an isosceles triangle measures 41°. Find the measure of the third angle.
b. Find the area of a 30°, 60°, 90°-triangle whose hypotenuse is 6 centimeters.
c. Find the side length of an equilateral triangle whose area is $25\sqrt{3}$ inches.
d. One triangle has base 16 inches and corresponding height 4 inches. A similar triangle has base 8 inches. Find the corresponding height.

11.3 More on Circles

Refer to Figure 11B. Assume O is the center of the circle. Find:

a. the measure of $\angle ABC$ b. the length of $\widehat{AC}$ c. the area of sector OCA

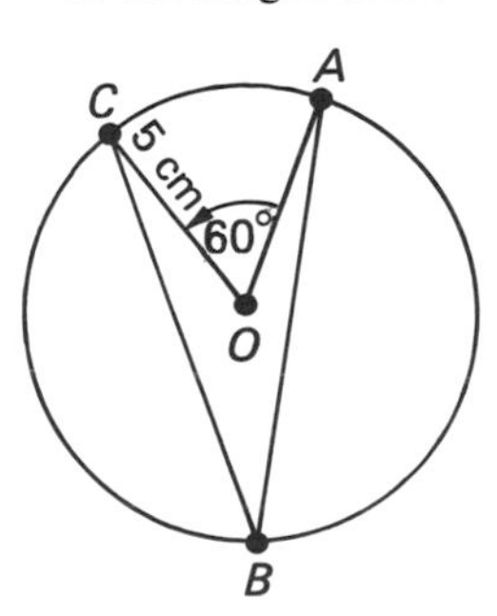

Fig. 11B

11.1 INTERSECTING LINES

MEASURING ANGLES

Recall that *angles* are formed when lines intersect. The intersecting lines are called the *sides* of the angle, and the point of intersection is called the *vertex*. To *identify* an angle, we often use points on their sides. Thus, in Figure 11.1,

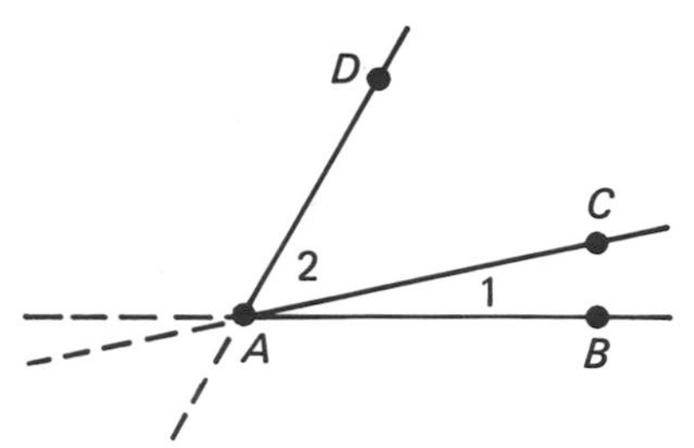

Fig. 11.1 $\angle 1 = \angle BAC$
$\angle 2 = \angle CAD$

$$\angle 1 = \angle BAC \qquad (\textit{Read: Angle 1 = Angle BAC})$$

$$\angle 2 = \angle CAD$$

Note that A, the *second* letter in each case, represents the vertex. We often speak of (the *line segments*) AB and AC as the *sides* of $\angle 1$. (Recall that line segment AB is the part of the line between points A and B.) In Figure 11.2, there is only one angle indicated, and we call it $\angle A$. The shaded region is called the **interior of** $\angle A$.

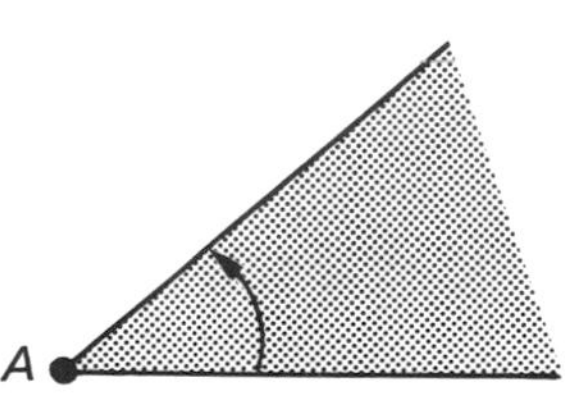

Fig. 11.2

An angle is *measured* in **degrees.** Finer measurements of angles employ degrees and **minutes.** *Each degree equals 60 minutes* (60′). Alternatively, decimals can be used. Thus, 20.3° means 20 and *three-tenths* degrees, whereas 20°03′ means 20 degrees and *three minutes*. To convert from decimal notation to degrees and minutes, observe that

$$.1° = .1 \times 60 \text{ minutes}$$

Thus,

$$\boxed{.1° = 6 \text{ minutes}}$$

Example 1 Convert into degrees and minutes. **a.** .4° **b.** 12.4°

Solution

a. $.4° = 4 \times .1°$
$= 4 \times 6$ minutes
$= 24$ minutes

b. $12.4° = 12° + .4°$
$= 12°24'$

Example 2 Express 17°33′ in decimal notation.

Solution

60 minutes = 1 degree

First, express the fraction $\frac{33}{60}$ as a decimal.

$$6.0\overline{)3.30}\quad = .55$$

Thus, $17°33' = 17.55°$

Write $\angle A = 40°$ when $\angle A$ measures 40°

CLASS EXERCISES

1. Convert 32.7° into degrees and minutes.
2. Convert 56°27′ into decimal notation.

CLASSIFYING ANGLES

As you know, a *right angle measures 90°*. The lines forming a right angle are said to be *perpendicular*. Write

$$L_1 \perp L_2$$

when lines L_1 and L_2 are perpendicular. An angle that measures 180° is called a **straight angle.** An **acute angle** is an angle that measures between 0° and 90°, whereas an **obtuse angle** is one that measures between 90° and 180°. Thus,

$$\angle A \text{ is acute if } 0° < \angle A < 90°; \qquad \angle B \text{ is obtuse if } 90° < \angle B < 180°$$

(See Figure 11.3.)

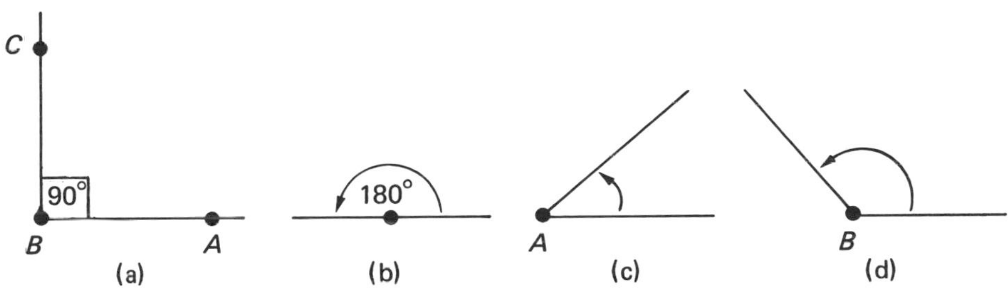

Fig. 11.3(a) A right angle measures 90°. $AB \perp BC$. **(b)** A straight angle measures 180°. **(c)** $\angle A$ is acute because $0° < \angle A < 90°$. **(d)** $\angle B$ is obtuse because $90° < \angle B < 180°$.

Example 3 Classify each angle as either a right angle, a straight angle, an acute angle, or an obtuse angle.

a. 72° **b.** 172° **c.** 180°

Solution **a.** acute **b.** obtuse **c.** straight

Write $\angle 3 = \angle 1 + \angle 2$

if (the measure of) $\angle 3$ is the sum of (the measures of) angles 1 and 2. In this case, $\angle 3$ is called the **sum of angles 1 and 2.** If $\angle 1 + \angle 2 = 90°$, each of these angles is the **complement** of the other. Thus, the complement of a 60° angle is a 30° angle. If $\angle 1 + \angle 2 = 180°$, each of these angles is the **supplement** of the other. The supplement of a 130° angle is a 50° angle.

Example 4 Suppose $\angle A = 53°20'$. Find the measure of **a.** its complement, **b.** its supplement.

Solution **a.** To find the measure of the complement of $\angle A$, consider:

$$\begin{array}{r} 90° \\ -53°20' \\ \hline \end{array}$$

In order to subtract *minutes*, borrow 60 minutes (1 degree) from 90°.

$$\begin{array}{r} 89°60' \\ \not{90}° \\ -53°20' \\ \hline 36°40' \end{array}$$

The complement of $\angle A$ measures 36°40'.

b.

$$\begin{array}{r} 179°60' \\ \not{180}° \\ -\ 53°20' \\ \hline 126°40' \end{array}$$

The supplement of $\angle A$ measures 126°40'.

Note that

measure of supplement of $\angle A$ = *90°* + *measure of complement of* $\angle A$

Example 5 Suppose the complement of $\angle A$ measures three times as much as $\angle A$. Find the measure of $\angle A$.

Solution Let x be the measure of $\angle A$. Reword the first sentence.

$$\underbrace{\text{The measure of the complement of } \angle A}_{90 - x} \ \underset{=}{\textit{is}} \ \underbrace{\textit{three times the measure of } \angle A}_{3x}.$$

Thus,

$$90 = 4x$$

$$\frac{90}{4} = x$$

$$22.5 = x$$

$\angle A = 22.5°$, or 22°30'

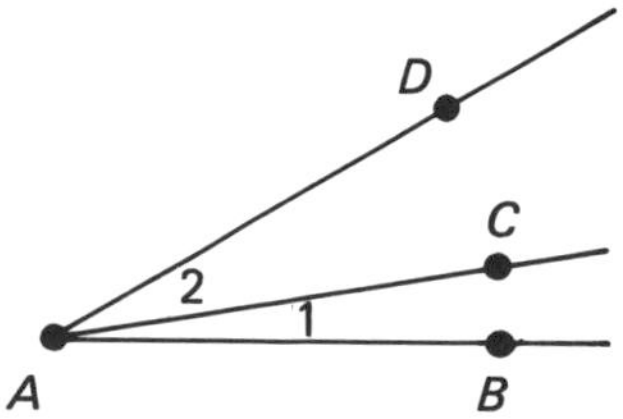

Fig. 11.4. $\angle 1$ and $\angle 2$ are adjacent angles. They share the common side AC. $\angle 1 + \angle 2 = \angle BAD$.

Angles that share a common side are known as **adjacent angles.** In Figure 11.4, $\angle 1$ and $\angle 2$ share a common side, AC. In Figure 11.5, $\angle 1$ and $\angle 2$ are

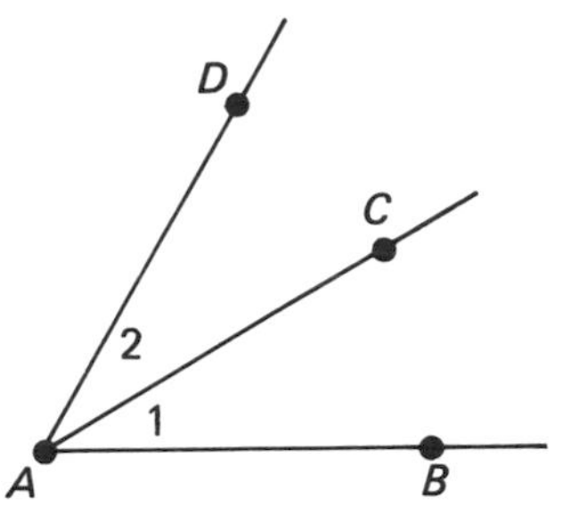

Fig. 11.5. $\angle 1 = \angle 2$ and $\angle 1 + \angle 2 = \angle BAD$. Here, AC bisects $\angle BAD$.

equal adjacent angles. Their common side, AC, is called the **bisector of** $\angle BAD$.

When lines intersect, four angles are formed. In Figure 11.6, $\angle 1$ and $\angle 3$ are called **vertical angles,** as are $\angle 2$ and $\angle 4$. Note that

$$\angle 1 + \angle 2 = 180°, \qquad \angle 3 + \angle 2 = 180°$$

From the first equation,

$$\angle 1 = 180° - \angle 2$$

And from the second equation,

$$\angle 3 = 180° - \angle 2$$

Thus,

$$\angle 1 = \angle 3$$

In general, *vertical angles are equal (in measure).*

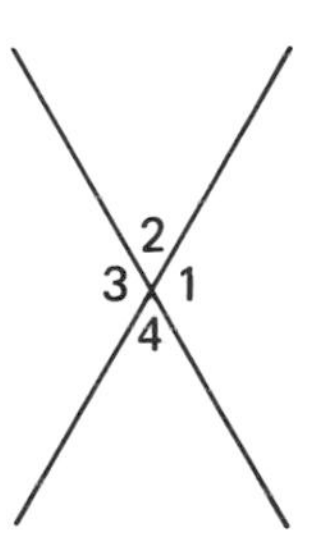

Fig. 11.6. $\angle 1$ and $\angle 3$ are vertical angles; $\angle 2$ and $\angle 4$ are vertical angles. $\angle 1 = \angle 3$; $\angle 2 = \angle 4$.

CLASS EXERCISES

In Exercises 3–5, suppose $\angle A = 85°30'$.

3. Is $\angle A$ acute or obtuse? **4.** Find the complement of $\angle A$. **5.** Find the supplement of $\angle A$. **6.** Suppose the supplement of $\angle B$ measures four times as much as the complement of $\angle B$. Find the measure of $\angle B$.

TRANSVERSALS

A line that intersects two or more other lines is known as the **transversal** of those lines. In Figure 11.7, L is the transversal of lines ℓ_1 and ℓ_2. The pairs of angles

$$\angle 1 \text{ and } \angle 5, \qquad \angle 2 \text{ and } \angle 6, \qquad \angle 3 \text{ and } \angle 7, \qquad \angle 4 \text{ and } \angle 8$$

are called **corresponding angles.** The pairs of angles

$$\angle 1 \text{ and } \angle 7, \qquad \angle 2 \text{ and } \angle 8$$

are called **alternate interior angles,** whereas the pairs of angles

$$\angle 3 \text{ and } \angle 5, \qquad \angle 4 \text{ and } \angle 6$$

are known as **alternate exterior angles.**

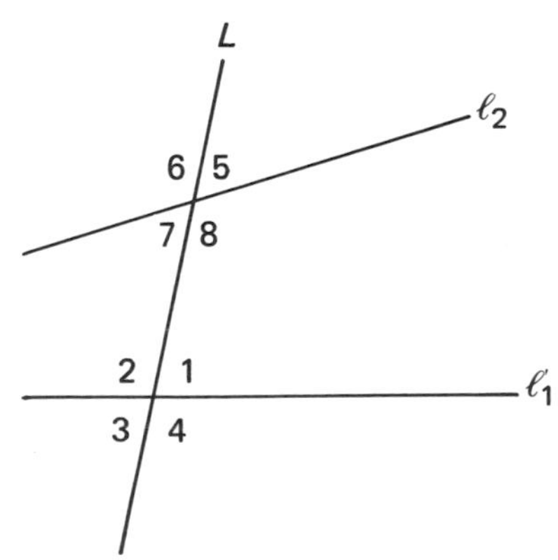

Fig. 11.7

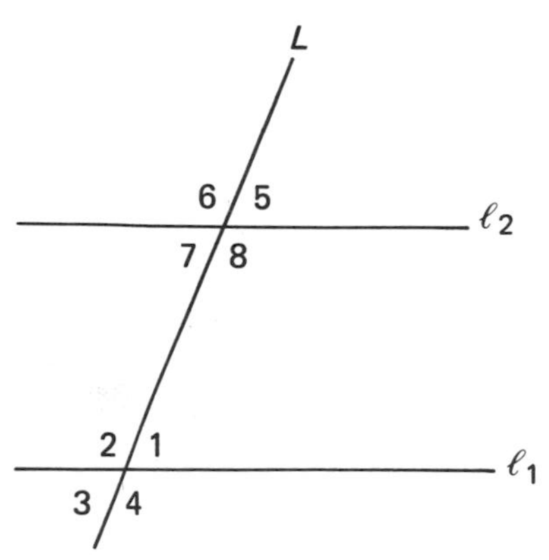

Fig. 11.8. $\ell_1 \parallel \ell_2$. $\angle 1 = \angle 5$, $\angle 2 = \angle 8$, $\angle 4 = \angle 6$.

Recall that *parallel lines* are lines on a plane that do not intersect. Write

$$\ell_1 \parallel \ell_2$$

when lines ℓ_1 and ℓ_2 are parallel. Consider a *transversal of parallel lines*, as in Figure 11.8. Then,

corresponding angles are equal (in measure)

alternate interior angles are equal

and *alternate exterior angles are equal*

In Figure 11.8, $\ell_1 \parallel \ell_2$. Therefore,

$\angle 1 = \angle 5$ (*Corresponding angles are equal.*)

$\angle 2 = \angle 8$ (*Alternate interior angles are equal.*)

$\angle 4 = \angle 6$ (*Alternate exterior angles are equal.*)

Now, suppose that two arbitrary lines, ℓ_1 and ℓ_2, are cut by a transversal.

If corresponding angles are equal, then $\ell_1 \parallel \ell_2$

Similarly, *if alternate interior angles are equal, then* $\ell_1 \parallel \ell_2$

And *if alternate exterior angles are equal, then* $\ell_1 \parallel \ell_2$

Example 6 In Figure 11.9, suppose that $\angle 4 = 100°$ and $\angle 5 = 80°$. Show that $\ell_1 \parallel \ell_2$.

Solution

$$\angle 1 = \text{supplement } \angle 4$$
$$= 180° - 100°$$
$$= 80°$$

Thus, $\angle 1 = \angle 5 = 80°$

Therefore, $\ell_1 \parallel \ell_2$ because corresponding angles are equal. (What other angles could you have used to show that $\ell_1 \parallel \ell_2$?)

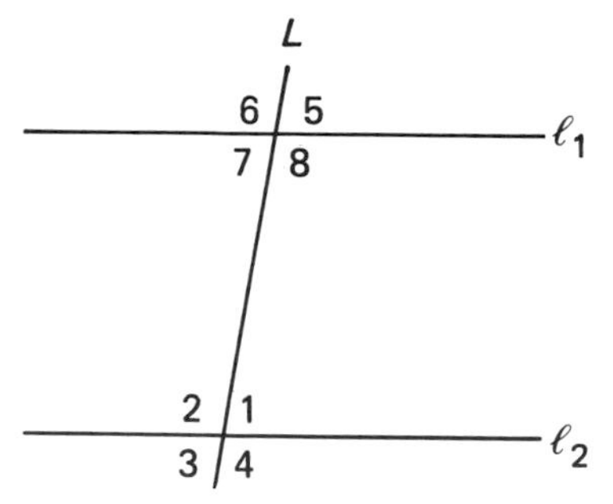

Fig. 11.9. $\angle 4 = 100°$, $\angle 5 = 80°$.

Write

$$AB = 7 \text{ inches}$$

if the length of the line segment AB is 7 inches.

When three (or more) parallel lines are cut by two (or more) transversals, the lengths of corresponding line segments of the transversals are proportional.

In Figure 11.10, ℓ_1, ℓ_2, and ℓ_3 are parallel. (Write $\ell_1 \parallel \ell_2 \parallel \ell_3$.) Thus,

$$\frac{AB}{BC} = \frac{DE}{EF}$$

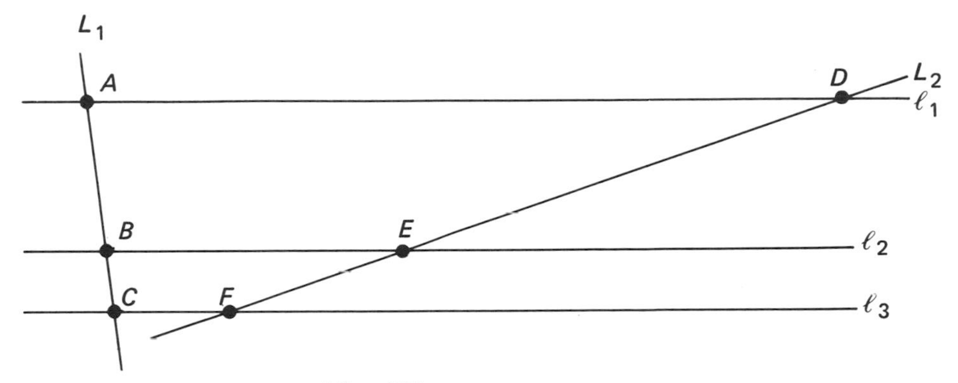

Fig. 11.10 $\ell_1 \parallel \ell_2 \parallel \ell_3$. $\dfrac{AB}{BC} = \dfrac{DE}{EF}$.

Example 7 In Figure 11.10, $\ell_1 \parallel \ell_2 \parallel \ell_3$. Suppose $AB = 5$ inches, $BC = 2$ inches, and $DE = 15$ inches. Find the length of EF.

Solution

$$\frac{AB}{BC} = \frac{DE}{EF}$$

$$\frac{5 \text{ in.}}{2 \text{ in.}} = \frac{15 \text{ in.}}{x \text{ in.}}$$

Cross-multiply: $5x = 30$

$x = 6$

Thus, $EF = 6$ inches.

CLASS EXERCISES *Exercises 7–10 refer to Figure 11.11. Suppose $\angle 1 = 75°$. Find the measure of each angle.*

7. $\angle 2$
8. $\angle 3$
9. $\angle 4$
10. $\angle 5$

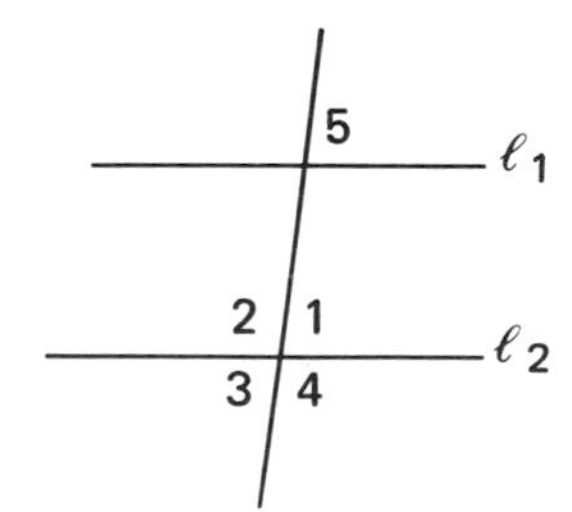

Fig. 11.11. $\ell_1 \parallel \ell_2$.

11. In Figure 11.12, assume $\ell_1 \parallel \ell_2 \parallel \ell_3$. Suppose $AB = 4$ inches, $DE = 12$ inches, and $EF = 15$ inches. Find the length of BC.

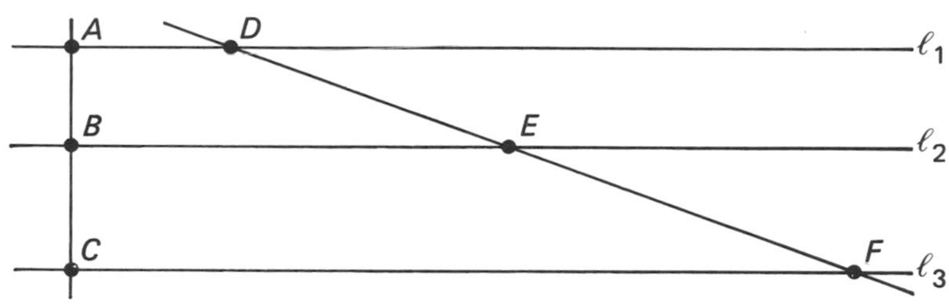

Fig. 11.12. $\ell_1 \parallel \ell_2 \parallel \ell_3$.

SOLUTIONS TO CLASS EXERCISES

1. 32°42′ **2.** 56.45° **3.** acute **4.** 4°30′ **5.** 94°30′

6. Let x be the measure of $\angle$B.

The measure of the supplement of $\angle$B | is | four times | that of the complement of $\angle$B.

Thus,

$$\begin{aligned} 180 - x &= 4\,(90 - x) \\ 180 - x &= 360 - 4x \\ 3x &= 180 \\ x &= 60 \end{aligned}$$

$\angle A = 60°$

7. 105° **8.** 75° **9.** 105° **10.** 75° **11.** 5 in.

HOME EXERCISES

In Exercises 1–5, convert into degrees and minutes.

1. .2° **2.** 40.2° **3.** 19.5° **4.** 100.8° **5.** 120.05°

In Exercises 6–10, write in decimal notation.

6. 5°30′ **7.** 12°12′ **8.** 39°48′ **9.** 80°03′ **10.** 118°57′

In Exercises 11–13, classify as either a right angle, a straight angle, an acute angle, or an obtuse angle.

11. 79° **12.** 180° **13.** 91°

In Exercises 14–18, find the measure of the complement of each angle.

14. 10° **15.** 45° **16.** 7°37′ **17.** 19.2° **18.** 19°02′

In Exercises 19–23, find the measure of the supplement of each angle.

19. 50° **20.** 37.4° **21.** 112°15′ **22.** 39 minutes **23.** a right angle

24. An angle measures twice as much as its complement. Find the measure of the angle.

25. An angle measures one-fifth as much as its complement. Find the measure of the angle.

26. An angle measures four-fifths as much as its supplement. Find the measure of the angle.

27. An angle measures 30° more than its supplement. Find the measure of the angle.

28. The supplement of angle A measures three times as much as the complement of angle A. Find the measure of angle A.

29. The complement of angle B measures one-tenth as much as the supplement of angle B. Find the measure of angle B.

30. In Figure 11.13, $\angle BAD = 45°$ and AC is the bisector of $\angle BAD$. Find the measure of $\angle 1$.

31. In Figure 11.14, $\angle FEH = 20°28'$ and EG is the bisector of $\angle FEH$. Find the measure of $\angle 2$.

Exercises 32–34 refer to Figure 11.15. Find the measure of each angle.

32. $\angle 2$ **33.** $\angle 3$ **34.** $\angle 4$

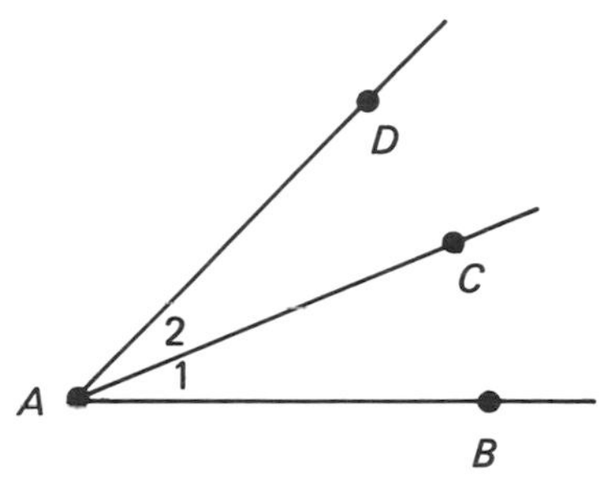

Fig. 11.13

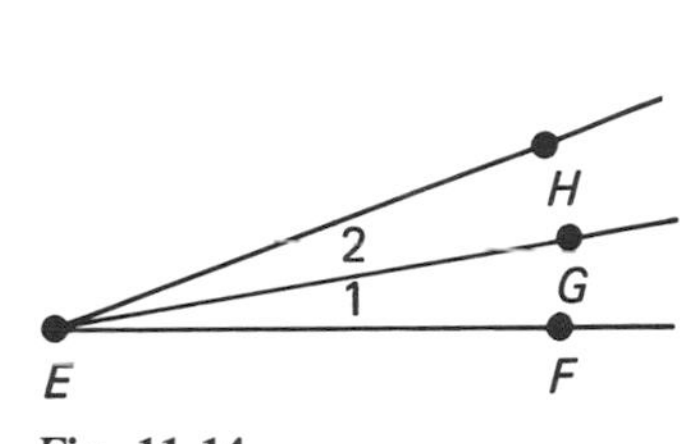

Fig. 11.14

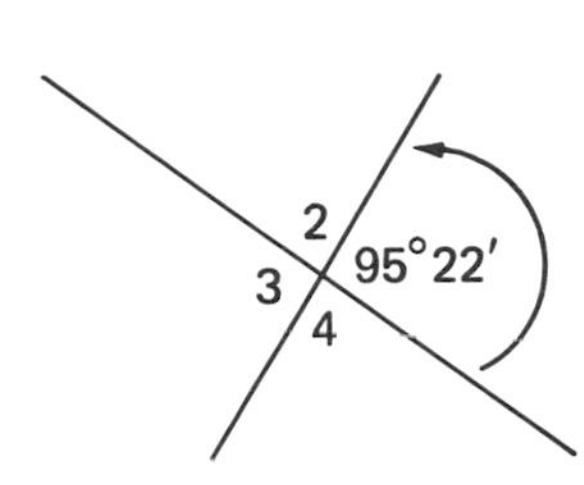

Fig. 11.15

Exercises 35–39 refer to Figure 11.16. Assume $\ell_1 \parallel \ell_2$ and $\angle 5 = 110°$. Find the measure of each angle.

35. $\angle 1$ **36.** $\angle 2$ **37.** $\angle 3$ **38.** $\angle 4$ **39.** $\angle 6$

Exercises 40–49 refer to Figure 11.17. Assume $\ell_1 \parallel \ell_2 \parallel \ell_3 \parallel \ell_4$. Also, $\angle BAE = 88°$, $\angle EFB = 117°$, $AB = 3$ inches, $BC = 6$ inches, $EF = 4$ inches, $FH = 24$ inches. Find the measure of each angle.

40. $\angle CBF$ **41.** $\angle GCB$ **42.** $\angle HDC$ **43.** $\angle AEF$ **44.** $\angle CGH$ **45.** $\angle GHD$

Find the length of each line segment.

46. FG **47.** BD **48.** CD **49.** GH

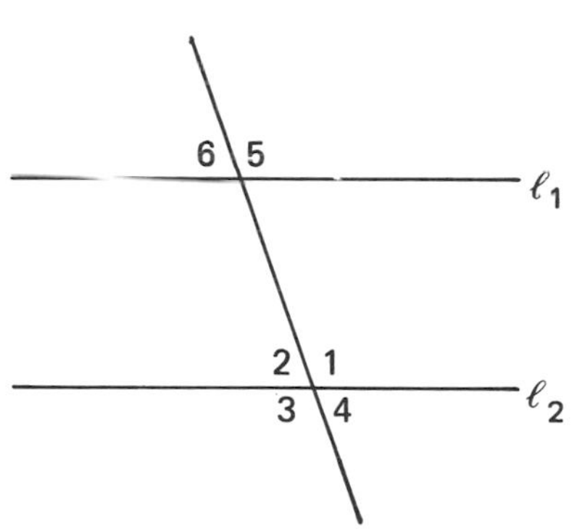

Fig. 11.16

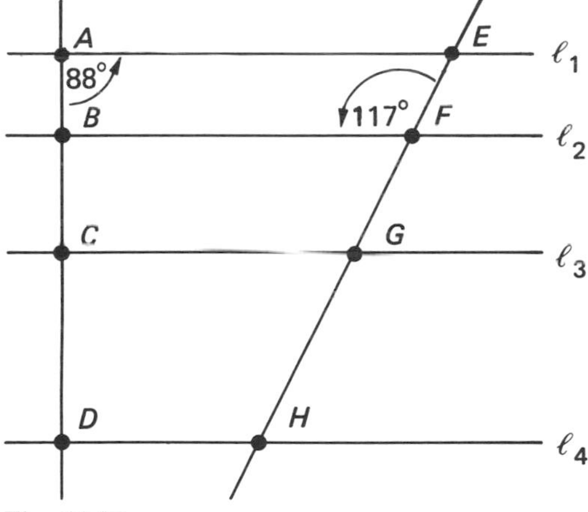

Fig. 11.17

11.2 MORE ON TRIANGLES

ANGLES AND TRIANGLES In Figure 11.18 on page 362, the three vertices of the triangle are labeled A, B and C, and the triangle is denoted by

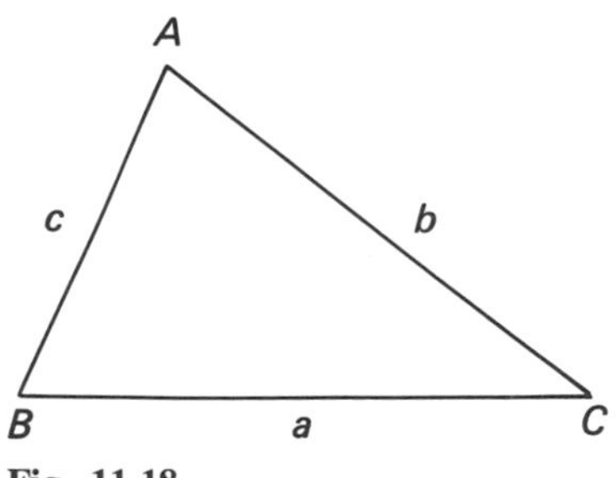

Fig. 11.18

$$\Delta ABC$$

The side opposite $\angle A$ is labeled a, the side opposite $\angle B$ is labeled b, and the side opposite $\angle C$ is labeled c.

The sum of the measures of the angles of a triangle is 180°

Thus, in ΔABC of Figure 11.19, $\angle A = 55°40'$ and $\angle B = 70°$. It follows that $\angle C$ is the supplement of $\angle A + \angle B$, or $54°20'$.

Recall that a *right triangle* is a triangle in which one angle is a right angle. An **acute triangle** is a triangle in which all three angles are acute. An **obtuse triangle** is a triangle in which one angle is obtuse. (See Figure 11.20 .)

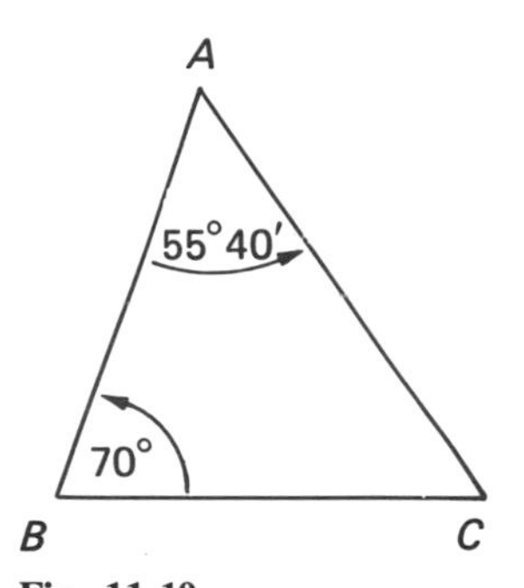

Fig. 11.19 .
$\angle C = 180° - (55°40' + 70°)$
$= 180° - 125°40'$
$= 54°20'$

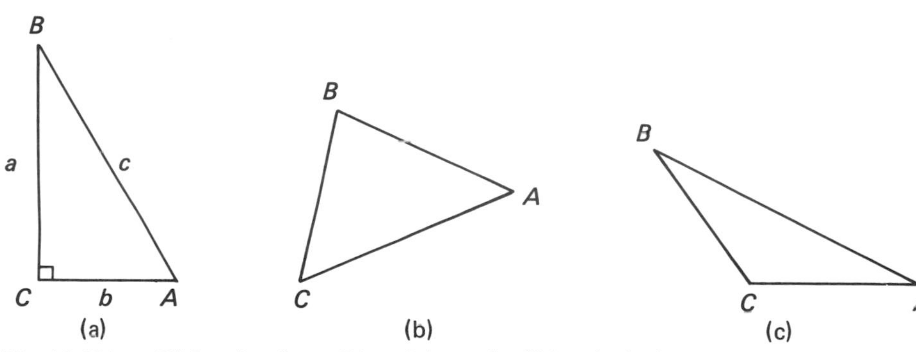

Fig. 11.20(a) Right triangle. $\angle C$ is a right angle. Side c is the hypotenuse. **(b)** Acute triangle. **(c)** Obtuse triangle. $\angle C$ is obtuse.

CLASS EXERCISES

1. **a.** In Figure 11.21, determine the measure of $\angle C$.
   b. Is this a right triangle, an acute triangle, or an obtuse triangle?
2. In Figure 11.22, $\angle C$ is a right angle. Determine the measure of $\angle A$.

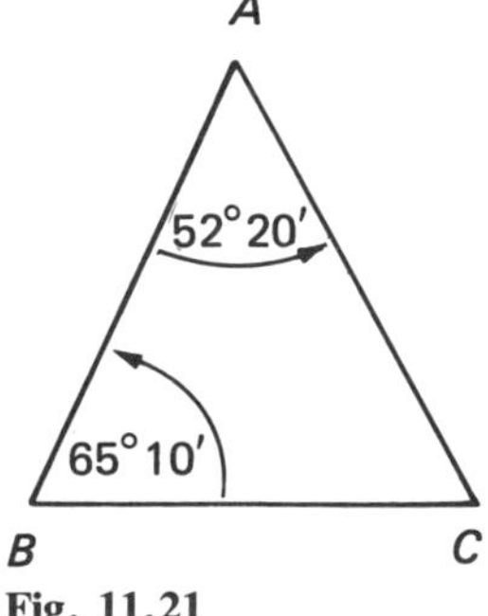

Fig. 11.21

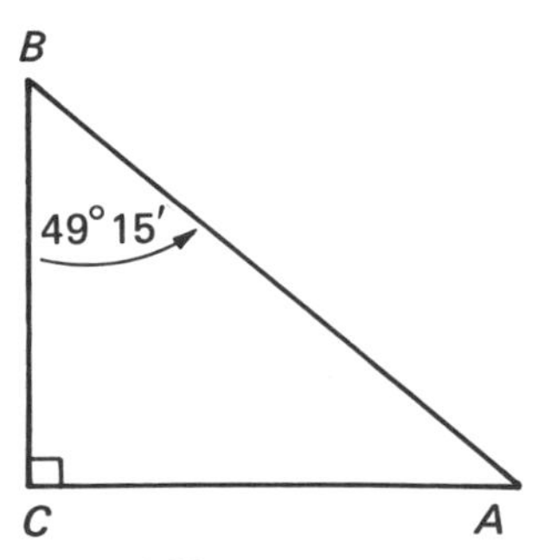

Fig. 11.22

ISOSCELES TRIANGLES An **isosceles triangles** is a triangle in which exactly two sides are equal (in length). The equal sides are called its **legs** and the third side is usually called the **base**. *The angles opposite the legs are also equal* (*in measure*) and are called the **base angles**. Also, *the height corresponding to the base bisects the base, that is, divides it into two equal line segments*. Figure 11.23 shows an isosceles triangle with legs b and c. Note that $\angle B = \angle C$. Also, $BD = DC$

If exactly two angles of a triangle are equal, the triangle is isosceles. Thus, Figure 11.24 depicts an isosceles triangle with legs b and c.

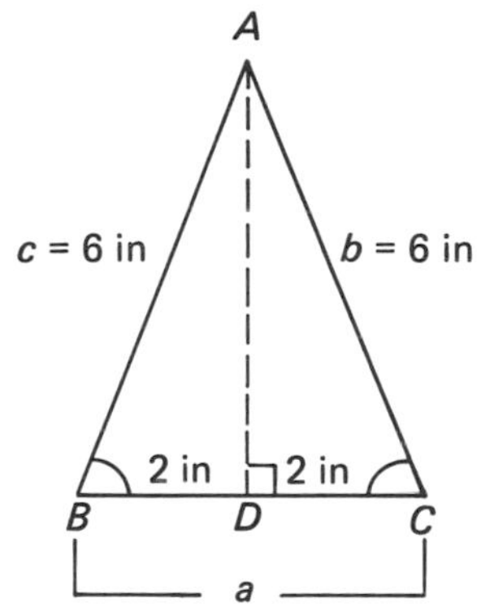

Fig. 11.23. $b = c$. Therefore, $\angle B = \angle C$. Also, $BD = DC$.

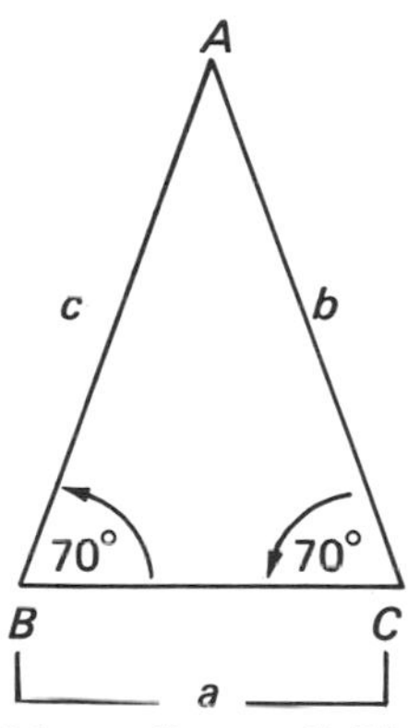

Fig. 11.24. $\angle B = \angle C$. Therefore, $b = c$.

In an ***isosceles right triangle***, *the equal base angles must each measure* 45°. In fact, if x is the number of degrees in each of these equal angles, then

$$x + x + 90 = 180$$
$$2x = 90$$
$$x = 45$$

Thus, an *isosceles* right triangle is often called a **45°, 45°, 90° - triangle**.

Example 1 Find **a.** the hypotenuse and **b.** the area of ΔABC of Figure 11.25 on page 364.

Solution **a.** $\angle B = 45°$, and thus, ΔABC is an isosceles right triangle. It follows that $b = 5$ inches. By the Pythagorean Theorem,

$$a^2 + b^2 = c^2$$
$$(5 \text{ in.})^2 + (5 \text{ in.})^2 = c^2$$
$$25 \text{ in.}^2 + 25 \text{ in.}^2 = c^2$$

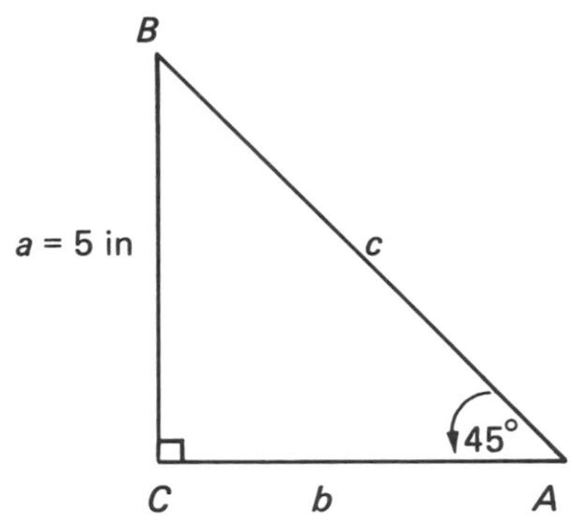

Fig. 11.25

$$50 \text{ in.}^2 = c^2$$
$$\sqrt{50} \text{ in.} = c$$
$$5\sqrt{2} \text{ in.} = c$$

b. The area of a triangle is given by the formula

$$\text{Area} = \frac{\text{base} \cdot \text{height}}{2}$$
$$A = \frac{5 \text{ in.} \times 5 \text{ in.}}{2}$$
$$A = \frac{25}{2} \text{ in.}^2$$

Example **1a** illustrates that *in an isosceles right triangle with legs a and b and hypotenuse c, the sides are in the ratio of*

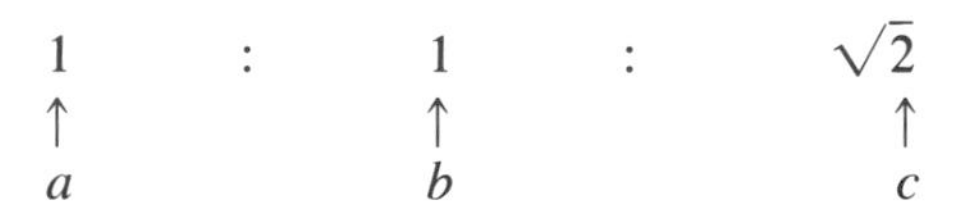

In other words, sides a and b are of equal length and

$$\textit{length of } c = \sqrt{2} \times (\textit{length of } a)$$

CLASS EXERCISES

3. Consider Figure 11.26. **a.** Show that $\triangle ABC$ is an isosceles triangle. **b.** Find the length of side BC.

4. In Figure 11.27, suppose ΔABC is an isosceles right triangle. Find **a.** its hypotenuse and **b.** its area.

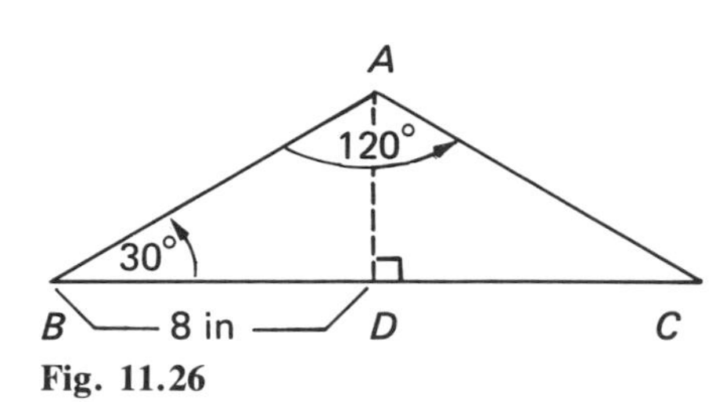

Fig. 11.26

Fig. 11.27

EQUILATERAL TRIANGLES

An **equilateral triangle** is one in which all three side lengths (and all three angle measurements) are equal. *Each angle of an equilateral triangle measures 60°. Each height bisects the corresponding base as well as the corresponding angle.* [See Figure 11.28(a).]

In Figure 11.28(b), ΔADC is known as a **30°, 60°, 90° - triangle**. Note that DC, the side opposite the 30° angle, is $\frac{1}{2}$ as long as side AC, because

$$DC = \frac{1}{2} BC = \frac{1}{2} AC$$

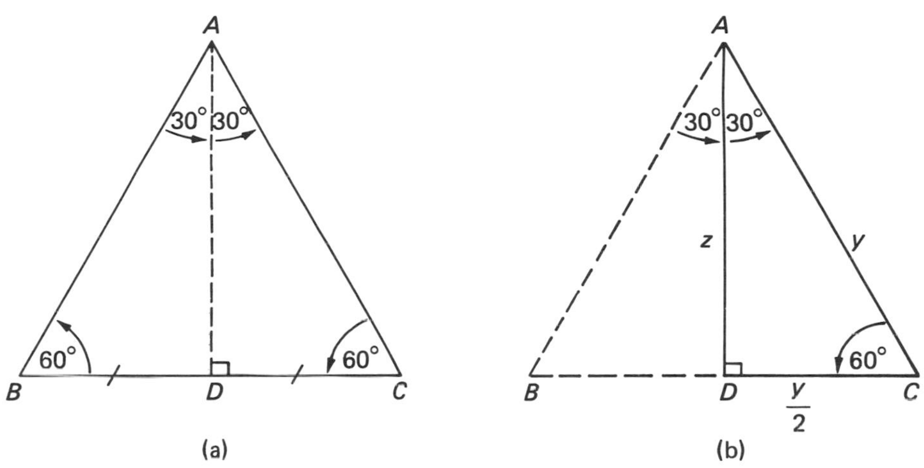

Fig. 11.28 An equilateral triangle. $\angle DAB = \angle CAD = 30°$.

Let $z = AD$ and $y = AC$

Then, $\frac{y}{2} = DC$, and by the Pythagorean Theorem,

$$z^2 + \left(\frac{y}{2}\right)^2 = y^2$$

$$z^2 + \frac{y^2}{4} = y^2$$

$$z^2 = \frac{3y^2}{4}$$

$$z = \frac{\sqrt{3}y}{2}$$

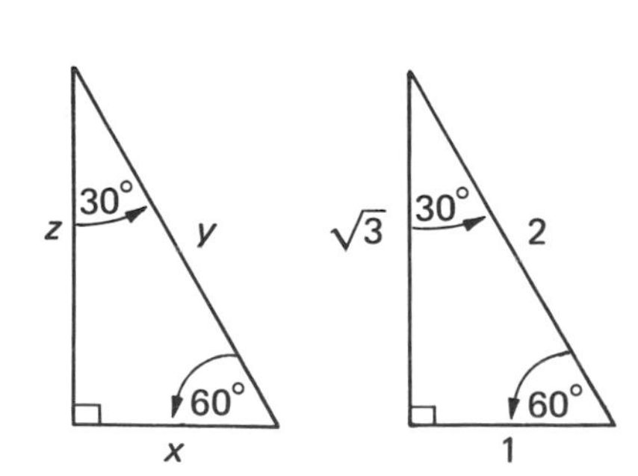

Fig. 11.29

Thus, *the ratio of the side lengths of a 30°, 60°, 90° - triangle is indicated in Figure 11.29 . Here, let x be the side opposite the 30° angle. Then,*

$$x : y : z$$
as $$1 : 2 : \sqrt{3}$$

Example 2 Consider the triangle in Figure 11.30. Find the length of sides b and c.

Solution This is a 30°, 60°, 90° - triangle. To find b, use:

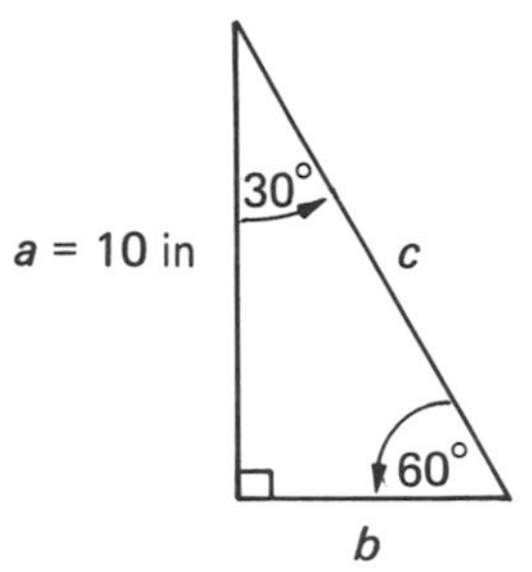

Fig. 11.30

$$\frac{a}{b} = \frac{\sqrt{3}}{1}$$

$$\frac{10 \text{ in.}}{b} = \sqrt{3}$$

$$10 \text{ in.} = b \cdot \sqrt{3}$$

$$\frac{10}{\sqrt{3}} \text{ in.} = b$$

It is easier to work with a fraction whose denominator is a rational number. *Rationalize the denominator* on the left by multiplying the fraction by $\frac{\sqrt{3}}{\sqrt{3}}$, which equals 1, so that the new fraction is equivalent to the old one.

$$b = \frac{10 \cdot \sqrt{3}}{\sqrt{3} \cdot \sqrt{3}} \text{ in.} = \frac{10\sqrt{3}}{3} \text{ in.}$$

To find c, use:

$$c = 2b = \frac{2 \cdot 10\sqrt{3}}{3} \text{ in.} = \frac{20\sqrt{3}}{3} \text{ in.}$$

Now, let us return to *an equilateral triangle with side length s and height h*, as in Figure 11.31. Observe that ΔADC is a 30°, 60°, 90° - triangle. Because

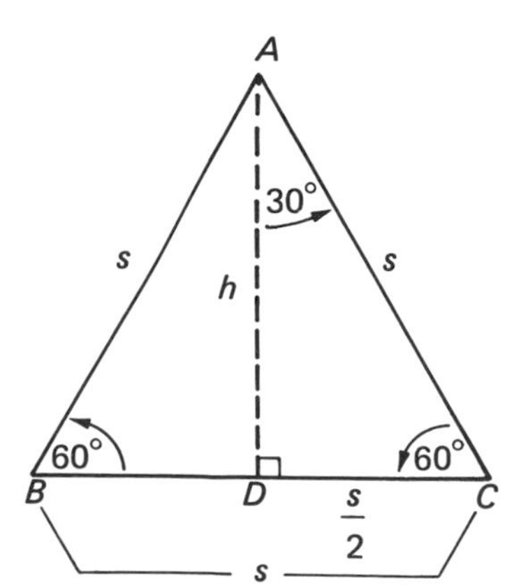

Fig. 11.31

$$DC = \frac{s}{2}$$

it follows that

$$\boxed{h = \frac{s\sqrt{3}}{2}}$$

To find the area, A, of this equilateral triangle, recall that

$$A = \frac{1}{2}\text{ base} \cdot \text{height}$$

$$A = \frac{1}{2} \cdot s \cdot \frac{s\sqrt{3}}{2} = \frac{s^2\sqrt{3}}{4}$$

Thus, *the area, A, of an equilateral triangle with side length s is given by the formula*

$$\boxed{A = \frac{s^2\sqrt{3}}{4}}$$

Example 3 Find **a.** the height and **b.** the area of an equilateral triangle with side length 6 decimeters.

Solution **a.** $h = \frac{s\sqrt{3}}{2}\text{ dm.} = \frac{6\sqrt{3}}{2}\text{ dm.}$
$= 3\sqrt{3}\text{ dm.}$

The height is $3\sqrt{3}$ decimeters.

b. $A = \frac{s^2\sqrt{3}}{4}$
$= \frac{6^2\sqrt{3}}{4}\text{ dm.}^2$
$= \frac{36\sqrt{3}}{4}\text{ dm.}^2$
$= 9\sqrt{3}\text{ dm.}^2$

The area is $9\sqrt{3}$ square decimeters.

Example 4 Find **a.** the side length and **b.** the height of an equilateral triangle whose area is $25\sqrt{3}$ square feet.

Solution **a.** $A = \frac{s^2\sqrt{3}}{4}$

$25\sqrt{3}\text{ ft.}^2 = \frac{s^2\sqrt{3}}{4}$

$100\text{ ft.}^2 = s^2$

$10\text{ ft.} = s$

The side length is 10 feet.

b. $h = \frac{s\sqrt{3}}{2} = \frac{10\sqrt{3}}{2}\text{ ft.}$
$= 5\sqrt{3}\text{ ft.}$

The height is $5\sqrt{3}$ feet.

CLASS EXERCISES *If necessary, leave your answers in terms of* $\sqrt{3}$. See Fig. 11.32, p. 368.

5. Consider the triangle in Figure 11.32(**a**). Find the lengths of sides a and b.

6. Consider the triangle in Figure 11.32(**b**). Find the lengths of sides b and c.

7. Find **a.** the height and **b.** the area of an equilateral triangle with side length 5 meters.

8. Find **a.** the side length and **b.** the height of an equilateral triangle whose area is $100\sqrt{3}$ square centimeters.

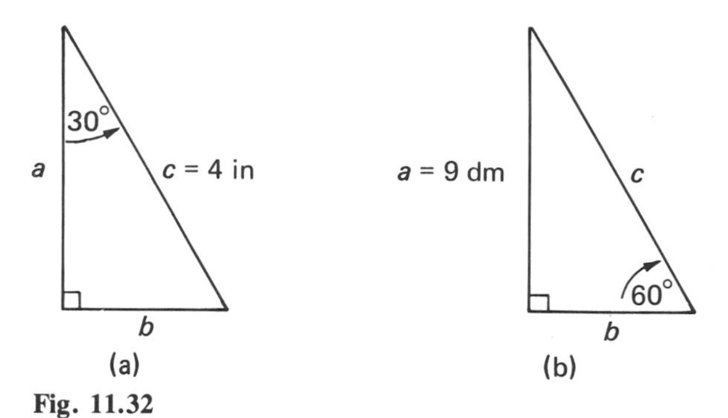

Fig. 11.32

SIMILAR TRIANGLES Intuitively, two geometric figures are said to be *similar* if they have the same shape. (Their sizes may possibly be different.) If the figures have the same size as well as the same shape, they are said to be *congruent*. We will be concerned with *similar triangles* that are *not* congruent. The triangles in Figure 11.33 are similar; those in Figure 11.34 are not. Here is a more precise definition of "similar triangles."

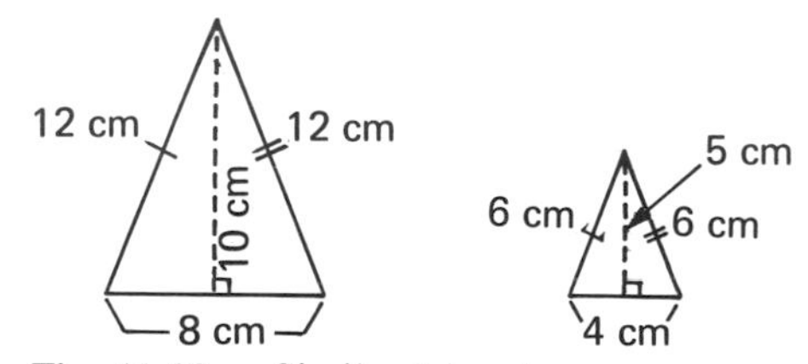

Fig. 11.33. Similar triangles.

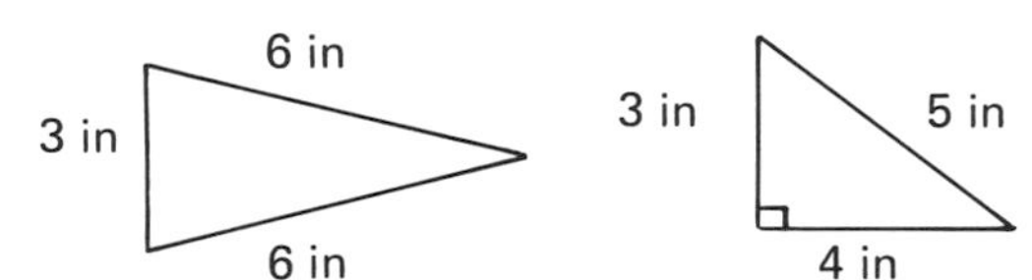

Fig. 11.34. These triangles are not similar.

DEFINITION

Let b be a base and h be the corresponding height of one triangle, and let b' be a base and h' be the corresponding height of another triangle. Then, the two triangles are **similar** if

$$\frac{b}{b'} = \frac{h}{h'}$$

In this case, $\frac{b}{b'}$ is called the **ratio of similarity.** (*We will always assume that* $\frac{b}{b'} > 1$.)

For the similar triangles of Figure 11.33, the ratio of similarity is

$$\frac{8 \text{ cm.}}{4 \text{ cm.}}, \text{ or } 2$$

The area, A, of the larger triangle is

$$\frac{8 \text{ cm.} \times 10 \text{ cm.}}{2}, \text{ or } 40 \text{ cm.}$$

and the arca, A', of the smaller triangle is

$$\frac{4 \text{ cm.} \times 5 \text{ cm.}}{2}, \text{ or } 10 \text{ cm.}$$

Observe the so-called **ratio of the areas:**

$$\frac{A}{A'} = \frac{40 \text{ cm.}}{10 \text{ cm.}} = 4 = 2^2$$

Thus, the ratio of similarity is 2 and the ratio of the area is 2^2.

In general, *if the ratio of similarity of two triangles is R, then the ratio of their areas is R^2.*

It can be shown that *two triangles are similar if the lengths of their corresponding sides are proportional.*

In Figure 11.35, the triangles are similar. Corresponding sides are marked with the same number of bars. The sides of length a, b, c of one triangle correspond, respectively, to the sides of length a', b', c' of the other. The lengths of corresponding sides of similar triangles are proportional. Thus,

$$\frac{a}{a'} = \frac{b}{b'} = \frac{c}{c'} = \text{ratio of similarity}$$

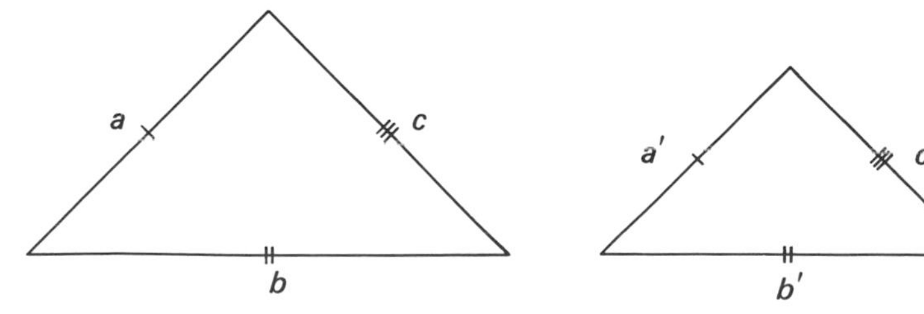

Fig. 11.35. Similar triangles.

Example 5 Suppose the triangles in Figure 11.36 are similar.

a. Find b'.
b. What is the ratio of similarity?
c. What is the ratio of the areas?
d. Find c'.

Solution **a.** $\frac{a}{a'} = \frac{b}{b'} = \frac{c}{c'}$

$$\frac{4 \text{ in.}}{2 \text{ in.}} = \frac{8 \text{ in.}}{b'} = \frac{9 \text{ in.}}{c'}$$

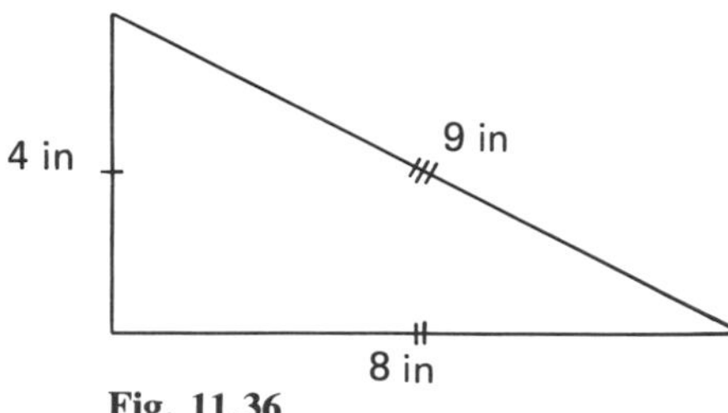

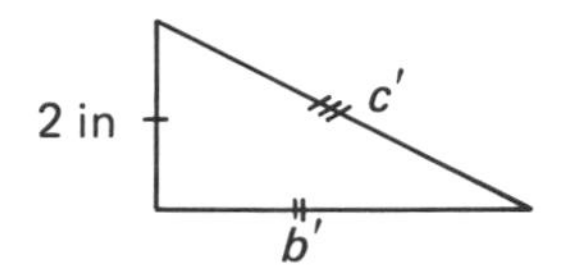

Fig. 11.36

Solve for b': First, divide numerator and denominator on the left by 2 in.

$$\frac{\overset{2}{\cancel{4 \text{ in.}}}}{\underset{1}{\cancel{2 \text{ in.}}}} = \frac{8 \text{ in.}}{b'}$$ *Now, cross-multiply.*

$$2b' = 8 \text{ in.}$$
$$b' = 4 \text{ in.}$$

b. The ratio of similarity is 2:

$$\frac{a}{a'} = \frac{4 \text{ in.}}{2 \text{ in.}} = 2$$

c. The ratio of the areas is 2^2, or 4.
d. Use part **b.** to find c'.

$$\frac{a}{a'} = \frac{c}{c'}$$
$$2 = \frac{9 \text{ in.}}{c'}$$ *Cross-multiply.*
$$2c' = 9 \text{ in.}$$
$$c' = \frac{9}{2} \text{ in., or } 4\frac{1}{2} \text{ in.}$$

Similarity of triangles can also be described in terms of angles. *Two triangles are similar if their corresponding angles are equal (in measure). Whenever corresponding side lengths are proportional, corresponding angles are equal, and vice versa. If two angles of one triangle are equal, respectively, to two angles of another triangle, the third angles must also be equal and the triangles are then similar.* (See Class Exercise **10**.)

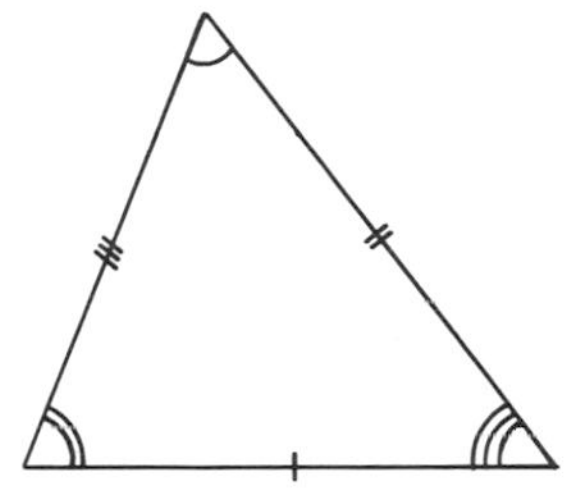

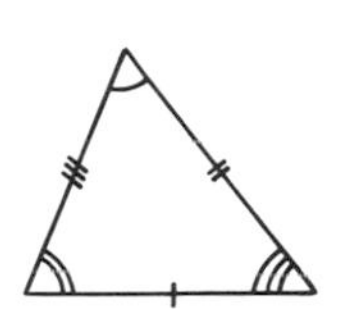

Fig. 11.37 Similar triangles. Corresponding sides and corresponding angles are indicated.

Example 6 In Figure 11.38(a), assume $AB \parallel DE$.

a. Show that ΔABC is similar to ΔCDE. **b.** What is the ratio of similarity?
c. Find the length of AB. **d.** Find the length of CE.

Solution **a.** $\angle BCA = \angle DCE$ because vertical angles are equal. $\angle A = \angle E$ and $\angle B = \angle D$ because $AB \parallel DE$ and $\angle A$ and $\angle E$ are alternate interior angles, as are $\angle B$ and $\angle D$. The two triangles are therefore similar.

In Figure 11.38(b), corresponding angles are indicated. Note that the sides opposite corresponding angles also correspond.

b. ratio of similarity $= \dfrac{DC}{BC}$

$$= \frac{10 \text{ in.}}{5 \text{ in.}}$$

$$= 2$$

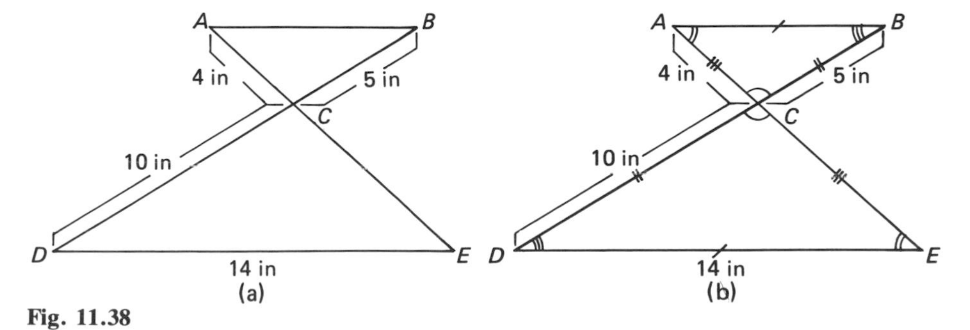

Fig. 11.38

c. $\frac{DE}{AB} = 2$

$\frac{14 \text{ in.}}{AB} = 2$

$\frac{14 \text{ in.}}{2} = AB$

$7 \text{ in.} = AB$

d. $\frac{CE}{AC} = 2$

$\frac{CE}{4 \text{ in.}} = 2$

$CE = 8 \text{ in.}$

CLASS EXERCISES

9. Assume the triangles in Figure 11.39 are similar.
a. Find b'. **b.** What is the ratio of similarity?
c. What is the ratio of the areas?

10. In Figure 11.40, show that ΔABC is similar to ΔDEF.

11. In Figure 11.41: **a.** Show that ΔABC is similar to ΔCDE.
b. Find the ratio of similarity. **c.** Find area ΔABC.

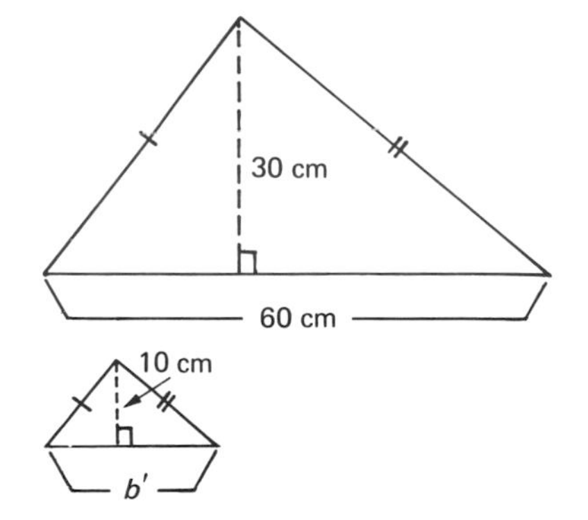

Fig. 11.39

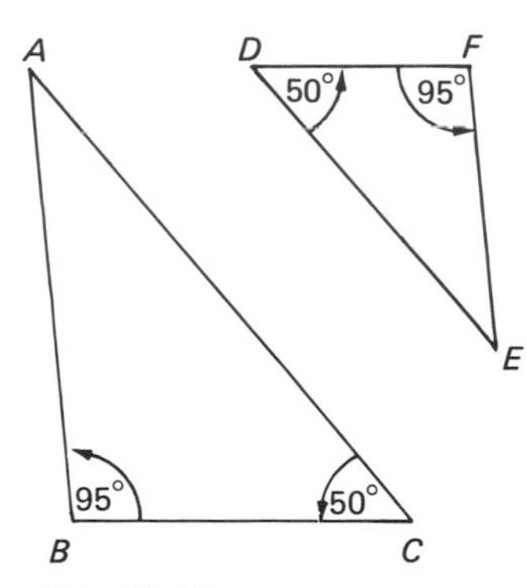

Fig. 11.40

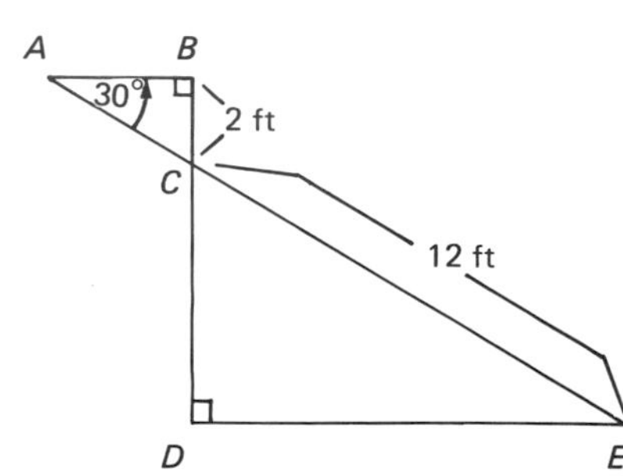

Fig. 11.41

SOLUTIONS TO CLASS EXERCISES

1. a. $62°30'$ **b.** an acute triangle **2.** $40°45'$

3. a. $\angle A + \angle B = 120° + 30° = 150°$

$\angle C = 180° - 150° = 30°$

Thus, $\angle C = \angle B$, and therefore, ΔABC is isosceles.

b. AD bisects BC. Thus, $BC = 2 \cdot BD$. Therefore, $BD = 16$ inches

4. a. $6\sqrt{2}$ cm. **b.** 18 cm.^{2} **5.** $b = 2$ in., $a = 2\sqrt{3}$ in. **6.** $b = 3\sqrt{3}$ dm., $c = 6\sqrt{3}$ dm.

7. a. $\frac{5\sqrt{3}}{2}$ m. **b.** $\frac{25\sqrt{3}}{4}$ m.^{2} **8. a.** $A = \frac{s^2\sqrt{3}}{4}$ **b.** $10\sqrt{3}$ cm.

$100\sqrt{3} \text{ cm.}^2 = \frac{s^2\sqrt{3}}{4}$

$400 \text{ cm.}^2 = s^2$

$20 \text{ cm.} = s$

9. a. 20 cm. **b.** 3 **c.** 9

10. $\angle B = \angle F = 95°$
$\angle C = \angle D = 50°$
$\angle A = 180° - (95° + 50°) = 180° - 145° = 35°$

And $\angle E = 180° - (95° + 50°) = 35°$
Thus, $\angle A = \angle E$
Therefore, ΔABC is similar to ΔDEF.

11. a. $\angle B = \angle D = 90°$
Therefore, $AB \| DE$ because alternate interior angles are equal.
$\angle A = \angle E = 30°$ (*alternate interior angles*)
$\angle BCA = \angle DCE = 60°$
Therefore, ΔABC is similar to ΔCDE.

b. $AC = 4$ ft. (*30°, 60°, 90° – triangle*)

ratio of similarity $= \dfrac{12 \text{ ft}}{4 \text{ ft}} = 3$

c. $AB = 2\sqrt{3}$ ft.

Area $\Delta ABC = \dfrac{2 \text{ ft.} \times 2\sqrt{3} \text{ ft.}}{2} = 2\sqrt{3} \text{ ft.}^2$

HOME EXERCISES

In Exercises 1–5, find the measure of the third angle of each triangle.

1. Fig. 11.42
2. Fig. 11.43
3. Fig. 11.44
4. Fig. 11.45
5. Fig. 11.46

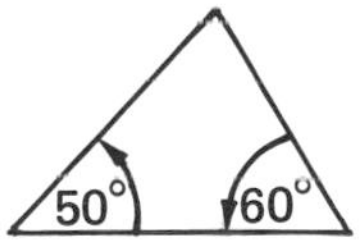

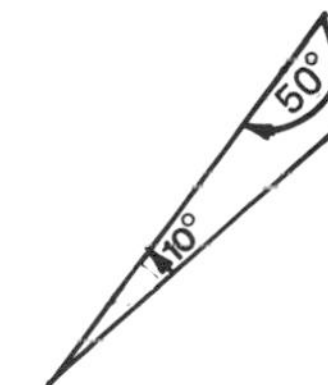

Fig. 11.42

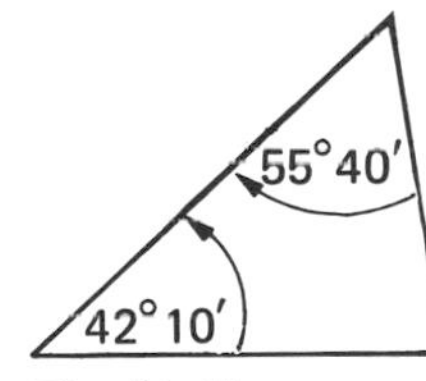

Fig. 11.43

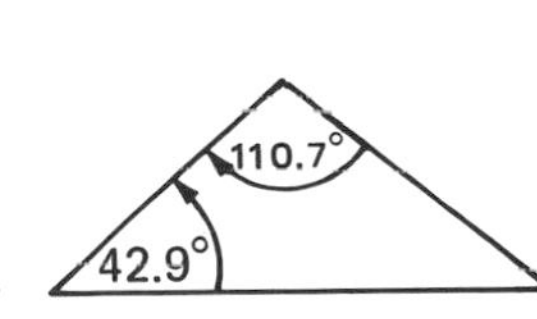

Fig. 11.44

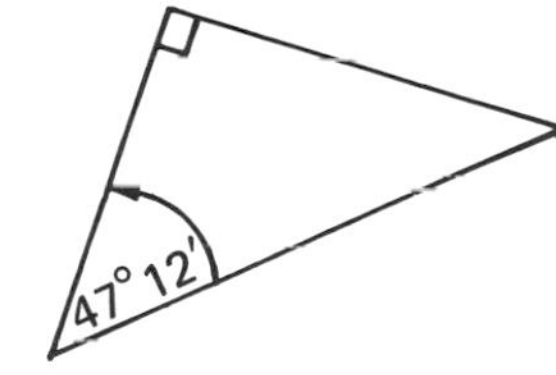

Fig. 11.45

Fig. 11.46

6. In a right triangle, one acute angle measures twice as much as the other. Find the measure of each acute angle.
7. In a right triangle, one acute angle measures 10° more than the other. Find the measure of each acute angle.
8. The largest angle of a triangle measures 20° more than the second largest angle and 25° more than the smallest angle. Find the measure of each angle.
9. The largest angle of a triangle measures twice another angle and three times the third angle. Find the measure of each angle.
10. In Figure 11.47, which triangles are isosceles?

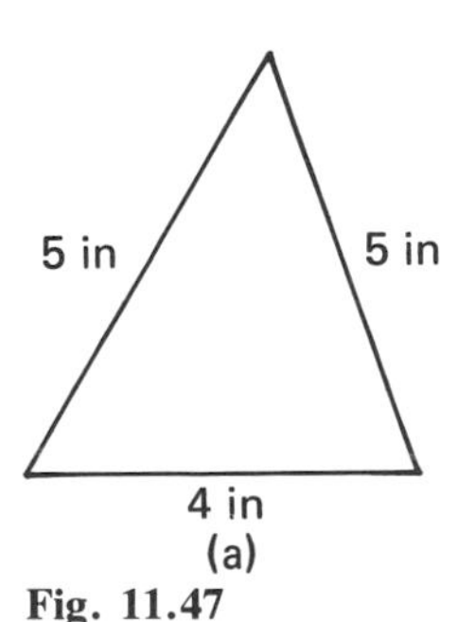

(a)

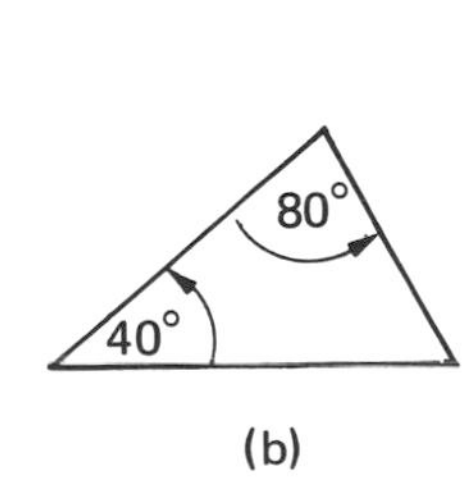

(b)

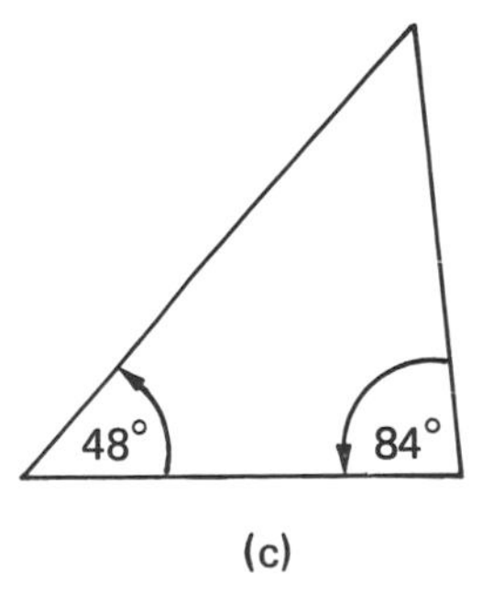

(c)

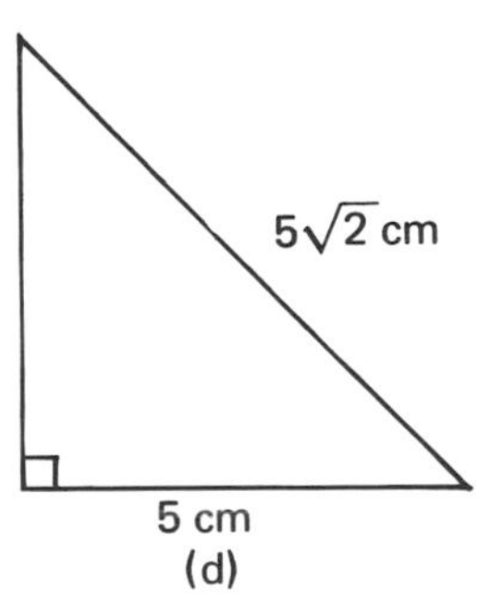

(d)

Fig. 11.47

11. Each of the base angles of an isosceles triangle measures 37°30′. Find the measure of the third angle.
12. In Figure 11.48 on page 374, $AB = AC$ and AD is the height corresponding to base BC.
    a. Find the length of BD. **b.** Find the area of ΔABC.

13. In Figure 11.49, $AB = AC$ and AD is the height corresponding to side BC.
    a. Find the length of base BC.
    b. Find the perimeter of ΔABC.
    c. Find the perimeter of ΔABD.
    d. Find the area of ΔABC.
    e. Find the area of ΔABD.
14. In Figure 11.50, suppose ΔABC is an isosceles right triangle. Find:
    a. the hypotenuse
    b. the area
15. In Figure 11.51, suppose ΔABC is an isosceles right triangle. Find:
    a. the length of each leg
    b. the area

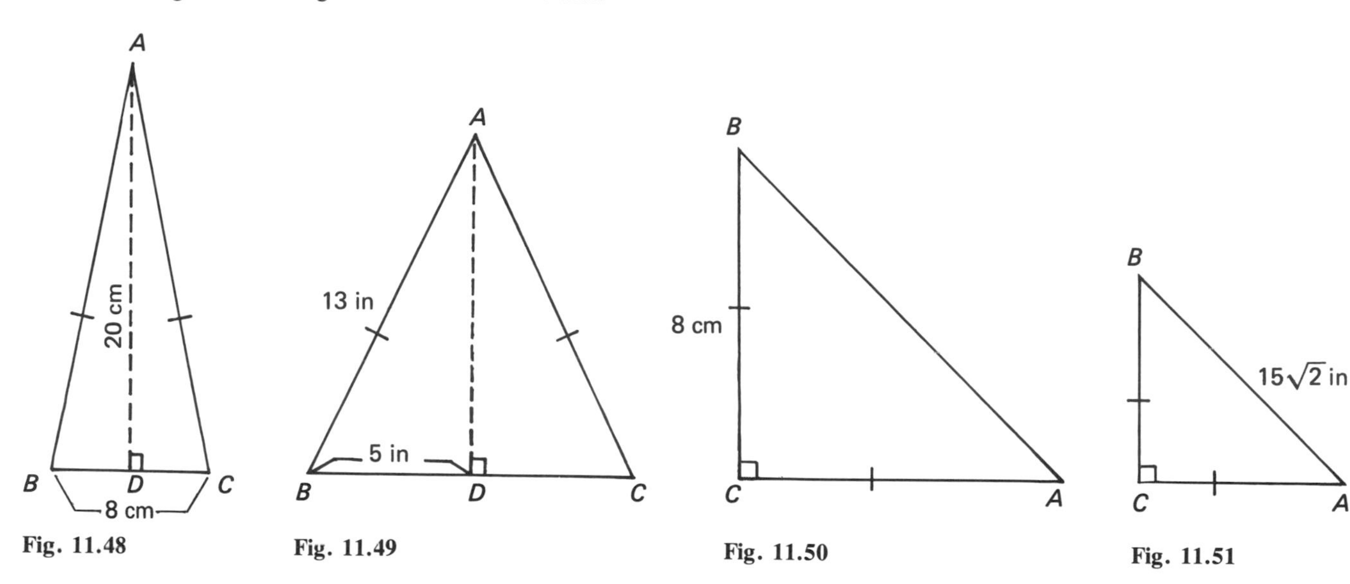

Fig. 11.48 Fig. 11.49 Fig. 11.50 Fig. 11.51

16. Consider the triangle in Figure 11.52. Find the lengths of sides a and b.
17. Consider the triangle in Figure 11.53. Find the lengths of sides a and c.
18. Consider the triangle in Figure 11.54. Find the lengths of sides b and c.
19. Find the area of a 30°, 60°, 90°-triangle whose hypotenuse is 4 inches.

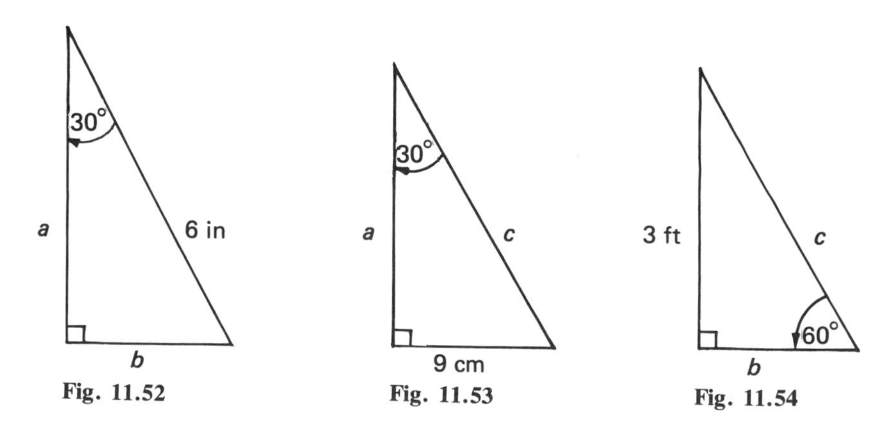

Fig. 11.52 Fig. 11.53 Fig. 11.54

20. Find **a.** the height, **b.** the perimeter, and **c.** the area of an equilateral triangle with side length 7 inches.

21. Find **a.** the side length and **b.** the area of an equilateral triangle with height $10\sqrt{3}$ centimeters.

22. Find **a.** the side length and **b.** the height of an equilateral triangle with area $9\sqrt{3}$ square inches.

23. An equilateral triangle has perimeter 27 decimeters. Find its area.

Exercises 24–34 refer to Figure 11.55. Find:

24. the measure of $\angle C$
25. the measure of $\angle BAD$
26. the measure of $\angle BAC$
27. the length of DC
28. the length of AC
29. the length of BD
30. the length of BC
31. the length of AB
32. the area of ΔADC
33. the area of ΔABC
34. the area of ΔABD

Exercises 35–41 refer to Figure 11.56. Find:

35. the measure of $\angle BAC$
36. the measure of $\angle DAC$
37. the length of BC
38. the length of DC
39. the length of AD
40. the area of ΔABC
41. the area of ΔADC

42. In Figure 11.57, find the area of ΔABC.

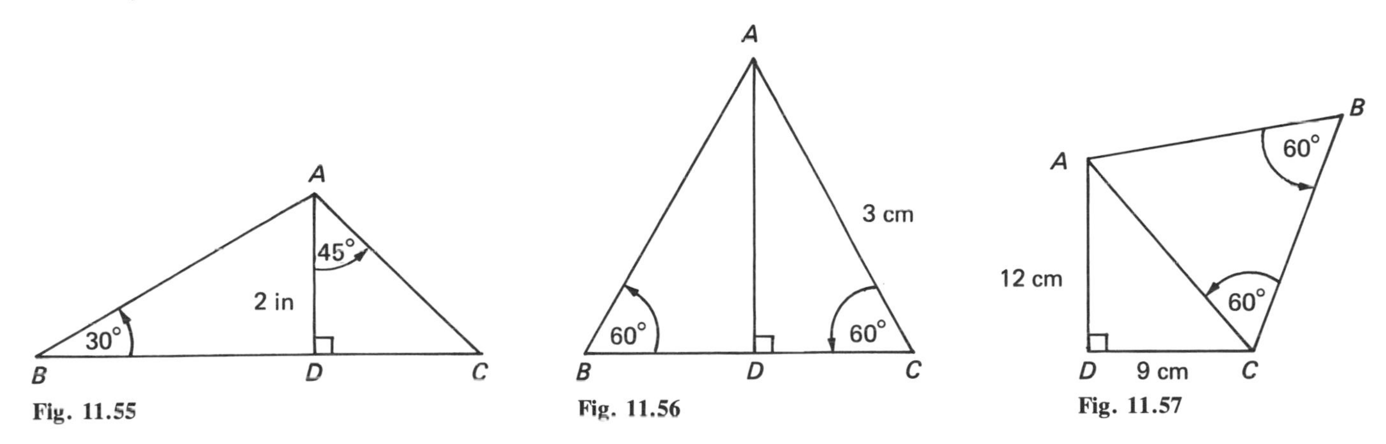

Fig. 11.55

Fig. 11.56

Fig. 11.57

In Exercises 43 and 44, assume each pair of triangles is similar.

a. Find c'.
b. What is the ratio of similarity?
c. Find b.
d. Find h.
e. What is the ratio of the areas?

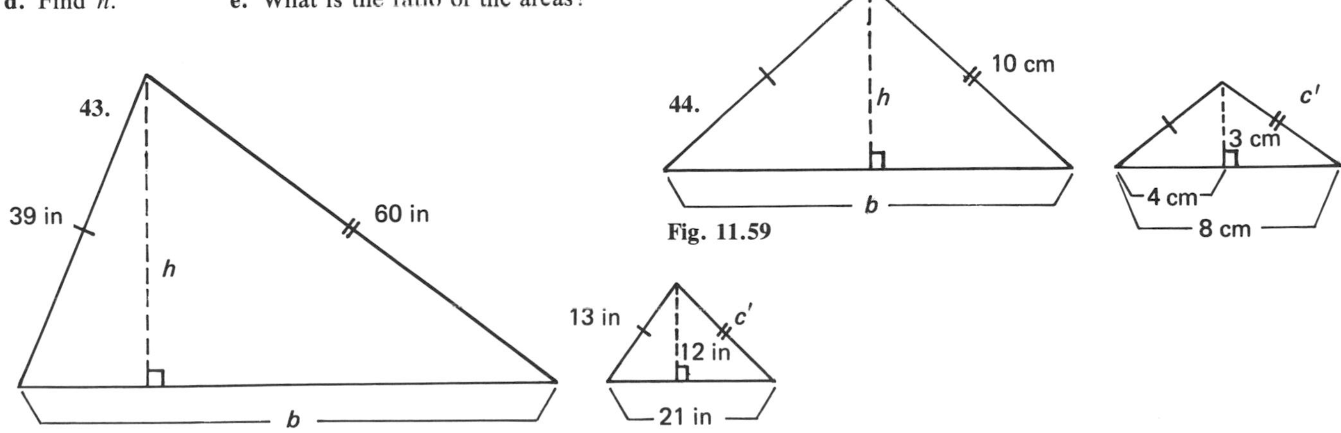

Fig. 11.58

Fig. 11.59

45. In Figure 11.60, show that ΔABC and ΔDEF are similar.
46. In Figure 11.61, show that ΔABC and ΔDEF are similar.

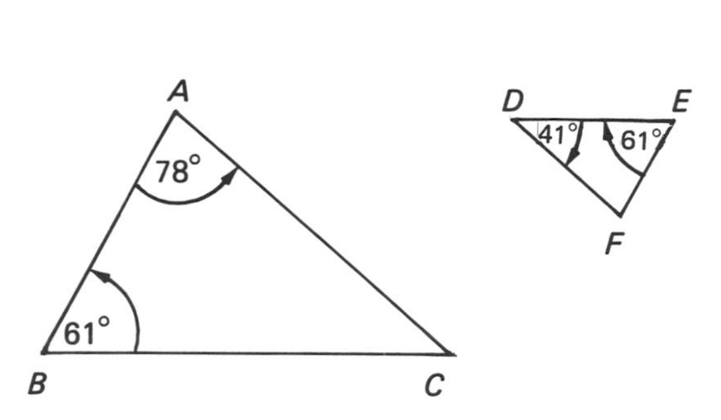

Fig. 11.60

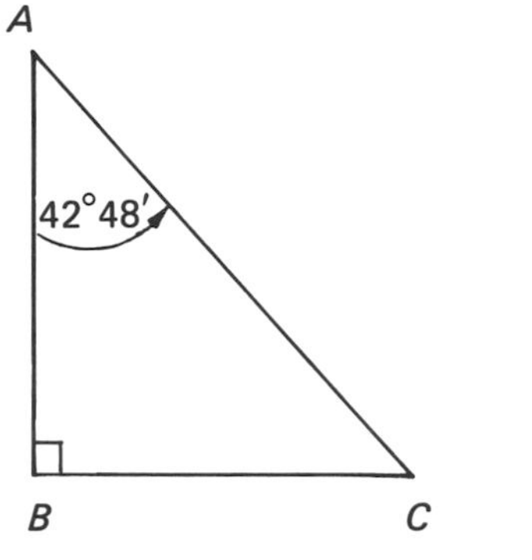

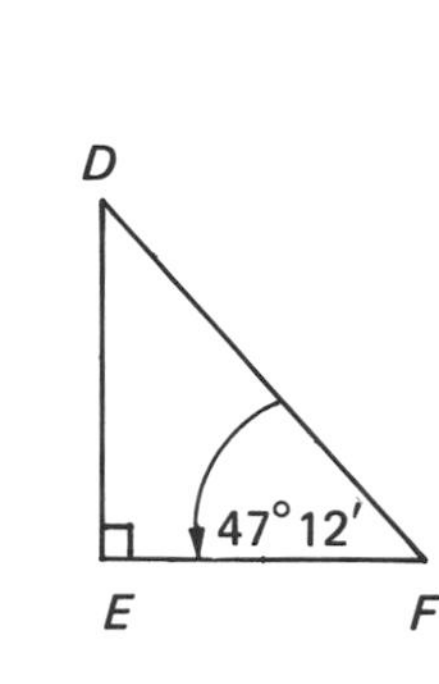

Fig. 11.61

47. In Figure 11.62, suppose $AB \parallel DE$.
   a. Show that ΔABC and ΔCDE are similar.
   b. What is the ratio of similarity?
   c. Find the length of AC.
   d. Find the length of BD.
48. In Figure 11.63, suppose $BC \parallel DE$.
   a. Show that ΔABC and ΔADE are similar.
   b. What is the ratio of similarity?
   c. Find the length of AC.
   d. Find the length of BD.
49. In Figure 11.64, suppose $BD = 18$ centimeters.
   a. Find the area of ΔCDE.
   b. Show that $AB \parallel DE$.
50. Suppose the ratio of similarity of two similar triangles is 4. If the height of the larger triangle is 20 centimeters, find the height of the smaller one.

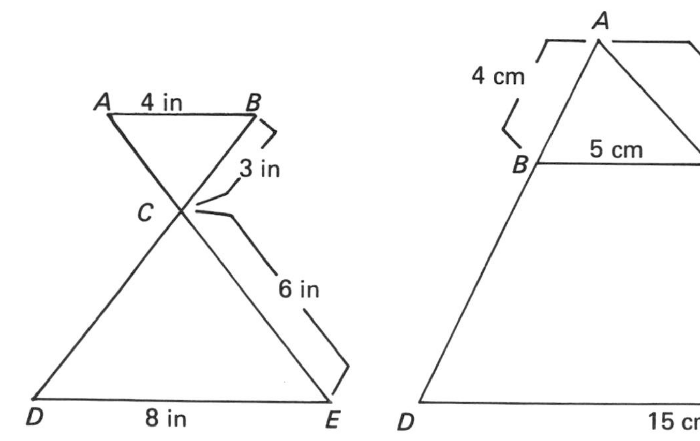

Fig. 11.62

Fig. 11.63

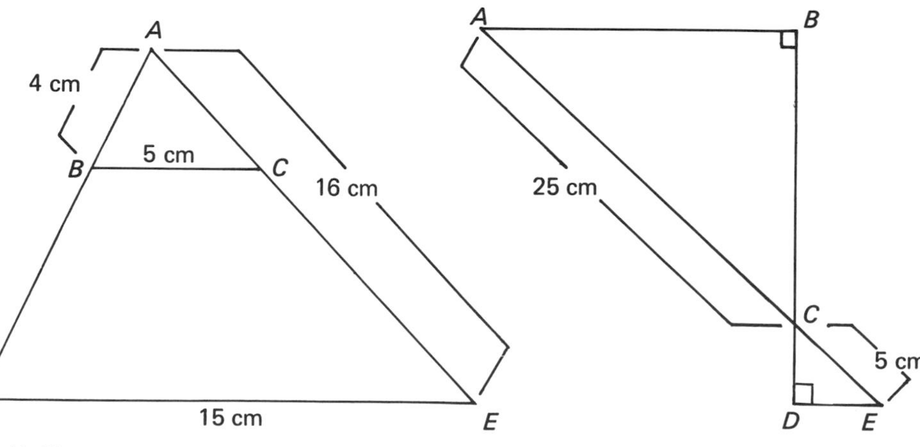

Fig. 11.64

51. Suppose the ratio of similarity of two similar triangles is 3. If the smaller triangle has base 4 inches and height 6 inches, find the area of the larger triangle.

52. A 6-foot man casts a 4-foot shadow. How long is the shadow cast by his 4-foot son?

53. A 16-foot ladder leaning against a wall makes a 30° angle with the wall. **a.** How far is the foot of the ladder from the bottom of the wall? **b.** To one decimal digit, how high up the wall does the ladder reach?

54. As an airplane rises, it makes a 45° angle with the runway. To one decimal digit, how many meters does the plane rise while traveling 800 meters in the air?

11.3 MORE ON CIRCLES

SECANTS, CHORDS, TANGENTS

Consider a circle in which

$d = diameter,$ $r = radius,$ $C = circumference,$ and $A = area$

Then, recall that

$$d = 2r$$
$$C = \pi d = 2\pi r$$
and $$A = \pi r^2$$

A **secant** to a circle is a line that intersects the circle at two points, P and Q. In Figure 11.65, ℓ_1 is a secant to the circle. The line segment PQ is known as a **chord.** A **tangent** to a circle is a line that intersects the circle at only one

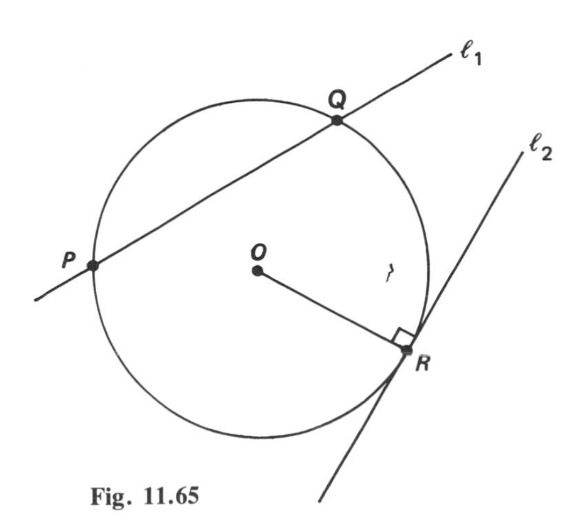

Fig. 11.65

point. Thus, ℓ_2 is tangent to the circle. *A tangent to a circle is perpendicular to the radius at the point of tangency.* In Figure 11.65, $OR \perp \ell_2$

A radius that bisects a chord is perpendicular to the chord. And a radius that intersects but does not bisect the chord is not perpendicular to the chord. Thus, in Figure 11.66, $OR \perp PQ$ and $PS = SQ$. But OA is not perpendicular to BC, and BD is larger than DC.

Example 1 In Figure 11.67, ℓ is tangent to the circle at S and O is the center of the circle.

a. Show that $\angle TSO$ is a right angle. **b.** Show that $\angle ORP$ is a right angle.

Solution **a.** The tangent, ℓ, is perpendicular to the radius of the circle at the point of tangency. Thus, $\angle TSO$ is a right angle.

b.

$$RQ = PQ - PR = 8 \text{ cm.} - 4 \text{ cm.} = 4 \text{ cm.}$$

Thus, $RQ = PR$, and radius OA bisects chord PQ. Therefore, $OA \perp PQ$, and $\angle ORP$ is a right angle.

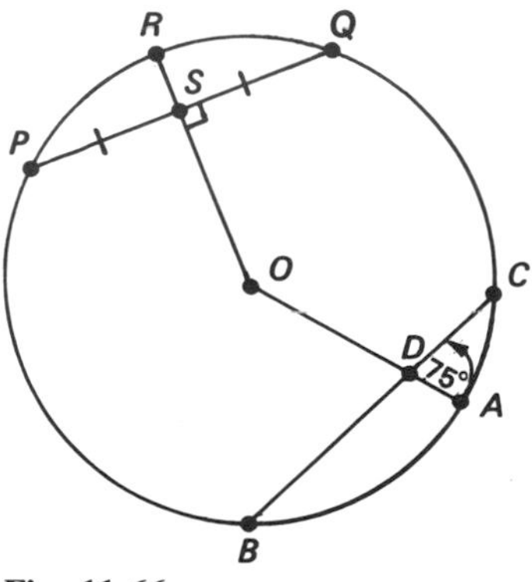

Fig. 11.66

Fig. 11.67

CLASS EXERCISES

1. In Figure 11.68, ℓ is tangent to the circle at R, O is the center of the circle, and $OP \perp RS$. Find the length of **a.** OP, **b.** RT, **c.** OT.

CENTRAL ANGLES

Let A and B be two points on a circle whose center is at O. $\angle AOB$ is called a **central angle.** (See Figure 11.69 .) The portion of the circle between A and B *in the interior of* $\angle AOB$ is called the **arc intercepted by** $\angle AOB$, and is written $\widehat{AB}$. The arc $\widehat{AB}$ is also said to be intercepted by the chord AB and by the secant through A and B. Now, consider *any* point C on the circle but not on arc $\widehat{AB}$. Then, $\angle ACB$ is called an **inscribed angle.** We also say that $\angle ACB$ **intercepts** $\widehat{AB}$. *When a central angle, $\angle AOB$, and an inscribed angle, $\angle ACB$, intercept the same arc, then*

$$\angle AOB = 2 \angle ACB$$

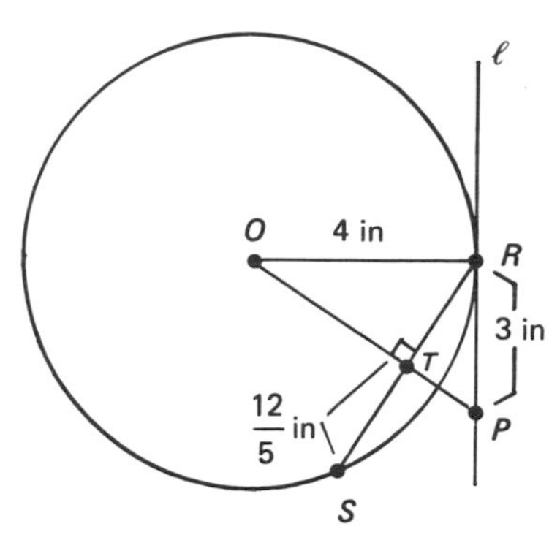

Fig. 11.68

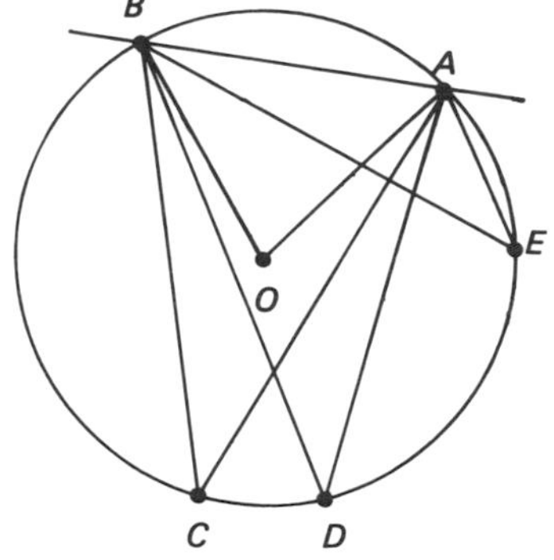

Fig. 11.69. $\angle AOB = 2\angle ACB$.

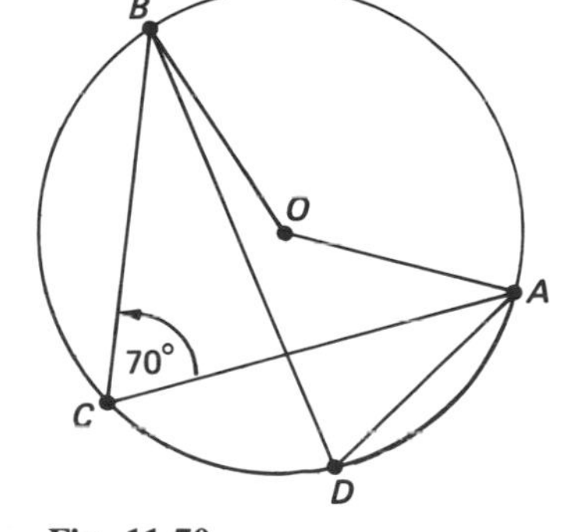

Fig. 11.70

In Figure 11.69, note that $\angle ADB$, an inscribed angle, also intercepts arc $\widehat{AB}$. Thus,

$$\angle AOB = 2\angle ADB = 2\angle ACB$$

and therefore,

$$\angle ADB = \angle ACB$$

Similarly, $\angle AEB$ is an inscribed angle that intercepts arc $\widehat{AB}$.

$$\angle AEB = \angle ACB$$

Any two inscribed angles that intercept the same arc have equal measures.

Example 2 In Figure 11.70, O is the center of the circle. Find the measure of

a. $\angle AOB$, **b.** $\angle ADB$.

Solution **a.** $\angle AOB$ is a central angle that intercepts the same arc $\widehat{AB}$, as does $\angle ACB$, an inscribed angle. Therefore,

$$\angle AOB = 2\angle ACB = 2 \times 70^\circ = 140^\circ$$

b. $\angle ADB$ is another inscribed angle that intercepts $\widehat{AB}$. Therefore,

$$\angle ADB = \angle ACB = 70^\circ$$

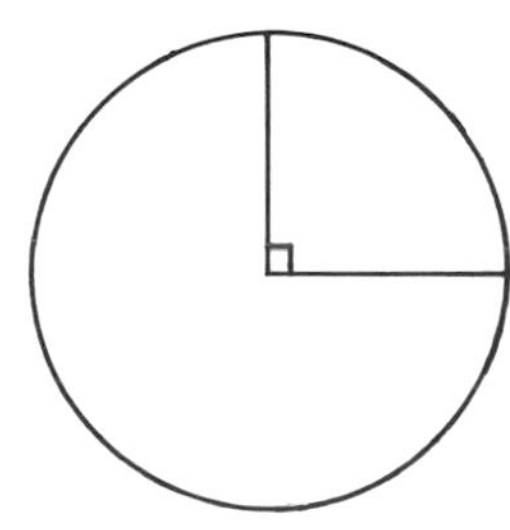

Fig. 11.71

A central angle of 360° intercepts the entire circle. A central angle of 90° intercepts an arc whose length is $\frac{1}{4}$ of the circumference. (See Figure 11.71.) Note that

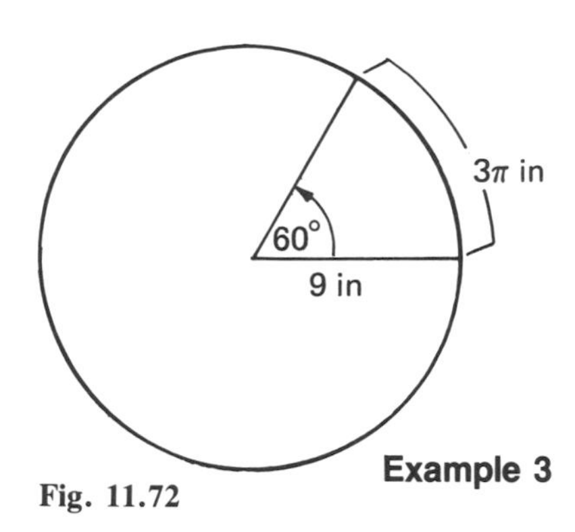

Fig. 11.72

$$\frac{90°}{360°} = \frac{1}{4}$$

In general, the following proportion holds.

$$\boxed{\frac{\text{central angle}}{360°} = \frac{\text{arc length}}{\text{circumference}}}$$

Example 3 A circle has radius 9 inches. Determine the length of an arc intercepted by a central angle of 60°.

Solution Let C = circumference, r = radius, Then,

$$C = 2\pi r = 2\pi \times 9 \text{ in.} = 18\pi \text{ in.}$$

$$\frac{\text{central angle}}{360°} = \frac{\text{arc length}}{\text{circumfcrence}}$$

$$\frac{60°}{360°} = \frac{\text{arc length}}{18\pi \text{ in.}}$$

$$\frac{1}{6} = \frac{\text{arc length}}{18\pi \text{ in.}}$$

$$\frac{18\pi \text{ in.}}{6} = \text{arc length}$$

$$3\pi \text{ in.} = \text{arc length}$$

Note that 3π ($\approx 3 \times 3.14$) is slightly more than 9. Thus, in Figure 11.72, the arc intercepted is slightly longer than the radius of the circle.

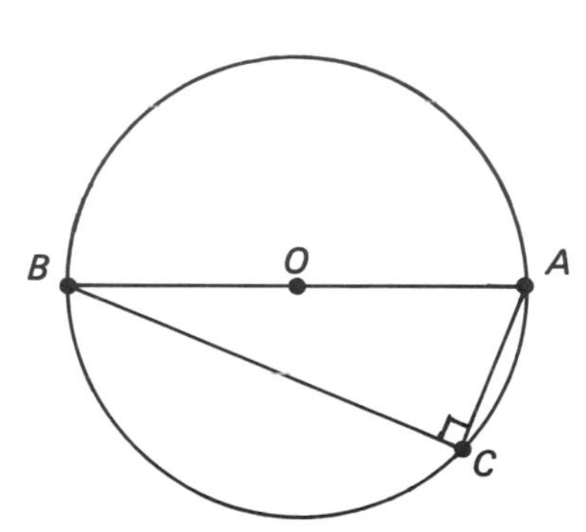

Fig. 11.73. $\angle AOB$ is a straight angle. $\widehat{AB}$ is a semicircle. $\angle ACB$ is a right angle.

A central angle of 180° (that is, a straight angle) intecepts an arc whose length is one-half the circumference. Such an arc is known as a **semicircle.** Observe, in Figure 11.73, that *a chord that intercepts a semicircle is a diameter. And an inscribed angle that intercepts a semicircle is a right angle* $\left(\frac{1}{2} \times 180°\right)$.

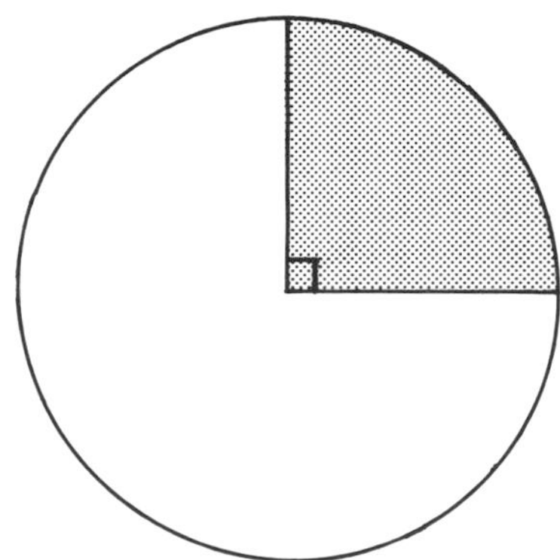
Fig. 11.74. The sector is shaded.

The portion of a circular region determined by central angle $\angle AOB$ is known as **sector** AOB. In Figure 11.74, the sector determined by a right angle (90°) is indicated. It appears that the area of the sector is one-fourth that of the circle. As previously noted,

$$\frac{90^\circ}{360^\circ} = \frac{1}{4}$$

In general,

$$\boxed{\frac{\text{central angle}}{360^\circ} = \frac{\text{area of sector}}{\text{area of circle}}}$$

Example 4 The diameter of a circle is 8 centimeters. Find the area of a sector of the circle determined by a 135° central angle.

Solution Let r = radius, d = diameter, A = area of circle.

$$r = \frac{1}{2} \cdot d = \frac{1}{2} \times 8 \text{ cm.} = 4 \text{ cm.}$$

$$A = \pi \times (4 \text{ cm.})^2 = 16\pi \text{ cm.}^2$$

$$\frac{\text{central angle}}{360^\circ} = \frac{\text{area of sector}}{\text{area of circle}}$$

$$\frac{135^\circ}{360^\circ} = \frac{\text{area of sector}}{16\pi \text{ cm.}^2}$$

$$\frac{3}{8} = \frac{\text{area of sector}}{16\pi \text{ cm.}^2}$$

$$\frac{3}{8} \times 16\pi \text{ cm.}^2 = \text{area of sector}$$

$$6\pi \text{ cm.}^2 = \text{area of sector}$$

(See Figure 11.75 .)

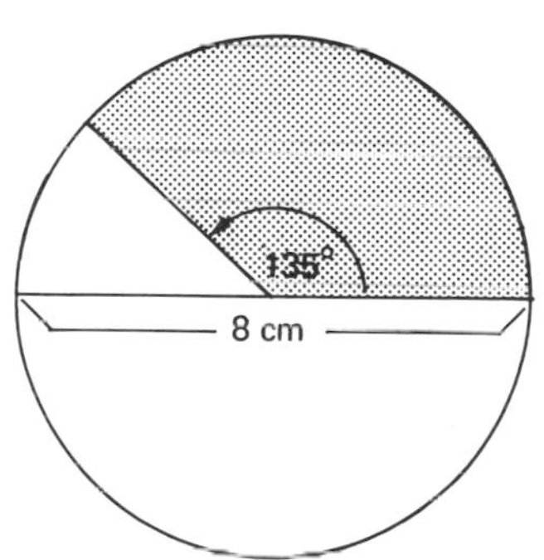

Fig. 11.75 . The circle has area 16π square centimeters. The (shaded) sector has area 6π square centimeters.

Fig. 11.76

CLASS EXERCISES

2. In Figure 11.76, O is the center of the circle. Find:
   a. the measure of $\angle ACB$
   b. the measure of $\angle OBA$
   c. the length of $\overset{\frown}{AB}$
   d. the area of sector AOB

SOLUTIONS TO CLASS EXERCISES

1. a. $OR \perp \ell$

Therefore, ΔORP is a right triangle with hypotenuse OP. By the Pythagorean Theorem, OP has length 5 inches.

b. $OP \perp RS$

Therefore, OP bisects RS. Thus, RT has length $\frac{12}{5}$ inches.

c. ΔOTR is a right triangle.

$$(OT)^2 + \left(\frac{12}{5} \text{ in.}\right)^2 = (4 \text{ in.})^2$$

$$(OT)^2 + \frac{144}{25}\text{ in.}^2 = 16\text{ in.}^2$$

$$(OT)^2 = \frac{256}{25}\text{ in.}^2$$

$$OT = \frac{16}{5}\text{ in.}$$

2. a. $\angle ACB = \frac{1}{2} \times 45° = 22.5°$ (or $22°30'$)

b. OB and OA are radii. Therefore, ΔOBA is isosceles. Thus,

$$\angle OBA = \angle BAO$$

$$2\angle OBA = 180° - 45°$$

$$2\angle OBA = 135°$$

$$\angle OBA = 67.5°$$

c.

$$\frac{45°}{360°} = \frac{\text{length } \widehat{AB}}{\text{circumference}}$$

$$\frac{1}{8} = \frac{\text{length } \widehat{AB}}{20\pi\text{ cm.}}$$

$$\frac{1}{8} \times 20\pi\text{ cm.} = \text{length } \widehat{AB}$$

$$2.5\pi\text{ cm.} = \text{length } \widehat{AB}$$

d.

$$\frac{45°}{360°} = \frac{\text{area of sector}}{\text{area of circle}}$$

$$\frac{1}{8} = \frac{\text{area of sector}}{100\pi\text{ cm.}^2}$$

$$\frac{1}{8} \times 100\pi\text{ cm.}^2 = \text{area of sector}$$

$$12.5\pi\text{ cm.}^2 = \text{area of sector}$$

HOME EXERCISES

Throughout these exercises, assume O is the center of each circle.

1. Refer to Figure 11.77, in which $\angle BOD$ is a straight angle. Match each entry in column (a) with its description in column (b).

Column (a)	*Column* (b)
$\widehat{AB}$	chord
$\angle AOB$	semicircle
$\angle ACB$	an arc that is not a semicircle
BC	tangent
ℓ_1	diameter
ℓ_2	secant
BD	central angle
$\widehat{BD}$	inscribed angle

Exercises 2–9 refer to Figure 11.78, in which ℓ is tangent to the circle at C. The radius of the circle is 5 centimeters and $AB = BC = \frac{5\sqrt{2}}{2}$ *centimeters. Find:*

2. the measure of $\angle OBC$
3. the length of OB
4. the area of ΔOBC
5. the measure of $\angle DCO$
6. the length of BD
7. the length of OD
8. the area of ΔOCD
9. the area of ΔBCD

10. In Figure 11.79, find: **a.** the measure of $\angle ACB$ **b.** the length of $\widehat{AB}$ **c.** the area of sector AOB

11. In Figure 11.80, find: **a.** the measure of $\angle AOB$ **b.** the measure of $\angle ADB$

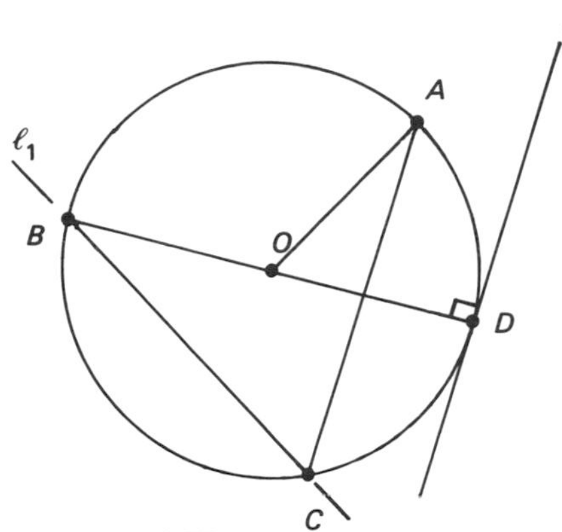

Fig. 11.77

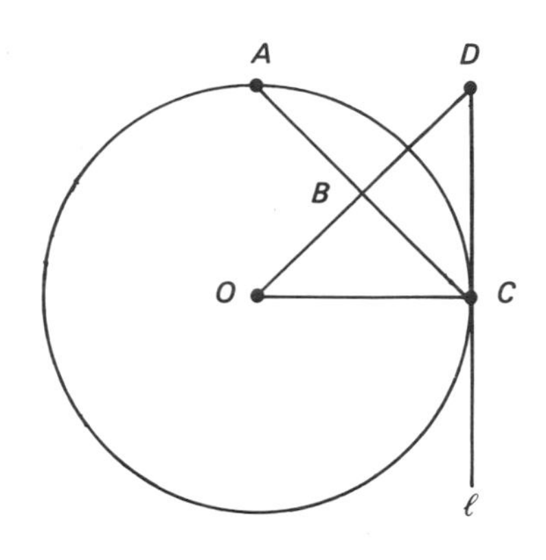

Fig. 11.78

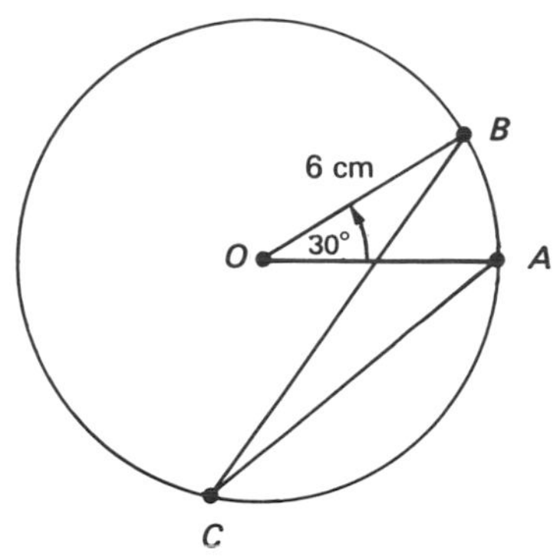

Fig. 11.79

12. In Figure 11.81, the area of the circle is 36π square inches and of sector AOB is 6π square inches. Find:
a. the measure of $\angle AOB$ **b.** the measure of $\angle ACB$ **c.** the length of $\widehat{AB}$

13. In Figure 11.82, $\widehat{AB}$ is of length 2π inches. Find:
a. the circumference of the circle **b.** the radius **c.** the area of sector AOB

14. In Figure 11.83, the area of sector AOB is 12π square centimeters. Find:
a. the area of the circle **b.** the radius **c.** the length of $\widehat{AB}$

15. In Figure 11.84, suppose that AB is a diameter of the circle. Find:
a. the measure of $\angle BCA$ **b.** the measure of $\angle ADB$

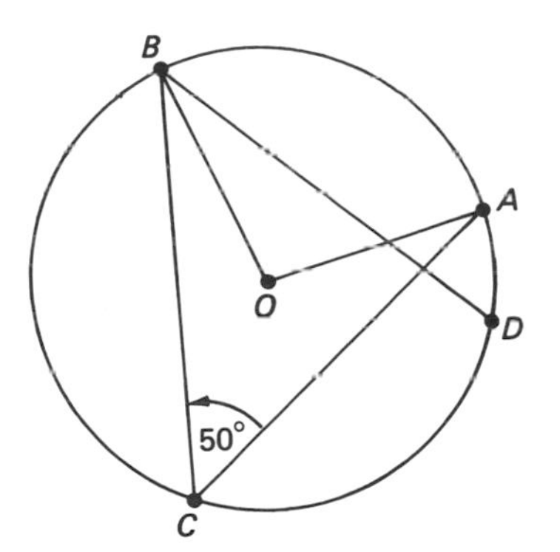

Fig. 11.80

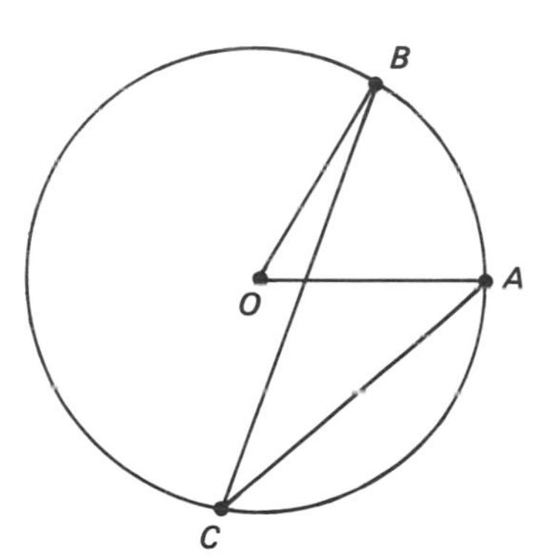

Fig. 11.81

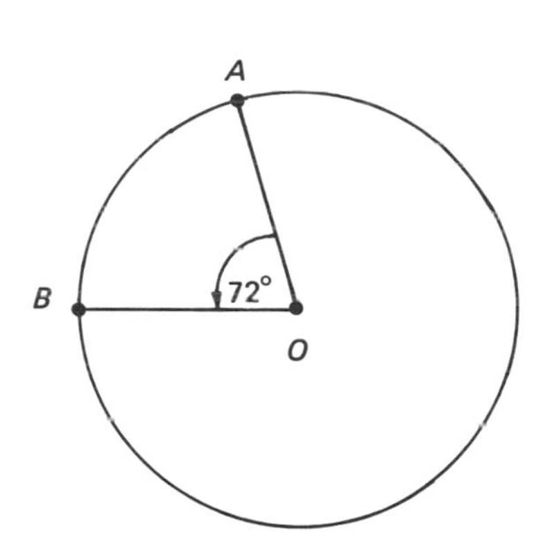

Fig. 11.82

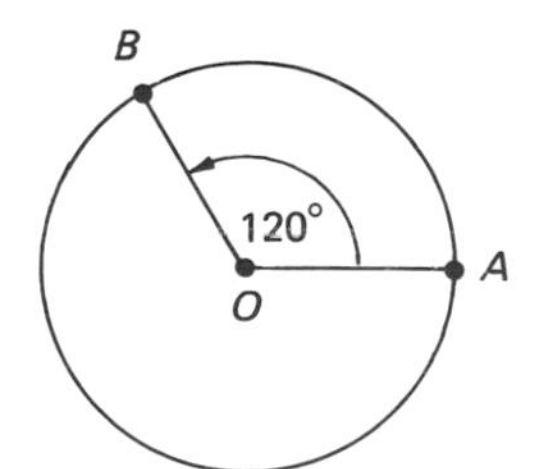

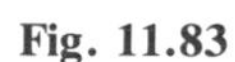

Fig. 11.83

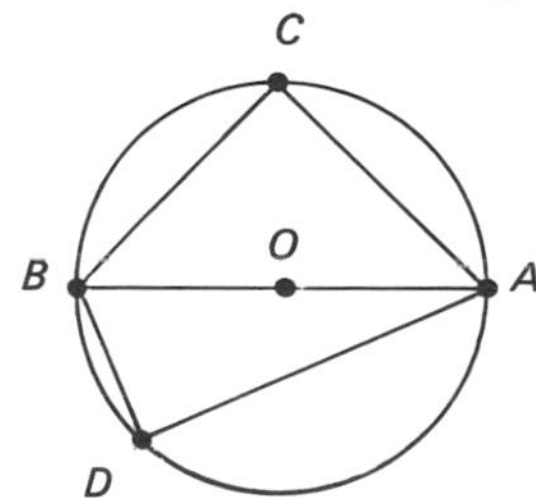

Fig. 11.84

For Exercises 16–21, find:
a. the length of the boundary and **b.** the area of each shaded region

16. Fig. 11.85
17. Fig. 11.86
18. Fig. 11.87
19. Fig. 11.88
20. Fig. 11.89
21. Fig. 11.90

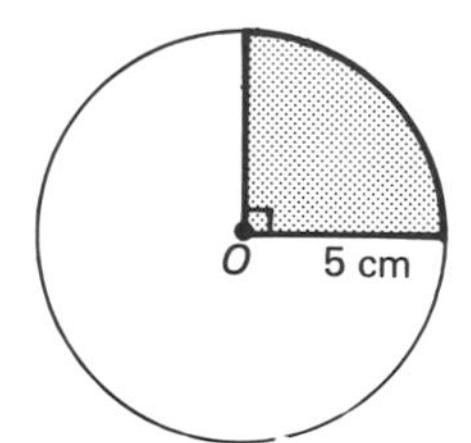

Fig. 11.85

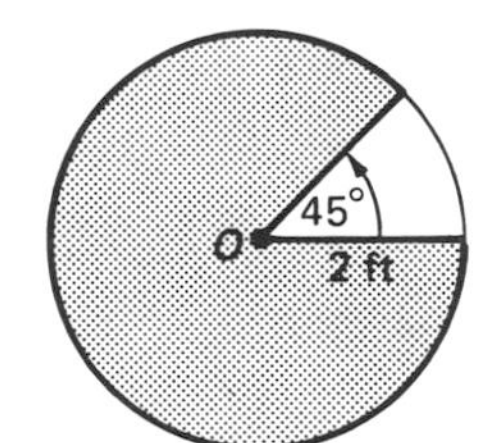

Fig. 11.86

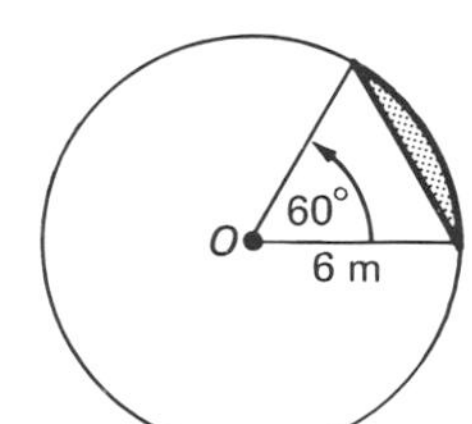

Fig. 11.87

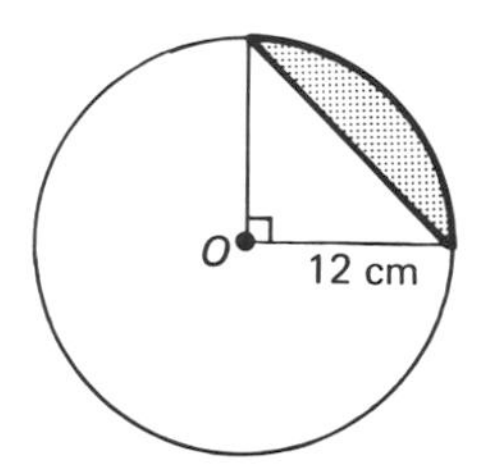

Fig. 11.88

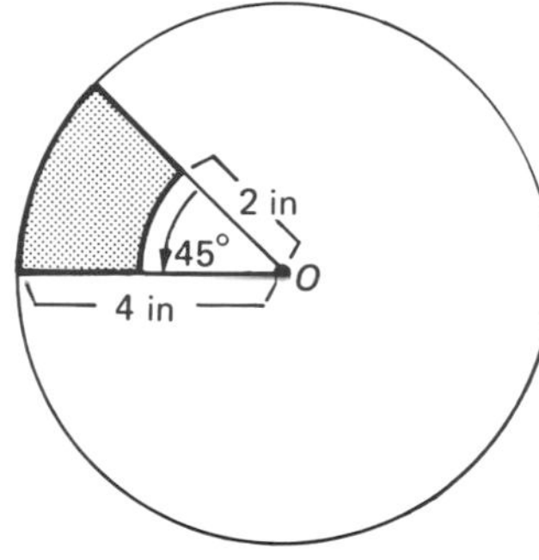

Fig. 11.89

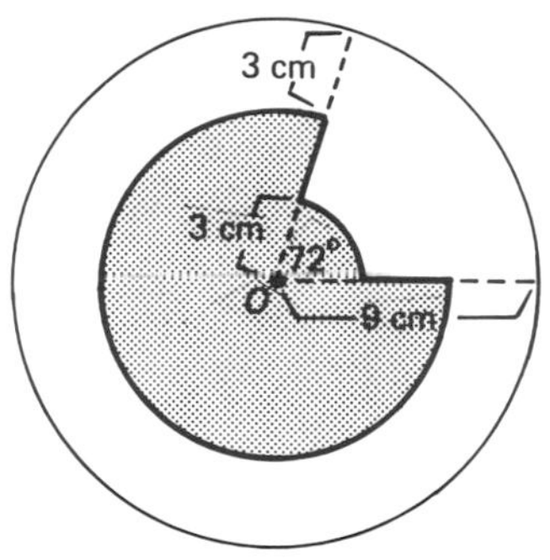

Fig. 11.90

Exercises 22–25 refer to Figure 11.91. Find:

22. the area of ΔAOB
23. the length of $\widehat{BC}$
24. the measure of $\angle CBO$
25. the measure of $\angle ADC$

Exercises 26–31 refer to Figure 11.92. Find:

26. the measure of $\angle ABO$
27. the measure of $\angle OAC$
28. the measure of $\angle AOB$
29. the length of AO
30. the area of ΔBOA
31. the area of sector OCD

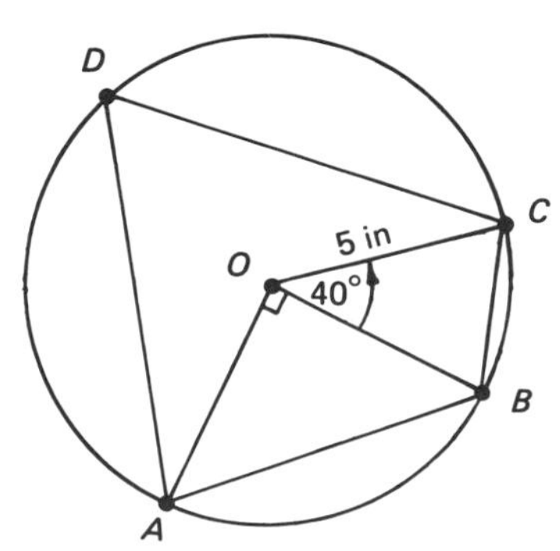

Fig. 11.91

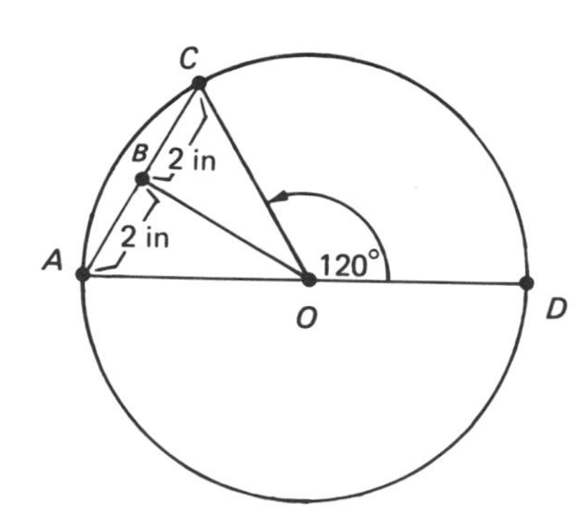

Fig. 11.92

Let's Review Chapter 11

1. Convert 28.6° into degrees and minutes.
2. Write 66° 12′ in decimal notation.
3. Classify as either a right angle, a straight angle, an acute angle, or an obtuse angle.
   a. 62° **b.** 145° **c.** 180°
4. Find the measure of **a.** the complement, **b.** the supplement, of an angle of 52°49′.
5. An angle measures 20° more than its supplement. Find the measure of the angle.
6. In Figure 11.93, assume $\ell_1 \parallel \ell_2$. Find the measure of: **a.** $\angle 1$, **b.** $\angle 2$
7. One acute angle of a right triangle measures 43°15′. Find the measure of the other acute angle.
8. In Figure 11.94, $AB = AC$ and AD is the height corresponding to BC. **a.** Find the length of BC. **b.** Find the area of ΔABC.
9. Find the area of a 30°, 60°, 90°-triangle whose hypotenuse is 10 inches.
10. Find **a.** the height, and **b.** the area of an equilateral triangle whose side length is 5 centimeters.
11. Suppose the ratio of similarity of two triangles is 3. If the area of the smaller triangle is 4 square feet, find the area of the larger triangle.

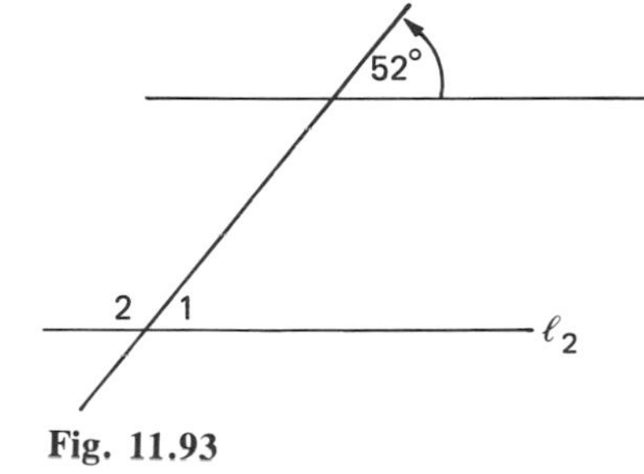

Fig. 11.93

In Problems 12–15, refer to Figure 11.95 . Assume O is the center of the circle. Find:

12. the measure of $\angle ACB$
13. the measure of $\angle ADB$
14. the length of arc $\widehat{AB}$
15. the area of sector AOB

Fig. 11.94

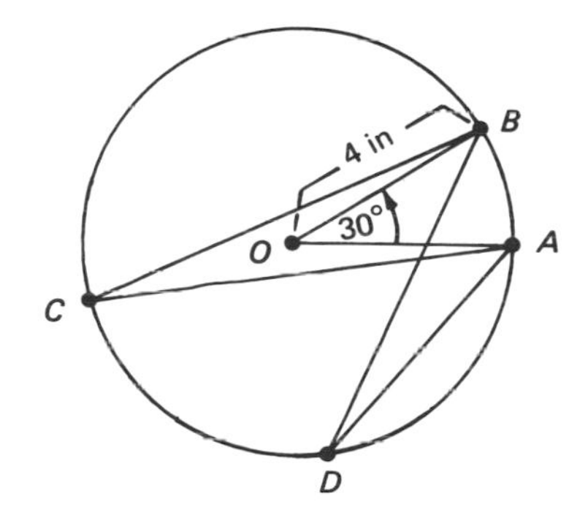

Fig. 11.95

TABLE A SQUARES AND SQUARE ROOTS

Number	Square	Square Root	Number	Square	Square Root	Number	Square	Square Root
1	1	1.000	35	1225	5.916	69	4761	8.307
2	4	1.414	36	1296	6.000	70	4900	8.367
3	9	1.732	37	1369	6.083			
4	16	2,000	38	1444	6,164	71	5041	8.426
5	25	2.236	39	1521	6,245	72	5184	8.485
6	36	2.449	40	1600	6.325	73	5329	8.544
7	49	2.646				74	5476	8.602
8	64	2.828	41	1681	6.403	75	5625	8,660
9	81	3.000	42	1764	6,481	76	5776	8,718
10	100	3.162	43	1849	6.557	77	5929	8.775
			44	1936	6.633	78	6084	8.832
11	121	3.317	45	2025	6.708	79	6241	8.888
12	144	3.464	46	2116	6,782	80	6400	8.944
13	169	3.606	47	2209	6.856			
14	196	3.742	48	2304	6.928	81	6561	9.000
15	225	3.873	49	2401	7.000	82	6724	9,055
16	256	4.000	50	2500	7.071	83	6889	9.110
17	289	4.123				84	7056	9.165
18	324	4.243	51	2601	7.141	85	7225	9.220
19	361	4.359	52	2704	7.211	86	7396	9.274
20	400	4.472	53	2809	7.280	87	7569	9,327
			54	2916	7.348	88	7744	9.381
21	441	4.583	55	3025	7.416	89	7921	9.434
22	484	4.690	56	3136	7.483	90	8100	9.487
23	529	4.796	57	3249	7.550			
24	576	4.899	58	3364	7.616	91	8281	9.539
25	625	5.000	59	3481	7.681	92	8464	9.592
26	676	5.099	60	3600	7.746	93	8649	9.644
27	729	5.196				94	8836	9.695
28	784	5.292	61	3721	7.810	95	9025	9.747
29	841	5.385	62	3844	7.874	96	9216	9.798
30	900	5.477	63	3969	7.937	97	9409	9.849
			64	4096	8.000	98	9604	9.899
31	961	5.568	65	4225	8.062	99	9801	9.950
32	1024	5.657	66	4356	8.124	100	10,000	10.000
33	1089	5.745	67	4489	8.185			
34	1156	5.831	68	4624	8.246			

ANSWERS

Answers to Diagnostic Tests, Let's Review Chapter questions, and odd-numbered Home Exercises are given below. Answers to even-numbered Home Exercises are given in the Instructor's Manual.

CHAPTER 1

DIAGNOSTIC TEST, p. 1

1.1 a. 740 **b.** five thousand nine hundred two **c.** $Q = 2, R - 5, S = 6, T = 9$ **d.** 80,000
1.2 a. $Q = -1, R = -4, S = -6, T = -8$ **b.** $<$ **c.** $-10 < -7 < -3 < 0 < 7$ **d.** 9 **e.** 9 and -9
1.3 a. 12,489 **b.** 683 **c.** 28,100 **d.** \$43
1.4 a. 3 **b.** -8 **c.** 14 **d.** 10 **e.** 10
1.5 a. 70,488 **b.** -6 **c.** 81 **d.** 32 **e.** -1 **f.** \$72
1.6 a. 0 **b.** not defined **c.** 8 **d.** 43 with remainder 88 *Check.* $(43)(93) + 88 = 3999 + 88 - 4089$
e. 6 pounds and 4 ounces
1.7 a. 39 **b.** 1 **c.** 7

HOME EXERCISES

1.1, p. 8

1. a. 5 **b.** 4 **c.** 2 **3. a.** 9 **b.** 0 **c.** 5 **5. a.** 7 **b.** 2 **c.** 6
7. a. 3 **b.** 0 **c.** 0 **9.** $4 \cdot 1000 + 2 \cdot 100 + 8 \cdot 10 + 3 \cdot 1$
11. $2 \cdot 1000 + 6 \cdot 100 + 7 \cdot 10 + 0 \cdot 1$
13. $5 \cdot 1{,}000{,}000 + 8 \cdot 100{,}000 + 3 \cdot 10{,}000 + 4 \cdot 1000 + 0 \cdot 100 + 0 \cdot 10 + 3 \cdot 1$ **15.** 524
17. 4,002,000 **19.** 60,055 **21.** four hundred nine
23. one hundred sixty-one thousand, two hundred eighty-four **25.** 250,000 **27.–32.** (See Figure 1A.)

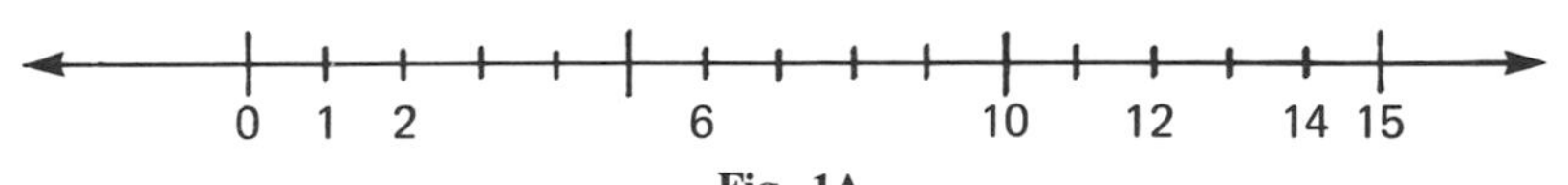

Fig. 1A

33. $Q = 3, R = 5, S = 8, T = 11, U = 12$ **35.** 6000 **37.** 7000 **39.** 20,000,000 **41.** 20,000,000
43. a. 536,200 **b.** 536,000 **c.** 500,000 **45.** 700,000 **47.** 2,000,000 **49.** \$50,000

HOME EXERCISES

1.2, p. 14

1.–6. (See Figure 1B on page A2.) **7.** $Q = -2, R = -5, S = -10, T = -15, U = -20$ **9.** $<$ **11.** $<$
13. $>$ **15.** left **17.** left **19.** right **21.** $1 < 3 < 5 < 8$ **23.** $-6 < -3 < 3 < 6$

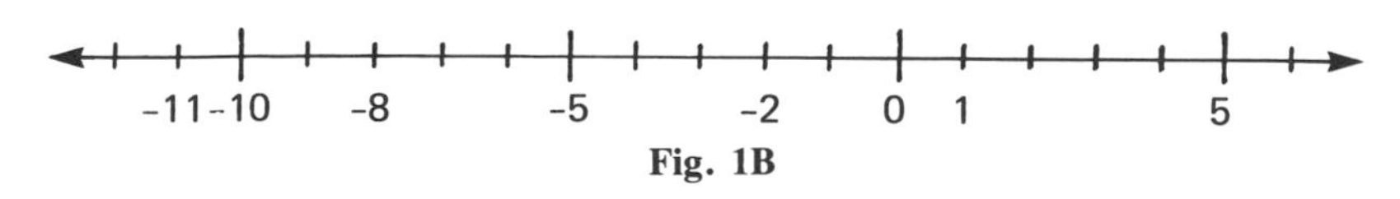

Fig. 1B

25. $-7 < -5 < -3 < 1 < 7$ **27.** -20 **29.** 0 **31.** 75 **33.** 8 **35.** 0 **37.** 100
39. -7 **41.** 7 **43.** 0 **45.** 8 and -8 **47.** 11 **49.** -44

HOME EXERCISES

1.3, p. 21

1. a. $(9 + 6) + 5 = 15 + 5 = 20$ **b.** $9 + (6 + 5) = 9 + 11 = 20$ **3.** 9947 **5.** 348,033
7. 1,795,592 **9.** 71,000,000 **11. a.** 8000 **b.** 8443 **c.** 8000 **d.** yes
13. 45 cents **15.** $164 **17.** 44 **19.** 5904 **21.** 14,710 **23.** 49,698
25. 849 (The checks follow.) **27.** 999 **29.** 1000 **31.** 7000 **33.** $18 **35.** $53
37. 821 **39.** 15°F *Checks.* **25.** $\begin{array}{r} 849 \\ -351 \\ \hline 498 \end{array}$ **27.** $\begin{array}{r} 999 \\ -284 \\ \hline 715 \end{array}$

HOME EXERCISES

1.4, p. 30

1. -5870 **3.** $-11,553$ **5.** 0 **7.** 3 **9.** 5 **11.** -13 **13.** 12 **15.** -35
17. 31 **19.** -3 **21.** -28 **23.** 1014 **25.** 45 **27.** 4 **29.** 7 **31.** 19
33. -23 **35.** 4 **37.** 11 **39.** 6°F **41.** $16 (won) **43.** $1

HOME EXERCISES

1.5, p. 38

1. 59,000 **3.** 1,200,000 **5. a.** $(4 \cdot 2) \cdot 6 = 8 \cdot 6 = 48$ **b.** $4 \cdot (2 \cdot 6) = 4 \cdot 12 = 48$
7. 150,000 **9.** 180,000 **11.** 16 **13.** 15 **15.** 299 **17.** 70,547 **19.** 21,882,200
21. -320 **23.** -32 **25.** 64 **27.** 49 **29.** -100 **31.** 125 **33.** -1
35. -1000 **37.** 0 **39.** 1 **41.** -128 **43.** 3 **45.** fourth **47.** sixth **49.** 480
51. $32,292 **53.** $9,317,440

HOME EXERCISES

1.6, p. 46

1. 0 **3.** -10 **5.** not defined **7.** -9 **9.** not defined **11.** 880 **13.** 7 ounces **15.** $12
17. quotient = 1203, remainder = 1 **19.** quotient = 599, remainder = 3 **21.** quotient = 583, remainder = 8
23. 18 yards and 2 feet **25. a.** 26 **b.** 4 **27.** 4 **29.** 5 **31.** 8 **33.** 22
35. 1080 **37.** 21 *Check.* $21 \times 25 = 525$ **39.** 60 *Check.* $60 \times 48 = 2880$
41. quotient = 21, remainder = 31 **43.** quotient = 100, remainder = 85

45. **a.** quotient = 30, remainder = 6 **b.** $(30)(33) + 6 = 990 + 6 = 996$
47. **a.** quotient = 86, remainder = 62 **b.** $86 \times 112 + 62 = 9632 + 62 = 9694$ **49.** \$12,009

HOME EXERCISES

1.7, p. 49

1. 23 **3.** 16 **5.** 3 **7.** -16 **9.** 0 **11.** 4 **13.** 48 **15.** 18
17. 14 **19.** 1 **21.** -27 **23.** -16 **25.** 35 **27.** 0 **29.** 16 **31.** 1
33. 9 **35.** 29 **37.** **a.** 3 **b.** -1 **39.** 20 **41.** 4 **43.** \$102

LET'S REVIEW CHAPTER 1, p. 49

1. **a.** 4 **b.** 0 **c.** 4
2. $8 \times 1{,}000{,}000 + 5 \times 100{,}000 + 0 \times 10{,}000 + 6 \times 1000 + 3 \times 100 + 9 \times 10 + 9 \times 1$
3. 40,601 **4.** $Q = 3, R = 8, S = 10, T = 14, U = 15$
5. **a.** 7,000,000 **b.** 7,300,000 **c.** 7,285,000
6. $Q = 4, R = -4, S = -6, T = -7, U = -10$ **7.** $<$ **8.** right **9.** $-8 < -4 < -2 < 0 < 2 < 8$
10. **a.** 4 **b.** -45 **c.** 0 **11.** **a.** 13 **b.** 13 **c.** 0
12. 11 and -11 **13.** 8504 **14.** 780,000 **15.** \$36 **16.** -9 **17.** -4 **18.** -5
19. -17 **20.** 2,880,480 **21.** 20 **22.** 49 **23.** -125 **24.** 16 **25.** 1
26. -7 **27.** 0 **28.** 0
29. **a.** quotient = 123, remainder = 44 **b.** $(123)(76) + 44 = 9348 + 44 = 9392$
30. 0 **31.** 4 **32.** 50 **33.** 6 **34.** -11

CHAPTER 2

DIAGNOSTIC TEST, p. 51

2.1 **a.** -5 **b.** no **c.** yes **d.** $7x + 3y + 5z$
2.2 **a.** $5a - 2c$ **b.** $x - 3y$ **c.** $3a - 2b$
2.3 **a.** 36 **b.** -7 **c.** 168 square centimeters
2.4 **a.** $-18x^3y^4$ **b.** $-3a^3$ **c.** $10x^3y^2 - 5x^2y^3$ **d.** $2b + 3 - 6a$
2.5 **a.** $2(7) - 1 \stackrel{?}{=} 13$ **b.** 8 **c.** 6 *Check.* $6 - 3 \stackrel{?}{=} 9 - 6$ **d.** 3

$14 - 1 \stackrel{?}{=} 13$ $3 \stackrel{\checkmark}{=} 3$

$13 \stackrel{\checkmark}{=} 13$

HOME EXERCISES

2.1, p. 55

1. $x \cdot x \cdot x \cdot x$ **3.** $10 \cdot x \cdot x \cdot y \cdot z \cdot z$ **5.** $-1 \cdot u$ **7.** 7 **9.** 1 **11.** 12
13. 6 **15.** like **17.** like **19.** unlike **21.** like **23.** like **25.** unlike
27. $7y$ **29.** $4xy$ **31.** $6xy$ **33.** 0 **35.** $8r^2st^2$ **37.** $6u - 3v$ **39.** $6a + 12b$
41. $6x^2y + 10z^2$ **43.** $3y^2 - 2z^2$ **45.** 14 **47.** $4n - 10p$ **49.** $14x$ **51.** $5a$

HOME EXERCISES

2.2, p. 58

1. (i) **3.** (i) **5.** (iii) **7.** (i) **9.** (ii) **11.** $3x + 5y$ **13.** $8x + 2$
15. $12x - 6y + z$ **17.** $7a - b + 2c + 6d + 2e$ **19.** $2a + b$ **21.** $5c - 11d$ **23.** $21y - 33z$
25. $3a + b + 2c - 2d$ **27.** $a - 4b$ **29.** $7x - 2y + 2z$ **31.** $6x - y$ **33.** $7 + 3a - 6b + c$
35. $10a - b$ **37.** $4x + y$ **39.** $7a + 2b$

HOME EXERCISES

2.3, p. 61

1. 12 **3.** 7 **5.** −8 **7.** −27 **9.** −3 **11.** 1 **13.** 3 **15.** −12
17. 25 **19.** 32 **21.** 10 **23.** 8 **25.** 1,009,702 **27. a.** 12 **b.** 8 **c.** 0
29. a. 1 **b.** 2 **c.** 2 **31. a.** 21 **b.** 45 **c.** 55 **33. a.** 1 **b.** 13 **c.** 73
35. a. 121 square inches **b.** 225 square inches **c.** 900 square inches **37.** 124 inches
39. a. 94 square feet **b.** 312 square feet

HOME EXERCISES

2.4, p. 68

1. b^3 **3.** b^5 **5.** $-5y^4$ **7.** a^7 **9.** $10c^3d^3$ **11.** $30x^5y^2$ **13.** $6x^3y^4z^6$
15. $10a^4b^2c^4$ **17.** $4x$ **19.** a^3 **21.** m^2n^4 **23.** $2x$ **25.** $4a^9b^2c$ **27.** $2x$
29. $x + 2$ **31.** $9x^2 + 9y$ **33.** $-3m + 3n$ **35.** $-10a^2 + 20b$ **37.** $x^3y + x^4$
39. $-4y^3 - 4y^3z$ **41.** $-2m^3n + 3mn^3$ **43.** $50x^6y^2z^6 + 80x^3y^3z^5$ **45.** $4x^3yz^3 - 16x^2y^2z^2 + 12x^2yz^3$
47. $14 - 4a$ **49.** $a^2 - 2ab + b^2$ **51.** $4a + 27b$ **53.** $x + y$ **55.** $a + 1$
57. $xy + 1$ **59.** $4x^2 + 6x$

HOME EXERCISES

2.5, p. 75

1. 6 is a root. **3.** 6 is not a root. **5.** −1 is a root. **7.** −3 is a root. **9.** 0 is a root.
11. 7 **13.** 14 **15.** −3 **17.** 15 **19.** 12 **21.** 5 **23.** 6 **25.** −1
27. 8 **29.** 72 **31.** −24 **33.** 4 **35.** −4 **37.** $\frac{-2}{5}$ **39.** 2

41. a. 4 **b.** $3\,(4) \stackrel{\checkmark}{=} 12$
$12 \stackrel{\checkmark}{=} 12$

43. a. 2 **b.** $7 - 3(2) \stackrel{?}{=} -1 + 2$
$7 - 6 \stackrel{?}{=} 1$
$1 \stackrel{\checkmark}{=} 1$

45. a. −2 **b.** $2(-2) + 7 \stackrel{?}{=} 7(-2) + 17$
$-4 + 7 \stackrel{?}{=} -14 + 7$
$3 \stackrel{\checkmark}{=} 3$

47. a. 1 **b.** $-3(1 + 2) \stackrel{?}{=} 1 - 10$
$-3(3) \stackrel{?}{=} -9$
$-9 \stackrel{\checkmark}{=} -9$

49. a. 7 **b.** $\frac{6(7) - (7 - 5)}{5.} \stackrel{?}{=} 8$ **51.** 4 **53.** 3

$$\frac{42 - 2}{5} \stackrel{?}{=} 8$$

$$\frac{40}{5} \stackrel{?}{=} 8$$

$$8 \stackrel{\checkmark}{=} 8$$

LET'S REVIEW CHAPTER 2, p. 76

1. a. $5 \cdot x \cdot y \cdot y \cdot y$ **b.** $\frac{1}{2} \cdot u \cdot v \cdot v \cdot w \cdot w \cdot w$ **2.** -12 **3.** yes **4.** no
5. $2m + 15n$ **6.** $5x + 2y - 3z$ **7.** $4w + x + 4y - z$ **8.** $3x + 3z$ **9.** $13a$
10. $a + b + c$ **11.** 50 **12.** 45 **13. a.** 4 **b.** -12 **14.** 400 square inches
15. $6x^3y^4$ **16.** $8m^4n^4$ **17.** $x^3 - 5x^2y$ **18.** $8a^4b^2c + 12a^3b^2c^2$ **19.** $2x^2y$ **20.** $8a^3$
21. $2a + 2b$ **22.** $4x - 3y$

23. $3(3) + 2 \stackrel{?}{=} 11$
$9 + 2 \stackrel{?}{=} 11$
$11 \stackrel{\checkmark}{=} 11$
3 is a root.

24. $5(2) - 2 \stackrel{?}{=} 2(2) + 3$
$10 - 2 \stackrel{?}{=} 4 + 3$
$8 \stackrel{x}{=} 7$
2 is not a root.

25. $5 - 2(-2) \stackrel{?}{=} -2 + 1$
$5 + 4 \stackrel{?}{=} -1$
$9 \stackrel{x}{=} -1$
-2 is not a root.

26. 39 **27.** 1 **28.** 12 **29.** 6 **30.** 2 *Check.* $2(2) + 3 \stackrel{?}{=} 15 - 4(2)$
$4 + 3 \stackrel{?}{=} 15 - 8$
$7 \stackrel{\checkmark}{=} 7$

CHAPTER 3

DIAGNOSTIC TEXT, p. 78

3.1 a. $2^3 \cdot 3 \cdot 5$ **b.** 14 **c.** $4xy(2x + 5y - 3)$
3.2 a. $a^2 + 6a + 8$ **b.** $x^2 - 2x - 35$ **c.** $m^3 - 3m^2n + 4mn - 12n^2$ **d.** $7a + 2$
3.3 a. $(x + 8)(x - 8)$ **b.** $(a^2 + 3b)(a^2 - 3b)$ **c.** $5(x + 2)(x - 2)$
3.4 a. $(x + 4)(x + 1)$ **b.** $(a + 6)(a - 7)$ **c.** $m(m + 3)^2$
3.5 a. $(4x + 1)(x + 1)$ **b.** $(4a + 1)(2a - 1)$ **c.** $(3x + y)(2x - 3y)$

HOME EXERCISES

3.1, p. 85

1. 1 and 5 **3.** 1, 2, 4, 5, 10, and 20 **5.** 1, 2, 4, 7, 14, and 28 **7.** prime **9.** prime
11. prime **13.** $2 \cdot 11$ **15.** $2^2 \cdot 3^2$ **17.** $2^4 \cdot 3$ **19.** $2^4 \cdot 5$ **21.** $2 \cdot 3^2 \cdot 5$
23. $2^4 \cdot 3^2$ **25.** 30 **27.** 9 **29.** 12 **31.** 2 **33.** 6 **35.** 4 **37.** 1 **39.** 6

41. $2(a + 1)$ **43.** $3(m^3 - 3)$ **45.** $16a + 25$ **47.** $x(2x + 3)$ **49.** $a^7(a^3 - 1)$
51. $4a^3(8a + 3)$ **53.** $2ab(ac + 2b)$ **55.** $x^3z^4(x^4y^6z - x^2y^6 + 1)$ **57.** $12yz^2(3yz^3 + 5yz - 2)$
59. $24x^2y^4z^2(4yz - 6y^6 + 3xz^3)$

HOME EXERCISES

3.2, p. 89

1. $z^2 + 9z + 8$ **3.** $a^2 + a - 6$ **5.** $b^2 - 6b + 8$ **7.** $m^2 - 7m - 30$ **9.** $5z^2 + 5z - 10$
11. $-3b^2 + 9b - 6$ **13.** $y^2 + 4y + 4$ **15.** $c^2 + 12c + 36$ **17.** $x^2 - 64$ **19.** $x^2 - 24x + 144$
21. $x^4 - 9$ **23.** $u^8 + 3u^4 + 2$ **25.** $x^6 - 6x^3 + 9$ **27.** $b^4 + 3b^2 - 10$ **29.** $c^5 + c^3 - c^2 - 1$
31. $x^2 - a^2$ **33.** $x^2 + 5ax + 6a^2$ **35.** $x^2 - xy - 2y^2$ **37.** $2x^2 - 6xy + x - 8y^2 - 4y$
39. $6a^2 - 4ab + 9a - 2b^2 + 3b$ **41.** $2x^2 + 10x + 10$ **43.** $2y^2 + 15y + 11$ **45.** $2x^2 + 7x + 5$

HOME EXERCISES

3.3, p. 91

1. $(y + 2)(y - 2)$ **3.** $(a + 5)(a - 5)$ **5.** $(y + 9)(y - 9)$ **7.** $(1 + x)(1 - x)$ **9.** $(11 + c)(11 - c)$
11. $(m + n)(m - n)$ **13.** $(3a + 5)(3a - 5)$ **15.** $(1 + 3u)(1 - 3u)$ **17.** $(3x + 4a)(3x - 4a)$
19. $(12y + 5x)(12y - 5x)$ **21.** $5(y + 3)(y - 3)$ **23.** $7(1 + t)(1 - t)$ **25.** $5(6 + y)(6 - y)$
27. $3(y + 3z)(y - 3z)$ **29.** $7(x + 5y)(x - 5y)$ **31.** $11(2a + 3b)(2a - 3b)$ **33.** $x(x + 2)(x - 2)$
35. $x^3(1 + x)(1 - x)$ **37.** $a(a + 2b)(a - 2b)$ **39.** $s^2(2s + 3t)(2s - 3t)$ **41.** $5a(a + 2)(a - 2)$
43. $2x(2x + 5y)(2x - 5y)$ **45. a.** $(x + 3)(x - 3)$ **b.** $(x + 1)(x - 1)$ **c.** does not factor
d. $(x + a)(x - a)$ **47. a.** $(y + 4)(y - 4)$ **b.** $3(y + 4)(y - 4)$ **c.** $16(y^2 + 1)$ **d.** does not factor

HOME EXERCISES

3.4, p. 96

1. $(x + 3)(x + 1)$ **3.** $(b + 1)^2$ **5.** $(a + 2)(a - 1)$ **7.** $(b + 1)(b - 2)$ **9.** $(s + 3)(s - 4)$
11. $(a - 6)^2$ **13.** $(x - 7)(x - 2)$ **15.** $(a + 7)(a - 2)$ **17.** $(c + 5)(c + 4)$
19. $(x + 8)(x + 5)$ **21.** $(x + 12)(x - 3)$ **23.** $2(x + 2)(x + 1)$ **25.** $3(x + 4)(x - 2)$
27. $-(y - 5)(y - 4)$ **29.** $a(a + 5)(a - 2)$ **31.** $4(m + 5)^2$
33. $3(6 - x)(4 + x)$, or $-3(x - 6)(x + 4)$ **35.** $v^2(u + 9)^2$ **37.** $2t(t - 7)(t - 5)$ **39.** $(x + a)^2$
41. $(x + 4a)(x + a)$ **43.** $(m + 2n)^2$ **45.** $(x + 3y)^2$ **47.** $2(x + 6y)^2$ **49. a.** $(x + 4y)(x + 2y)$
51. a. $(x + 3)(x + 2)$ **b.** $(x - 3)(x - 2)$ **c.** $(x - 5)(x - 1)$ **d.** does not factor

HOME EXERCISES

3.5, p. 100

1. $(3x + 1)(x + 1)$ **3.** $(3x - 1)(x + 1)$ **5.** $(2x + 1)(x + 3)$ **7.** $(2y + 3)(y + 1)$
9. $(2a + 5)(a + 1)$ **11.** $(2m - 1)(m - 2)$ **13.** $(2a - 1)(a + 4)$ **15.** $(2a + 5)(a + 5)$
17. $(3y - 1)^2$ **19.** $(2a + 1)^2$ **21.** $(3y - 7)(y + 1)$ **23.** $(2m + 3)(2m + 1)$ **25.** $(3x - 2)(2x + 3)$

27. $(7a+2)(a+1)$ **29.** $(5x+4)(2x-1)$ **31.** $(3x+1)(3x-13)$ **33.** $(5a-3)(a+4)$
35. $2(2x+1)(x+1)$ **37.** $(2b-1)(b+12)$ **39.** $4(2c+3)(c+1)$ **41.** $2x(2x+3)(x+1)$
43. $(2x+y)^2$ **45.** $(2y+z)(y+z)$ **47.** $(2x+y)(x+2y)$ **49.** $(3a+2b)(3a-10b)$
51. $2(3x+y)(x+2y)$

LET'S REVIEW CHAPTER 3, p. 100

1. 1, 2, 3, 6, 9, and 18 **2.** composite **3.** prime **4.** composite **5.** $2^2 \cdot 7$ **6.** $2 \cdot 5 \cdot 7$
7. 12 **8.** 4 **9.** $2(2x+3)$ **10.** $3a^4(3a^2-5)$ **11.** $x^2y^3z(x^3yz^2+xz^2+1)$
12. x^2+6x+5 **13.** z^4-3z^2-4 **14.** $m^2-mn-6n^2$ **15.** $(a+5)(a-5)$
16. $(2y+7z)(2y-7z)$ **17.** $3x(y+2)(y-2)$ **18.** $(c+4)(c+3)$ **19.** $(m+4)(m-5)$
20. $(x-3)(x-2)$ **21.** $(2x+5)(x+1)$ **22.** $(2y-1)(y-3)$ **23.** $(4x-3y)(x+y)$
24. $5(m-2n)$ **25.** $(6+a)(6-a)$ **26.** $(x+7)(x+2)$ **27.** $7(a+2b)(a-2b)$
28. $x(x+2)(x-2)$ **29.** $x(3x+1)(x-2)$ **30.** $(m+5n)(m+3n)$ **31.** $4(m+2n^2)(m-2n^2)$
32. $s(s+7)(s-1)$ **33.** $t^2(t+4)(t-4)$ **34.** $(a-8)(a-2)$ **35.** does not factor
36. $(5x+2)(x-2)$

CHAPTER 4

DIAGNOSTIC TEST, p. 102

4.1 a. $\frac{-2}{5}$ **b.** $Q = -\frac{4}{5}, R = -\frac{1}{5}, S = \frac{1}{5}, T - \frac{2}{5}, U = \frac{7}{5}$ **c.** $\frac{3}{4}$ **d.** $\frac{-3}{5}$ **4.2 a.** $\frac{9}{4}$ **b.** $\frac{6}{5}$
4.3 a. $\frac{-5x}{8z^3}$ **b.** $\frac{3(a+b)^3}{5(a-b)}$ **c.** $\frac{5x+4}{x}$ **d.** $\frac{c^2}{6a-5c}$ **4.4 a.** $\frac{3}{a-b}$ **b.** $\frac{2x+3y}{x+y}$
4.5 a. x^2+5x+2 **b.** quotient x^2-x+2, remainder -1

HOME EXERCISES

4.1, p. 109

1. numerator 4 and denominator 9 **3.** numerator -3 and denominator 2 **5.** 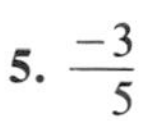

7.–10.

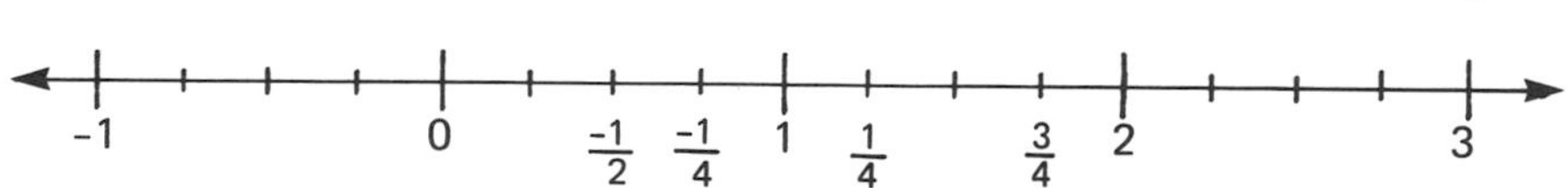

Fig. 4A

11. $Q = \frac{3}{7}, R = \frac{9}{7}, S = \frac{-6}{7}, T = \frac{-10}{7}, U = \frac{-8}{7}, V = \frac{1}{7}, W = \frac{-1}{7}$ **13.** $\frac{1}{1}$

15. $\frac{0}{D}$ for any integer $D \neq 0$; in particular $\frac{0}{1}$

17. Does $(-4)(3) = 6(-2)$?
Yes. $-12 = -12$
Thus, $\frac{-4}{6} = \frac{-2}{3}$

19. Does $(-8)(-7) = 14(4)$?
Yes. $56 = 56$
Thus, $\frac{-8}{14} = \frac{4}{-7}$

21. $\frac{1}{5}$ **23.** $\frac{1}{5}$ **25.** $\frac{-1}{3}$ **27.** $\frac{1}{4}$ **29.** $\frac{-1}{2}$ **31.** $\frac{4}{5}$ **33.** $\frac{2}{5}$ **35.** $\frac{1}{6}$
37. $\frac{-5}{3}$ **39.** $\frac{-3}{10}$ **41.** $\frac{1}{11}$ **43.** $\frac{6}{7}$ **45.** **a.** and **c.** **47.** $\frac{3}{4}$

HOME EXERCISES

4.2, p. 115

1. a. x **b.** $x + 1$ **3. a.** $y + 3$ **b.** $y - 4$ **5. a.** 5 **b.** $y^2 + 2y$
7. a. $t^4 - 3t^2 + 1$ **b.** $t^3 + t + 2$ **9.** $\frac{20}{11}$ **11.** $\frac{5}{2}$ **13.** $\frac{5}{2}$ **15.** $\frac{8}{7}$ **17.** -1 **19.** $\frac{17}{15}$
21. 1 **23.** 3 **25.** $\frac{1}{5}$ **27.** $\frac{15}{4}$ **29.** 5 **31.** $\frac{1}{2}$ **33.** $\frac{10}{9}$ **35. a.** $\frac{1}{3}$ **b.** $\frac{1}{5}$ **c.** $\frac{1}{7}$
37. a. $\frac{7}{4}$ **b.** 4 **c.** $\frac{5}{2}$ **39. a.** 0 **b.** $\frac{4}{5}$ **c.** $\frac{117}{50}$ **41.** 2

43. Does $x(x + 4) = 1(x^2 + 4x)$?

Yes. $x^2 + 4x = x^2 + 4x$

Thus, $\frac{x}{x^2 + 4x} = \frac{1}{x + 4}$

45. Does $(x^2 + x)5 = (5x + 5)x$?

Yes. $5x^2 + 5x = 5x^2 + 5x$

Thus, $\frac{x^2 + x}{5x + 5} = \frac{x}{5}$

HOME EXERCISES

4.3, p. 118

1. 5 **3.** $\frac{49}{11}$ **5.** $\frac{1}{y^3}$ **7.** $\frac{1}{a}$ **9.** 1 **11.** $\frac{-8}{r^2s}$ **13.** $\frac{-2xy^3}{z}$ **15.** $\frac{3m^9p}{2n}$
17. $\frac{2a^3}{3}$ **19.** $\frac{1}{m + n}$ **21.** $\frac{8}{(c - d)^3}$ **23.** $\frac{(a + b)^6}{2}$ **25.** $\frac{1}{25x(a + 2)}$ **27.** $a - b$
29. $x + y$ **31.** $x - y$ **33.** $3x + 6y$ **35.** $\frac{5x^2 + 4y^2}{2}$ **37.** $\frac{x^2 + y^2}{x}$ **39.** $x + 2y$
41. $\frac{5(a^2 - b^2)}{a}$ **43.** $\frac{x(x - 2)}{2}$ **45.** $\frac{a - 2b + c}{5}$ **47.** $\frac{s + 2t - u}{6}$ **49.** $\frac{a - 2x + 3ax}{4ax}$
51. $\frac{4bc + 5a^2 - 6c}{5b}$ **53.** $\frac{1}{2 + 3x}$ **55.** $\frac{yz}{2y^2 - 3z^2}$ **57.** $\frac{20xy}{2y - 5x + 10xy}$

HOME EXERCISES

4.4, p. 121

1. $\frac{a + b}{x + y}$ **3.** $\frac{x + y}{x - y}$ **5.** $\frac{x + y}{x - y}$ **7.** $\frac{x + y}{x - 2y}$ **9.** $\frac{6}{5}$ **11.** $\frac{a}{b}$ **13.** $\frac{x - y}{3}$

15. $\frac{3}{x+y}$ **17.** $\frac{a-3}{4}$ **19.** $\frac{3}{5(m-n)}$ **21.** $\frac{a+2b}{2}$ **23.** $\frac{2x-3y}{6}$ **25.** $\frac{x+1}{x+3}$ **27.** $\frac{y-3}{y+1}$

29. $\frac{u-6}{u+1}$ **31.** $\frac{m+4}{m+2}$ **33.** $\frac{x-2}{x-4}$ **35.** $\frac{a-b}{2(a+2b)}$ **37.** $\frac{x+4}{2(x-2)}$ **39.** $\frac{2x+5}{x+3}$

HOME EXERCISES

4.5, p. 129

1. a. $x^2 + 2x + 1$ **b.** 2 **3. a.** $-t^5 + t^3 + t$ **b.** 5 **5. a.** $z^{10} + 2z^7 - z^4 + z$ **b.** 10

7. $x + 4$ *Check.* $(x + 4)(x + 1) = x^2 + 5x + 4$ **9.** $x + 6$ *Check.* $(x + 6)(x - 1) = x^2 + 5x - 6$

11. $x + 5$ *Check* $(x + 5)(x - 8) = x^2 - 3x - 40$ **13.** $3t^2 - 1$ *Check.* $(3t^2 - 1)(t + 4) = 3t^3 + 12t^2 - t - 4$
$= 12t^2 + 3t^3 - 4 - t$

15. $z^2 + 1$ **17.** $a + 3$ **19.** $a^2 + 5$ **21.** $x + 2$ **23.** $x^2 + 2x + 1$

25. a. 4 **b.** 3 **c.** $\underbrace{\text{degree }(3)}_{0} < \underbrace{\text{degree }(2x)}_{1}$

27. a. $x + 1$ **b.** 5 **c.** $\underbrace{\text{degree }(5)}_{0} < \underbrace{\text{degree }(x+1)}_{1}$

29. a. $2x$ **b.** 5 **c.** $\underbrace{\text{degree }(5)}_{0} < \underbrace{\text{degree }(x^2-2)}_{2}$

31. a. 3 **b.** -1 **c.** $3(x + 1) + (-1) = 3x + 2$

33. a. $x + 3$ **b.** 2 **c.** $(x + 3)(x + 2) + 2 = (x^2 + 5x + 6) + 2$
$= x^2 + 5x + 8$

35. a. $x^2 + 2x + 1$ **b.** 4 **c.** $(x^2 + 2x + 1)(x + 1) + 4 = (x^3 + 3x^2 + 3x + 1) + 4 = x^3 + 3x^2 + 3x + 5$

37. a. $3x + 5$ **b.** $10x + 2$ **c.** $(3x + 5)(x^2 - x) + (10x + 2) = (3x^3 + 2x^2 - 5x) + (10x + 2) = 3x^3 + 2x^2 + 5x + 2$

39. a. $x + 5$ **b.** 5 **c.** $x + 5 + \frac{5}{x+3}$ **41. a.** $x + 5$ **b.** 25 **c.** $x + 5 + \frac{25}{x-5}$

43. $4x - 3$ **b.** $-3x + 13$ **c.** $4x - 3 + \frac{-3x+13}{x^2+x+1}$

45. a. $x^2 + 3$ **b.** 6 **c.** $x^2 + 3 + \frac{6}{x^2-2}$

47. a. $x^2 - 4x + 6$ **b.** -5 **c.** $x^2 - 4x + 6 + \frac{-5}{x+1}$

49. a. $x^2 - 4x + 9$ **b.** -16 **c.** $x^2 - 4x + 9 + \frac{-16}{x+2}$

LET'S REVIEW CHAPTER 4, p. 130

1. numerator 3 and denominator 8 **2.** $\frac{-2}{5}$ **3.** $Q = \frac{-4}{5}, R = \frac{-1}{5}, S = \frac{1}{5}, T = \frac{3}{5}, U = \frac{6}{5}, V = \frac{9}{5}$

4. $\frac{-8}{1}$ **5.** Does $4(-3) = 6(-2)$? Yes. $-12 = -12$ Thus, $\frac{4}{6} = \frac{-2}{-3}$ **6.** $\frac{-1}{3}$ **7.** $\frac{3}{5}$ **8. a.** $2y - 1$ **b.** $y^2 + 3y - 5$

9. 32 **10.** $\frac{3}{4}$ **11. a.** -2 **b.** -2 **c.** $\frac{5}{2}$ **12.** $\frac{a^2}{b^4}$ **13.** $\frac{50z^3}{x^3}$ **14.** $\frac{4a^3}{3c^2}$

15. $\frac{(x+y)^2}{27(x-y)}$ **16.** $2(x - y)$ **17.** $\frac{a+b}{x}$ **18.** $\frac{a+2c}{3}$ **19.** $\frac{2}{1-5y}$ **20.** $\frac{a+b}{5(a+2b)}$

21. $\frac{m-3}{m+3}$ **22.** $\frac{1}{s}$ **23.** $\frac{2(x+y)}{x-y}$ **24. a.** $7x^6 + 5x^5 - x^4 + 2x^2$ **b.** 6

25. $y + 5$ *Check.* $(y + 5)(y + 2) = y^2 + 7y + 10$ **26.** $x^2 - x + 1$ **27.** $a^2 - 1$

28. a. 5 **b.** 2 **c.** $\underbrace{\text{degree }(2)}_{0} < \underbrace{\text{degree }(x+1)}_{1}$

29. a. $4x - 15$ **b.** 39 **c.** $(4x - 15)(x + 2) + 39 = (4x^2 - 7x - 30) + 39 = 4x^2 - 7x + 9$

30. a. $x^2 - 5x + 20$ **b.** -79 **c.** $x^2 - 5x + 20 + \frac{-79}{x+4}$

CHAPTER 5

DIAGNOSTIC TEST, p. 132

5.1 a. $\frac{1}{36}$ **b.** $\frac{(x-1)(x+y)}{4}$ **c.** $\frac{2ax}{3b}$ **d.** $\frac{x+3a}{2(x+2)}$

5.2 a. 60 **b.** $x^3(x - 1)$ **c.** $(x + 1)(x - 3)$; $\frac{x}{x+1} = \frac{x^2 - 3x}{(x+1)(x-3)}$, $\frac{3}{x-3} = \frac{3x+3}{(x+1)(x-3)}$

5.3 a. $\frac{1}{2}$ **b.** $\frac{1}{x-2}$ **c.** $\frac{9}{20}$ **d.** $\frac{x^2 + 2x - 4}{(x+1)(x+4)(x-4)}$

5.4 a. $>$ **b.** $<$ **c.** $<$

5.5 a. $\frac{1}{6}$ **b.** $\frac{6}{7}$ **c.** $\frac{x}{x+3}$

5.6 a. $\frac{5}{4}$ **b.** -24 **c.** 5

HOME EXERCISES

5.1, p. 139

1. $\frac{1}{6}$ **3.** $\frac{3}{2}$ **5.** $\frac{1}{12}$ **7.** $\frac{5}{14}$ **9.** $\frac{-4}{15}$ **11.** $\frac{2}{3}$ **13.** $\frac{ax}{y}$ **15.** $2b$ **17.** $\frac{20xb}{ay}$

19. $\frac{48a^2}{x^2}$ **21.** $\frac{5y}{x}$ **23.** $\frac{2(a-2)}{3(a+2)}$ **25.** $\frac{3}{a+3}$ **27.** 24 **29.** $\frac{2}{3}$ **31.** 2 **33.** $\frac{-1}{3}$

35. 25 **37.** 6 **39.** $\frac{22}{25}$ **41.** $\frac{b}{a}$ **43.** $\frac{1}{a}$ **45.** $\frac{b^2}{ac^2}$ **47.** $\frac{2}{x-a}$ **49.** 6 **51.** $\frac{x+2}{a^2b}$

53. a^2 **55.** $2a^2$ **57.** $\frac{5}{2}$ **59.** $\frac{3}{x}$

HOME EXERCISES

5.2, p. 145

1. 6 **3.** 12 **5.** 30 **7.** 108 **9.** 30 **11.** 12 **13.** 300 **15.** $5x$

17. x^2 **19.** $a(a+b)$ **21.** $8a^2bc$ **23.** $24m^2n^2$ **25.** $(x+2)(x-2)$ or x^2-4

27. $(a+5)^2(a-5)$ **29.** $8(x-1)^2$ **31.** $a^2b^2(x-y)$ **33.** $a^2b^7x^4y^3z$ **35.** $25(a+2)(a-2)$

37. a. 3 **b.** $\frac{1}{3}=\frac{1}{3}$, $1=\frac{3}{3}$ **39. a.** 6 **b.** $\frac{1}{2}=\frac{3}{6}$, $\frac{5}{6}=\frac{5}{6}$ **41. a.** 20 **b.** $\frac{5}{4}=\frac{25}{20}$, $\frac{7}{10}=\frac{14}{20}$

43. a. 140 **b.** $\frac{7}{20}=\frac{49}{140}$, $\frac{3}{28}=\frac{15}{140}$ **45. a.** 108 **b.** $\frac{5}{36}=\frac{15}{108}$, $\frac{1}{27}=\frac{4}{108}$

47. a. xy **b.** $\frac{1}{x}=\frac{y}{xy}$, $\frac{1}{x}=\frac{x}{xy}$ **49. a.** $x+a$ **b.** $\frac{2}{x+a}=\frac{2}{x+a}$, $-1=\frac{-(x+a)}{x+a}$

51. a. $x^2y^4z^2$ **b.** $\frac{x-a}{x^2y^3z^2}=\frac{y(x-a)}{x^2y^4z^2}$, $\frac{a}{xy^4z}=\frac{axz}{x^2y^4z^2}$

53. a. $(x+1)(x-1)^2$ **b.** $\frac{1}{x^2-1}=\frac{x-1}{(x+1)(x-1)^2}$, $\frac{x}{(x-1)^2}=\frac{x(x+1)}{(x+1)(x-1)^2}$

55. a. $a(x+2)(x-2)$ **b.** $\frac{1}{ax-2a}=\frac{x+2}{a(x+2)(x-2)}$, $\frac{a}{x^2-4}=\frac{a^2}{a(x+2)(x-2)}$, $\frac{-1}{ax+2a}=\frac{2-x}{a(x+2)(x-2)}$

HOME EXERCISES

5.3, p. 151

1. $\frac{3}{7}$ **3.** 1 **5.** $\frac{5}{6}$ **7.** $\frac{8}{9}$ **9.** $\frac{6}{5}$ **11.** $\frac{7}{x}$ **13.** $\frac{11}{x+1}$ **15.** $\frac{9}{a}$ **17.** 1

19. 3 **21.** -1 **23.** $\frac{1}{x+2}$ **25.** $\frac{b^2+1}{b+1}$ **27.** $y+1$ **29.** $\frac{5}{6}$ **31.** $\frac{3}{4}$ **33.** $\frac{23}{24}$

35. $\frac{5}{192}$ **37.** $\frac{25}{18}$ **39.** $\frac{71}{50}$ **41.** $\frac{x+y}{xy}$ **43.** $\frac{a(x+1)}{x^2}$ **45.** $\frac{x+a+2}{(x+a)^2}$ **47.** $\frac{1-a^2}{a(t+1)}$

49. $\frac{x+1}{x(x+2)}$ **51.** $\frac{10a+9}{a(a+9)(a-9)}$ **53.** $\frac{3x^2-12x+11}{(x-1)(x-2)(x-3)}$ **55.** $\frac{ax+4x+a}{a^3(a+4)(a-4)}$ **57.** $\frac{1}{x-1}$

59. $\frac{3}{5}$ **61.** $\frac{3}{2}$ **63.** $-2a$

HOME EXERCISES

5.4, p. 154

1. $<$ **3.** $<$ **5.** $>$ **7.** $<$ **9.** $<$ **11.** $<$ **13.** $<$ **15.** $>$ **17.** $>$

19. $>$ 21. $<$ 23. $<$ 25. $<$ 27. $>$ 29. $>$ 31. $<$ 33. $<$

35. Jerry 37. the $\frac{44}{3}$-inch side 39. the 8-ounce box

HOME EXERCISES

5.5, p. 158

1. $\frac{1}{4}$ 3. $\frac{-1}{4}$ 5. 3 7. $\frac{-1}{6}$ 9. $\frac{3}{4}$ 11. 3 13. 6 15. 5 17. $\frac{20}{23}$

19. $\frac{25}{3}$ 21. $\frac{a}{xy}$ 23. $\frac{-2}{xy}$ 25. $\frac{a^3}{b^3}$ 27. $-b$ 29. $\frac{x^2}{2y^3}$ 31. $\frac{4ab^2}{3d^2}$ 33. $\frac{x-1}{x-2}$

35. $\frac{1+a}{a^3}$ 37. -1 39. $\frac{-(1+a-b)}{a}$

HOME EXERCISES

5.6, p. 164

1. $\frac{9}{2}$ 3. $\frac{1}{3}$ 5. -81

7. $\frac{3}{4}$ *Check.* $\frac{3}{4}-\frac{1}{4} \stackrel{?}{=} \frac{1}{2}$

$$\frac{2}{4} \stackrel{?}{=} \frac{1}{2}$$

$$\frac{1}{2} \stackrel{\checkmark}{=} \frac{1}{2}$$

9. $\frac{2}{5}$ *Check.* $1+5\left(\frac{2}{5}\right) \stackrel{?}{=} 3$

$$1+2 \stackrel{?}{=} 3$$

$$3 \stackrel{\checkmark}{=} 3$$

11. 6 13. 2 15. -6 17. \$8.10 19. 22 21. 120 23. 4 25. -3

27. 10 29. 5 31. 9 33. 2 35. 2 37. 0 39. -3 41. -3 43. $\frac{6}{5}$

45. 1 47. -16

49. 7 *Check.* $\frac{7+3}{7-2} \stackrel{?}{=} 2$

$$\frac{10}{5} \stackrel{?}{=} 2$$

$$2 \stackrel{\checkmark}{=} 2$$

51. $\frac{-8}{7}$ *Check.* $\dfrac{2}{\frac{-8}{7}+2} \stackrel{?}{=} \dfrac{5}{1-\frac{-8}{7}}$

$$2\left(1+\frac{8}{7}\right) \stackrel{?}{=} 5\left(\frac{-8}{7}+2\right)$$

$$2+\frac{16}{7} \stackrel{?}{=} -\frac{40}{7}+10$$

$$\frac{30}{7} \stackrel{\checkmark}{=} \frac{30}{7}$$

LET'S REVIEW CHAPTER 5, p. 165

1. $\frac{3}{8}$ 2. $\frac{4}{45}$ 3. ab^2 4. $\frac{24(x+3)}{z(x-3)}$ 5. -3 6. 8 7. axy

8. $(x+1)(x-y)$ **9.** 30 **10.** 12 **11.** x^2y **12.** $(x-1)(x+2)(x-2)$

13. a. 15 **b.** $\frac{1}{3}=\frac{5}{15}$, $\frac{2}{5}=\frac{6}{15}$ **14. a.** $(x-2)(x+3)$ **b.** $\frac{1}{x-2}=\frac{x+3}{(x-2)(x+3)}$, $\frac{x}{x+3}=\frac{x^2-2x}{(x-2)(x+3)}$

15. $\frac{1}{3}$ **16.** 2 **17.** $\frac{1}{x+3}$ **18.** $\frac{5}{4}$ **19.** $\frac{3x^2+1}{x^3}$ **20.** $\frac{x^2-ax+a}{(x+a)(x-a)}$

21. $<$ **22.** $<$ **23.** $>$ **24.** $>$ **25.** $\frac{3}{2}$ **26.** $\frac{4}{5}$ **27.** $\frac{ab}{c}$ **28.** $\frac{1}{(x-1)(x+a)}$

29. $\frac{13}{15}$ **30.** -12 **31.** 6 **32.** $\frac{1}{2}$

CHAPTER 6

DIAGNOSTIC TEST, p. 167

6.1 a. 7 **b.** .007 **c.** 2.5 **d.** $<$ **e.** $>$
6.2 a. 2.325 **b.** $-.29$
6.3 a. 37,250 **b.** .045 **c.** .003 02 **d.** 40
6.4 a. 120 **b.** 400 **c.** .501
6.5 a. 2.5 **b.** .18
6.6 a. .625 **b.** .222 222 ... **c.** $\frac{3}{4}$ **d.** .08, $\frac{2}{25}$ **e.** 210%
6.7 a. 8 **b.** 10,000
6.8 a. \$6250 **b.** \$75.56

HOME EXERCISES

6.1, p. 176

1. .5 **3.** .127 **5.** $-.7$ **7.** .07 **9.** .0009 **11.** .000 07 **13.** .007 77
15. 9.0 **17.** -16.03 **19.** b and c **21.** a and c **23.** three hundred eighty-five thousandths
25. fifty-nine thousandths. **27.** .29 **29.** $<$ **31.** $>$ **33.** $<$ **35.** $>$ **37.** $<$
39. $>$ **41.** $>$ **43.** $<$ **45.** $>$ **47.** $>$ **49.** $<$ **51.** $.2 < .23 < .3 < .302 < .32$
53. $-.101 < -.1 < -.02 < -.011 < -.01$ **55.** the steel rod **57. a.** 1 **b.** .7 **c.** .68 **d.** .676
59. a. 4 **b.** 3.6 **c.** 3.56 **d.** 3.565 **61.** \$8

HOME EXERCISES

6.2, p. 179

1. 1.4 **3.** 1.916 **5.** 1.6901 **7.** 47.5416 **9.** $-.9$ **11.** -1.636 **13.** $-.76$
15. 8.484 **17.** .2 **19.** .622 **21.** .0888 **23.** .055 **25.** 1 **27.** -1.2 **29.** 3.81
31. .4 **33.** .03 **35.** $-.8$ **37.** \$67.88 **39.** \$8.05

HOME EXERCISES

6.3, p. 185

1. 3.8 **3.** 238 **5.** 53 **7.** 3724 **9.** 9.5 **11.** .3 **13.** .008 **15.** .001
17. .000 93 **19.** .000 001 **21.** .2 **23.** .0002 **25.** .0001 **27.** −.0003 **29.** $30.75
31. 1.4 inches **33.** 4.94 **35.** .2704 **37.** .000 43 **39.** .0012 **41.** .000 012 **43.** 20
45. −.002 **47.** .9 **49.** .0002 **51.** 2500 **53.** −2500 **55.** $57.20 **57.** 22
59. 144 **61.** 4 **63.** $12

HOME EXERCISES

6.4, p. 187

1. 10 **3.** 22 **5.** 41 **7.** 20 **9.** 2 **11.** .11 **13.** .387 **15.** .16
17. .000 98 **19.** .15 **21.** .05 **23.** .76 **25.** .09 **27.** 4 **29.** 6.6 **31.** .004
33. .02 **35.** .2

HOME EXERCISES

6.5, p. 189

1. 1 **3.** 2 **5.** 2.14 **7.** 1.4 **9.** 1 **11.** 3 **13.** 120 **15.** 200
17. 1.2 **19.** 2 **21.** 20 **23.** 15.1 **25.** $\frac{-2}{9}$ **27.** 4 **29.** $\frac{3}{59}$

31. 10 *Check.* $10 - .02 \stackrel{?}{=} 9.98$
$9.98 \stackrel{\checkmark}{=} 9.98$

33. 12 *Check.* $\frac{12}{.06} \stackrel{?}{=} 200$
$12 \stackrel{?}{=} .06\,(200)$
$12 \stackrel{\checkmark}{=} 12$

35. 5 **37.** 50

HOME EXERCISES

6.6, p. 196

1. .4 **3.** .18 **5.** .005 **7.** .0055 **9.** .111 111 ... **11.** 1.333 333 ...
13. $\frac{29}{100}$ **15.** $-\frac{1}{5}$ **17.** $\frac{3}{25}$ **19.** $\frac{3}{4}$ **21. a.** $\frac{11}{10}$ **b.** $1\frac{1}{10}$ **23. a.** $\frac{5}{4}$ **b.** $1\frac{1}{4}$
25. 63.72 **27.** .0427 **29.** .93 **31.** 3.00 **33.** 1.75 **35.** .0075 **37.** 43% **39.** 3%
41. 500% **43.** 1.25% **45.** $\frac{17}{100}$ **47.** $\frac{1}{5}$ **49.** $\frac{21}{200}$ **51.** 81% **53.** 7% **55.** 62.5%
57. 550% **59.** > **61.** > **63.** >

HOME EXERCISES

6.7, p. 204

1. $35\% \times 200$ **3.** $x\% \times 120$ **5.** $x = 35\% \times 200$ **7.** $x\% \times 120 = 30$ **9.** $x = 40\% \times 80$
11. $25\% \cdot x = 4$; $x = 16$ **13.** $x\% \times 80 = 40$; $x = 50$; *Check.* $50\% \times 80 \stackrel{?}{=} 40$, $.50 \times 80 \stackrel{?}{=} 40$, $40.0 \stackrel{\checkmark}{=} 40$ **15.** 8.8
17. $6.5\% \times 200 = x$; $x = 13$; *Check.* $6.5\% \times 200 \stackrel{?}{=} 13$, $.065 \times 200 \stackrel{?}{=} 13$, $13.0 \stackrel{\checkmark}{=} 13$ **19.** 60% **21.** 40 **23.** 180,000 **25.** \$550
27. a. \$720 **b.** \$18,720 **29.** 25% **31.** \$22,155 **33.** \$90,000 **35.** 17,460 **37.** 5%
39. a. 360 **b.** 160

HOME EXERCISES

6.8, p. 210

1. \$42 **3.** 6% *Check.* $.06(750) \stackrel{?}{=} 45$, $45.00 \stackrel{\checkmark}{=} 45$ **5.** \$1200 *Check.* $.09(1200) \stackrel{?}{=} 108$, $108.00 \stackrel{\checkmark}{=} 108$ **7.** \$1000 at $7\frac{1}{2}\%$
9. \$210 **11.** $\frac{1}{4}$ **13.** \$3000 at 8% for $\frac{3}{4}$ of a year **15.** \$532.48 **17.** \$302.25

LET'S REVIEW CHAPTER 6, p. 211

1. .17 **2.** .033 **3.** c and d **4.** > **5.** > **6.** > **7.** .778 **8.** .175
9. −.9 **10.** 4.09 **11.** .001 05 **12.** .001 98 **13.** .03 **14.** 68 **15.** .49
16. 74.5 **17.** .3252 **18.** .0048 **19.** 216 **20.** 2 **21.** 7 **22.** .08 **23.** .8
24. .1 666 666 … **25.** $\frac{12}{25}$ **26.** .085 **27.** $\frac{9}{25}$ **28.** 15% **29.** $75\% \times x$ **30.** 60%
31. 20% **32.** 4% **33.** \$1250 **34.** \$36.27

CHAPTER 7

DIAGNOSTIC TEST, p. 212

7.1 a. 28, 29, 30 **b.** 28 **c.** 12 feet, 8 feet, 4 feet
7.2 a. 3 hours **b.** 480 miles **c.** 15 miles

7.3 **a.** $\frac{3b+10}{2}$ **b.** $\frac{1}{2x-1}$

7.4 P: **a.** 2 **b.** 1 **c.** $P = (2, 1)$
Q: **a.** 1 **b.** 5 **c.** $Q = (1, 5)$
R: **a.** -4 **b.** 3 **c.** $R = (-4, 3)$
S: **a.** 0 **b.** $-\frac{3}{2}$ **c.** $S = \left(0, -\frac{3}{2}\right)$

7.5 **a.** Figure 7A. **b.** Figure 7B. **c.** Figure 7C.

7.6 **a.** Figure 7D. **b.** (6, 1) **c.** (5, 3)

7.7 **a.** 24 **b.** $-\frac{1}{2}$

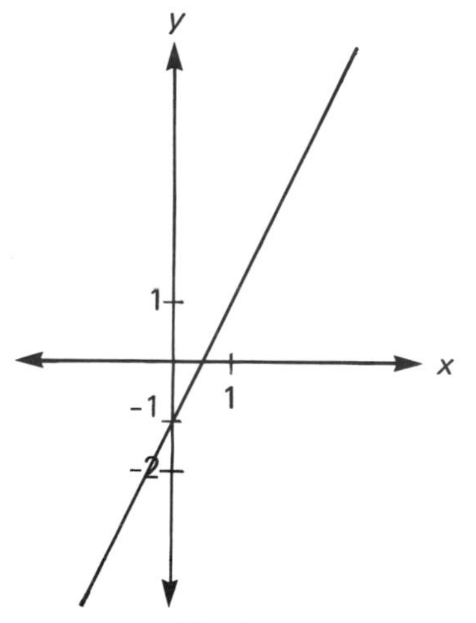

Fig. 7A

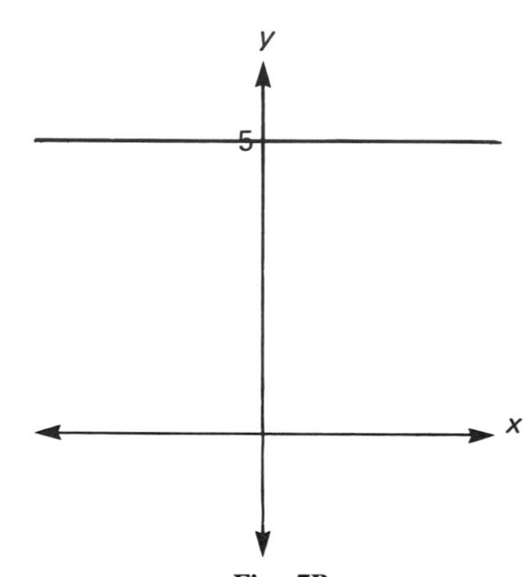

Fig. 7B

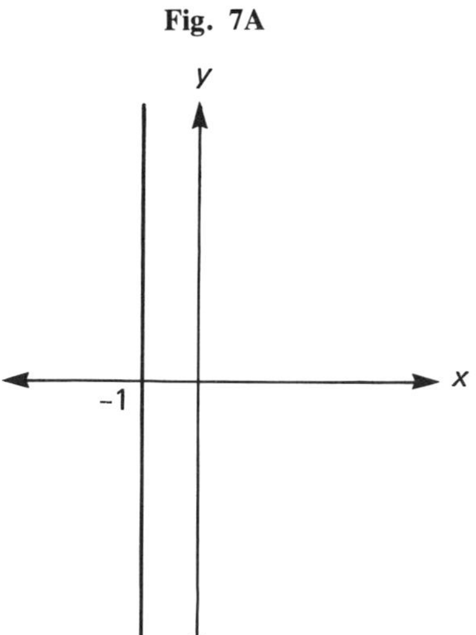

Fig. 7C

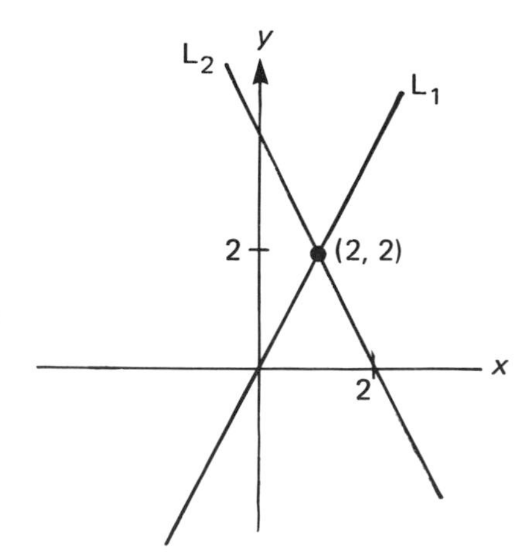

Fig. 7D

HOME EXERCISES

7.1, p. 219

1. a. $x + 2$ **b.** $x - 6$ **c.** $3x$ **d.** $\frac{x}{3}$ **3. a.** $x + 9$ **b.** $x - 2$ **c.** $\frac{x}{2}$ **d.** $x + 1$

5. $x - 19$ **7.** $2x - 5$ **9.** $x + 6$ **11. a.** $x + 5$ **b.** $\frac{x}{2}$ **13. a.** $y + 3$ **b.** $y + 10$

15. 13 **17.** 32 **19.** 15 and 16 **21.** 7 *Check.* $2(7) + 1 \stackrel{?}{=} 15$
$15 \stackrel{\checkmark}{=} 15$

23. 22 and 28 *Check.* $22 + 28 \stackrel{?}{=} 50$ and $28 \stackrel{?}{=} 22 + 6$
$50 \stackrel{\checkmark}{=} 50$ $28 \stackrel{\checkmark}{=} 28$

25. 13 *Check.* $28 - 3 \stackrel{?}{=} 5(8 - 3)$
$25 \stackrel{?}{=} 5 \cdot 5$
$25 \stackrel{\checkmark}{=} 25$

27. 28

29. 25 *Check.* $25 - 15 \stackrel{?}{=} 2[(25 - 5) - 15]$
$10 \stackrel{?}{=} 2[5]$
$10 \stackrel{\checkmark}{=} 10$

31. 17 *Check.* $31 + 17 \stackrel{?}{=} 2(7 + 17)$
$48 \stackrel{?}{=} 2(24)$
$48 \stackrel{\checkmark}{=} 48$

33. 8 feet and 4 feet

35. 20 **37.** $22\frac{1}{2}$ minutes **39.** $25

HOME EXERCISES

7.2, p. 227

1. 90 miles **3.** 180 miles **5.** 300 miles per hour **7. a.** 25 miles **b.** 75 miles

9. 280 miles **11.** a mile and a half *Check.* $(8 - 6)\left(\frac{3}{4}\right) \stackrel{?}{=} \frac{3}{2}$
$2\left(\frac{3}{4}\right) \stackrel{?}{=} \frac{3}{2}$
$\frac{3}{2} \stackrel{\checkmark}{=} \frac{3}{2}$

13. 18 miles *Check.* $\frac{18}{6} + \frac{18}{9} \stackrel{?}{=} 5$
$3 + 2 \stackrel{?}{=} 5$
$5 \stackrel{\checkmark}{=} 5$

HOME EXERCISES

7.3, p. 232

1. $\frac{y}{5}$ **3.** $2y + 3$ **5.** $\frac{2y - 1}{3}$ **7.** $\frac{2t - 3}{4}$ **9.** $\frac{-2z}{3}$ **11.** $\frac{5 - 2x}{3}$ **13.** $\frac{2u - 6}{3}$

15. $\frac{12 - b + 3c}{2}$ **17.** $\frac{3}{x - 1}$ **19. a.** $\frac{y + 20}{5}$ **b.** $5x - 20$ **21. a.** $\frac{4c - 1}{10}$ **b.** $\frac{10b + 1}{4}$

23. a. $\frac{11z - 1}{3}$ **b.** $\frac{3y + 1}{11}$ **25. a.** $\frac{1 - 4y}{y}$ **b.** $\frac{1}{x + 4}$ **27. a.** $\frac{10 + y - 2z}{5}$ **b.** $5x + 2z - 10$

29. a. b **b.** a **31. a.** $\frac{y+10}{2}$ **b.** $2\left(\frac{y+10}{2}\right) - y \stackrel{?}{=} 10$

$y + 10 - y \stackrel{?}{=} 10$

$10 \stackrel{\checkmark}{=} 10$

33. a. $\frac{5a-4}{2}$ **b.** $5a - 2\left(\frac{5a-4}{2}\right) \stackrel{?}{=} 4$

$5a - (5a - 4) \stackrel{?}{=} 4$

$4 \stackrel{\checkmark}{=} 4$

35. $\frac{A}{w}$ **37. a.** $\frac{2A}{b}$ **b.** 50 inches

39. a. $\frac{d}{t}$ **b.** 60 miles per hour

HOME EXERCISES

7.4, p. 236

1. Figure 7E.

3.

P:	**a.** 5	**b.** 1	**c.** P = (5, 1)
Q:	**a.** 2	**b.** 8	**c.** Q = (2, 8)
R:	**a.** −3	**b.** 2	**c.** R = (−3, 2)
S:	**a.** −3	**b.** −2	**c.** S = (−3, −2)
T:	**a.** 3	**b.** −2	**c.** T = (3, −2)

5.–27. Figure 7F.

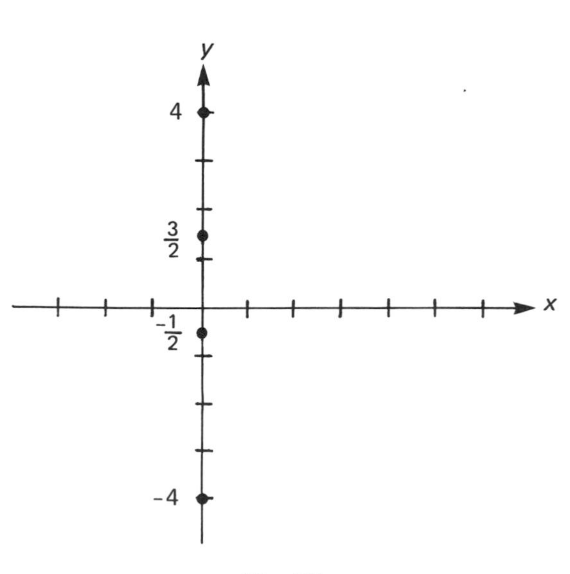

Fig. 7E

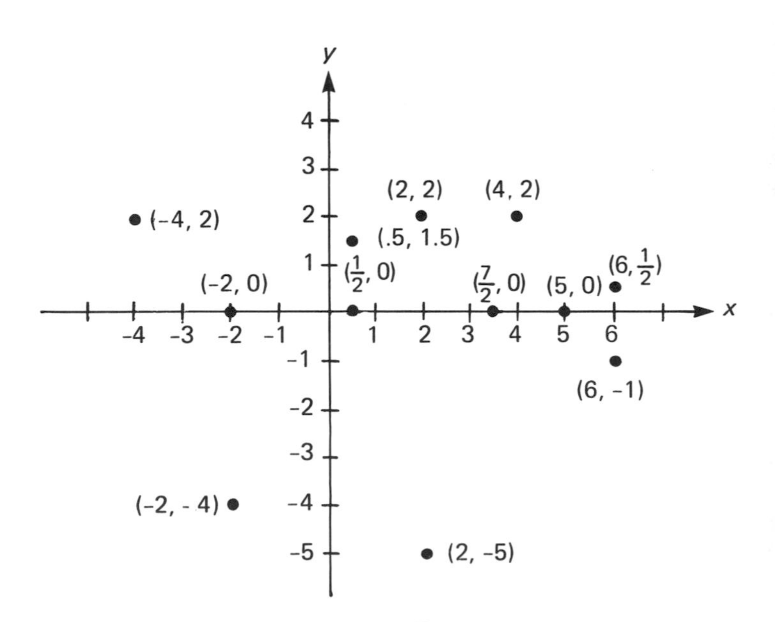

Fig. 7F

HOME EXERCISES

7.5, p. 241

1.

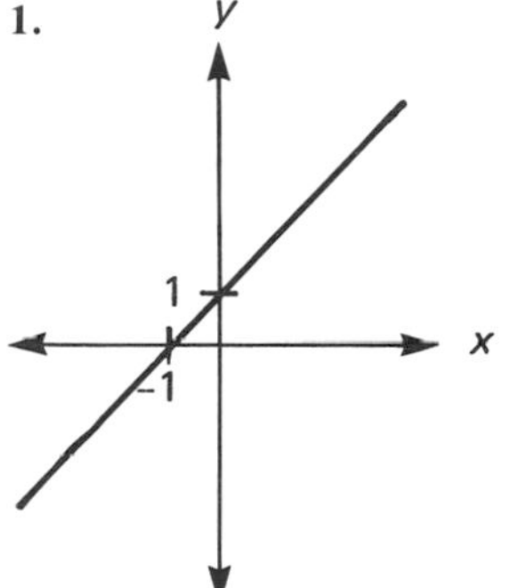

Fig. 7G The graph of $y = x + 1$

3.

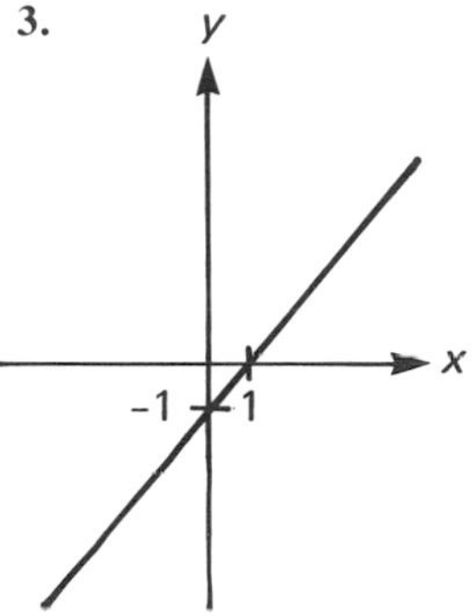

Fig. 7H The graph of $y = x - 1$

5.

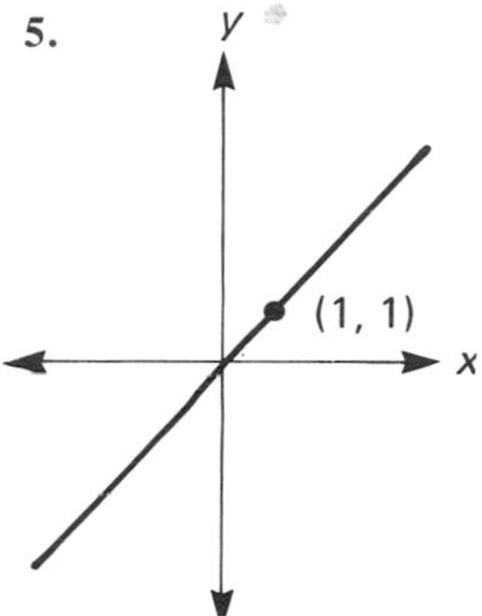

Fig. 7I The graph of $y = x$

7.

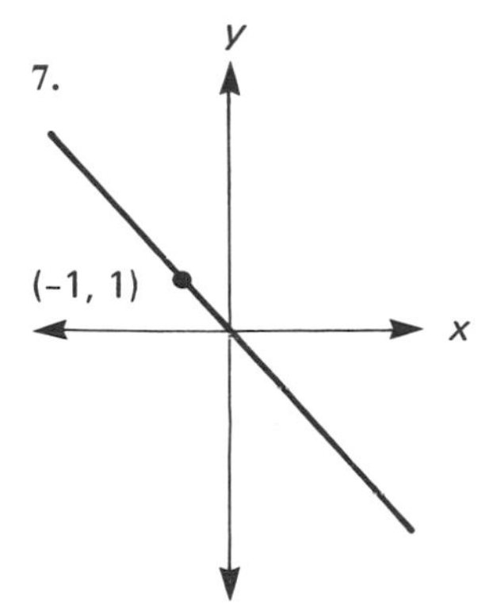

Fig. 7J The graph of $y = -x$

9.

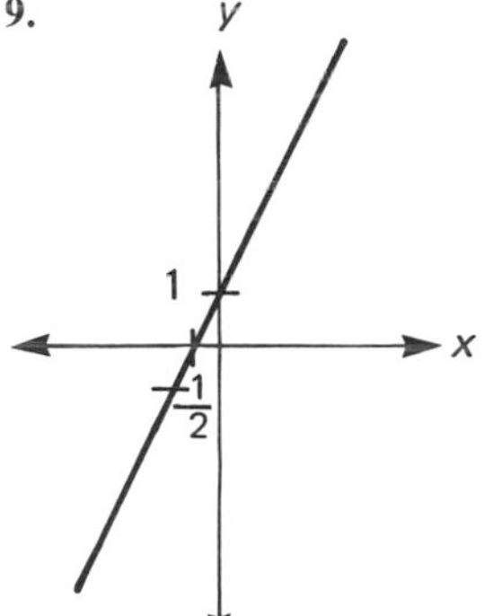

Fig. 7K The graph of $y = 2x + 1$

11.

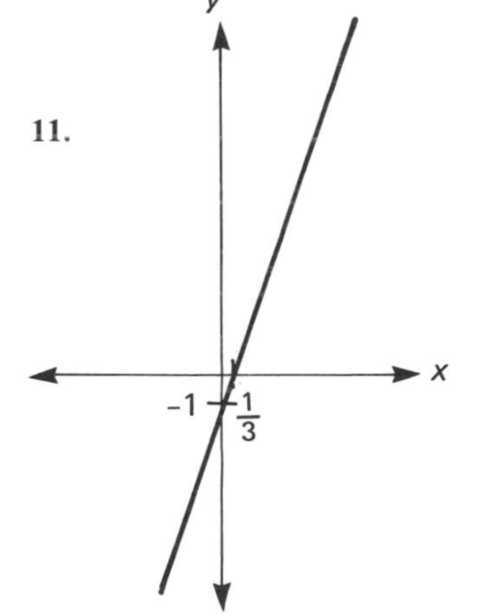

Fig. 7L The graph of $y = 3x - 1$

13.

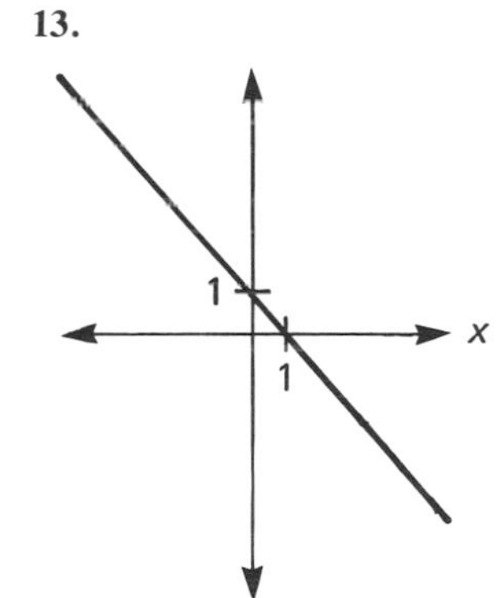

Fig. 7M The graph of $y = 1 - x$

15.

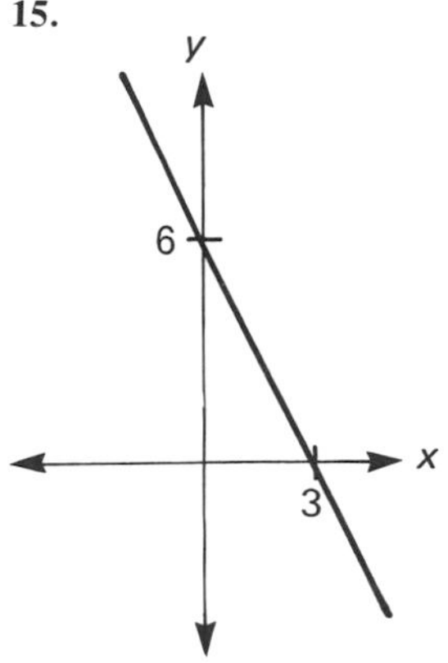

Fig. 7N The graph of $y = 6 - 2x$

17.

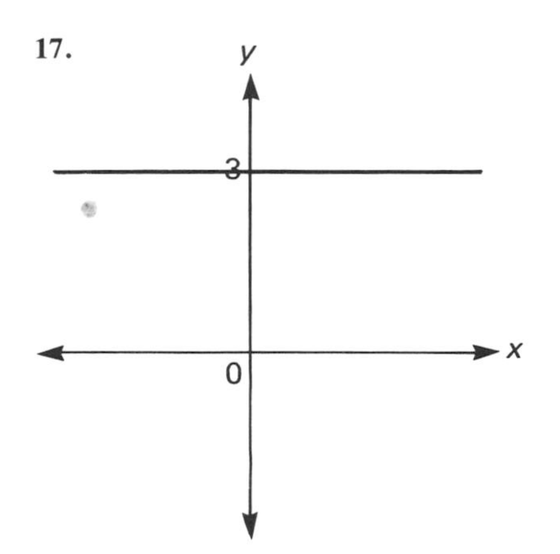

Fig. 7O The graph of $y = 3$

19.

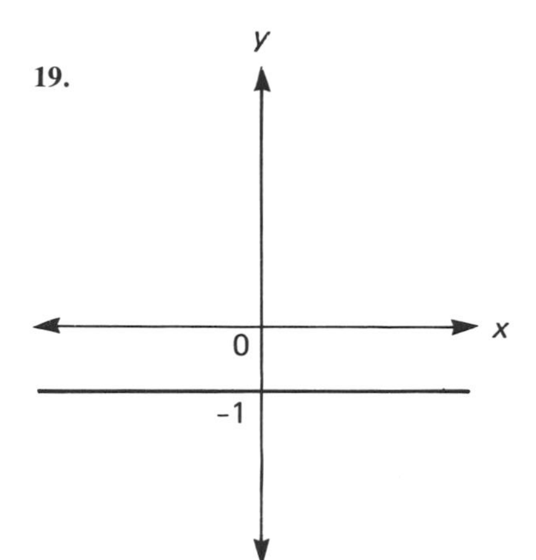

Fig. 7P The graph of $y = -1$

21. x **23.** y

HOME EXERCISES

7.6, p. 252

(The checks for 11, 13, 17, and 19 are on page A21.)

1. (Figure 7Q) *Check.* : $2 \overset{\checkmark}{=} 2$ L_2: $2 \overset{?}{=} 4 - 2$
$2 \overset{\checkmark}{=} 2$

3. (Figure 7R) *Check.* L_1: $4 - 4 \overset{?}{=} 0$ L_2: $4 - 2 \overset{?}{=} 2(0 + 1)$
$0 \overset{\checkmark}{=} 0$ 2 $\overset{\checkmark}{=} 2$

5. (Figure 7S on page A21) **7.** (Figure 7T on page A21) **9.** (Figure 7U on page A21)

11. (3, 4) **13.** (4, 5) **15.** (1, 1) **17.** (3, 1) **19.** (2, 3) **21.** (2, 1) **23.** (2, 2) **25.** (18, 24) **27.** (3, 1) **29.** (−3, −2) **31.** (5, 1) **33.** $\frac{6}{5}, \frac{14}{5}$ **35.** 9

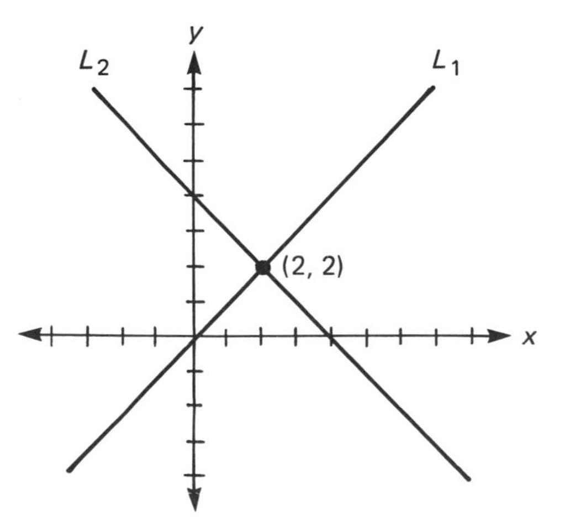

Fig. 7Q

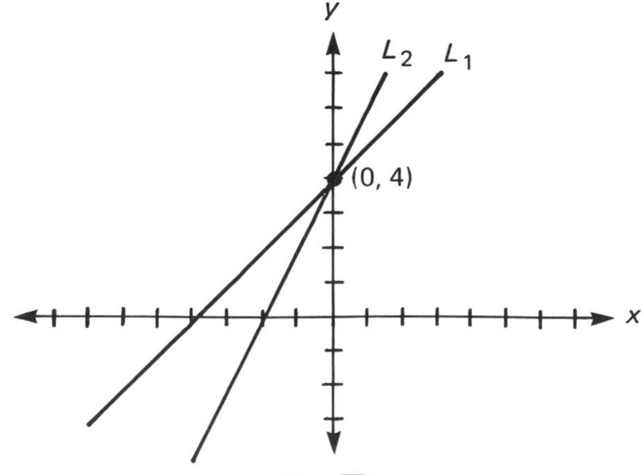

Fig. 7R

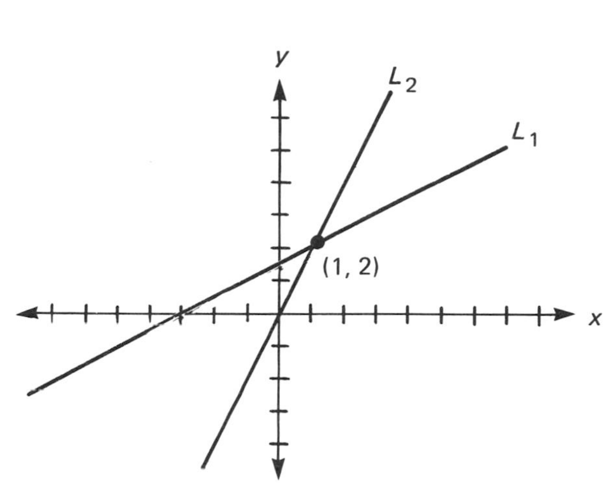

Fig. 7S

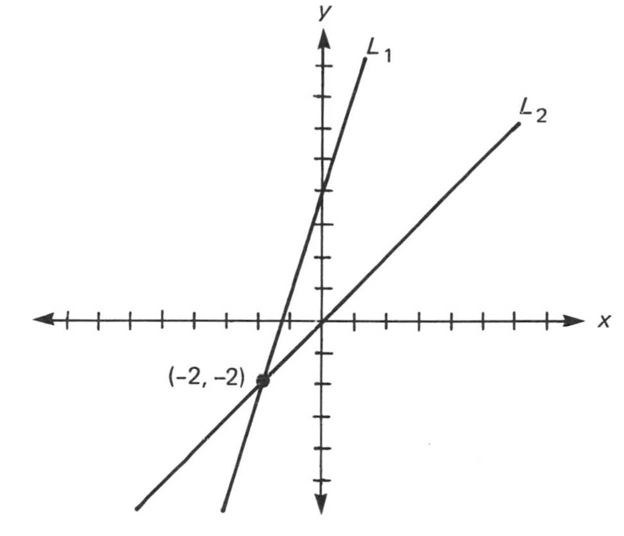

Fig. 7T

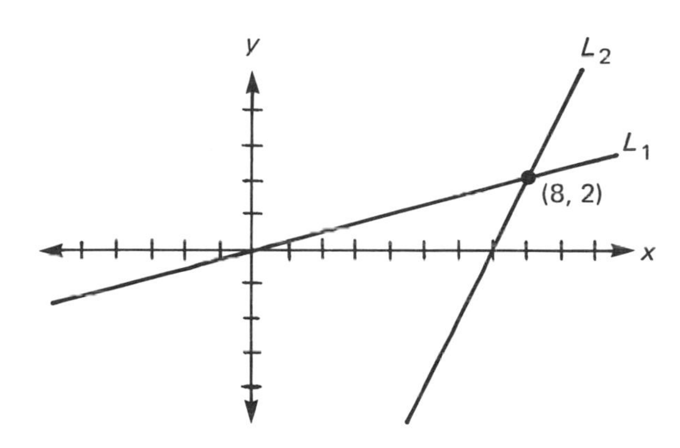

Fig. 7U

Checks.

11. $4 \stackrel{?}{=} 3 + 1$ $\quad 4 \stackrel{\checkmark}{=} 4$

$4 \stackrel{?}{=} 3(3) - 5$ $\quad 4 \stackrel{?}{=} 9 - 5$ $\quad 4 \stackrel{\checkmark}{=} 4$

13. $5 \stackrel{?}{=} 2(4) - 3$ $\quad 5 \stackrel{?}{=} 8 - 3$ $\quad 5 \stackrel{\checkmark}{=} 5$

$5 \stackrel{?}{=} 3(4) - 7$ $\quad 5 \stackrel{?}{=} 12 - 7$ $\quad 5 \stackrel{\checkmark}{=} 5$

17. $3 + 1 \stackrel{?}{=} 4$ $\quad 4 \stackrel{\checkmark}{=} 4$

$3 - 1 \stackrel{?}{=} 2$ $\quad 2 \stackrel{\checkmark}{=} 2$

19. $2 + 2(3) \stackrel{?}{=} 8$ $\quad 2 + 6 \stackrel{?}{=} 8$ $\quad 8 \stackrel{\checkmark}{=} 8$

$4(2) - 3 \stackrel{?}{=} 5$ $\quad 8 - 3 \stackrel{?}{=} 5$ $\quad 5 \stackrel{\checkmark}{=} 5$

HOME EXERCISES

7.7, p. 258

1. 2 **3.** 3 **5.** -2 **7.** 2 **9.** 10 **11.** 10 **13.** 40 **15.** -144 **17.** 8
19. 2 **21.** 2 **23.** 32 **25.** 1 **27.** 2 **29.** -12 **31.** 1 **33.** directly
35. directly **37.** inversely **39.** \$2.10 **41.** 1000

LET'S REVIEW CHAPTER 7, p. 259

1. 7 **2.** 28, 30, 32 *Check.* $28 + 30 + 32 \stackrel{?}{=} 90$
$90 = 90$
3. 22 **4.** 50 **5.** 3½ hours **6.** 2 hours

7. 8 miles **8.** $\frac{3 + 2x}{7}$ **9. a.** $\frac{-2b}{3}$ **b.** $\frac{-3a}{2}$ **10. a.** $\frac{7u}{u+3}$ **b.** $7u - 3\left(\frac{7u}{u+3}\right) \stackrel{?}{=} u\left(\frac{7u}{u+3}\right)$

$$7u - \frac{21u}{u+3} \stackrel{?}{=} \frac{7u^2}{u+3}$$

$$\frac{7u^2 + 21u - 21}{u+3} \stackrel{?}{=} \frac{7u^2}{u+3}$$

$$\frac{7u^2}{u+3} \stackrel{\checkmark}{=} \frac{7u^2}{u+3}$$

11. P: **a.** 1 **b.** 4 **c.** $P = (1, 4)$
Q: **a.** -2 **b.** 3 **c.** $Q = (-2, 3)$
R: **a.** -4 **b.** -2 **c.** $R = (-4, -2)$
S: **a.** 3 **b.** -3 **c.** $S = (3, -3)$

12.–16.

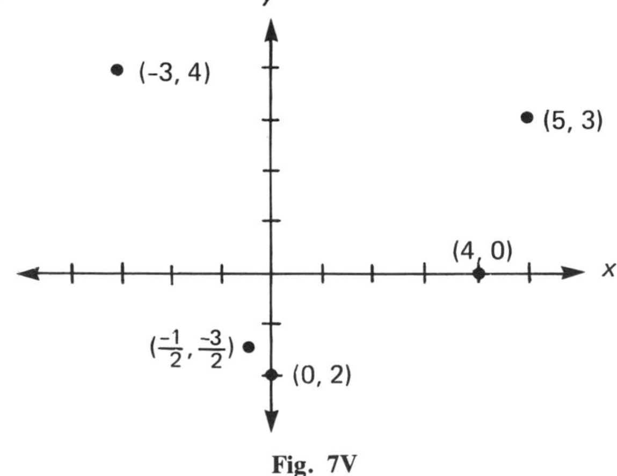

Fig. 7V

17.

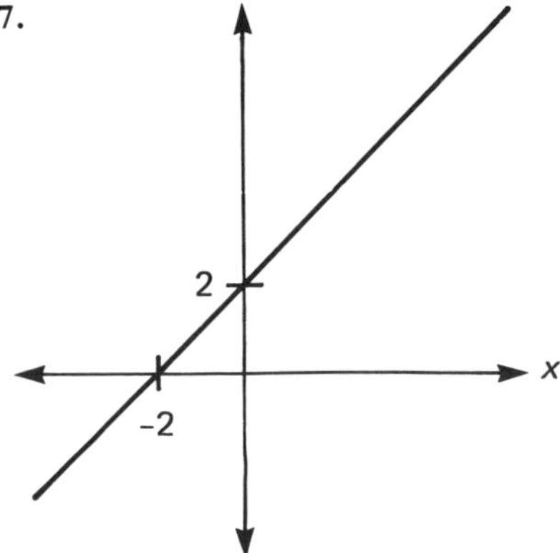

Fig. 7W The graph of $y = x + 2$

18.

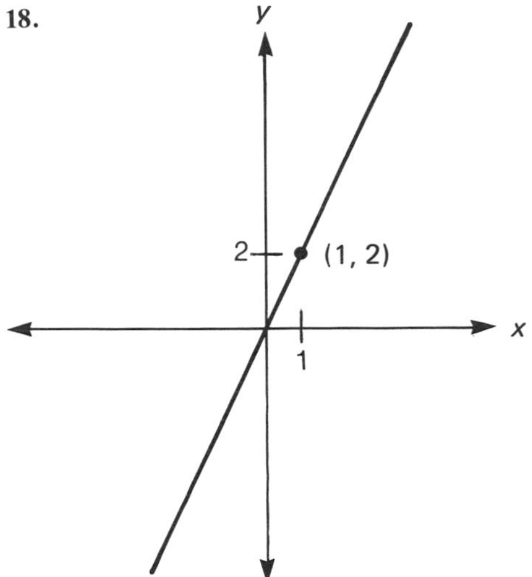

Fig. 7X The graph of $y\ -2x$

19.

Fig. 7Y The graph of $y = 2 - x$

20.

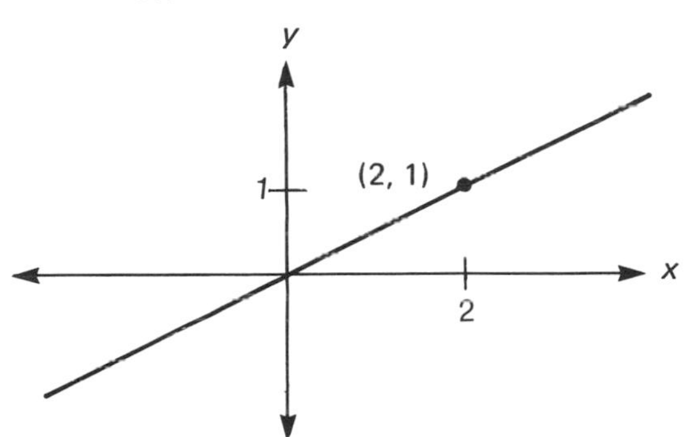

Fig. 7Z The graph of $y = \frac{x}{2}$

21.

Fig. 7AA

22. $\left(\frac{5}{3}, \frac{13}{3}\right)$ **23.** (2, 1) **24.** (10, −7) **25.** −2 **26.** 15 **27.** 32 **28.** inversely

CHAPTER 8

DIAGNOSTIC TEST, p. 261

8.1 a. 11 **b.** $|ab|$, or $-ab$ **c.** $10 < \sqrt{111} < 11$ **d.** .3

8.2 a. 700 **b.** $\frac{6a}{b^2}$ **c.** $2a\sqrt{7a}$

8.3 a. $2\sqrt{x}$ **b.** $\frac{-\sqrt{3}}{10}$

8.4 **a.** 6 **b.** $\sqrt{15}$ **c.** 1 **d.** $3(1 + \sqrt{2})$ **e.** $\frac{1 + \sqrt{2}}{3}$

8.5 **a.** $\frac{1}{36}$ **b.** 1 **c.** $\frac{64}{27}$

8.6 **a.** 1.96×10^6 **b.** .000 003 89 **c.** 6×10^6

HOME EXERCISES

8.1, p. 269

1. 4 **3.** 7 **5.** 9 **7.** -12 **9.** 50 **11.** 0 **13.** 3 **15.** -3 **17.** 6
19. x **21.** ab **23.** x^4 **25.** a^6 **27.** x^9 **29.** m **31.** $-m$ **33.** $-n$
35. mn **37.** n^2 **39.** $3 < \sqrt{14} < 4$ **41.** $8 < \sqrt{65} < 9$ **43.** $12 < \sqrt{145} < 13$
45. $40 < \sqrt{1670} < 41$ **47.** .4 **49.** 3.5 **51.** 3.1 **53.** 14.5

HOME EXERCISES

8.2, p. 274

1. 70 **3.** 110 **5.** 500 **7.** 10,000 **9.** $4a^2$ **11.** $10a^9$ **13.** xy^2 **15.** a^4x^5
17. a^2bc^3 **19.** $11yz^6$ **21.** $\frac{3}{5}$ **23.** $\frac{9}{2}$ **25.** $\frac{10}{11}$ **27.** $\frac{x}{y^4}$ **29.** $\frac{5}{d^4}$ **31.** $\frac{12x^2}{yz^5}$
33. **a.** $3\sqrt{2}$ **b.** 4.2 **35.** **a.** $6\sqrt{2}$ **b.** 8.5 **37.** **a.** $5\sqrt{3}$ **b.** 8.7 **39.** $a\sqrt{a}$
41. $\frac{a^2\sqrt{a}}{b^3}$ **43.** $3b^4\sqrt{3b}$ **45.** $\frac{3a^4\sqrt{2a}}{2b}$ **47.** $\frac{5a\sqrt{a}}{7b^4}$ **49.** .3 **51.** .02 **53.** .001
55. $.9x^4y^2$

HOME EXERCISES

8.3, p. 277

1. $2\sqrt{2}$ **3.** $4\sqrt{2}$ **5.** $5\sqrt{6}$ **7.** $8\sqrt{11}$ **9.** $7\sqrt{5}$ **11.** $3\sqrt{2}$ **13.** $2\sqrt{2} + 3\sqrt{3}$
15. $5\sqrt{7} + 2\sqrt{5}$ **17.** $3\sqrt{2}$ **19.** 0 **21.** $\sqrt{7}$ **23.** $7\sqrt{3}$ **25.** $5\sqrt{a}$ **27.** $19y\sqrt{y}$
29. $23b\sqrt{a}$ **31.** $3\sqrt{3a}$ **33.** $12x\sqrt{2x}$ **35.** $6b\sqrt{7ab}$ **37.** $\frac{2 + \sqrt{5}}{5}$ **39.** $\frac{\sqrt{2} - \sqrt{5} + \sqrt{7}}{3}$
41. $\frac{7\sqrt{2} - 5\sqrt{3}}{35}$ **43.** $\frac{2\sqrt{2} + 3\sqrt{3}}{24}$

HOME EXERCISES

8.4, p. 282

1. $\sqrt{10}$ **3.** $2\sqrt{14}$ **5.** 7 **7.** 10 **9.** a **11.** b^3 **13.** $10x$ **15.** $xy^2z^2\sqrt{x}$

17. $2 + \sqrt{6}$ 19. 3 21. 47 23. $38 + 16\sqrt{5}$ 25. $-23 - 5\sqrt{55}$ 27. $3 + 2\sqrt{2}$ 29. 3

31. -2 33. $2\sqrt{3}$ 35. $\frac{\sqrt{2}}{3}$ 37. $\frac{4\sqrt{y}}{3z}$ 39. $\frac{3}{10}$ 41. a. $3(1 + \sqrt{2})$ b. $\frac{1 + \sqrt{2}}{3}$

43. a. $5(1 + \sqrt{2})$ b. $\frac{-(1 + \sqrt{2})}{3}$ 45. a. $\sqrt{2}(1 + \sqrt{3})$ b. $\frac{1 + \sqrt{3}}{2}$

47. a. $\sqrt{ab}(a + \sqrt{c})$ b. $\frac{a + \sqrt{c}}{ab}$ 49. a. $-ab$ b. 5 51. $\frac{5}{6}$ 53. 2 55. 1 57. $\frac{5st\sqrt{t}}{12}$

HOME EXERCISES

8.5, p. 285

1. $\frac{1}{4}$ 3. $-\frac{1}{2}$ 5. $\frac{1}{64}$ 7. $-\frac{1}{16}$ 9. $-\frac{1}{27}$ 11. $\frac{1}{125}$ 13. $\frac{1}{10{,}000}$ 15. 1

17. 1 19. $\frac{5}{3}$ 21. 4 23. $\frac{25}{16}$ 25. $\frac{64}{27}$ 27. 100,000 29. $\frac{1}{16}$ 31. $\frac{3}{4}$ 33. a

35. y^{-10} 37. b^{-3} 39. b^{-10} 41. 1 43. -2 45. -3 47. -1 49. 0 51. -2

HOME EXERCISES

8.6, p. 290

1. 1.25×10^1 3. 7.028×10^3 5. 4.97×10^{-1} 7. 8×10^{-1} 9. 10^2 11. 3.61×10^5
13. 5×10^{-6} 15. 6 17. 65.2 19. 3,230,000 21. .5 23. .000 901 25. 878,000,000
27. 200.1 29. 54,314 31. 10^3 33. 4×10^{11} 35. 1.24×10^{-7} 37. 2.5×10^{11}
39. 1.08×10^{15} miles 41. 1.60×10^{-13} joules 43. 8.8×10^{-5}

LET'S REVIEW CHAPTER 8, p. 291

1. 8 2. -30 3. x^2 4. $-xy$ 5. $7 < \sqrt{55} < 8$ 6. 9.9 7. $5a$ 8. a^2b^3

9. $\frac{9}{10}$ 10. $5\sqrt{3} + 2\sqrt{2}$ 11. $3\sqrt{5}$ 12. $\sqrt{a}$ 13. $17ab^2$ 14. $\frac{2\sqrt{3}}{3}$ 15. 10 16. 2

17. 2 18. 2 19. $\frac{2}{5}$ 20. $\frac{4\sqrt{3}}{3}$ 21. a. $7(1 + \sqrt{2})$ b. $\frac{1 + \sqrt{2}}{3}$ 22. $\frac{1}{49}$ 23. 1

24. $\frac{9}{4}$ 25. x^{-7} 26. 2.65×10^5 27. 2.57×10^{-2} 28. 000 345 29. 4×10^2

CHAPTER 9

DIAGNOSTIC TEST, p. 292

9.1 a. 1, 7 b. $\frac{1}{4}, -2$

9.2 **a.** $\pm 10\sqrt{2}$ **b.** $-1, \frac{-5}{3}$

9.3 **a.** $-2 \pm \sqrt{2}$ **b.** $\frac{1}{3}$ **c.** $-2 \pm \sqrt{2}$

9.4 **a.** 10, −12 **b.** 11 seconds

HOME EXERCISES

9.1, p. 298

1. b, d **3.** −2 **5.** 1 **7.** −2 *Check.*

$$(-2+4)^2 \stackrel{?}{=} (-2)^2$$
$$2^2 \stackrel{?}{=} 4$$
$$4 \stackrel{\checkmark}{=} 4$$

9. 3 **11.** 2, 3

13. −3, 8 *Check.* (for −3)

$$2(-3+3)(-3-8) \stackrel{?}{=} 0$$
$$2(0)(-11) \stackrel{?}{=} 0$$
$$0 \stackrel{\checkmark}{=} 0$$

(for 8)

$$2(8+3)(8-8) \stackrel{?}{=} 0$$
$$2(11)(0) \stackrel{?}{=} 0$$
$$0 \stackrel{\checkmark}{=} 0$$

15. −2 **17.** −2, −3 **19.** 3. −4 **21.** −4, 2 **23.** −3, −6 **25.** 5 **27.** 3, −7

29. 1, −9 **31.** 1, −2 *Check.* (for 1)

$$2(1)^2 + 2(1) - 4 \stackrel{?}{=} 0$$
$$2 + 2 - 4 \stackrel{?}{=} 0$$
$$0 \stackrel{\checkmark}{=} 0$$

(for − 2)

$$2(-2)^2 + 2(-2) - 4 \stackrel{?}{=} 0$$
$$2(4) - 4 - 4 \stackrel{?}{=} 0$$
$$8 - 4 - 4 \stackrel{?}{=} 0$$
$$0 \stackrel{\checkmark}{=} 0$$

33. $-1, -\frac{1}{2}$ **35.** $\frac{1}{2}, -4$ **37.** $-\frac{1}{2}, -\frac{3}{2}$ **39.** $-\frac{5}{2}$ **41.** $2, \frac{3}{2}$ **43.** 0, 2 **45.** $0, \frac{1}{2}$

HOME EXERCISES

9.2, p. 302

(The checks are given below.)

1. ± 4 **3.** ± 7 **5.** $\pm\sqrt{6}$ **7.** $\pm\sqrt{13}$ **9.** $\pm 2\sqrt{2}$ **11.** $\pm 3\sqrt{2}$ **13.** $\pm 3\sqrt{5}$

15. $\pm\frac{2}{3}$ **17.** $\pm\sqrt{\frac{2}{3}}$ or $\frac{\pm\sqrt{6}}{3}$ **19.** 3, −4 **21.** $\frac{6}{5}, \frac{-2}{5}$ **23.** $\frac{4}{3}, -4$ **25.** $\frac{2 \pm \sqrt{3}}{3}$

27. $\frac{-4 \pm \sqrt{7}}{5}$ **29.** $\frac{-5 \pm \sqrt{6}}{3}$ **31.** $\frac{1 \pm 2\sqrt{2}}{2}$ **33.** $-3 \pm 4\sqrt{2}$ **35.** $1 \pm 14\sqrt{2}$ **37.** 2 **39.** $\frac{3}{5}$

Checks.

7. for $\sqrt{13}$:

$$(\sqrt{13})^2 - 13 \stackrel{?}{=} 0$$
$$13 - 13 \stackrel{?}{=} 0$$
$$0 \stackrel{\checkmark}{=} 0$$

for $-\sqrt{13}$:

$$(-\sqrt{13})^2 - 13 \stackrel{?}{=} 0$$
$$13 - 13 \stackrel{?}{=} 0$$
$$0 \stackrel{\checkmark}{=} 0$$

19. for 3:

$$[2(3)+1]^2 \stackrel{?}{=} 49$$
$$7^2 \stackrel{?}{=} 49$$
$$49 \stackrel{\checkmark}{=} 49$$

for −4:

$$[2(-4)+1]^2 \stackrel{?}{=} 49$$
$$[-7]^2 \stackrel{?}{=} 49$$
$$49 \stackrel{\checkmark}{=} 49$$

25. for $\dfrac{2+\sqrt{3}}{3}$:

$$\left[3\left(\frac{2+\sqrt{3}}{3}\right)-2\right]^2 \stackrel{?}{=} 3$$
$$[2+\sqrt{3}-2]^2 \stackrel{?}{=} 3$$
$$[\sqrt{3}]^2 \stackrel{?}{=} 3$$
$$3 \stackrel{\checkmark}{=} 3$$

for $\dfrac{2-\sqrt{3}}{3}$:

$$\left[3\left(\frac{2-\sqrt{3}}{3}\right)-2\right]^2 \stackrel{?}{=} 3$$
$$[2-\sqrt{3}-2]^2 \stackrel{?}{=} 3$$
$$[-\sqrt{3}]^2 \stackrel{?}{=} 3$$
$$3 \stackrel{\checkmark}{=} 3$$

HOME EXERCISES

9.3, p. 309

(The checks are given below.)

1. a. 8 **b.** 2 **c.** $-2 \pm \sqrt{2}$ **3. a.** 1 **b.** 2 **c.** $1, \frac{3}{2}$

5. a. 13 **b.** 2 **c.** $\frac{3 \pm \sqrt{13}}{2}$ **7. a.** 12 **b.** 2 **c.** $\frac{3 \pm \sqrt{3}}{2}$

9. a. 0 **b.** 1 **c.** $\frac{1}{2}$ **11. a.** 320 **b.** 2 **c.** $\frac{5 \pm 2\sqrt{5}}{2}$

13. a. 5 **b.** 2 **c.** $\frac{-1 \pm \sqrt{5}}{2}$ **15. a.** -35 **b.** 0 **17.** $\frac{-5 \pm \sqrt{33}}{2}$ **19.** $\frac{5 \pm \sqrt{13}}{6}$

21. $-5 \pm 2\sqrt{5}$ **23.** $\frac{-3 \pm \sqrt{13}}{2}$ **25.** $-1, \frac{-3}{2}$ **27.** $3, \frac{1}{3}$ **29.** 1, 2 **31.** 4, -5

33. $-1 \pm \sqrt{5}$ **35.** $\frac{1}{2}, \frac{-1}{3}$

Checks.

17. for $\dfrac{-5+\sqrt{33}}{2}$:

$$\left(\frac{-5+\sqrt{33}}{2}\right)^2 + 5\left(\frac{-5+\sqrt{33}}{2}\right) - 2 \stackrel{?}{=} 0$$
$$\frac{58-10\sqrt{33}}{4} + \frac{-25+5\sqrt{33}}{2} - 2 \stackrel{?}{=} 0$$
$$\frac{58-10\sqrt{33}-50+10\sqrt{33}-8}{4} \stackrel{?}{=} 0$$
$$0 \stackrel{\checkmark}{=} 0$$

for $\dfrac{-5-\sqrt{33}}{2}$:

$$\left(\frac{-5-\sqrt{33}}{2}\right)^2 + 5\left(\frac{-5-\sqrt{33}}{2}\right) - 2 \stackrel{?}{=} 0$$
$$\frac{58+10\sqrt{33}}{4} + \frac{-25-5\sqrt{33}}{2} - 2 \stackrel{?}{=} 0$$
$$\frac{58+10\sqrt{33}-50-10\sqrt{33}-8}{4} \stackrel{?}{=} 0$$
$$0 \stackrel{\checkmark}{=} 0$$

19. For $\dfrac{5+\sqrt{13}}{6}$:

For $\dfrac{5-\sqrt{13}}{6}$:

$$3\left(\frac{5+\sqrt{13}}{6}\right)^2 - 5\left(\frac{5+\sqrt{13}}{6}\right) + 1 \stackrel{?}{=} 0$$

$$3\left(\frac{38+10\sqrt{13}}{36}\right) + \frac{-25-5\sqrt{13}}{6} + 1 \stackrel{?}{=} 0$$

$$\frac{38+10\sqrt{13}-50-10\sqrt{13}+12}{12} \stackrel{?}{=} 0$$

$$0 \stackrel{\checkmark}{=} 0$$

$$3\left(\frac{5-\sqrt{13}}{6}\right)^2 - 5\left(\frac{5-\sqrt{13}}{6}\right) + 1 \stackrel{?}{=} 0$$

$$3\left(\frac{38-10\sqrt{13}}{36}\right) + \frac{-25+5\sqrt{13}}{6} + 1 \stackrel{?}{=} 0$$

$$\frac{38-10\sqrt{13}-50+10\sqrt{13}+12}{12} \stackrel{?}{=} 0$$

$$0 \stackrel{\checkmark}{=} 0$$

HOME EXERCISES

9.4, p. 312

1. 4 and 5 **3.** 10 and -7 **5.** ± 6 **7.** 5 **9.** 6 seconds
11. a. 1 second **b.** 7 seconds **13.** 6 or 10 **15.** 10 feet

LET'S REVIEW CHAPTER 9, p. 313

(The checks are given below.)

1. $-3, -4$ **2.** 3 **3.** $\frac{1}{3}, -2$ **4.** $\pm\sqrt{17}$ **5.** $\pm 4\sqrt{5}$ **6.** $1, -2$ **7.** $\frac{5 \pm \sqrt{2}}{3}$
8. a. -11 **b.** 0 **9.** $\frac{-5 \pm \sqrt{5}}{2}$ **10.** $\frac{-1}{2}$ **11.** $-3 \pm \sqrt{10}$ **12.** 9 and -5
13. $3\sqrt{2}$ seconds or about 4.2 seconds

Checks.

1. for 3:

$$(-3)^2 + 7(-3) + 12 \stackrel{?}{=} 0$$

$$9 - 21 + 12 \stackrel{?}{=} 0$$

$$0 \stackrel{\checkmark}{=} 0$$

for -4:

$$(-4)^2 + 7(-4) + 12 \stackrel{?}{=} 0$$

$$16 - 28 + 12 \stackrel{?}{=} 0$$

$$0 \stackrel{\checkmark}{=} 0$$

6. for 1:

$$[2(1) + 1]^2 \stackrel{?}{=} 9$$

$$(3)^2 \stackrel{?}{=} 9$$

$$9 \stackrel{\checkmark}{=} 9$$

for -2:

$$[2(-2) + 1]^2 \stackrel{?}{=} 9$$

$$(-3)^2 \stackrel{?}{=} 9$$

$$9 \stackrel{\checkmark}{=} 9$$

9. for $\frac{-5+\sqrt{5}}{2}$:

$$\left(\frac{-5+\sqrt{5}}{2}\right)^2 + 5\left(\frac{-5+\sqrt{5}}{2}\right) + 5 \stackrel{?}{=} 0$$

$$\frac{30-10\sqrt{5}}{4} + \frac{-25+5\sqrt{5}}{2} + 5 \stackrel{?}{=} 0$$

$$\frac{30-10\sqrt{5}-50+10\sqrt{5}+20}{4} \stackrel{?}{=} 0$$

$$0 \stackrel{\checkmark}{=} 0$$

for $\frac{-5-\sqrt{5}}{2}$:

$$\left(\frac{-5-\sqrt{5}}{2}\right)^2 + 5\left(\frac{-5-\sqrt{5}}{2}\right) + 5 \stackrel{?}{=} 0$$

$$\frac{30+10\sqrt{5}}{4} + \frac{-25-5\sqrt{5}}{2} + 5 \stackrel{?}{=} 0$$

$$\frac{30+10\sqrt{5}-50-10\sqrt{5}+20}{4} \stackrel{?}{=} 0$$

$$0 \stackrel{\checkmark}{=} 0$$

CHAPTER 10

DIAGNOSTIC TEST, p. 314

10.1 **a.** 84 inches **b.** .06 meter **c.** 25 miles
10.2 **a.** 14 centimeters **b.** 25 inches **c.** 35 square inches **d.** 7 meters by 5 meters
10.3 **a.** 15 square centimeters **b.** 4 square feet **c.** 300 square centimeters
10.4 **a.** 9π square centimeters **b.** 36π square inches **c.** 14π meters
10.5 **a.** volume = 6 cubic feet, surface area = 22 square feet
b. volume = $\frac{500\pi}{3}$ cubic inches, surface area = 100π square inches
c. volume = 45 cubic centimeters, lateral surface area = 30π cubic centimeters **d.** 10 feet

HOME EXERCISES

10.1, p. 319

1. 8 feet **3.** 6 yards **5.** 7 miles **7.** $\frac{8}{9}$ yard **9.** 4.5 yards **11.** 15,840 feet
13. 28 feet **15.** 528 feet **17.** 5000 meters **19.** 4.05 meters **21.** 2.44 kilometers
23. 358 centimeters **25.** 1969 inches **27.** 25 centimeters **29.** 31 miles **31.** 48 kilometers
33. 338 kilometers **35.** **a.** 3600 inches **b.** 9144 centimeters **37.** 2 gallons

HOME EXERCISES

10.2, p. 325

1. 25 inches **3.** 38 inches **5.** 22 inches **7.** 9 miles **9.** 6 miles **11.** 10 inches
13. 280 feet **15.** 360 feet **17.** 12 square inches **19.** $\frac{3}{2}$ square inches **21.** 4 inches
23. 6 yards **25.** 81 square inches **27.** 24 square feet **29.** 5333 yards **31.** $28
33. 15 inches by 20 inches **35.** 18 feet by 20 feet **37.** 100 square inches

HOME EXERCISES

10.3, p. 335

1. 8 square inches **3.** $\frac{9}{4}$ square inches **5.** 12 centimeters **7.** 8 feet **9.** $8\sqrt{2}$ meters
11. $\sqrt{15}$ centimeters **13.** $3\sqrt{3}$ feet **15.** 36 square inches **17.** 9 inches **19.** 8 inches
21. 25 square inches **23.** 110 square inches **25.** $\frac{70}{9}$ inches **27.** 4 inches **29.** 30 square centimeters
31. 42 square inches **33.** 128 square centimeters **35.** 52 square centimeters **37.** 180 square centimeters
39. 180 square centimeters

HOME EXERCISES

10.4, p. 341

1. 20 inches **3.** 2 inches **5. a.** π inches **b.** 3 inches **7. a.** 9π yards **b.** 28 yards
9. a. 2π feet **b.** 6 feet **11. a.** π miles **b.** 3 miles **13. a.** 25π square feet **b.** 79 square feet
15. a. 2500π square centimeters **b.** 7854 square centimeters **17. a.** 400π square feet **b.** 1256 square feet
19. 9π square inches **21. a.** 10 centimeters **b.** 5 centimeters **c.** 25π square centimeters
23. 188 feet **25.** 21π square inches **27.** $8(8 - \pi)$ square centimeters **29.** 4 times as fast

HOME EXERCISES

10.5, p. 349

1. a. 20 cubic inches **b.** 48 square inches **3. a.** 180 cubic centimeters **b.** 222 square centimeters
5. a. 800 cubic centimeters **b.** 600 square centimeters
7. a. 8 cubic centimeters **b.** 24 square centimeters
9. a. 216 cubic centimeters **b.** 216 square centimeters **11.** 5 inches **13.** 4 inches
15. 1260 cubic feet **17. a.** 198 cubic inches **b.** 234 square inches
19. a. $\frac{256\pi}{3}$ cubic kilometers **b.** 268 cubic kilometers **c.** 64π square kilometers
21. a. 972π cubic meters **b.** 3054 cubic meters **c.** 324π square centimeters
23. 2 yards **25. a.** 125π cubic centimeters **b.** 393 cubic centimeters **c.** 50π square centimeters
27. a. 800π cubic centimeters **b.** 2514 cubic centimeters **c.** 80π square centimeters
29. 2 centimeters **31.** 2 centimeters
33. a. 98π cubic centimeters **b.** 308 cubic centimeters **c.** $7\pi\sqrt{85}$ square centimeters
35. a. $\frac{128\pi}{3}$ cubic centimeters **b.** 134 cubic centimeters **c.** $16\pi\sqrt{3}$ square inches **37.** 3 centimeters
39. $3\sqrt{3}$ centimeters **41.** 300π cubic centimeters **43.** 20π cubic inches **45.** $\frac{32\pi}{3}$ cubic inches

LET'S REVIEW CHAPTER 10, p. 351

1. 42 inches **2.** 567 centimeters **3.** 393.7 inches **4.** 52 kilometers
5. a. 18 centimeters **b.** 20 square centimeters **6. a.** 36 inches **b.** 81 square inches **7.** 7 inches
8. 108 square inches **9.** 45 square centimeters **10.** 42 square inches **11.** 12 meters
12. 54 square inches **13.** $\frac{9}{2}$ centimeters **14. a.** 4π square feet **b.** 12.6 square feet
15. a. 8 centimeters **b.** 4 centimeters **c.** 16π square centimeters
16. a. 60 cubic inches **b.** 94 square inches **17. a.** 64 cubic centimeters **b.** 96 square centimeters
18. a. $\frac{256\pi}{3}$ cubic inches **b.** 268 cubic inches **c.** 64π square inches
19. a. 36π cubic inches **b.** 113 cubic inches **c.** 24π square inches
20. a. 32π cubic inches **b.** 100 cubic inches **c.** $8\sqrt{13} \cdot \pi$ square inches **21.** 2 centimeters

CHAPTER 11

DIAGNOSTIC TEST, p. 353

11.1 **a.** 25°24′ **b.** 53°38′ **c.** 143°38′ **d.** 41°

11.2 **a.** 98° **b.** $\frac{9\sqrt{3}}{2}$ square centimeters **c.** 10 inches **d.** 2 inches

11.3 **a.** 30° **b.** $\frac{5\pi}{3}$ centimeters **c.** $\frac{25\pi}{6}$ square centimeters

HOME EXERCISES

11.1, p. 360

1. 0°12′ **3.** 19°30′ **5.** 120°03′ **7.** 12.2° **9.** 80.05° **11.** acute **13.** obtuse **15.** 45° **17.** 70.8° **19.** 130° **21.** 67°45′ **23.** 90° **25.** 15° **27.** 105° **29.** 80° **31.** 10°14′ **33.** 95°22′ **35.** 110° **37.** 110° **39.** 70° **41.** 92° **43.** 63° **45.** 117° **47.** 18 inches **49.** 16 inches

HOME EXERCISES

11.2, p. 373

1. 70° **3.** 82°10′ **5.** 42°48′ **7.** 50° and 40° **9.** $\left(\frac{1080}{11}\right)^\circ$, $\left(\frac{540}{11}\right)^\circ$, and $\left(\frac{360}{11}\right)^\circ$ **11.** 105°

13. **a.** 10 inches **b.** 36 inches **c.** 30 inches **d.** 60 square inches **e.** 30 square inches

15. **a.** 15 inches **b.** $\frac{225}{2}$ square inches **17.** a = $9\sqrt{3}$ centimeters, c = 18 centimeters

19. $2\sqrt{3}$ square inches **21.** **a.** 20 centimeters **b.** $100\sqrt{3}$ square centimeters

23. $\frac{81\sqrt{3}}{4}$ square decimeters **25.** 60° **27.** 2 inches **29.** $2\sqrt{3}$ inches **31.** 4 inches

33. $2(1 + \sqrt{3})$ square inches **35.** 60° **37.** 3 centimeters **39.** $\frac{3\sqrt{3}}{2}$ centimeters

41. $\frac{9\sqrt{3}}{8}$ square centimeters **43.** **a.** 20 inches **b.** 3 **c.** 63 inches **d.** 36 inches **e.** 9

45. $\angle B = \angle E = 61°$, $\angle A = \angle F = 78°$, $\angle C = \angle D - 41°$

47. **a.** $\angle ACB = \angle DCE$, $\angle A = \angle E$, $\angle B = \angle D$ **b.** 2 **c.** 3 inches **d.** 9 inches

49. **a.** 6 square centimeters **b.** $\angle ABC = \angle CDE$ **51.** 108 square inches **53.** **a.** 8 feet **b.** 13.9 feet

HOME EXERCISES

11.3, p. 382

1.

column (a)	column (b)
$\widehat{AB}$	an arc that is not a semicircle
$\angle AOB$	central angle
$\angle ACB$	inscribed angle
BC	chord
ℓ_1	secant
ℓ_2	tangent
BD	diameter
$\widehat{BD}$	semicircle

3. 3 centimeters **5.** 90° **7.** 3 centimeters

9. 6 square centimeters **11. a.** 100° **b.** 50°

13. a. 10π inches **b.** 5 inches **c.** 5π square inches **15. a.** 90° **b.** 90°

17. a. $\frac{8 + 7\pi}{2}$ feet **b.** $\frac{7\pi}{2}$ square feet

19. a. $(12\sqrt{2} + 6\pi)$ centimeters **b.** $(36\pi - 72)$ square centimeters

21. a. $\frac{30 + 54\pi}{5}$ centimeters **b.** $\frac{153\pi}{5}$ square centimeters **23.** $\frac{10\pi}{9}$ inches **25.** 65° **27.** 60°

29. 4 inches **31.** $\frac{16\pi}{3}$ square inches

LET'S REVIEW CHAPTER 11, p. 385

1. 28°36′ **2.** 66.2° **3. a.** acute **b.** obtuse **c.** straight **4. a.** 37°11′ **b.** 127°11′

5. 100° **6. a.** 52° **b.** 128° **7.** 46°45′ **8. a.** 8 inches **b.** 64 square inches

9. $\frac{25\sqrt{3}}{2}$ square inches **10. a.** $\frac{5\sqrt{3}}{2}$ centimeters **b.** $\frac{25\sqrt{3}}{4}$ square centimeters **11.** 36 square feet

12. 15° **13.** 15° **14.** $\frac{2\pi}{3}$ inches **15.** $\frac{4\pi}{3}$ square inches

INDEX

LIST OF SYMBOLS

Symbol	Use	Page
$=$	is equal to	3
\$28,898	28,898 dollars	9
0°C	0 degrees Celsius	9
$a < b$	a is less than b	10
$b > a$	b is greater than a	11
$-a$	the (additive) inverse of a	12
$\|a\|$	the absolute value of a	12
$a + b$	a plus b	15
$a - b$	a minus b	19
$a \cdot b$, ab, $a \times b$	a times b	31
a^2	the square of a	36
a^3	thc cubc of a	36
a^n	the nth power of a	37
$\frac{a}{c}$, $a \div c$, $c\overline{)a}$	a divided by c	39
$\neq$	is not equal to	39
x, y, z	variables	52
$\stackrel{?}{=}$, $\stackrel{\checkmark}{=}$, $\stackrel{\times}{=}$		70
gcd	greatest common divisor	81
lcm	least common multiple	140
lcd	least common denominator	142
.7	seven tenths	168
17%	17 percent	193
212°F	212 degrees Fahrenheit	230
(a, b)	ordered pair	234
$\sqrt{a}$	the square root of a	262
$\approx$	is approximately equal to	267
a^{-n}	$\frac{1}{a^n}$	283
a^0	1	283

Symbol	Use	Page
$\pm n$	both n and $-n$	299
$\ell_1 \perp \ell_2$	ℓ_1 and ℓ_2 are perpendicular.	321
π	pi	338
$\angle 1$	angle 1	354
$20°03'$	20 degrees and 3 minutes	354
$\ell_1 \parallel \ell_2$	ℓ_1 and ℓ_2 are parallel.	358
$\triangle ABC$	the triangle with vertices A, B, and C	361–362
$\widehat{AB}$	the arc AB	378